Symptoms, Diagnosis and Treatment of Thrombosis and Hemostasis

Symptoms, Diagnosis and Treatment of Thrombosis and Hemostasis

Editor: Nicholle Newman

hayle medical

New York

Hayle Medical,
750 Third Avenue, 9th Floor,
New York, NY 10017, USA

Visit us on the World Wide Web at:
www.haylemedical.com

ISBN 978-1-64647-567-4 (Hardback)

Cataloging-in-Publication Data

Symptoms, diagnosis and treatment of thrombosis and hemostasis / edited by Nicholle Newman.
p. cm.
Includes bibliographical references and index.
ISBN 978-1-64647-567-4
1. Thrombosis. 2. Hemostasis. 3. Blood coagulation disorders. 4. Symptoms. 5. Thrombosis--Diagnosis.
6. Hemostasis--Diagnosis. 7. Thrombosis--Treatment. 8. Hemostasis--Treatment. I. Newman, Nicholle.
RC647.C55 S96 2023
616.157--dc23

Contents

Preface

Hemostasis is a physiological process in the body that helps in stopping the bleeding and starting the repair process. This process prevents excessive blood loss and infections in the body. The formation of a blood clot inside a blood vessel, which obstructs the blood flow through the circulatory system is referred to as thrombosis. It can be categorized into venous thrombosis and arterial thrombosis. Venous thromboembolism (VTE) refers to blood clots that start in the veins. Deep vein thrombosis (DVT) and pulmonary embolism are the two sub-categories of VTE. Some common symptoms of thrombosis include sudden change in mental state, swelling in leg or arm, weakness in one side of the body, and chest pain. Diagnosis of thrombosis is performed through tests such as ventilation-perfusion scans, computed tomography (CT) scan, venography, and ultrasound. Anticoagulants such as heparin are the most effective medications used for the treatment of thrombosis. This book outlines the symptoms, diagnosis and treatment of thrombosis and hemostasis in detail. It consists of contributions made by international experts. The book is appropriate for students as well as for experts seeking detailed information in the area of hematology.

This book is a comprehensive compilation of works of different researchers from varied parts of the world. It includes valuable experiences of the researchers with the sole objective of providing the readers (learners) with a proper knowledge of the concerned field. This book will be beneficial in evoking inspiration and enhancing the knowledge of the interested readers.

In the end, I would like to extend my heartiest thanks to the authors who worked with great determination on their chapters. I also appreciate the publisher's support in the course of the book. I would also like to deeply acknowledge my family who stood by me as a source of inspiration during the project.

Editor

1

An illustrative case of endocardial fibroelastosis and recalcitrant intracardiac thrombosis

Denizhan Ozdemir[1], Isabel Oliva Cortopassi[2] and Robert Lawrence McNamara[3*]

Abstract

Background: Endocardial Fibroelastosis is diffuse, accentuated proliferation of ventricular endocardium causing a rare form of restrictive cardiomyopathy in both children and adults. It is an incompletely understood cause of heart failure predominantly in Sub-Saharan Africa associated with high morbidity and mortality. Atrial fibrillation and thrombus formation are common accompanying complications and portend a poor prognosis. Due to rarity of the condition in the developed countries and lack of evidence based options, the optimal strategy for anticoagulation is unclear.

Case presentation: Herein, we describe a relatively asymptomatic patient with endocardial fibroelastosis who has been found to have atrial fibrillation and a large thrombus in the right atrium. Currently, there is no evidence-based strategy in the management of endocardial fibroelastosis-associated intracardiac thrombus. This case report illustrates a scenario by which the use of apixaban potentially benefited or prevented the thrombus formation compared with warfarin as demonstrated by imaging findings.

Conclusions: The patients with endocardial fibroelastosis are at risk of developing intracardiac thrombus due to sticky substrate lining cardiac chambers while being relatively asymptomatic. No directed therapy is known for the management of heart failure and any complications of subsequent arrhythmias. The general recommendations follow those of same conditions in other hosts. Novel oral anticoagulant agents can be considered in the treatment of atrial thrombus in the appropriate settings.

Keywords: Endocardial fibroelastosis, Atrial thrombus, Apixaban, Novel oral anticoagulant agent

Background

Endocardial fibroelastosis (EFE) is defined as thickening of endocardial surface by proliferation of fibrous and elastic tissue [1, 2]. Although it mainly presents in pediatric population in Sub-Saharan Africa, EFE can also affect adults [3]. The etiology of EFE is unclear but proposed to be related to infectious agents and the climate due to variation with geographical distribution [4]. Histologically, the abnormal endocardium is composed of multiple layers of elastic and fibrous tissue that can functionally impair the contractility of cardiac chambers [5]. Due to underlying contractility dysfunction and propensity to coagulation in endocardial microenvironment, patients are susceptible to heart failure, atrial fibrillation and thrombotic complications [6–9]. Currently, the management follows the general rule of heart failure and atrial fibrillation from different etiologies [6–8].

Case presentation

Herein, we describe a case report of EFE that developed atrial fibrillation and aggressive thrombosis of right atrium. A 68-year-old Nigerian man is brought into the emergency room with asymptomatic heart rate of 30 per minute, noted before a cardiopulmonary stress test which he was undergoing for evaluation of prognosis and candidacy for cardiac transplantation. His medical history is significant for endocardial fibroelastosis diagnosed 15 years ago, persistent atrial fibrillation with slow ventricular response, hypertension, moderate sleep apnea, pre-diabetes,

* Correspondence: robert.mcnamara@yale.edu
[3]Section of Cardiovascular Medicine, Yale New Haven Hospital, Yale School of Medicine, Dana Clinic Building, 3rd Floor, 789 Howard Avenue, New Haven, CT 06519, USA
Full list of author information is available at the end of the article

and non-sustained ventricular tachycardia on ambulatory Holter monitoring for which he had previously declined an implantable cardioverter defibrillator (ICD). His medications included loop diuretics and warfarin. His social history is notable for immigration to the United States at mid-adult age and remote tobacco use. Family history is remarkable for multiple first-degree family members with sudden cardiac death before 50 years of age. He transferred his care to our health care system a few years prior to current presentation. Review of systems was notable for fatigue.

Physical exam demonstrated jugular venous distention with marked Y descent present, irregularly irregular heart rate, hepatomegaly, and ascites. LDL level was borderline elevated. Otherwise, laboratory values were all within normal limits, including hematologic evaluation, IgE levels and stool ova and parasites.

An electrocardiogram (EKG) revealed atrial fibrillation with slow ventricular response, rightward axis deviation, borderline low voltage in limb leads, and Q waves in lead III.

His first available transthoracic echocardiogram in our health system was performed in 2015. It revealed normal left ventricular cavity size, diastolic and systolic function with estimated ejection fraction of 55%; severely enlarged right atrium without thrombus; diffuse right ventricular endocardial thickening with normal systolic function and cavity obliteration; mild to moderate tricuspid regurgitation; and trace pericardial effusion. Cardiac catheterization outside this health care system reportedly showed restrictive pattern; however, details are not known. Initial cardiac magnetic resonance imaging (cMRI) in 2015 showed massive dilation of right atrium, distortion of right ventricle and mass-like thickening of mid-septum demonstrating diffuse and mildly heterogeneous enhancement on delayed post-contrast imaging with no evidence of intracardiac thrombus. From 2015 to 2017, the patient remained asymptomatic without hospitalization or any major illness and was able to work part-time. He was seen in scheduled intervals for the adjustments in his management and monitoring. His restrictive physiology and volume status were managed with low dose diuretics. He did not tolerate beta blocker due to bradycardia in the past. To reduce the risk of thromboembolic complications from atrial fibrillation, he was anticoagulated with apixaban for two years (2015 through 2017) until three weeks before presentation with no evidence of intracardiac thrombus on periodic cardiac MRI imaging. Three weeks prior to presentation, he was switched from apixaban to warfarin for unknown reasons.

During his hospitalization a cMRI was obtained to monitor any changes associated with EFE. It revealed a 29 × 22 mm thrombus in the right atrial cavity (Fig. 1a). There was no evidence of medication non-adherence, and his International Normalized Ratio (INR) was 2.6 at the time of detection of thrombus. He continued to take warfarin for anticoagulation to treat large thrombus in the right atrium for the next 3 months. Notably, his INR remained within the therapeutic range (2.0–3.0) nine out of eleven measurements over 3-month time period since thrombus was first detected, with only two sub-therapeutic levels of 1.7 and 1.8. However, the size of thrombus increased to 71 × 21 mm on the following repeat imaging (Fig. 1b). Due to progression of atrial thrombus, the apixaban was re-instated, and warfarin was discontinued. Repeat imaging 5 months after apixaban was restarted demonstrated the reduction in thrombus size to 31 × 5 mm (Fig. 1c). Subsequent imaging demonstrated further regression. From the first encounter the large thrombus was detected until the last clinic visit, he continued to remain asymptomatic and at his usual state of health without any evidence of thromboembolic or bleeding complications.

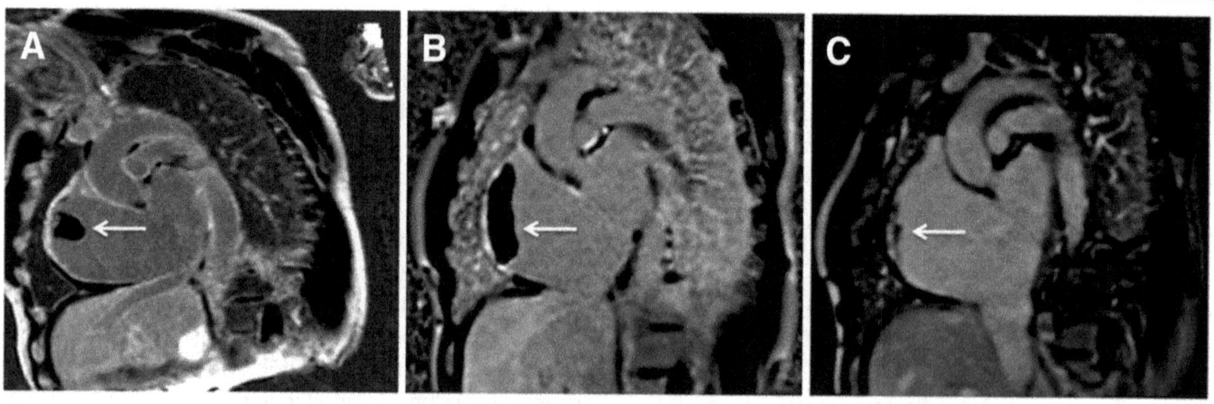

Fig. 1 **a**, arrow: Short axis late gadolinium enhancement images through the right atrium demonstrate thrombus formation measuring 29 × 22 mm three weeks after switch from apixaban to warfarin. **b**, arrow: The thrombus enlarged to 71 × 21 mm 3 months after continuing use of warfarin. **c**, arrow: The thrombus diminished to 31 × 5 mm 5 months after discontinuation of warfarin and re-instating apixaban

Discussion and conclusions

Endocardial fibroelastosis (EFE) constitutes a type of restrictive cardiomyopathy predominantly seen in Sub-Saharan Africa [10]. The literature is scarce on EFE, and the etiology is not fully understood. However, infection, climate, and genetic factors are thought to be implicated [11, 12]. EFE is a distinct entity from endomyocardial fibrosis where hypereosinophilia is considered to be the original insult [13]. Also, it should not be confused with left ventricular non-compaction cardiomyopathy, where pronounced ventricular trabeculations are characteristic [14].

EFE manifests as excessive thickening of endocardium with subsequent myocardial dysfunction. The diagnosis is usually made with non-invasive testing including clinical features, cardiac imaging findings and exclusion of associated conditions such as hypereosinophilic syndrome, infections or storage diseases [15]. It often eventually leads to typical restrictive heart failure manifestations and cardiac arrhythmias, predominantly atrial fibrillation [7–9]. Due to altered endothelium lining associated with disorganized tissue growth, patients with EFE are also predisposed to thrombus formation [16, 17]. Unfortunately, no directed therapy for EFE is known because of rarity of the condition, and current recommended management follows those of heart failure in general such as addressing heart failure symptoms with optimization of volume status, managing rate control in case of atrial fibrillation with rapid ventricular rates, and risk stratification for conduction disturbances for candidacy of implantable devices among others [6, 18, 19]. Of particular note, recommendations for thromboembolic risk prevention and management of detected thrombus follow general guidelines as limited specific evidence exists in patients with EFE. In this case, no thrombus was noted on multiple cMRIs during treatment with a direct anticoagulant, apixaban, but a thrombus was noted and continued to enlarge on warfarin therapy despite therapeutic INR values. Subsequent switch back to apixaban demonstrated successful treatment. One explanation of the improved efficacy of the direct anticoagulant may be the predictable pattern of anticoagulant effect compared with the warfarin. In addition, the abnormal lining of endocardium in EFE may provide a particularly thrombogenetic surface, particularly in the hemodynamic setting of atrial fibrillation and high right atrial pressures due to restrictive heart disease. One recent case report is consistent with our findings. It discusses a potentially related scenario, a patient with left ventricular non-compaction cardiomyopathy who was successfully treated with a direct oral anticoagulant instead of warfarin [20]. Further studies may provide deeper understanding of thrombus formation and specific anticoagulation strategies in EFE.

Abbreviations

cMRI: cardiac magnetic resonance imaging; EFE: Endocardial fibroelastosis; EKG: Electrocardiogram; ICD: Implantable cardioverter defibrillator; LDL: Low density lipoprotein; NOAC: Novel oral anticoagulant agent

Acknowledgements

Not applicable

Author details

[1]Department of Internal Medicine, Yale New Haven Hospital, Yale School of Medicine, New Haven, CT, USA. [2]Section of Radiology and Biomedical Imaging, Yale New Haven Hospital, Yale School of Medicine, New Haven, CT, USA. [3]Section of Cardiovascular Medicine, Yale New Haven Hospital, Yale School of Medicine, Dana Clinic Building, 3rd Floor, 789 Howard Avenue, New Haven, CT 06519, USA.

References

1. Letteer CR. Endocardial fibroelastosis. Ann Intern Med. 1953;38(3):573–82.
2. Ino T, Benson LN, Freedom RM, Rowe RD. Natural history and prognostic risk factors in endocardial fibroelastosis. Am J Cardiol. 1988;62(7):431–4.
3. Raman SV, Mehta R, Walker J, Pennell DJ. Cardiovascular magnetic resonance in endocardial fibroelastosis. J Cardiovasc Magn Reson. 2005;7(2):391–3.
4. Adams FH, Katz B. Endocardial fibroelastosis: case reports with special emphasis on the clinical findings. J Pediatr. 1952;41(2):141–52.
5. Halliday WR. Endomyocardial fibroelastosis; a study of thirty cases. Dis Chest. 1954;26(1):27–40.
6. Manning JA, Sellers FJ, Bynum RS, Keith JD. The medical management of clinical endocardial FIBROELASTOSIS. Circulation. 1964;29:60–5.
7. Hung W, Walsh BJ. Congenital auricular fibrillation in a newborn infant with endocardial fibroelastosis. Report of a case with necropsy J Pediatr. 1962;61:65–9.
8. Thomas WA, Randall RV, Bland EF, Castleman B. Endocardial fibroelastosis; a factor in heart disease of obscure etiology: a study of 20 autopsied cases in children and adults. N Engl J Med. 1954;251(9):327–38.
9. Radford DJ, Izukawa T. Atrial fibrillation in children. Pediatrics. 1977;59(2):250–6.
10. Stein H, Shnier MH, Wayburne S, Isaacson C. Cardiomyopathy in African children. Arch Dis Child. 1964;39(208):610–7.
11. Chan JL, Rosing DR, Klion AD, Horvath KA. Surgical management of adult endocardial fibroelastosis. J Thorac Cardiovasc Surg. 2017;154(5):e81–4.
12. Lurie PR. Changing concepts of endocardial fibroelastosis. Cardiol Young. 2010;20(2):115–23.
13. Cheung CC, Constantine M, Ahmadi A, Shiau C, Chen LYC. Eosinophilic myocarditis. Am J Med Sci. 2017;354(5):486–92.
14. Towbin JA, Lorts A, Jefferies JL. Left ventricular non-compaction cardiomyopathy. Lancet. 2015;386(9995):813–25.
15. Sobieh A, Bolen MA, Rodriguez ER, Tan CD, Rodriquez LL, Renapurkar RD. Multimodality imaging and pathologic assessment in an adult with endocardial Fibroelastosis. CASE (Phila). 2017;1(5):176–8.
16. Becker RC. Thrombogenesis in atrial fibrillation contributing mechanisms and natural history. J Thromb Thrombolysis. 2009;27(1):119–21.
17. Branch GL, Castle RF. Thromboembolic complications in primary endocardial fibroelastosis. J Pediatr. 1966;69(2):250–8.
18. Muchtar E, Blauwet LA, Gertz MA. Restrictive cardiomyopathy: genetics, pathogenesis, clinical manifestations, diagnosis, and therapy. Circ Res. 2017;121(7):819–37.
19. Chan K, Harper AR, Ashrafian H, Yavari A. Cardiomyopathies. Medicine. 2018;46(10):606–17.
20. Sun H, Zhao Q, Wang Y, Lakin R, Feng H, Fan X, Luo H, Gao D, Liu L. He Y and others. Daily 10 mg rivaroxaban as a therapy for ventricular thrombus related to left ventricular non-compaction cardiomyopathy: a case report. Medicine (Baltimore). 2018;97(4):e9670.

2

Effects of glycemic control and hypoglycemia on Thrombus formation assessed using automated microchip flow chamber system

Kiyoaki Yamamoto[1], Takashi Ito[2*], Tomoka Nagasato[2,3], Atsushi Shinnakasu[1], Mihoko Kurano[1], Aiko Arimura[1], Hiroshi Arimura[1], Hiroshi Hashiguchi[1], Takahisa Deguchi[1], Ikuro Maruyama[2] and Yoshihiko Nishio[1]

Abstract

Background: Thrombus formation is an important factor affecting cardiovascular events and venous thromboembolism in type 2 diabetes. However, it is unclear whether glycemic control reduces thrombogenicity. We investigated the effect of short-term glycemic control (STUDY 1) and hypoglycemia (STUDY 2) on thrombus formation using an automated microchip flow chamber system.

Methods: For STUDY 1, we recruited 10 patients with type 2 diabetes. Before and after 2 weeks of treatment, blood glucose was analyzed with a continuous glucose monitoring system, and thrombogenicity was analyzed with an automated microchip flow chamber system. For STUDY 2, we recruited 10 subjects without diabetes who underwent an insulin tolerance test. We evaluated the change in thrombogenic potential with hypoglycemia.

Results: STUDY1: The mean blood glucose level reduced from 10.1 ± 2.6 to 6.9 ± 0.97 mM ($P < 0.01$). T10, an indicator of thrombogenicity, significantly attenuated after glycemic control (338 ± 65 vs. 425 ± 117 s, $P < 0.05$). The attenuation in T10 was significantly correlated with changes in mean blood glucose level after treatment ($r = -0.718$, $P < 0.05$). STUDY 2: Platelet function was enhanced with decreasing blood glucose; increased platelet function was strongly correlated with an increase in epinephrine.

Conclusions: We demonstrated attenuation in thrombogenicity with short-term comprehensive diabetes care and enhancement in thrombogenicity with hypoglycemia, using a new flow chamber system.

Trial registration: UMIN-CTR UMIN 000019899, registered 26-Jan-2015 (STUDY 2).

Keywords: Glycemic control, Hyperglycemia, Hypoglycemia

Background

Atherothrombotic disease is a major cause of death in patients with diabetes [1]. The mechanism of increased thrombosis in patients with diabetes involves multiple pathways. Thrombus formation, the last step in the atherothrombotic process, is an important factor influencing the risk and prognosis associated with cardiovascular events [2]. Moreover, diabetes is a risk factor for arterial and venous thrombosis. A recent meta-analysis found that patients with diabetes were at increased risk of developing venous thromboembolism (VTE), which can occur in a hypercoagulable state [3].

Several studies have shown that patients with diabetes have increased thrombogenicity owing to platelet hyperreactivity, activation of coagulation factors and hypo-fibrinolysis [4–6]. However, it is unclear whether glycemic control reduces thrombus formation. Especially, the

* Correspondence: takashi@m3.kufm.kagoshima-u.ac.jp
[2]Department of Systems Biology in Thromboregulation, Kagoshima University Graduate School of Medical and Dental Sciences, 8-35-1, Sakuragaoka, Kagoshima 890-8544, Japan
Full list of author information is available at the end of the article

influence of glycemic control over short periods of time, such as one or two weeks, has not been assessed, even though it is a clinically-important issue for preventing post-surgical complications [7].

On the other hand, evidence is accumulating that hyperglycemia and hypoglycemia, which can occur when trying to achieve strict glycemic control, is associated with increased risk of cardiovascular and cerebrovascular death. In the ACCORD study, the mortality rate was higher in the strict-treatment than in the standard-treatment group [8]. The involvement of hypoglycemia as a risk factor for thrombosis is a topic of active discussion.

A previous study evaluated thrombogenicity using plasma markers [9] and platelet function tests with an agonist, above physiological concentrations; however, the effects of blood flow were not considered and blood flow is a well-known influence on the thrombus formation process [10]. Therefore, it is necessary to evaluate thrombogenicity comprehensively, using close-to-physiological conditions and considering blood flow. Recently, a Total Thrombus formation Analysis System (T-TAS, Fujimori Kogyo Co., Yokohama, Kanagawa) [11] was found useful [12] for quantitative analysis of thrombus formation. The T-TAS continuously monitors the pressure on microchips with thrombogenic surfaces that simulate pathological blood vessels [11, 13]. Compared with conventional systems, the T-TAS can evaluate thrombogenicity under blood flow conditions; therefore, it is possible to perform analyses that more closely resemble in vivo conditions.

This pilot study aimed at investigating the influence of short-term glycemic control (STUDY 1) and change in hypoglycemia (STUDY 2) on thrombus formation using T-TAS.

Materials and methods

Study 1

Glycemic control efforts were started immediately after hospitalisation. Blood glucose levels were controlled with insulin, GLP-1 analogues and oral hypoglycemic agents. The goal of treatment was to achieve a fasting plasma glucose (FPG) level < 7.2 mM and a postprandial PG level < 11.1 mM. A continuous glucose monitoring (CGM) device (iPro2: Medtronic Minimed, Northridge, CA, USA) was attached to each patient from the day of hospitalization, and monitoring was performed for 3 days. Blood collection and evaluation for T-TAS were carried out on the day after hospitalization. These evaluations were performed again after glycemic control was achieved. The exclusion criteria were antiplatelet drug or anticoagulant use, infectious diseases, malignant tumors, drug-induced diabetes, ketoacidosis and possible pregnancy.

Blood samples

Following overnight fasting, blood was drawn for the measurement of blood glucose, hematocrit (Ht), platelet, plasminogen activator inhibitor 1 (PAI-1), fibrinogen, factor VII (FVII), factor VIII (FVIII) and prothrombin fragment 1 + 2 (F1 + 2), and for assessment using T-TAS.

Glucose levels were measured using a GA echo buffer solution, Glucose BP standard solution and automated glucose analyzer GA08III (A&T corporation, Kanagawa, Japan). The LPIA·tPAI test with the automated immunoanalyzer STACIA (LSI Medience Corporation, Tokyo, Japan) for PAI-1, Thrombo check Fib (L) with CS-5100 Automated Coagulation Analyzer (Sysmex Corporation, Hyogo, Japan) for fibrinogen, HemosIL RecombiPlasTin or HemosIL Synth Asil with ACL TOP automated analyzer (Instrumentation Laboratory, Bedford, MA) for FVII or FVII and Enzygnost with Microplate Reader Emax (Molecular Devices Japan, Tokyo, Japan) for F1 + 2 were applied respectively.

CGM

We used a CGM device and considered the result as an indicator of short-term glycemic control. The mean blood glucose level, peak blood glucose levels and standard deviation (SD) and mean amplitude of glycemic excursion (MAGE) were calculated from the results.

T-TAS

T-TAS evaluations were carried out using two kinds of microchips (PL chip and AR chip). The PL chip, which is coated with collagen, represents platelet thrombus formation. The AR chip, which is coated with collagen plus tissue thromboplastin, represents thrombus formation mediated by the coagulo-fibrinolysis system and platelets. Whole blood anticoagulated with hirudin was perfused through the PL chip at a shear rate of $1500\,S^{-1}$; then, PL-T10 (time to 10 kPa), PL-OT (occlusion time: time to 60 kPa) and PL-AUC10 (area under the flow curve in 10 min) were used to evaluate thrombus formation. Recalcified whole blood containing corn trypsin inhibitor was perfused through the AR chip at a sheer rate of $600\,S^{-1}$; then, AR-T10 (time to 10 kPa), AR-OT (occlusion time: time to 80 kPa) and AR-AUC30 (area under the flow curve in 30 min) were evaluated. The intra-assay coefficient of variability of PL-AUC and AR-AUC are 5.85 and 1.24% respectively (Additional file 1: Figure S1). The reference ranges (mean ± SD) of PL-T10, PL-OT and PL-AUC in healthy volunteers were reported to be 163 ± 51 s, 347 ± 96 s, and 369.1 ± 71.8 kPa × min, respectively [14].

Study 2

Human regular insulin (Eli Lilly, Indianapolis, IN) was injected intravenously at a dose of 0.05–0.10 U/kg body weight. Blood samples for glucose and epinephrine were collected before the injection and at 15, 30, 45, 60, 90 and 120 min after the injection; for T-TAS blood samples were collected before the injection and at 15, 30, 60, 90 and 120 min; and for WBC blood samples were collected before the injection and at 30 and 60 min after the injection. For measurement of epinephrine, the CA Test 「TOSOH」 (Tosoh Corporation, Tokyo, Japan) and HPLC system (SHIMADZU CORPORATION, Kyoto, Japan) were used.

Statistical analysis

Normally distributed data are presented as means ± SDs, and the paired *t*-test was used to assess the presence of significant differences. Non-normally distributed data are presented as medians (quartile range), and the Wilcoxon signed rank test was used for assessment. Spearman's rank correlation was used for assessing correlation. IBM SPSS 20 statistical package (IBM Corp., Armonk, NY) was used to perform all the statistical analyses. A P-value < 0.05 was considered statistically significant.

Results

Study 1

We included 10 patients with type 2 diabetes who were admitted to Kagoshima University Hospital for glycemic control (Table 1) (Additional file 1: Table S1). The mean duration of the period between pre- and post-glycemic control was 10.4 days. The levels of fasting plasma glucose (FPG) reduced from 9.6 ± 2.3 mM to 6.3 ± 1.4 mM ($P < 0.01$), mean blood glucose reduced from 10.1 ± 2.5 mM to 6.8 ± 0.9 mM ($P < 0.01$), and peak blood glucose reduced from 15.6 ± 4.3 mM to 11.5 ± 4.1 mM ($P < 0.05$). The SD of blood glucose and MAGE were unchanged (2.1 ± 1.3 mM vs. 1.6 ± 1.2 mM, $P = 0.24$ and 87.5 ± 42.4 vs. 68.5 ± 51.2, $P = 0.31$, respectively) (Table 2). In one case, mild hypoglycemia (3.1 mM) was recognized in the night during CGM; however, this was not observed in other patients.

Table 1 Patient characteristics in STUDY 1

Age (years)	47.5 ± 16.6
Sex (male/female)	5 / 5
BMI (kg/m^2)	29.6 ± 7.9
FPG (mM)	10.2 ± 3.0
HbA1c (%)	10.5 ± 3.1
Duration of diabetes (years)	4.8 ± 4.0
Systolic blood pressure (mmHg)	121.1 ± 13.3
Diastolic blood pressure (mmHg)	77.7 ± 12.0
LDL cholesterol (mM)	2.9 ± 1.3
HDL cholesterol (mM)	1.3 ± 0.4
Triglyceride (mM)	1.8 ± 1.1
eGFR (mL/min/1.73 m^2)	99.4 ± 21.1

Values are presented as mean ± standard deviation (SD)
FPG fasting plasma glucose

Hematologically, no significant changes were observed in WBC count, Ht level, platelet count, fibrinolysis marker (PAI-1) level and coagulation marker (fibrinogen, FVII, FVIII, F1 + 2) levels (Table 2).

In STUDY 1, the T-TAS and two microchips were used to evaluate thrombus formation. Inside the PL chip, activation of the platelet is triggered on the surface of collagen [13]. Inside the other AR chip, activation of both platelets and coagulation is triggered by collagen and tissue thromboplastin [11]. We evaluated thrombogenicity by assessing the changes in flow pressure during thrombus formation.

AR-T10 and AR-OT increased from 337.8 ± 64.7 s to 425.2 ± 116.9 s ($P = 0.005$) and from 441.2 ± 73.0 s to 527.7 ± 115.4 s ($P = 0.006$), respectively, while AR-AUC30 reduced from 1890.3 ± 91.3 to 1775.9 ± 154.5 kPa × min (P = 0.006) (Fig. 1). No PL chip parameters changed (Table 2).

We compared the change in the mean PG (Δmean PG) and SD (ΔSD) with the change in T-TAS parameters (Fig. 2). The change in AR-T10 (ΔAR-T10) was significantly correlated with Δmean PG after treatment (r = −0.718, $P < 0.05$), but not with FPG, SD and MAGE.

Although the correlation between Δmean PG and ΔAR-OT or ΔAR-AUC30 was not statistically significant, non-significant trend was observed. Additional file 1: Table S2 presents individual data on mean PG, AR-T10, AR-OT and AR-AUC30 in STUDY 1.

Study 2

We included 10 patients who underwent an insulin tolerance test for hypothalamic pituitary evaluation (Table 3). After insulin injection, PG decreased from 5.2 ± 0.7 mM (baseline) to 1.7 ± 0.5 mM (45 min), and Ht increased from 37.5 ± 3.5% to 39.1 ± 3.7% ($P < 0.01$) and epinephrine increased from 19.5 ± 11.2 pg/mL to 449.5 ± 289.6 pg/mL (P < 0.01). However, the platelet count did not change (24.5 × 10^4/μL to 25.8 × 10^4/μL, $P = 0.053$).

PL-T10 decreased from 156.4 ± 55.5 s to 109.7 ± 15.2 s (P < 0.05), while PL-AUC10 increased from 380.3 ± 72.4 s to 428.3 ± 31.9 s (P < 0.05). PL-T10 was significantly lower, and PL-AUC10 was significantly higher, at 60 min compared to the baseline values (Fig. 3c, d).

Along with hypoglycemia, epinephrine and WBC count significantly changed (Fig. 3e). In addition, there was a correlation between the decrease in PG (ΔPG) and the change in epinephrine (Δepinephrine). Furthermore, Δepinephrine was correlated with ΔPL-T10 and ΔPL-AUC10 (Fig. 4). We found the significant correlations between ΔPL-T10 and change in Ht (r = − 0.772, $P = 0.009$), change in WBC count ($r = -0.915$, $P < 0.001$) or change in platelet

Table 2 Parameters of haematologic data, CGMs and T-TAS before and after glycemic control in STUDY 1

	Baseline	After control	*P*-value
FPG (mM)	9.6 ± 2.3	6.3 ± 1.4	0.004
Hematologic parameters			
WBC count (/μL)	5752 ± 1440	5468 ± 1342	0.191
Ht (%)	41.1 ± 4.8	40.7 ± 4.9	0.523
Plt count (10^4 /μL)	19.3 (16.6–24.2)	20.2 (18.9–24.1)	0.683
PAI-1 (ng/mL)	24.0 (16.3–45.3)	21.0 (14.5–28.5)	0.211
Fibrinogen (g/dL)	3.3 ± 0.7	3.6 ± 0.3	0.128
Factor VII (%)	85.4 ± 8.6	83.6 ± 11.2	0.522
Factor VIII (%)	101.1 ± 33.4	92.3 ± 28.7	0.288
F1 + 2 (pM)	135.5 (105.0–169.3)	139.5 (107.3–173.5)	0.580
CGMs			
Mean PG (mM)	10.1 ± 2.5	6.8 ± 0.9	0.005
Peak PG (mM)	16.3 (11.3–19.1)	10.2 (8.8–12.4)	0.017
SD (mM)	1.8 (1.2–2.4)	1.1 (0.9–1.7)	0.242
MAGE (mM)	4.3 (2.9–7.0)	2.8 (2.2–4.1)	0.310
T-TAS			
PL-T10 (s)	125.5 (110.8–141.0)	126.0 (120.5–135.3)	0.881
PL-OT (s)	327.8 ± 67.6	308.7 ± 56.5	0.166
PL-AUC10 (kPa × min)	393.2 ± 45.3	398.3 ± 32.2	0.647

Values are presented as mean ± standard deviation (SD) or median (quartile range)
WBC white blood cell, *Ht* hematocrit, *Plt* platelet, *PAI-1* plasminogen activator inhibitor-1, *F1 + 2* prothrombin fragment 1 + 2, *FPG* fasting plasma glucose, *MAGE* mean amplitude of glycemic excursions, *T10* time to 10 kPa, *OT* occlusion time, *AUC10* area under the curve in 10 min

count (r = – 0.648, $P = 0.043$) (Additional file 1: Figure S2). Additional file 1: Table S3 shows individual data on the PG, epinephrine, PL-T10 and PL-AUC10 in STUDY2.

Discussion

We aimed to elucidate the effects of short-term glycemic control on blood thrombogenicity in patients with diabetes (STUDY1) and change of platelet function during hypoglycemia (STUDY2) using a newly-developed, microchip-based flow chamber system. We used two types of microchips [13]: the PL chip to evaluate thrombogenicity, mainly involving platelets, and the AR chip to evaluate platelet activation and the coagulo-fibrinolytic reaction [11].

Although it has been reported that thrombogenic potential improves as glycemic control improves, for at least several months [15], it was not clear whether short-term control reduces thrombus formation. Our study (STUDY1) demonstrated that short-term glycemic control attenuates thrombogenicity. Of note, improvements

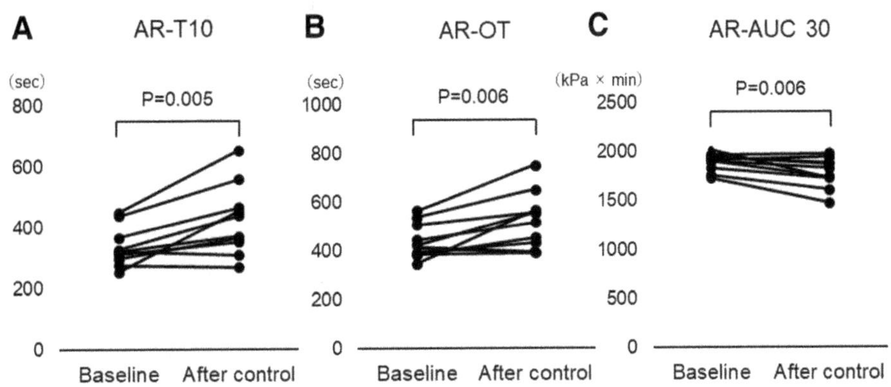

Fig. 1 Changes in time for the flow pressure to reach 10 kPa (AR-T10) (**a**), time to occlusion (AR-OT) (**b**) and area under the flow curve at 30 min (AR-AUC 30) (**c**) before and after glycemic control in STUDY 1. Data ($n = 10$) were evaluated by two-tailed t-test

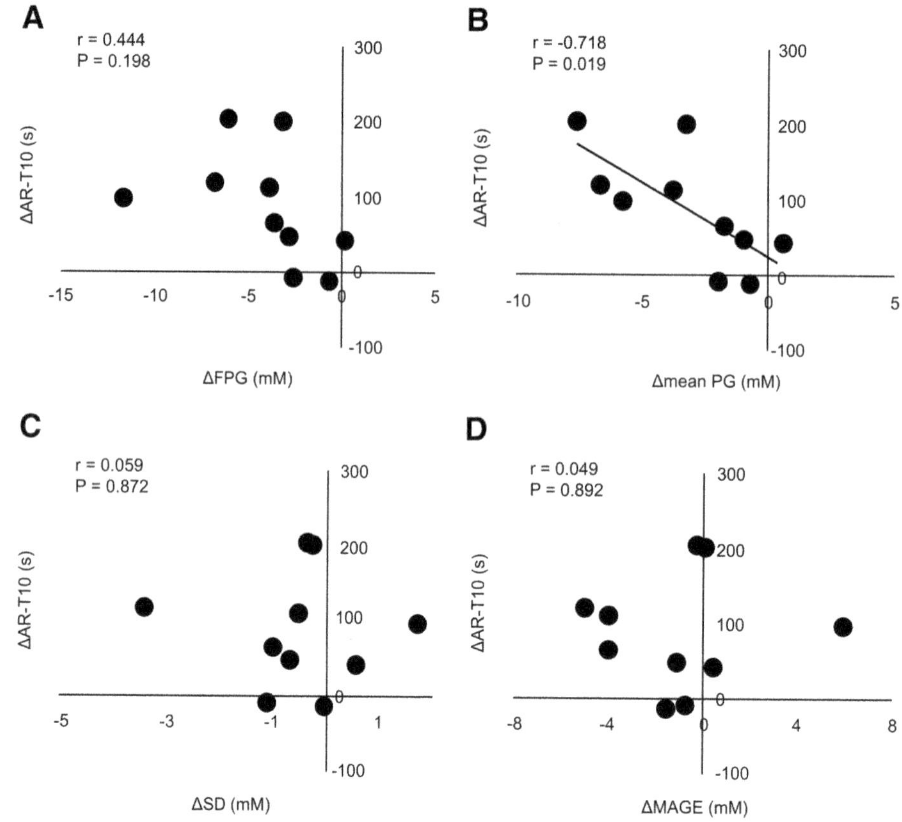

Fig. 2 Correlation between the change in AR-T10 and the change in FPG (**a**), the change in mean BG (**b**), the change in SD (**c**), or the change in MAGE (**d**) in STUDY 1. Pearson's rank correlation was used for assessing correlation (*n* = 10)

in mean blood glucose level, but not glucose excursion, were strongly correlated with reduced thrombogenicity. In impaired glucose tolerance patients, it has been reported that suppression of postprandial hyperglycemia can prevent the onset of cardiovascular events [16]. But in type 2 diabetic patients, the protective effect of the attenuation of glucose spikes remains controversial [17]. Furthermore, there is a large-scale clinical research showing that the cardiovascular risk is lower in patients with low HbA1c values, which represent low mean blood glucose levels [18]. These results underscore the clinical significance of short-term pre-operative glycemic control to improve patients' prognosis.

Table 3 Patient characteristics in STUDY 2

Age (years)	46.8 ± 14.8
Sex (male/female)	3 / 7
BMI (kg/m^2)	24.1 ± 3.3
HbA1c (%)	5.7 ± 0.5
Systolic blood pressure (mmHg)	121.7 ± 14.5
Diastolic blood pressure (mmHg)	75.6 ± 11.5
eGFR (mL/min/1.73 m^2)	97.3 ± 18.6

Values are presented as mean ± standard deviation (SD)
FPG fasting plasma glucose; BMI, body mass index

The mean value of AR-OT before glycemic control was 425.2 s, shorter than the mean value in healthy adults (489 +/− 96 s, unpublished data). After glycemic control, AR-OT improved to 527.7 s. The other AR chip parameters (AR-T10 and AR-AUC30) also improved after glycemic control. The PL chip parameters, which represent platelet-mediated thrombogenicity, showed no attenuation.

One possible explanation for the reduced thrombogenicity of the AR chip, without attenuation of PL chip thrombogenicity and coagulo-fibrinolysis markers, is glycation of fibrinogen and plasminogen because of high blood glucose. In the studies that used the purified-fibrinogen model, chronic hyperglycemia accelerated glycation of fibrinogen non-enzymatically. Glycated fibrinogen rapidly constructed a fibrin network, increasing the resistance to the fibrinolytic system, compared to non-diabetic patients [19, 20], due to changes in the kinetics of fibrin polymerization [20]. Past reports found that patients with diabetes and rapidly-improving glycemic control had significantly decreased glycated fibrinogen over 3 days [21]. In patients with diabetes, glycation of plasminogen reduced plasmin generation and impaired plasminogen function, both of which were

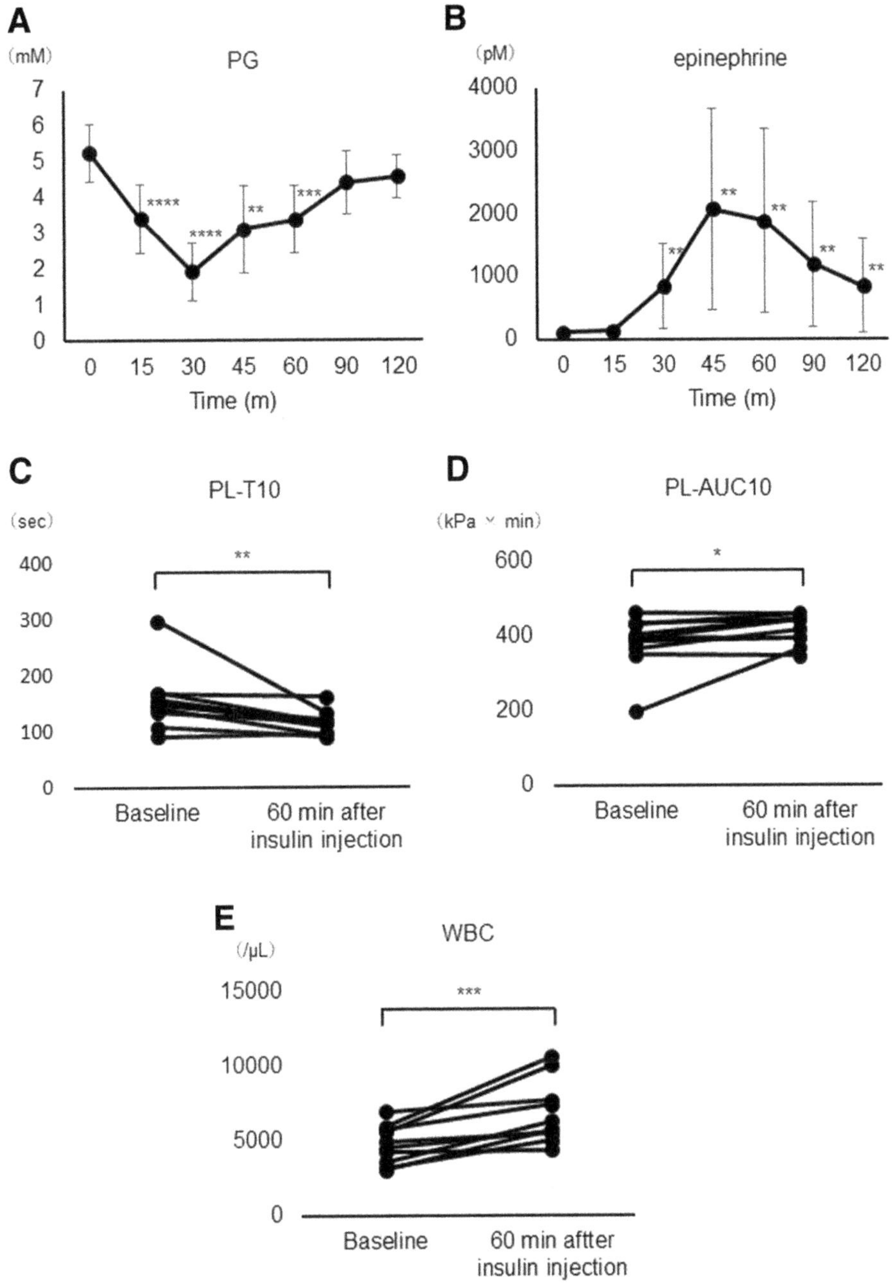

Fig. 3 Time course of PG (**a**) and epinephrine (**b**) and changes in PL-T10 (**c**), PL-AUC 10 (**d**), WBC (**e**) before and 60 min after insulin injection in insulin tolerance test (STUDY 2). Data are presented as mean ± SD. *$P < 0.05$ vs basal, **$P < 0.01$ vs basal, ***$P < 0.005$ vs basal, **** $P < 0.001$ vs basal. The change in WBC was evaluated by two-tailed t-test, the others were evaluated by Wilcoxon signed rank test ($n = 10$)

ameliorated by glycemic control [22]. Based on these results, the hypo-fibrinolytic state may have improved secondary to decreases in glycated fibrinogen and plasminogen, even during short-term glycemic control.

It is not clear what factor, among all those altered by comprehensive diabetes care, contributed to the attenuation of thrombotic occlusion. Since the decrease in the mean PG change showed a relatively strong negative correlation with the change in AR-T10, mean PG might be an important factor. However, it is also possible that some specific treatment options, such as metformin, glucagon-like peptide-1 receptor agonists or statins, contribute to reduction of thrombogenic potentials. A large-scale study with a stratified analysis is necessary to examine this issue further.

The findings of this study differ from the findings of previous studies [23, 24] of the relationship between glycemic control and thrombogenicity. Several reported

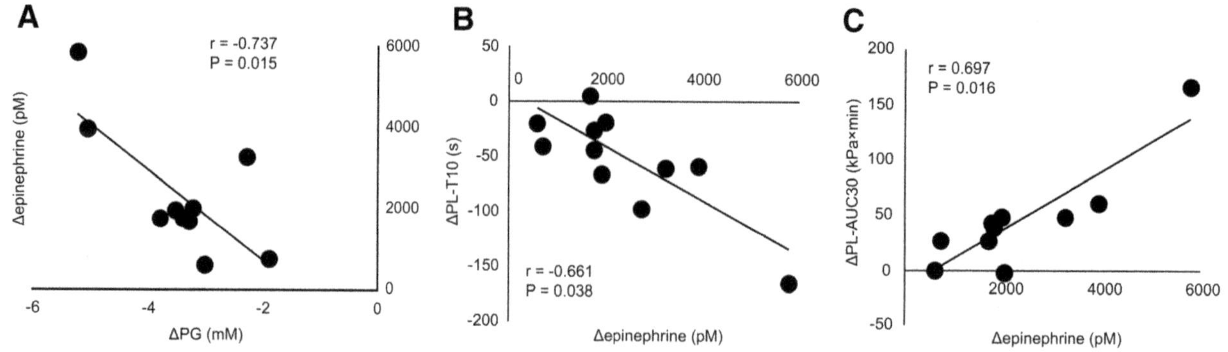

Fig. 4 Correlation between the change in PG and the change in epinephrine (**a**), the change in epinephrine and the change in PL-T10 (**b**) and the change in epinephrine and the change in PL-AUC10 (c) in STUDY 2. Each data is the difference between the value before administration of insulin and the bottom value (PG and PL-T10) or the peak value (epinephrine and PL-AUC10) at any time point after administration of insulin. Correlation between the change in PG and the change in epinephrine was analzsed by Pearson's rank correlation, and others were analyzed by Spearman's rank correlation ($n = 10$)

that levels of fibrinogen, PAI-1, FVII, FVIII and F1 + 2 are increased in patients with diabetes [25]; however, in the present study, the levels were within normal limits and did not change after glycemic control. These markers are influenced by various factors, such as low-grade inflammation [26], insulin resistance [27] and hyperglycemia itself [28]; therefore, our results may reflect differences in patients' backgrounds.

The relationship between glycemic control and platelet function is controversial [29, 30]. Factors other than the glucose level may be involved in platelet function among patients with type 2 diabetes [31]. In a previous clinical study, aspirin administered for primary prevention in patients with type 2 diabetes failed to reduce the risk of macrovascular events [32]. The role of platelets in thrombus formation, among patients with type 2 diabetes, may be limited.

Conventionally, there is a recognized relationship between hypoglycaemia and increased platelet function [33]. However, prior results were verified in the absence of blood flow or with supra-physiological concentrations. Moreover, the samples consisted of platelet-rich plasma (PRP) and not whole blood. To the best of our knowledge, this is the first study to evaluate thrombus formation associated with hypoglycaemia using a flow chamber system with microchips simulated pathological blood vessels.

Platelet function was enhanced with decreasing blood glucose, and increased platelet function was strongly correlated with an increase in epinephrine (STUDY2). Several studies have shown that plasma epinephrine may have an important role in platelet activation during hypoglycemia in patients with type 2 diabetes [34, 35]. Epinephrine is secreted from the adrenal medulla through the activation of the sympathetic nervous system as a physiological defence when PG levels fall [36]. Increased epinephrine levels cause platelet aggregation through the activation of α2-adrenergic receptors [37].

Additionally, the changes in WBC count and Ht levels correlated with changes in PL chip parameters. Increased WBC count, partially that induced by epinephrine, may contribute to thrombus formation and progress. WBCs, especially monocytes and neutrophils, induce thrombus formation in collaboration with platelets [38]. Furthermore, increased Ht level after insulin injection could have affected blood viscosity inside the PL chip.

Our patients included one with acromegaly and one with adrenal insufficiency; however, there was no association of cortisol and growth hormone levels with platelet function [39].

It has been reported that, in patients with acute coronary syndrome, hyperglycemia is associated with poor prognosis and high rates of subsequent ischaemic events [40]. However, there is no firm evidence that glycemic control improves prognosis in patients who have experienced cardiovascular events. Our results suggest that glycemic control improves prognosis in critical situations and underscore the importance of avoiding hypoglycemia.

Although there is no study on the relationship between glycemic control and development of VTE, glycemic control-related improvements in the hypo-fibrinolysis state may lower the risk of VTE while improving prognosis.

Earlier studies investigated the effects of anti-platelet or anti-coagulant drugs on T-TAS measurements [41, 42]. One study that evaluated the effects of warfarin or direct-acting oral coagulants (DOACs) with T-TAS [43] showed a decrease in AR-AUC30, similar to the results of STUDY 1. Changes in T-TAS parameters, obtained by comprehensive diabetes care involving short-term glycemic control, appear physiologically-relevant.

Our study has several limitations. First, the number of participants was small. Nevertheless, we think that our preliminary findings are worth reporting and may inform future, more scientifically-rigorous studies. Second, we enrolled patients without diabetes in STUDY 2, because patients with diabetes have increased risk of ischaemic events associated with hypoglycaemia. Third, we did not evaluate thrombus formation with the AR chip in STUDY2. Interpretation of AR chip results might be difficult if significant changes in platelet function are observed because the AR chip results reflected either platelet aggregation or coagulation factor. Finally, we could not obtain direct evidence of glycated state alteration.

Conclusions

In summary, we demonstrated attenuation in thrombogenicity with comprehensive diabetes care and enhancement in thrombogenicity with hypoglycemia, using a new flow chamber system. To confirm these findings, further large-scale clinical studies are needed.

Abbreviations

AR-AUC30: area under the flow curve in 30 min; AR-OT: occlusion time: time to 80 kPa; AR-T10: time to 10 kPa; CGM: Continuous glucose monitoring; DOAC: Direct-acting oral coagulants; F1 + 2: Prothrombin fragment 1 + 2; FPG: Fasting plasma glucose; FVII : Factor VII; FVIII: Factor VIII; Ht: Hematocrit; MAGE: Mean amplitude of glycemic excursion; PAI-1: Plasminogen activator 1; PL-AUC10: area under the flow curve in 10 min; PL-OT: occlusion time: time to 60 kPa; PL-T10: time to 10 kPa; PRP: Platelet-rich plasma; SD: Standard deviation; T-TAS: Total Thrombus formation Analysis System; WBC: White blood cell

Acknowledgements

None.

Authors' contributions

KY, TI, AA, HA, HH, TD, IM and NY participated in the design and conception of the design. KY, TI and TN collected the data. KY and TI analyzed the data. SGD, AS, MK, AA and HA participated in recruitment and study conduction. TI and YN interpreted the data and edited the manuscript. YN was the principal investigator of the study who was primarily responsible for the conception and design of the study and oversaw the entire conduction, data collection and analysis the study. All authors read and approved the final manuscript.

Author details

[1]Department of Diabetes and Endocrine Medicine, Kagoshima University Graduate School of Medical and Dental Sciences, 8-35-1, Sakuragaoka, Kagoshima 890-8544, Japan. [2]Department of Systems Biology in Thromboregulation, Kagoshima University Graduate School of Medical and Dental Sciences, 8-35-1, Sakuragaoka, Kagoshima 890-8544, Japan. [3]Fujimori Kogyo Co., Research Institute, 1-10-1, Yokohama, Kanagawa 236-0003, Japan.

References

1. Sarwar N, Gao P, Seshasai SR, Gobin R, Kaptoge S, Di Angelantonio E, Ingelsson E, Lawlor DA, Selvin E, Stampfer M, et al. Diabetes mellitus, fasting blood glucose concentration, and risk of vascular disease: a collaborative meta-analysis of 102 prospective studies. Lancet. 2010;375:2215–22.
2. Furie B, Furie BC. Mechanisms of thrombus formation. N Engl J Med. 2008;359:938–49.
3. Gariani K, Mavrakanas T, Combescure C, Perrier A, Marti C. Is diabetes mellitus a risk factor for venous thromboembolism? A systematic review and meta-analysis of case-control and cohort studies. Eur J Intern Med. 2016;28:52–8.
4. Colwell JA, Nesto RW. The platelet in diabetes: focus on prevention of ischemic events. Diabetes Care. 2003;26:2181–8.
5. Boden G, Vaidyula VR, Homko C, Cheung P, Rao AK. Circulating tissue factor procoagulant activity and thrombin generation in patients with type 2 diabetes: effects of insulin and glucose. J Clin Endocrinol Metab. 2007;92:4352–8.
6. Seljeflot I, Larsen JR, Dahl-Jorgensen K, Hanssen KF, Arnesen H. Fibrinolytic activity is highly influenced by long-term glycemic control in type 1 diabetic patients. J Thromb Haemost. 2006;4:686–8.
7. Newman JD, Wilcox T, Smilowitz NR, Berger JS. Influence of diabetes on trends in perioperative cardiovascular events. Diabetes Care. 2018.
8. Gerstein HC, Miller ME, Byington RP, Goff DC Jr, Bigger JT, Buse JB, Cushman WC, Genuth S, Ismail-Beigi F, Grimm RH Jr, et al. Effects of intensive glucose lowering in type 2 diabetes. N Engl J Med. 2008;358:2545–59.
9. Fonseca VA, Reynolds T, Hemphill D, Randolph C, Wall J, Valiquet TR, Graveline J, Fink LM. Effect of troglitazone on fibrinolysis and activated coagulation in patients with non-insulin-dependent diabetes mellitus. J Diabetes Complicat. 1998;12:181–6.
10. Brass LF, Diamond SL. Transport physics and biorheology in the setting of hemostasis and thrombosis. J Thromb Haemost. 2016;14:906–17.
11. Hosokawa K, Ohnishi T, Kondo T, Fukasawa M, Koide T, Maruyama I, Tanaka KA. A novel automated microchip flow-chamber system to quantitatively evaluate thrombus formation and antithrombotic agents under blood flow conditions. J Thromb Haemost. 2011;9:2029–37.
12. Hosokawa K, Ohnishi T, Sameshima H, Miura N, Ito T, Koide T, Maruyama I. Analysing responses to aspirin and clopidogrel by measuring platelet thrombus formation under arterial flow conditions. Thromb Haemost. 2013;109:102–11.
13. Hosokawa K, Ohnishi T, Fukasawa M, Kondo T, Sameshima H, Koide T, Tanaka KA, Maruyama I. A microchip flow-chamber system for quantitative assessment of the platelet thrombus formation process. Microvasc Res. 2012;83:154–61.
14. Yamaguchi Y, Moriki T, Igari A, Matsubara Y, Ohnishi T, Hosokawa K, Murata M. Studies of a microchip flow-chamber system to characterize whole blood thrombogenicity in healthy individuals. Thromb Res. 2013;132:263–70.
15. Osende JI, Badimon JJ, Fuster V, Herson P, Rabito P, Vidhun R, Zaman A, Rodriguez OJ, Lev EI, Rauch U, et al. Blood thrombogenicity in type 2 diabetes mellitus patients is associated with glycemic control. J Am Coll Cardiol. 2001;38:1307–12.
16. Chiasson JL, Josse RG, Gomis R, Hanefeld M, Karasik A, Laakso M. Acarbose for prevention of type 2 diabetes mellitus: the STOP-NIDDM randomised trial. Lancet. 2002;359:2072–7.
17. Raz I, Wilson PW, Strojek K, Kowalska I, Bozikov V, Gitt AK, Jermendy G, Campaigne BN, Kerr L, Milicevic Z, Jacober SJ. Effects of prandial versus fasting glycemia on cardiovascular outcomes in type 2 diabetes: the HEART2D trial. Diabetes Care. 2009;32:381–6.
18. Stratton IM, Adler AI, Neil HA, Matthews DR, Manley SE, Cull CA, Hadden D, Turner RC, Holman RR. Association of glycaemia with macrovascular and microvascular complications of type 2 diabetes (UKPDS 35): prospective observational study. Bmj. 2000;321:405–12.
19. Dunn EJ, Ariens RA, Grant PJ. The influence of type 2 diabetes on fibrin structure and function. Diabetologia. 2005;48:1198–206.
20. Pieters M, Covic N, van der Westhuizen FH, Nagaswami C, Baras Y, Toit Loots D, Jerling JC, Elgar D, Edmondson KS, van Zyl DG, et al. Glycaemic control improves fibrin network characteristics in type 2 diabetes - a purified fibrinogen model. Thromb Haemost. 2008;99:691–700.
21. Hammer MR, John PN, Flynn MD, Bellingham AJ, Leslie RD: Glycated fibrinogen: a new index of short-term diabetic control. Ann Clin Biochem 1989, 26 (Pt 1):58–62.

22. Ajjan RA, Gamlen T, Standeven KF, Mughal S, Hess K, Smith KA, Dunn EJ, Anwar MM, Rabbani N, Thornalley PJ, et al. Diabetes is associated with posttranslational modifications in plasminogen resulting in reduced plasmin generation and enzyme-specific activity. Blood. 2013;122:134–42.
23. Derosa G, Dangelo A, Ragonesi PD, Ciccarelli L, Piccinni MN, Pricolo F, Salvadeo S, Montagna L, Gravina A, Ferrari I, et al. Effects of rosiglitazone and pioglitazone combined with metformin on the prothrombotic state of patients with type 2 diabetes mellitus and metabolic syndrome. J Int Med Res. 2006;34:545–55.
24. Bruno G, Cavallo-Perin P, Bargero G, Borra M, D'Errico N, Pagano G. Association of fibrinogen with glycemic control and albumin excretion rate in patients with non-insulin-dependent diabetes mellitus. Ann Intern Med. 1996;125:653–7.
25. Barazzoni R, Kiwanuka E, Zanetti M, Cristini M, Vettore M, Tessari P. Insulin acutely increases fibrinogen production in individuals with type 2 diabetes but not in individuals without diabetes. Diabetes. 2003;52:1851–6.
26. Sam S, Haffner S, Davidson MH, D'Agostino RB Sr, Feinstein S, Kondos G, Perez A, Mazzone T. Relation of abdominal fat depots to systemic markers of inflammation in type 2 diabetes. Diabetes Care. 2009;32:932–7.
27. Juhan-Vague I, Thompson SG, Jespersen J. Involvement of the hemostatic system in the insulin resistance syndrome. A study of 1500 patients with angina pectoris. The ECAT angina pectoris study group. Arterioscler Thromb. 1993;13:1865–73.
28. Ceriello A, Giugliano D, Quatraro A, Dello Russo P, Torella R. Blood glucose may condition factor VII levels in diabetic and normal subjects. Diabetologia. 1988;31:889–91.
29. Konya H, Hasegawa Y, Hamaguchi T, Satani K, Umehara A, Katsuno T, Ishikawa T, Miuchi M, Kohri K, Suehiro A, et al. Effects of gliclazide on platelet aggregation and the plasminogen activator inhibitor type 1 level in patients with type 2 diabetes mellitus. Metabolism. 2010;59: 1294–9.
30. Singer J, Weissler Snir A, Leshem-Lev D, Rigler M, Kornowski R, Lev EI. Effect of intensive glycemic control on platelet reactivity in patients with long-standing uncontrolled diabetes. Thromb Res. 2014;134:121–4.
31. Schneider DJ. Factors contributing to increased platelet reactivity in people with diabetes. Diabetes Care. 2009;32:525–7.
32. Ogawa H, Nakayama M, Morimoto T, Uemura S, Kanauchi M, Doi N, Jinnouchi H, Sugiyama S, Saito Y. Low-dose aspirin for primary prevention of atherosclerotic events in patients with type 2 diabetes: a randomized controlled trial. Jama. 2008;300:2134–41.
33. Trovati M, Anfossi G, Cavalot F, Vitali S, Massucco P, Mularoni E, Schinco P, Tamponi G, Emanuelli G. Studies on mechanisms involved in hypoglycemia-induced platelet activation. Diabetes. 1986;35:818–25.
34. Mikhailidis DP, Barradas MA, Jeremy JY, Dandona P. Effect of alpha a-adrenoceptor antagonist on platelet activation during insulin-induced hypoglycaemia in type 2 (non-insulin-dependent) diabetes mellitus. Diabetologia. 1989;32:147–9.
35. Takeda H, Kishikawa H, Shinohara M, Miyata T, Suzaki K, Fukushima H, Ichinose K, Shichiri M. Effect of alpha 2-adrenoceptor antagonist on platelet activation during insulin-induced hypoglycaemia in type 2 (non-insulin-dependent) diabetes mellitus. Diabetologia. 1988;31:657–63.
36. Cryer PE. Mechanisms of hypoglycemia-associated autonomic failure in diabetes. N Engl J Med. 2013;369:362–72.
37. Kishikawa H, Takeda H, Kiyota S, Sakakida M, Fukushima H, Ichinose K, Matsuda H, Nakamura N, Uzawa H. Role of alpha 2-adrenergic receptor in platelet activation during insulin-induced hypoglycemia in normal subjects. Diabetes. 1987;36:407–12.
38. Gawaz M, Langer H, May AE. Platelets in inflammation and atherogenesis. J Clin Invest. 2005;115:3378–84.
39. Hutton RA, Mikhailidis D, Dormandy KM, Ginsburg J. Platelet aggregation studies during transient hypoglycaemia: a potential method for evaluating platelet function. J Clin Pathol. 1979;32:434–8.
40. De Caterina R, Madonna R, Sourij H, Wascher T. Glycaemic control in acute coronary syndromes: prognostic value and therapeutic options. Eur Heart J. 2010;31:1557–64.
41. Arima Y, Kaikita K, Ishii M, Ito M, Sueta D, Oimatsu Y, Sakamoto K, Tsujita K, Kojima S, Nakagawa K, et al. Assessment of platelet-derived thrombogenicity with the total thrombus-formation analysis system in coronary artery disease patients receiving antiplatelet therapy. J Thromb Haemost. 2016;14:850–9.
42. Idemoto Y, Miura SI, Norimatsu K, Suematsu Y, Hitaka Y, Shiga Y, Morii J, Imaizumi S, Kuwano T, Iwata A, et al. Evaluation of the antithrombotic abilities of non-vitamin K antagonist oral anticoagulants using the Total Thrombus-formation analysis system((R)). Heart Vessel. 2017;32:309–16.
43. Ishii M, Kaikita K, Ito M, Sueta D, Arima Y, Takashio S, Izumiya Y, Yamamoto E, Yamamuro M, Kojima S, et al. Direct Oral anticoagulants form Thrombus different from warfarin in a microchip flow chamber system. Sci Rep. 2017;7:7399.

3

Down-regulation of platelet adhesion receptors is a controlling mechanism of thrombosis, while also affecting post-transfusion efficacy of stored platelets

Ehteramolsadat Hosseini[1], Maryam Mohtashami[1] and Mehran Ghasemzadeh[1,2*]

Abstract

Physiologically, upon platelet activation, uncontrolled propagation of thrombosis is prevented by regulating mechanisms which affect the expression and function of either platelet adhesion receptors or integrins. Receptor ectodomain shedding is an elective mechanism which is mainly involved in down-regulation of adhesion receptors GPIbα and GPVI. Platelet integrin $\alpha_{IIb}\beta_3$ can also be modulated with a calpain-dependent proteolytic cleavage. In addition, activating signals may induce the internalization of expressed receptors to selectively down-regulate their intensity. Alternatively, further activation of platelets is associated with microvesiculation as a none-selective mechanism which leads to the loss of membrane- bearing receptors. In a non-physiological condition, the storage of therapeutic platelets has also shown to be associated with the unwilling activation of platelets which triggers receptors down-regulation via aforementioned different mechanisms. Notably, herein the changes are time-dependent and not controllable. While the expression and shedding of pro-inflammatory molecules can induce post-transfusion adverse effects, stored-dependent loss of adhesion receptors by ectodomain shedding or microvesiculation may attenuate post-transfusion adhesive functions of platelets causing their premature clearance from circulation. In its first part, the review presented here aims to describe the mechanisms involved in down-regulation of platelet adhesion receptors. It then highlights the crucial role of ectodomain shedding and microvesiculation in the propagation of "platelet storage lesion" which may affect the post-transfusion efficacy of platelet components.

Keywords: $\alpha_{IIb}\beta_3$, Adhesion, Ectodomain shedding, GPIbα, GPVI, Microparticle, Platelet refractoriness, Transfusion, Thrombosis

Highlights

- Different mechanisms of internalization, microvesiculation and ectodomain shedding modulate platelet receptors
- Platelets lose the expression and function of adhesion receptors during storage
- Ectodomain shedding and microvesiculation are mainly involved in platelet storage lesion (PSL)
- Platelet transfusion can be complicated by platelet refractoriness
- PSL decreases post-transfusion efficacy of platelets

Introduction

Platelets typically circulate in the bloodstream for 7–10 days and their principal function is to survey the inner lining of blood vessels to detect and seal any breaches in the vasculature by the creation of thrombi. The firm adhesion of platelets to the site of injury forms a monolayer which serves as a reactive site for further recruitment of free-flowing platelets. Platelet

* Correspondence: mehran1476@yahoo.com
[1]Blood Transfusion Research Centre, High Institute for Research and Education in Transfusion Medicine, Iranian Blood Transfusion Organization Building, Hemmat Exp. Way, Next to the Milad Tower, PO Box: 14665-1157, Tehran, Iran
[2]Australian Center for Blood Diseases, Monash University, Melbourne, Victoria 3004, Australia

accumulation is dependent on platelet-bound Willebrand factor (vWF) and fibrinogen which serves to cross-link adjacent platelets and promote stable platelet aggregation [1, 2]. This process is essential for the formation of a primary hemostatic plug (white thrombus) and also for the development of pathological thrombi at site of atherosclerotic plaque rupture [3]. Of note, in a developing thrombus further activation converts platelets from a proaggregatory to a to pro-coagulant phenotype which enables the assembly of the coagulation reaction complexes (the Tenase and Prothrombinase complex) on the cell surface, necessary for thrombin and fibrin generation. This supports secondary hemostasis by the conversion of the white thrombus to a red clot containing a planner of fibrin network and trapped RBCs [4]. Nonetheless, physiological thrombus formation (mural thrombi) is tightly regulated to avoid excessive platelet accumulation at the injury site and vascular obstruction, the principal pathological process causing heart attacks and ischaemic stroke [5]. Figure 1 demonstrates different steps of platelet tethering, adhesion, aggregation and thrombus formation.

The important functions of platelet adhesion receptors

Classically, thrombus formation on the site of vascular injury is triggered by the interaction of main platelets adhesive receptors Glycoprotein Ib/V/IX and Glycoprotein VI (GPVI) with their specific ligands which are exposed to circulation followed by endothelial damage. The initial capturing of free flowing platelets occurs through the binding of GPIb/V/IX to immobilized von vWF expressed at sites of vascular injury. This interaction slows down platelet movement and allows other adhesion receptors with slower-binding kinetics, including integrin $\alpha2\beta1$ and GPVI, as the most potent adhesion receptor to be engaged with the exposed collagen in sub-endothelial matrix. Followed by GPVI ligation with collagen, the induction of strong inside-out signals induces platelet activation and release. These events are associated with integrin activation on the surface of both adhered and adjacent free flowing platelets while the interaction of these activated integrins with fibrinogen/VWF can crosslink platelets to make aggregation and thrombus formation (Fig. 1a). Integrin ligation induce potent outside-in signals which augment cytosolic calcium influx associated with the conversion of pro-aggregatory phenotype of platelets located on developing thrombi to pro-inflammatory and pro-coagulant platelets [3]. P-selectin expressing platelets recruit leukocytes to the site of vascular injury while pro-coagulant platelets provide a highly efficient scaffold for coagulation-cascade activity and fibrin generation which develop clot formation [6]. At this stage, polymerized fibrin has also shown to recruit circulating rest platelets lacking activated integrin through the interaction with adhesive receptors Glycoprotein Ibα (GPIbα) and GPVI [7]. This second phase of platelet recruitments may provide a new scaffold for thrombin generation enhancing coagulant function. Developing thrombus also expresses a verity of pro-inflammatory mediators and receptors (Fig. 1b). Leukocytes recruitment to thrombi is initiated by the interaction between P-selectin and P-selectin glycoprotein ligand-1 (PSGL-1) which respectively are expressed on activated platelets and circulating rest leukocytes. Via this interaction, leukocyte begins to tether and roll around the thrombus while being affected by chemokines and chemoattractant which are continuously released from activated platelets. This cellular crosstalk induces potent inside-out signals which activate leukocyte integrin Machrophage-1 antigen (MAC-1). Followed by MAC-1 interaction with its main ligand in platelets, GPIbα, leukocytes firmly adhere to thrombi while getting fully activated and polarized. These activated leukocytes may release chromatin neutrophil extracellular traps (NETs) which ensnare more circulating cells including platelets while providing an efficient negatively charged matrix that propagates pro-coagulant function [6, 8] (Fig. 1b).

Different mechanisms modulating platelet adhesion receptors

Changes of platelet receptors expression by different stimuli or during storage can influence their surface density and ligand binding capacity, signal transduction as well as receptor cross-linking and clustering. Regardless of platelets functional activity which can be affected by these alterations, physiologically, the changes may also control platelet aging or clearance. Studies have shown that the expression of adhesion receptors on the surface of platelets can be modulated by four different mechanisms including A) internalization, B) microvesiculation (microparticulation), C) secretion and D) ectodomain shedding. In addition, followed by ectodomain shedding, soluble forms of receptors lacking a transmembrane domain may also interfere with receptor functional activity while during microvesiculation, microparticle -bearing receptors may also play some important physio-pathological roles [9–11]. The controlled termination of signaling pathways after ligand-induced activation is an important mechanism to ensure appropriate signal intensity and the consequent cellular response [12]. Internalization, secretion and ectodomain shedding are three mechanisms involved in dampening ligation capacity and signal transduction via the selective removal of receptors from platelet membrane [13, 14]. These mechanisms which are ignited from early stage of platelet activation modulate different aspects of cellular

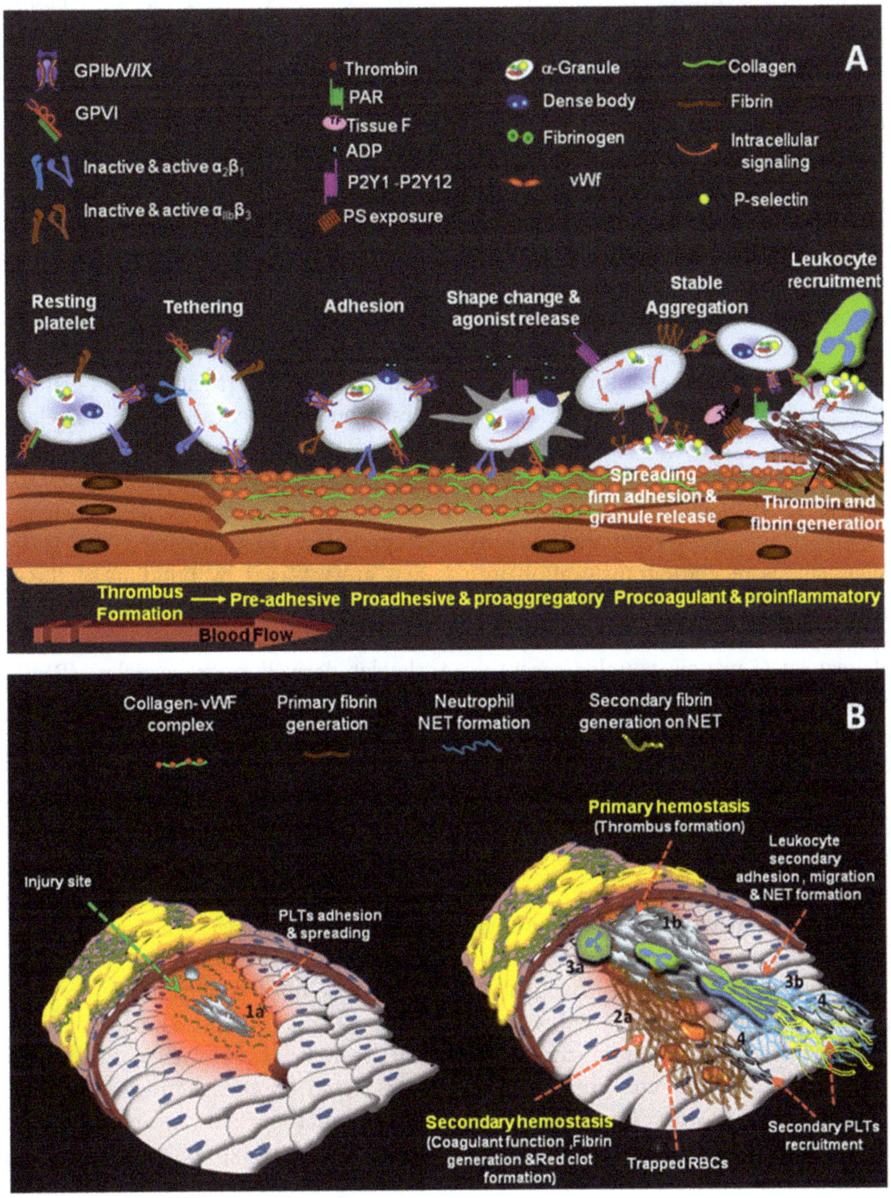

Fig. 1 a Different Stages of Platelet Adhesion to the Site of Vascular Injury. Resting platelets are initially captured on the site of vascular injury via the interaction between GPIb/V/IX and immobilized vWF ***(Tethering).*** Platelets firmly adhere to the sub-endothelial matrix through the engagement of collagen receptors $\alpha 2\beta 1$ and GPVI ***(Adhesion)***. This triggers potent inside out signals inducing ADP release from dense bodies (***Shape change & agonist release***) as well as activating platelet major integrin $\alpha_{IIb}\beta_3$. Integrins facilitate platelet ***spreading*** and subsequent ***aggregation*** through the binding to vWF and fibrinogen. Activating signals down-stream engaged receptors induce the ***release of granule*** contents including P-selectin which provides an efficient scaffold for ***leukocyte recruitment*** linking pro-aggregatory phase of platelet activation to pro-inflammatory function. On the other hand, the accumulative signals further activate platelets and induce sustained calcium influx which results in the surface exposure of phosphatidylserine (PS) and pro-coagulant function leading to thrombin production and fibrin generation at the site of injury. Interacting with PAR receptors, generated thrombin also acts as a potent agonist which supports more efficient ***pro-inflammatory and pro-coagulant*** function. **b** Primary and secondary hemostasis: mutual links between pro-inflammatory and pro-coagulant function 1- (**a**) Followed by the injury, platelet recruitment to the exposed sub-endothelial matrix leads to the formation of a developing thrombus(**b**) which express either pro-inflammatory molecules (mainly P-selectin) or pro-coagulant phospholipids (***primary hemostasis***). **2**- Pro-coagulant matrix presented by platelets provides an efficient scaffold for fibrin generation (**a**) which supports ***secondary hemostasis*** by the conversion of the white thrombus to a red clot containing a planner of fibrin network and trapped RBCs.**3-** Platelets recruits leukocyte(**a**) while during their crosstalk, neutrophils get fully activated and release their chromatin contents as extracellular NET(**b**). The negatively charged NET materials provide an efficient pro-coagulant scaffold for fibrin generation**. 4-**Platelets may also interact with generated fibrin while creating a secondary thrombus functions including platelets trafficking and adhesion to the site of vascular injury as well as their pro-inflammatory and pro-coagulant functions during thrombus formation. Damaged receptors specially affected by oxidant stress have also shown to be eliminated by such selective removal mechanisms [15]. On the other hand, microvesiculation is considered as a none-selective mechanism of receptor removal, usually from fully activated platelets during membrane shedding or from apoptotic platelets by membrane fragmentation. Although platelets continuously shed microparticles (MPs) even under resting condition, significant microvesiculation occurs in post-activated or apoptotic end stage platelets [16].

Internalization of platelet receptors

Internalization can also affect platelet receptors redistribution by different specific and nonspecific mechanisms. One of these mechanisms is receptor translocation into the open canalicular system (OCS). This is an important mechanism to regulate the redistribution of GPIb and GPIV [17] . GPVI, platelet G-protein coupled receptors (PAR-1, PAFR, P_2Y_1, P_2Y_{12}), integrins ($\alpha_2\beta_1$, $\alpha_{IIb}\beta_3$), Platelet-derived growth factor receptor α, thrombopoietin receptors and prostacyclin receptors are also other platelet receptors which are internalized by different mechanism [18–24].

General mechanisms of receptor internalization

Upon activation, internalization redistributes the receptors while shuttling them from the plasma membrane to endosomes where they may experience inactivation process through lysosomal degradation [25, 26]. Generally, endocytosis is triggered by ubiquitin molecules interaction with activated receptors and their related proteins which are associated with tyrosine kinase dependent Phosphorylation [12]. One of the important components of endocytic machinery is clathrin which is likely regulated by ubiquitination processes [27]. When a clathrin-coated pit forms on the inner surface of the cell membrane, it then buds inside the cell and forms a coated vesicle containing receptors and other membrane associated molecules. Cytoplasmic proteins including dynamin and adaptors such as adaptin may be also involved in endocytosis process [28]. Other molecular factors which may affect membrane receptors redistribution are included in cytoskeletal proteins and lipid composition [29].

Receptor endocytosis may also occur constitutively without cellular activation. This is a clathrin- independent pathway of endocytosis which is not affected by receptor antagonisms or associated with receptor signaling [26]. In some condition receptor redistribution may be induced even by non-molecular factors such as external physical stress which affects membrane integrity and cause none specific endocytosis [29].

Platelet Microvesiculation

Notably, since more than 80% of circulating microparticles express platelet antigens in healthy individuals, the physiological importance of platelet-originated microparticles and their relevance with the status of different diseases especially cardiovascular and inflammatory disorders, including arterial thrombosis, heparin-induced thrombocytopenia, immune thrombocytopenia, malaria infection, *acquired immune deficiency syndrome* (AIDS), and rheumatoid arthritis have been so far of interests in numerous researches [30, 31]. Platelets release two types of vesicles including exosomes (approximately 40–100 nm in diameter) which are shed by exocytosis from the multivesicular bodies as well as alpha-granules and microparticles (approximately 100–1000 nm in diameter) that are released by budding of the cytoplasmic membrane or by platelet fragmentation during apoptosis [10]. As a general marker, like other type of microparticles, platelet derived microparticles; (PMPs) usually express Annexin-V or tissue factor (TF). PMPs also express variety of platelet surface molecules including adhesive receptors (such as GPIbα, β_3 integrin (GPIIIa, CD61), α_{IIb}-integrin (GPIIb, CD41), GPVI, P-selectin, lysosomal-associated membrane protein3 (LAMP3, CD63), thrombospondin – 1 and Platelet endothelial cell adhesion molecule (PECAM)-1 [31, 32]. Therefore, membrane loss during microparticulation is associated with the loss of platelet surface proteins including adhesive receptors. Given this, microparticle formation can be considered as a down-regulating mechanism of platelet adhesiveness to reactive matrixes. However, PMPs show diverse phenotypes based on their activation states. Resting platelets-derived microparticles are phosphatidylserine (PS) negative, which not expressing pro-inflammatory molecules including P-selectin whereas activated platelets express PS, P-selectin and CD40 ligand (CD40L) [33]. In addition, as an another important pro-inflammatory molecule, PMPs also showed to express platelet-activating factor (PAF) [34]. Amongst adhesion molecules expressed by microparticle, GPVI also detected on megakaryocyte microparticles while not being expressed by microparticles released from activated platelets [35]. This may suggest that, contrary to other adhesion receptor GPIbα (which is significantly expressed by PMPs), GPVI down regulation on platelets cannot be affected by microparticulation and mainly modulated by other mechanisms including ectodomain shedding or internalization. Notably, regardless of expressed molecules, PMPs also comprise a vast array of biologic materials including growth factors, coagulation factors, enzymes, chemokines, cytokines, complement proteins, apoptosis regulators, bioactive lipids, miRNAs, β-amyloid precursor protein, Ca^{2+}-dependent protease) calpain (, which may play important roles in different biological functions [36].

Microvesiculation mechanism in platelets

Membrane integrity in resting platelets can be affected with intensive physical stress induced by some pathological states such as the presence of abnormal shear force in atherosclerotic vessels, mechanical artificial valve and microangiopathic conditions [37]. Contrary to agonist-induced microparticle generation or what occurs in activated platelets, Destabilization of the actin cytoskeleton in resting platelets can trigger microparticle shedding by the mechanisms which are not dependent to Ca^{2+} elevation and calpain activity [30]. However, the involvement of the phosphoinositide 3-kinase signaling pathway in the articulation of $\alpha_{IIb}\beta_3$-integrin with actin dynamics might be one of the possible mechanisms [38]. Upon platelet activation, the elevation of intracellular Ca^{2+} and ROS generation (through cyclooxygenase and NADPH oxidase) induce calpains activation and cytoskeletal changes. Clapains cleave skeleton associated proteins in platelet membrane while aberrantly destabilizing actin polymerization and cytoskeletal arrangement [39, 40]. Generally, Calpain activity in addition to the loss of membrane integrity induced by PS exposure, are the main mechanisms behind microparticle shedding in activated platelets. Moreover, shear stress-induced PMPs formation may also occur during platelet activation and adhesion through the interaction between GPIbα and vWf where the loosely adhered platelets are tethering off from the adhesive matrix. This means that the circulating detached platelets lose some parts of their membrane contents including adhesive receptors [41]. With fairly different mechanisms involved, platelet apoptosis is also associated with the significant microparticle formation during membrane fragmentation which might be mechanistically different from calcium dependent budding of cell membrane in activated platelets. Platelet apoptosis is associated with mitochondrial damage, cytochrome release and caspase activity which leads to both calpain dependent and independent cytoskeleton cleavage/reorganization resulting in microparticles release from remnant of the apoptotic cell [16, 42]. Activation based receptor modulation has been also shown to be relevant with microparticle formation. In addition to calpain, GPIbα modulation during platelet activation may also affect microvesiculation in stored platelets. A current study has indicated that upon platelet activation, cytosolic phospholipase A2 (PLA2) activity with subsequent AA production can result in dimerization of 14–3-3ζ and its interaction with the cytoplasmic domains of GPIbα in lipid rafts. While this interaction induces the stabilization and clustering of GPIbα, on the other hand, it causes the loss of the membrane anchoring interaction of Filamin A with this receptor, which affects platelet surface integrity and makes the membrane prone to microvesiculation [43].

Ectodomain shedding in platelets

Ectodomain shedding through limited proteolysis is a general mechanism to modulate the function of surface transmembrane proteins. This process usually occurs during cell activation and is associated with the proteolysis of a broad range of transmembrane macromolecules including adhesion receptors, numerous cytokines, growth factors and their receptors. However, resting (non-stimulated) cells also experience a basal level of constitutive ectodomain shedding which is increased in a time-dependent manner, while can be dramatically enhanced in response to activating stimuli or apoptotic conditions [44, 45]. Platelet adhesion and activation has been shown to be associated with surface protein shedding [46] and since platelets have not the translational capability to renew their full complement of surface proteins, ectodomain shedding is considered to be a rapid and irreversible key mechanism which effectively down-regulates the surface expression of platelet receptors and their signaling capacities [11, 13]. Endogenous platelet metalloproteinases (sheddases) including a disintegrin and metalloproteases (ADAMs) and matrix metalloproteinases (MMPs) are the key players in platelet ectodomain shedding [47]. Ectodomain shedding has been reported as an important mechanism for the regulation of GPIbα and GPV expressions on the platelet surface. In addition to down-regulation of the interaction between GPIb-V-IX and vWF, this phenomenon has been also suggested to be responsible for the clearance of aged platelets from the circulation [48]. Actually desialylation of GPIb-V-IX through sialidase activity also causes platelet clearance while inducing the irreversible formation of receptor clustering and the ectodomain shedding of GPIbα and GPV from the platelets [49, 50]. ADAM17 is the most prominent endogenous sheddas, which cleaves both GPIbα and GPV at specific sites releasing a 130- and an 80-kDa soluble fragment, respectively [48]. The 130 kDa fragment is referred as glycocalicin (GC) and circulates in plasma at a concentration of 1 to 3 μg/mL [51] . Generally, remarkable function of ADAM 17 occurs during platelet activation by a variety of agonists, however due to the presence of constitutive levels of mature ADAM17, even resting platelets experience basal levels of GPIbα and GPV shedding. Evidence for the physiological regulation of GPIbα by ADAM 17 can be derived from the observation that the concentration of glycocalicin in the plasma of the mice deficient in functional ADAM17 is considerably lower than wild type animals [52]. In addition to cell activation, GPIbα shedding also occurs during cell death. Treatment of platelets with carbonyl cyanide 3-chlorophenylhydrazone (CCCP) provokes mitochondrial injury and platelet apoptosis similar to that observed during the storage of platelets in vitro. In both scenarios, ADAM17 has been suggested as the responsible protease [45, 52, 53]. As the main collagen receptor, GPVI is another molecule which is shed from the platelet surface during platelet activation by agonists as well as shear stress. However, these shedding events are mainly modulated by ADAM10 while leading to the release of a soluble ~ 55-kDa fragment with a ~ 10 kDa platelet-associated remnant fragment. In addition, similar to GPIbα, the mitochondrial damage also shown to be associated with GPVI shedding in a mechanism which is mainly ADAM 17 dependent [54]. Resting platelets also constitutively present both ADAM 10 and ADAM 17. However contrary to what already mentioned for GPIb and despite the presence of ADAM10 with detectable proteolytic activity on the surface of resting platelets, GPVI expression has shown to be quite stable on circulating inactivated platelets. This might be due to cryptic nature of vulnerable site of GPVI which only expose to ADAM 10 proteolytic activity under the condition of ligand binding or membrane perturbation in elevated fluid shear stress [55].

Mechanistic features for the shedding of platelet adhesion receptors

Several lines of evidence suggested that ADAM-dependent shedding of platelet adhesion receptors can be modulated by Ca^{2+} elevation, protein kinase C (PKC) activation, PS exposure, and caspase activity [56]. A calcium-modulated protein, calmodulin is one of the important modulators of ectodomain shedding. It has been shown that the direct attachment of calmodulin to juxtamembrane cytoplasmic sequences of GPV, GPIbβ (from GPIb-V-IX complex) and GPVI receptor can interfere with the activity of the sheddase. This is supported by the observation that dissociation of calmodulin from the ectodomain cleavage sites, induced by calmodulin inhibitors such as W7, renders receptors more vulnerable to shedding. It has been shown that calmodulin disassociation from these inhibitors significantly increases the levels of sGPVI and GC. On the other hand, signaling events followed by platelet activation can increase intracellular calcium while its interaction with calmodulin can be associated with rapid conformational changes of the molecule and its dissociation from receptor cleavage sites which make them vulnerable to the proteolytic activities by ADAMs [57]. The another possibility is that the interaction between those affected calmodulin and some cytoskeleton modulators such as myosin light chain kinase (MLCK) causes cytoskeletal rearrangement which may induce shedding event by the effects on either receptors stability or ADAMs activity [40, 57, 58]. Additionally, calmodulin may affect receptor shedding through its involvement in other pathways including calcium/calmodulin-dependent protein kinase type I/II (CaMKI/II) while calmodulin antagonists also showed to induce receptor shedding by the activation of apoptotic pathways or calcium elevation [59, 60]. Generally, platelet activation and Ca^{2+} elevation is associated with increased levels of intracellular reactive oxygen species (ROS) [61]. ROS–induced oxidation of cysteine residues located on cysteine-rich domain of ADAMs can activate these proteolytic molecules, while ROS interaction with intracellular cytoplasmic domains of ADAMs may also increase their affinity with their substrate [58]. On the other hand, the direct or indirect interaction of ROS with cytoplasmic domains of the adhesive receptors GPVI may also induce shedding events. It has been indicated that following ligand binding of GPVI, rapid oxidation of an unpaired thiol in the cytoplasmic tail of this receptor results in the formation of GPVI homodimers, preceding down-stream signaling and ectodomain metalloproteolysis of GPVI [62, 63]. ROS also can indirectly affect the receptors and induce their shedding by the oxidation of cysteine residues of different related protein kinases including p38-MAPK (mitogen-activated protein kinase), PTP (protein tyrosine phosphatase), SFK (sarcoma family kinase) and PKs C, G, or A [64–66]. It is known that ADAM17 (as the main sheddas of platelet GPIbα and GPV) is activated via a p38 MAPK-dependent pathway which is per se modulated by accumulating ROS [67]. On the other hand, activated p38-MAPK can lead to the dimerization of the protein 14,3,3ζ attached to the cytoplasmic domain of GPIbα. This induces the receptor clustering and its disassociation from filamin which makes it unstable and susceptible to shedding events [40]. ROS-induced activation of p38-MAPK, PKC and SFK can enhance NADPH oxidase activity leading to further ROS generation [68]. The generated ROS can activate other signaling pathways such as PLC and PI3K which increase intracellular calcium by its either release from platelet DTS or entrance by specific channels [61]. Consequently, this ROS-induced calcium influx can enhance receptor shedding in the way already discussed.

Proteolytic modulation of platelet receptor integrin $\alpha_{IIb}\beta_3$

It has been currently shown that platelets can experience the ectodimain shedding of $\alpha_{IIb}\beta_3$ under non-physiological shear stress while the concentration of this receptor components has been increased in microparticle-free plasma [69]. However, more studies are required to characterize a specific sheddase for $\alpha_{IIb}\beta_3$

or define an in vivo pathophysiologic importance for this observation and its relevance to the mechanisms of integrin down-regulation. Additionally, a recent study also claimed that oxidative stress can induced modulation of platelet integrin $\alpha_{IIb}\beta_3$ expression and shedding in patients implanted with continuous flow left ventricular assist devices (CF-LVADs). However, in this study samples examined for the receptor shedding analysis were not microparticle-free and since PMPs are rich of $\alpha_{IIb}\beta_3$, both observed reduction in the expression of $\alpha_{IIb}\beta_3$ and higher levels of soluble integrin could be resulted from platelet microvesiculation rather than ectodomain shedding [70]. Regardless of these reports, so far all other studies have not reported ectodomain shedding for $\alpha_{IIb}\beta_3$ in any in vivo or in vitro experimental models. On the other hand, In different condition of either agonist-induced or storage-dependent PMPs formations, our western blot analysis also confirmed that using multi-steps ultra-centrifugation can totally reduce the soluble levels of $\alpha_{IIb}\beta_3$ in microparticle-free samples [54] . However, studies showed that, following platelet activation especially within sustained calcium influx, calpain-dependent proteolytic cleavage of cytoplasmic part of β3 may affect $\alpha_{IIb}\beta_3$ activation and clustering [48] .Several line of evidence have also reported calpain-mediated cleavage of integrin-dependent signaling molecules (including FAK, Src, PTP-1B) and some related cytoskeletal proteins (such as talin, filamin-1, and cortactin). This may also be considered as an alternative proteolytic pathway which can control the adhesive function of $\alpha_{IIb}\beta_3$ while regulating thrombus formation. However, the role of calpain in regulating integrin activity still remains controversial as shown that this protease may present either positive or negative effects on function of $\alpha_{IIb}\beta_3$ platelets [71].

Platelet receptors crosstalk and ectodomain shedding

Receptors crosstalk is also considered as another feature of shedding mechanism. The studies showed that the GPIb ligation causes its cross-interaction with FC receptors (FCR) γ and FCRIIa through the PI3K signaling pathway [72]. The interactions induce signaling cascades down-streams of these two receptors leading to the enhancement of GPVI shedding [73]. C-type lectin-like receptor 2 (CLEC-2) is a currently described molecules expressed by platelets, which can interact with Podoplanin as a widely expressed molecule outside of the vasculature while inducing activating signals through a hemimmunoreceptor tyrosine-based activation motif (hemITAM). It has been shown that the ligation of this receptor is associated with the proteolytic cleavage of GPVI and FcgRIIa. However, there is no evidence that shows CLEC-2 shedding during platelet activation [31] .Integrin involvement in ADAMs activation is another calcium-dependent mechanism which can control shedding events. It is known that followed by inside-out signaling, integrin gets primed while its ligation with fibrinogen can induce significant activating outside-in signals which ignite calcium influx, ROS generation and consequent receptors shedding [61, 74]. Still, there is one more intriguing mechanism by which integrin can modulate shedding through its direct interaction with ADAMs. In one side, integrin binding to the ADAMs can activate these molecules enhancing the shedding of adhesive receptors while this interaction also can modulate integrin ligation capacities attenuating outside-in signaling [75]. It seems that, this mechanism can efficiently down-regulate platelets adhesive function through the either decrease in platelet activation or increase of receptor shedding. On Receptor shedding could also be essentially Ca^{2+} independent. There are some PKC mimetic compounds, such as phorbol-myristate-acetat (PMA) which act independent of Ca^{2+} while significantly increasing ADAMs activities and shedding [56, 76].

Platelet apoptosis and receptor shedding

Platelet apoptosis is associated with the significant ectodomain shedding of surface receptors. Ex vivo studies showed that the treatment of human platelets with BH3 mimetic agents can decrease platelet adhesive function due to ectodomain shedding of GPIbα and GPVI. On the other hand, thrombocytopenic mice deficient in Bcl-xL also showed the reduced expression of GPIbα and GPVI in platelets [46, 77], confirming the important role of apoptotic pathway in shedding process. In addition to BH3 mimetic agents that induce a time- and dose-dependent decrease in platelet adhesive function resulted from ectodomain shedding of GPIbα and GPVI, the mitochondrial poison carbonyl cyanide 3-chlorophenylhydrazone (CCCP) also creates premature platelet ageing and shedding of these platelet receptors [78]. Although, apoptosis-induced shedding of receptors is mainly independent of Ca2+, its exact mechanism is not well characterized yet, as BH3 mimetic agents like ABT 737 also showed to induce a small increases in Ca2+ in a minor platelet population [79]. It has been shown that under apoptotic condition, metalloproteinase inhibitors can significantly reduce receptor shedding. Therefore, it is postulated that the activation of metalloproteinase downstream of mitochondrial injury and caspase activation can be the main event that regulates receptor shedding under apoptotic condition. Studies showed that target receptor shedding by activated ADAM17 is modulated in a p38-dependent manner while the inhibition of p38-MAPK can block the shedding of GPIbα in stored platelets [46]. On the other hand, studies also confirmed that p38-MAPK significantly plays a role in ABT 737-induced platelet apoptosis; however whether the activation of this kinase during apoptosis can be also involved in platelet receptor shedding requires further investigation [80]. Of note, in apoptotic condition, the possible mechanism by which MAP kinase regulating ADAMs activity and shedding cannot be ROS-dependent as already described under activation status because BH3 mimetic agents did not show to whether induce or increase intra platelet ROS generation [80].

Physiological significance for the shedding of platelet adhesive receptors

In a physiological basis, shedding of adhesion receptors is considered as part of a mechanistic process facilitating platelet spreading via the sequential events of platelet de-attachment and re-attachment at the tips of filopodia or the leading edge of lamellipodia. In addition receptor shedding is an important mechanism modulating thrombotic function through a controlled down-regulation of platelets reactivity [48]. However, in some cases, the uncontrolled shedding of adhesion receptors can molest platelet function and hemostasis [81]. Lowered surface density of GPIbα and GPVI due to the shedding can attenuate signaling capacity which affects platelet activation and secretion. This is in addition to the platelet adhesive malfunction which is in a direct correlation with lower expression of these receptors. On the other hand, ectodomain shedding of adhesion receptors can also decrease the number of platelet-matrix contacts, which potentiates thrombo-embolic events [48] . It has been shown that in thrombi formed ex vivo, procoagulant platelets undergoing ADAM-induced glycoprotein shedding are less-adhesive due to defects in GPIbα and GPVI dependent adhesion to VWF, collagen, and fibrin. In addition calpain-induced integrin cleavage in pro-coagulant platelets can be also considered as another reason aggravating platelet signaling capacity and adhesive capabilities [56]. Notably, pro-coagulant function is associated with the excessive microparticle formation leading to the loss of platelet membrane containing adhesion receptors [81]. Thus in a same scenario but with a different mechanism, platelets lose more adhesive capacities within post activation stage.

Platelet storage and increasing loss of adhesion receptors

Transfusion of platelet concentrates (PCs) is one of the most important therapeutic approaches in the management of thrombocytopenia and bleeding complications. Regardless of preparation techniques, stored platelets gradually experience inevitable deleterious changes so-called platelet storage lesion (PSL) that may lead to a progressive structural and functional damage. These changes which begin from the time of platelets isolation till the transfusion to a recipient are mainly induced by either reversible or irreversible increases in the basal levels of platelet activation. The most important phase of PSL is associated with irreversible changes in platelet morphology and function, which are significantly initiated with platelet α granules release and its conversion from pro-aggregatory status to pro-inflammatory phenotypes identified by the expression of P-selectin and CD40L [54] . It has been shown that along with platelet activation, the increased levels of intracellular calcium acts as the main modulator of signaling events which results in platelet receptor ectodomain shedding and membrane loss due to microparticulation. Both of these phenomena cause progressive loss of adhesive receptors during platelet storage [82]. Notably this extensive loss of platelet receptors may affect its proper function required for therapeutic uses [54, 83]. Decreasing levels of GPIbα expression during the storage of platelet concentrates (PCs) were reported by several studies [54]. Receptor ectodomain shedding and platelet membrane loss due to microparticulation are two important causes for this gradual decrease [82]. We already indicated that the continuous shedding can lead to significant decrease in GPIbα expression in 5-day stored platelets [54]. It is postulated that the down-regulation of GPIbα expression affect the adhesive function of transfused platelet in elevated arterial shear rates [84]. Obviously the older platelet products with the lower GPIbα expression may show less functionality after transfusion. This is in addition to the fact that metalloproteinase- dependent loss of surface GPIbα plays a role in the clearance of aged platelets from the circulation [52]. Therefore, the higher levels of shedding for this receptor in older PCs could be also associated with a rapid clearance of transfused platelets leading to reduced platelet recovery and survival. GPVI expression also showed to be modulated during storage. We already demonstrated a decreasing expression of GPVI during storage where was directly correlated with the GPIbα expression. A negative correlation between GPVI expression and shedding was also observed, a finding that verifies the main role of ectodomain shedding in the modulation of GPVI expression. It is worthy to know that given the absence of GPVI in the proteome of platelet microparticles, dislike GPIbα, the reducing levels of GPVI expression should be mainly due to ectodomain shedding rather than microvesiculation. A direct correlation between the shedding levels of P-selectin and GPVI during storage suggested this adhesion receptor as a valid marker of PSL [54] . This finding was also consistent with data from clinical studies in which the levels of soluble GPVI was correlated with soluble P-selectin in patients with acute coronary syndrome and or with

Acute Ischemic Stroke [85]. Further studies also suggested soluble GPVI as a helpful biomarker to monitor platelet activation during anti-platelet therapy [86]. Notably considerable shedding of GPVI and its association with decreasing expression of this receptor on the surface of stored platelets can affect GPVI dependent platelet function during storage. Some studies showed significant attenuation in platelet adhesion capacity to collagen in stored PCs [87, 88]. We also showed a direct correlation between this observed decreasing adhesive capacities with the increasing levels of GPVI shedding during storage. However, whether the observed loss of GPVI during storage can render platelet signaling capacities and integrin activation was another important question which has been addressed by the observation of attenuating platelet aggregation in responses to collagen during storage and its significant reverse correlation with GPVI shedding in stored PCs [89] .

Clinical significance of PSL

Although clinical research has not yet found any significant correlation between PSL and post-transfusion adverse effects or bleeding events in the patients, these studies have confirmed that storage duration can affect platelets survival in circulation. It is known that transfusion of fresher platelets is accompanied with higher CCI (corrected count increment) and more prolonged transfusion intervals in patients [90]. In other word, clinically, transfusion of the longer stored PCs has been associated with reduced CCI, platelet recovery and survival [91]. This might be due to the fact that in longer stored products, platelets with P-selectin exposure, GPIbα shedding and reduced membrane-associated sialic acid are selectively cleared from circulation [87, 92] .Using the mice models, in vivo studies showed that in stored platelets the inhibition of GPIbα shedding by metalloproteinase inhibitor or 5G6 Fab fragment can promote post transfusion recovery of platelets while improving their hemostatic function [87, 93]. Conclusively these studies suggest that GPIbα shedding reduces platelet reactivity while increasing its clearance in vivo. However, while lower CCI, platelet recovery, clearance and survival are much related to expression levels of adhesion receptors has not well defied yet.

Conclusion

Therapeutic platelet transfusion can be complicated by platelet refractoriness either modulated by immune or non-immune causes. Immune-related refractoriness especially those induced by antibodies against HLA antigens are the most studied causes of post transfusion platelet malfunctions. However platelet storage dependent lesion seems to play an important role in the premature clearance of transfused circulating platelets, which also results in lower efficacy for platelet transfusion. Evidently platelet storage is associated with the ectodomain shedding and microvesiculation which down-regulate platelet adhesion receptors, GPIbα and GPVI. This is in addition to the storage-dependent loss of integrin activity, which is also crucially involved in platelet aggregability and thrombus formation. Whereas these alterations have shown to irreversibly affect platelet function in vitro, clinical observations have indicated that the premature clearance or lower survivals rather than platelet malfunction is the only important adverse effect of platelet transfusion. Therefore given the improvement of hemostasis in the patients who transfused the longer-stored platelets, some studies suggested that platelets dysfunction due to PSL is generally restored after transfusion. However, this is a controversial viewpoint, as basic studies have confirmed that platelet adhesion during storage is affected by irreversible down-regulating mechanisms such as ectodomain shedding, micopaticulation and other proteolytic cleavage which permanently destroy some receptor adhesive and signaling function. Thereby highlighting detailed mechanisms involved in down-regulation of platelet adhesion receptor,This review may benefit the readership to better judge whether the functional capabilities of stored platelets can be properly recovered after transfusion or not.

Abbreviations

ADAMs: A disintegrin and metalloproteases; AIDS: Acquired immune deficiency syndrome; CaMKI/II: Calcium/calmodulin-dependent protein kinase type I/II; CCCP: Carbonyl cyanide 3-chlorophenylhydrazone; CCI: Corrected count increment; CD40L: CD40 ligand; CF-LVADs: Continuous flow left ventricular assist devices; CLEC-2: C-type lectin-like receptor 2; FCR: FC receptors; GC: Glycocalicin; GPIbα: Glycoprotein Ibα; GPVI: Glycoprotein VI; hemITAM: Hemimmunoreceptor tyrosine-based activation motif; LAMP3: Lysosomal-associated membrane protein 3; MAC-1: Machrophage-1 antigen; MAPK: Mitogen-activated protein kinase; MMPs: Matrix metalloproteinases; MPs: Microparticles; MLCK: Myosin light chain kinase; NETs: Neutrophil extracellular traps; OCS: Open canalicular system; PI3K: Phosphoinositide 3-kinases; PMA: Phorbol-myristate-acetat; PS: Phosphatidylserine; PLA2: Phospholipase A2; PAF: Platelet activating factor; PCs: Platelet concentrates; PECAM: Platelet endothelial cell adhesion molecule; PMPs: Platelet derived microparticles; PSL: Platelet storage lesion; PKC: Protein kinase C; PSGL-1: P-selectin glycoprotein ligand-1; PTP: Protein tyrosine phosphatase; ROS: Reactive oxygen species; SFK: Sarcoma family kinase; TF: Tissue factor; vWF: Von Willebrand factor

Acknowledgements

A few parts of this review are extracted from MG's doctorate thesis at Monash University (as referred). The authors wish to thank Prof. Shaun P. Jackson from Australian center for blood diseases for his constructive advice.

Authors' contributions

MG provided intellectual input designed and wrote the paper as main author, depicted the figures. EH provided intellectual input, co-wrote the paper and depicted the figures. MM provided intellectual input and helped with the writing of paper. All authors read and approved the final manuscript.

References

1. Kulkarni S, Dopheide SM, Yap CL, Ravanat C, Freund M, Mangin P, Heel KA, Street A, Harper IS, Lanza F, Jackson SP. A revised model of platelet aggregation. J Clin Invest. 2000;105(6):783–91.
2. Ruggeri ZM, factor v W. J Clin Invest. 1997;100(11 Suppl):S41–6.
3. Hosseini E, Ghasemzadeh M. Different stages of platelet adhesion to the site of vascular injury. Iranian J Blood Cancer. 2012;4(3):133–42.
4. Investigation of Signaling Cross-talk Between Platelets and Neutrophils. Melbourne: Monash University; 2009.
5. Gibbins JM. Platelet adhesion signalling and the regulation of thrombus formation. J Cell Sci. 2004;117(16):3415–25.
6. Ghasemzadeh M, Hosseini E. Platelet-leukocyte crosstalk: linking proinflammatory responses to procoagulant state. Thromb Res. 2013;131(3): 191–7.
7. Mammadova-Bach E, Ollivier V, Loyau S, Schaff M, Dumont B, Favier R, Freyburger G, Latger-Cannard V, Nieswandt B, Gachet C, Mangin PH, Jandrot-Perrus M. Platelet glycoprotein VI binds to polymerized fibrin and promotes thrombin generation. Blood. 2015;126(5):683–91.
8. Ghasemzadeh M, Hosseini E. Intravascular leukocyte migration through platelet thrombi: directing leukocytes to sites of vascular injury. Thromb Haemost. 2015;113(06):1224–35.
9. Gardiner EE, Al-Tamimi M, Andrews RK, Berndt MC. Platelet receptor shedding. Methods Mol Biol. 2012;788:321–39.
10. Melki I, Tessandier N, Zufferey A, Boilard E. Platelet microvesicles in health and disease. Platelets. 2017;28(3):214–21.
11. Chen Z, Mondal NK, Ding J, Koenig SC, Slaughter MS, Griffith BP, Wu ZJ. Activation and shedding of platelet glycoprotein IIb/IIIa under non-physiological shear stress. Mol Cell Biochem. 2015;409(1–2):93–101.
12. Holler D, Dikic I. Receptor endocytosis via ubiquitin-dependent and -independent pathways. Biochem Pharmacol. 2004;67(6):1013–7.
13. Hayashida K, Bartlett AH, Chen Y, Park PW. Molecular and cellular mechanisms of ectodomain shedding. Anat Rec (Hoboken). 2010;293(6): 925–37.
14. Watanabe S, Boucrot E. Fast and ultrafast endocytosis. Curr Opin Cell Biol. 2017;47:64–71.
15. Piper RC, Lehner PJ. Endosomal transport via ubiquitination. Trends Cell Biol. 2011;21(11):647–55.
16. Flaumenhaft R. Formation and fate of platelet microparticles. Blood Cells Mol Dis. 2006;36(2):182–7.
17. Lopez-Vilchez I, Hedner U, Altisent C, Diaz-Ricart M, Escolar G, Galan AM. Redistribution and hemostatic action of recombinant activated factor VII associated with platelets. Am J Pathol. 2011;178(6):2938–48.
18. Wang Z, Leisner TM, Parise LV. Platelet α2β1 integrin activation: contribution of ligand internalization and the α2-cytoplasmic domain. Blood. 2003;102(4): 1307–15.
19. Smith TH, Li JG, Dores MR, Trejo J. Protease-activated receptor-4 and purinergic receptor P2Y12 dimerize, co-internalize, and activate Akt signaling via endosomal recruitment of beta-arrestin. J Biol Chem. 2017; 292(33):13867–78.
20. Dupre DJ, Chen Z, Le Gouill C, Theriault C, Parent JL, Rola-Pleszczynski M, Stankova J. Trafficking, ubiquitination, and down-regulation of the human platelet-activating factor receptor. J Biol Chem. 2003;278(48):48228–35.
21. Avrov K, Kazlauskas A. The role of c-Src in platelet-derived growth factor α receptor internalization. Exp Cell Res. 2003;291(2):426–34.
22. Rabie T, Strehl A, Ludwig A, Nieswandt B. Evidence for a role of ADAM17 (TACE) in the regulation of platelet glycoprotein V. J Biol Chem. 2005; 280(15):14462–8.
23. Midgett C, Stitham J, Martin KA, Hwa J. Prostacyclin receptor regulation-from transcription to trafficking. Curr Mol Med. 2011;11(7):517–27.
24. Gao W, Shi P, Chen X, Zhang L, Liu J, Fan X, Luo X. Clathrin-mediated integrin alphaIIbbeta3 trafficking controls platelet spreading. Platelets. 2018; 29(6):610–21.
25. Koenig JA. Assessment of receptor internalization and recycling. In: Receptor signal transduction protocols, vol. 259; 2004. p. 249–73.
26. Segawaa JSK. Constitutive exposure of phosphatidylserine on viable cell. PNAS. 2012;108(48):19246–51.
27. Piper RC. Endosomal Transportation via Ubiquitination. Trends Cell Biol. 2011;21(11):647–55.
28. Ehrlich WBM. Endocytosis by random initiation and stabilizatio of clathrin-coated pits. Cell Press. 2004;118(5):591–605.
29. Nishimura NMT. Membrane re-modelling by BAR domain superfamily proteins via molecular and non-molecular factors. Biochem Soc Trans. 2018; 46(2):379–89.
30. Morel O, Jesel L, Freyssinet JM, Toti F. Cellular mechanisms underlying the formation of circulating microparticles. Arterioscler Thromb Vasc Biol. 2011; 31(1):15–26.
31. Gitz E, Pollitt AY, Gitz-Francois JJ, Alshehri O, Mori J, Montague S, Nash GB, Douglas MR, Gardiner EE, Andrews RK. CLEC-2 expression is maintained on activated platelets and on platelet microparticles. Blood. 2014;124(14):2262–70.
32. Badimon L, Suades R, Fuentes E, Palomo I, Padro T. Role of platelet-derived microvesicles as crosstalk mediators in Atherothrombosis and future pharmacology targets: a link between inflammation, atherosclerosis, and thrombosis. Front Pharmacol. 2016;7:293.
33. Flaumenhaft R, Dilks JR, Richardson J, Alden E, Patel-Hett SR, Battinelli E, Klement GL, Sola-Visner M, Italiano JE Jr. Megakaryocyte-derived microparticles: direct visualization and distinction from platelet-derived microparticles. Blood. 2009;113(5):1112–21.
34. Mitsios JV, Vini MP, Stengel D, Ninio E, Tselepis AD. Human platelets secrete the plasma type of platelet-activating factor acetylhydrolase primarily associated with microparticles. Arterioscler Thromb Vasc Biol. 2006;26(8):1907–13.
35. Lok CA, Nieuwland R, Sturk A, Hau CM, Boer K, Vanbavel E, Vanderpost JA. Microparticle-associated P-selectin reflects platelet activation in preeclampsia. Platelets. 2007;18(1):68–72.
36. Edelstein LC. The role of platelet microvesicles in intercellular communication. Platelets. 2017;28(3):222–7.
37. Boilard E, Duchez AC, Brisson A. The diversity of platelet microparticles. Curr Opin Hematol. 2015;22(5):437–44.
38. Platelet responsiveness and function during storage : implications for platelet transfusion therapy. Maastricht: Maastricht University; 2007.
39. Bhullar J, Bhopale VM, Yang M, Sethuraman K, Thom SR. Microparticle formation by platelets exposed to high gas pressures - an oxidative stress response. Free Radic Biol Med. 2016;101:154–62.
40. Ghasemzadeh M, Hosseini E, Roudsari ZO, Zadkhak P. Intraplatelet reactive oxygen species (ROS) correlate with the shedding of adhesive receptors, microvesiculation and platelet adhesion to collagen during storage: does endogenous ROS generation downregulate platelet adhesive function? Thromb Res. 2018;163:153–61.
41. Nomura MSS. Clinical significance of procoagulant microparticle. J Intensive Care. 2015;3(1):2.
42. Morel O, Toti F, Jesel L, Freyssinet JM. Mechanisms of microparticle generation: on the trail of the mitochondrion! Semin Thromb Hemost. 2010; 36(8):833–44.
43. Smethurst PA. Aging of platelets stored for transfusion. Platelets. 2016;27(6): 526–34.
44. Goth CK, Halim A, Khetarpal SA, Rader DJ, Clausen H, Schjoldager KT. A systematic study of modulation of ADAM-mediated ectodomain shedding by site-specific O-glycosylation. Proc Natl Acad Sci U S A. 2015;112(47): 14623–8.
45. Arribas J, Borroto A. Protein ectodomain shedding. Chem Rev. 2002;102(12): 4627–38.
46. Au AE, Josefsson EC. Regulation of platelet membrane protein shedding in health and disease. Platelets. 2017;28(4):342–53.
47. Andrews RK, Gardiner EE. Metalloproteolytic receptor shedding...Platelets "acting their age". Platelets. 2016;27(6):512–8.
48. Andrews RK, Karunakaran D, Gardiner EE, Berndt MC. Platelet receptor proteolysis: a mechanism for downregulating platelet reactivity. Arterioscler Thromb Vasc Biol. 2007;27(7):1511–20.
49. Quach ME, Chen W, Li R. Mechanisms of platelet clearance and translation to improve platelet storage. Blood. 2018;131(14):1512–21.
50. Jansen AJ, Josefsson EC, Rumjantseva V, Liu QP, Falet H, Bergmeier W, Cifuni SM, Sackstein R, von Andrian UH, Wagner DD, Hartwig JH, Hoffmeister KM. Desialylation accelerates platelet clearance after refrigeration and initiates GPIbalpha metalloproteinase-mediated cleavage in mice. Blood. 2012;119(5): 1263–73.
51. Coller BS, Kalomiris E, Steinberg M, Scudder LE. Evidence that glycocalicin circulates in normal plasma. J Clin Invest. 1984;73(3):794–9.
52. Bergmeier W, Piffath CL, Cheng G, Dole VS, Zhang Y, von Andrian UH, Wagner DD. Tumor necrosis factor-alpha-converting enzyme (ADAM17) mediates GPIbalpha shedding from platelets in vitro and in vivo. Circ Res. 2004;95(7):677–83.
53. Aktas B, Pozgajova M, Bergmeier W, Sunnarborg S, Offermanns S, Lee D, Wagner DD, Nieswandt B. Aspirin induces platelet receptor shedding via ADAM17 (TACE). J Biol Chem. 2005;280(48):39716–22.
54. Hosseini E, Ghasemzadeh M, Nassaji F, Jamaat ZP. GPVI modulation during

platelet activation and storage: its expression levels and ectodomain shedding compared to markers of platelet storage lesion. Platelets. 2017; 28(5):498-508.
55. Gardiner EE. Proteolytic processing of platelet receptors. Res Pract Thromb Haemost. 2018;2(2):240–50.
56. Baaten CC, Swieringa F, Misztal T, Mastenbroek TG, Feijge MA, Bock PE, Donners MM, Collins PW, Li R, van der Meijden PE. Platelet heterogeneity in activation-induced glycoprotein shedding: functional effects. Blood advances. 2018;2(18):2320–31.
57. Gardiner E, Arthur J, Berndt M, Andrews R. Role of calmodulin in platelet receptor function. Curr Med Chem Cardiovasc Hematol Agents. 2005;3(4):283–7.
58. Hartmann M, Herrlich A, Herrlich P. Who decides when to cleave an ectodomain? Trends Biochem Sci. 2013;38(3):111–20.
59. Andrews RK, Munday AD, Mitchell CA, Berndt MC. Interaction of calmodulin with the cytoplasmic domain of the platelet membrane glycoprotein Ib-IX-V complex. Blood. 2001;98(3):681–7.
60. Wang Z, Li S, Shi Q, Yan R, Liu G, Dai K. Calmodulin antagonists induce platelet apoptosis. Thromb Res. 2010;125(4):340–50.
61. Stalker TJ, Newman DK, Ma P, Wannemacher KM, Brass LF. Platelet signaling. Handb Exp Pharmacol. 2012;210:59–85.
62. Arthur JF, Shen Y, Kahn ML, Berndt MC, Andrews RK, Gardiner EE. Ligand binding rapidly induces disulfide-dependent dimerization of glycoprotein VI on the platelet plasma membrane. J Biol Chem. 2007;282(42):30434–41.
63. Qiao J, Arthur JF, Gardiner EE, Andrews RK, Zeng L, Xu K. Regulation of platelet activation and thrombus formation by reactive oxygen species. Redox Biol. 2018;14:126–30.
64. Giannoni E, Buricchi F, Raugei G, Ramponi G, Chiarugi P. Intracellular reactive oxygen species activate Src tyrosine kinase during cell adhesion and anchorage-dependent cell growth. Mol Cell Biol. 2005;25(15):6391–403.
65. Yoo SK, Freisinger CM, LeBert DC, Huttenlocher A. Early redox, Src family kinase, and calcium signaling integrate wound responses and tissue regeneration in zebrafish. J Cell Biol. 2012;199(2):225–34.
66. Ray PD, Huang BW, Tsuji Y. Reactive oxygen species (ROS) homeostasis and redox regulation in cellular signaling. Cell Signal. 2012;24(5):981–90.
67. Brill A, Chauhan AK, Canault M, Walsh MT, Bergmeier W, Wagner DD. Oxidative stress activates ADAM17/TACE and induces its target receptor shedding in platelets in a p38-dependent fashion. Cardiovasc Res. 2009;84(1):137–44.
68. Yamamori T, Inanami O, Nagahata H, Cui Y-D, Kuwabara M. Roles of p38 MAPK, PKC and PI3-K in the signaling pathways of NADPH oxidase activation and phagocytosis in bovine polymorphonuclear leukocytes. FEBS Lett. 2000;467(2–3):253–8.
69. Bender M, Stegner D, Nieswandt B. Model systems for platelet receptor shedding. Platelets. 2017;28(4):325–32.
70. Mondal NK, Chen Z, Trivedi JR, Sorensen EN, Pham SM, Slaughter MS, Griffith BP, Wu ZJ. Oxidative stress induced modulation of platelet integrin α2bβ3 expression and shedding may predict the risk of major bleeding in heart failure patients supported by continuous flow left ventricular assist devices. Thromb Res. 2017;158:140–8.
71. Kulkarni S, Jackson SP. Platelet factor XIII and calpain negatively regulate integrin alphaIIbbeta3 adhesive function and thrombus growth. J Biol Chem. 2004;279(29):30697–706.
72. Wu Y, Asazuma N, Satoh K, Yatomi Y, Takafuta T, Berndt MC, Ozaki Y. Interaction between von Willebrand factor and glycoprotein Ib activates Src kinase in human platelets: role of phosphoinositide 3-kinase. Blood. 2003; 101(9):3469–76.
73. Gardiner EE, Karunakaran D, Arthur JF, Mu FT, Powell MS, Baker RI, Hogarth PM, Kahn ML, Andrews RK, Berndt MC. Dual ITAM-mediated proteolytic pathways for irreversible inactivation of platelet receptors: de-ITAM-izing FcgammaRIIa. Blood. 2008;111(1):165–74.
74. Cosentino-Gomes D, Rocco-Machado N, Meyer-Fernandes JR. Cell signaling through protein kinase C oxidation and activation. Int J Mol Sci. 2012;13(9): 10697–721.
75. Bridges LC, Bowditch RD. ADAM-integrin interactions: potential integrin regulated ectodomain shedding activity. Curr Pharm Des. 2005;11(7):837–47.
76. Yanez-Mo M, Sanchez-Madrid F, Cabanas C. Membrane proteases and tetraspanins. Biochem Soc Trans. 2011;39(2):541–6.
77. Jackson SP, Schoenwaelder SM. Procoagulant platelets: are they necrotic? Blood. 2010;116(12):2011–8.
78. Andrews RK, Gardiner EE. Basic mechanisms of platelet receptor shedding. Platelets. 2017;28(4):319–24.
79. van Kruchten R, Mattheij NJ, Saunders C, Feijge MA, Swieringa F, Wolfs JL, Collins PW, Heemskerk JW, Bevers EM. Both TMEM16F-dependent and TMEM16F-independent pathways contribute to phosphatidylserine exposure in platelet apoptosis and platelet activation. Blood. 2013;121(10): 1850–7.
80. Rukoyatkina N, Mindukshev I, Walter U, Gambaryan S. Dual role of the p38 MAPK/cPLA2 pathway in the regulation of platelet apoptosis induced by ABT-737 and strong platelet agonists. Cell Death Dis. 2013;4:e931.
81. Chen Z, Mondal NK, Ding J, Gao J, Griffith BP, Wu ZJ. Shear-induced platelet receptor shedding by non-physiological high shear stress with short exposure time: glycoprotein Ibα and glycoprotein VI. Thromb Res. 2015; 135(4):692–8.
82. Gardiner EE, Andrews RK. Platelet receptor expression and shedding: glycoprotein Ib-IX-V and glycoprotein VI. Transfus Med Rev. 2014;28(2):56–60.
83. Ghasemzadeh M, Hosseini E. Platelet granule release is associated with reactive oxygen species generation during platelet storage: a direct link between platelet pro-inflammatory and oxidation states. Thromb Res. 2017; 156:101–4.
84. Jilma-Stohlawetz P, Horvath M, Eichelberger B, Koren D, Jilma B, Panzer S. Platelet function under high-shear conditions from platelet concentrates. Transfusion. 2008;48(1):129–35.
85. Pennings GJ, Yong AS, Wong C, Al-Tamimi M, Gardiner EE, Andrews RK, Kritharides L. Circulating levels of soluble EMMPRIN (CD147) correlate with levels of soluble glycoprotein VI in human plasma. Platelets. 2014;25(8):639–42.
86. Naitoh K, Hosaka Y, Honda M, Ogawa K, Shirakawa K, Furusako S. Properties of soluble glycoprotein VI, a potential platelet activation biomarker. Platelets. 2015;26(8):745–50.
87. Chen W, Liang X, Syed AK, Jessup P, Church WR, Ware J, Josephson CD, Li R. Inhibiting GPIbα shedding preserves post-transfusion recovery and hemostatic function of platelets after prolonged StorageHighlights. Arterioscler Thromb Vasc Biol. 2016;36(9):1821–8.
88. Boomgaard MN, Gouwerok CW, Homburg CH, de Groot G, MJ IJ, de Korte D. The platelet adhesion capacity to subendothelial matrix and collagen in a flow model during storage of platelet concentrates for 7 days. Thromb Haemost. 1994;72(4):611–6.
89. Hosseini E, Beshkar P, Ghasemzadeh M. Reverse correlations of collagen-dependent platelet aggregation and adhesion with GPVI shedding during storage. J Thromb Thrombolysis. 2018;46(4):534–40.
90. Aubron C, Flint AW, Ozier Y, McQuilten Z. Platelet storage duration and its clinical and transfusion outcomes: a systematic review. Crit Care. 2018;22(1):185.
91. Ng MS, Tung JP, Fraser JF. Platelet storage lesions: what more do we know now?. Transfusion medicine reviews. 2018;32(3):144-54.
92. Soslau G, Giles J. The loss of sialic acid and its prevention in stored human platelets. Thromb Res. 1982;26(6):443–55.
93. Bergmeier W, Burger PC, Piffath CL, Hoffmeister KM, Hartwig JH, Nieswandt B, Wagner DD. Metalloproteinase inhibitors improve the recovery and hemostatic function of in vitro–aged or–injured mouse platelets. Blood. 2003;102(12):4229–35.

4

Analysis of the association between resolution of disseminated intravascular coagulation (DIC) and treatment outcomes in post-marketing surveillance of thrombomodulin alpha for DIC with infectious disease and with hematological malignancy by organ failure

Noriaki Kawano[1*], Hideo Wada[2], Toshimasa Uchiyama[3], Kazuo Kawasugi[4], Seiji Madoiwa[5], Naoki Takezako[6], Kei Suzuki[7], Yoshinobu Seki[8], Takayuki Ikezoe[9], Tsuyoshi Hattori[10] and Kohji Okamoto[11]

Abstract

Background: Although disseminated intravascular coagulation (DIC) is life-threatening, any organ failure associated with DIC resolution and outcomes have been unclear.

Patients and methods: A total of 2795 DIC patients (infection: 1990, hematological malignancy: 805) were analyzed in the post-marketing surveillance of thrombomodulin alpha (TM-α). The background factors of sequential organ failure assessment (SOFA) and antithrombin (AT) were investigated in DIC with infectious disease for their association with DIC resolution and outcome using κ statistics, indicating DIC resolution and survival or DIC non-resolution and non-survival. The same analyses were performed for total bilirubin, creatinine, lactate dehydrogenase, and underlying disease in DIC with hematological malignancy.

Results: In DIC with infectious disease, higher SOFA score severity was closely correlated with lower overall survival in both the DIC resolution and non-resolution groups, but AT activity was not. κ coefficients were 0.234, 0.295, and 0.311 for the SOFA score 0–6, 7–12, and 13–24 groups, respectively. In DIC with hematological malignancy, κ coefficients of total bilirubin were 0.251 and 0.434, and those of creatinine were 0.283 and 0.437 in the normal and abnormal groups, respectively, showing better concordance in the abnormal group than in the normal. Other factors had poor concordance.

Conclusion: In DIC with infectious disease, DIC resolution is an important therapeutic target in patients who have higher SOFA score severity. In DIC with hematological malignancy, DIC resolution is similarly important in patients with abnormality of bilirubin and/or creatinine.

(Continued on next page)

* Correspondence: kawanoriaki@yahoo.co.jp
[1]Department of Internal Medicine, Miyazaki Prefectural Miyazaki Hospital, 5-30 Kitatakamatsu, Miyazaki 880-8510, Japan
Full list of author information is available at the end of the article

(Continued from previous page)
Trial registration: The clinical characteristics and treatment outcomes of patients with DIC treated with TM-α between May 2008 and April 2010 were retrospectively analyzed by subgroup analysis of the post-marketing surveillance data.

Keywords: DIC, Infectious disease, Hematological malignancy, SOFA, Bilirubin, Creatinine

Background

Disseminated intravascular coagulation (DIC) is a life-threatening clinical condition with high mortality due to the severe underlying disease, such as sepsis, hematological malignancy, and solid tumors, that is characterized by the systemic activation of coagulation pathways resulting in multiple organ failure [1–5]. Although the mechanism of DIC differs depending on the underlying disease, there is a common process across all cases, characterized by excessive production of thrombi that cause fibrin generation and deposition. Furthermore, fibrinolytic activation and overconsumption of anticoagulation factors can lead to systemic hemorrhage [1–5].

In the pathogenesis of DIC with sepsis, inflammation and coagulation have been closely linked to damage-associated molecular patterns (e.g. high mobility group box 1, histone), pathogen-associated molecular patterns (e.g. lipopolysaccharide), which facilitate secretion of neutrophil extracellular traps from activated neutrophils, and other inflammatory cytokines (e.g. IL-1β, TNF-α) [1–4].

In the pathogenesis of DIC with hematological disease, cancer procoagulant or tissue factor in leukemic cells and tissue plasminogen activator activation have been closely related to activation of excessive production of thrombi and fibrinolytic activation [1–3, 5].

Regarding the diagnosis and treatment of DIC, harmonization of guidelines for DIC was recently performed by the British Committee for Standards in Haematology (BCSH), the Japanese Society of Thrombosis and Hemostasis (JSTH), and the Italian Society for Thrombosis and Haemostasis (SISET) because the recommendations for diagnosis and treatment differed for each of the three guidelines [6]. In Japan, most emergency and hematological physicians make a diagnosis of and provide treatment for DIC according to the diagnostic criteria of the Japanese Association for Acute Medicine (JAAM) for infectious DIC and the Japanese Ministry of Health and Welfare (JMHW) for hematological DIC. Treatment for the underlying diseases of DIC is essential in DIC patients. Furthermore, supportive modalities such as the administration of platelet concentrates, fresh frozen plasma, heparin, and antithrombin play an important role in controlling DIC [6–8].

The hallmark of DIC treatment is the control of inflammation and coagulopathy in DIC with infectious disease and with hematological disease, and thrombomodulin alpha (TM-α) may be an appropriate anticoagulant and anti-inflammatory agent because of its two major effective sites, the lectin-like domain and epidermal growth factor-like domains [3–5]. These sites control inflammation and bind to thrombin to inactivate coagulation activity, forming a complex that activates protein C to generate activated protein C for the control of abnormal coagulopathy [3–5].

A phase 3 study and several retrospective studies including post-marketing surveillance (PMS) have reported the efficacy and safety of TM-α for DIC patients with infectious and hematological diseases [9–12], and it is generally prescribed in clinical practice. Wada H et al. reported the addition of recommendations for the use of TM-α to the "Expert consensus for the treatment of disseminated intravascular coagulation in Japan [13]. However, TM-α is still made no recommendation according to the international guidelines for management of sepsis and septic shock [14]. Furthermore, in the recent RCT, the 28-day mortality rate was not statistical significant but favorable in the TM-α group, including the DIC resolution [15–17].

With respect to DIC resolution, Okuda et al. reported that DIC resolution in patients treated with TM-α was related to a better outcome of DIC [18]. In contrast, Saito et al. reported that higher DIC resolution in patients treated with TM-α was not significantly related to the outcome of DIC [9]. To date, the relationship between DIC resolution and the outcome of DIC has not been fully examined because the etiologies and background of DIC vary in the underlying diseases.

To elucidate the clinical significance of organ failure for the association between DIC resolution and treatment outcome in DIC patients with infectious disease and hematological malignancy, the clinical impact of disease severity of backgrounds on the association between DIC resolution and treatment outcomes was retrospectively analyzed using subgroup analysis of PMS of TM-α. Furthermore, since DIC patients with infectious disease and those with hematological disease have rarely been both studied together, the above issues were analyzed in each subgroup and then discussed with respect to clinical practice.

Patients and methods

The clinical characteristics and treatment outcomes of patients with DIC treated with TM-α between May 2008 and April 2010 were retrospectively analyzed by subgroup analysis of the PMS data. Patient selection for the analyses is shown in Fig. 1.

The analyses were conducted with 2795 DIC patients (infection: 1990, hematological malignancy: 805) from the PMS of TM-α who fulfilled the DIC diagnostic criteria of the JAAM for infectious diseases (over 4 points) and the JMHW for hematological diseases (over 3 points).

After treatment with TM-α, resolution of DIC was defined as a score of ≤3 points using the diagnostic criteria of the JAAM and of ≤2 points using those of the JMHW for DIC with infectious and hematological diseases, respectively. Survival at 28 days from the beginning of TM-α treatment or at the end of observation was calculated.

In DIC with infectious disease, the severity of organ failure was assessed by the sequential organ failure assessment (SOFA) score. In hematological malignancy, to assess the severity of organ failure, total bilirubin (T. Bil) and creatinine, which are a subscore of the SOFA because the SOFA score was not clinically applied for DIC with hematological malignancy, were used. Furthermore, lactate dehydrogenase (LDH) was also used for the evaluation of the activity of hematological malignancy.

Statistical analysis

The following background factors were investigated to examine the association between DIC resolution and treatment outcome using κ statistics [19], indicating the concordance of DIC resolution and survival or DIC non-resolution and non-survival: SOFA and antithrombin (AT) activity for DIC with infectious disease; underlying disease, T. Bil, LDH, and creatinine for DIC with hematological malignancy. The strength of agreement for the κ statistics was described in a previous study [19]: κ < 0.00, poor; 0.00–0.20, slight; 0.21–0.40, fair; 0.41–0.60, moderate; 0.61–0.80, substantial; and 0.81–1.00, almost

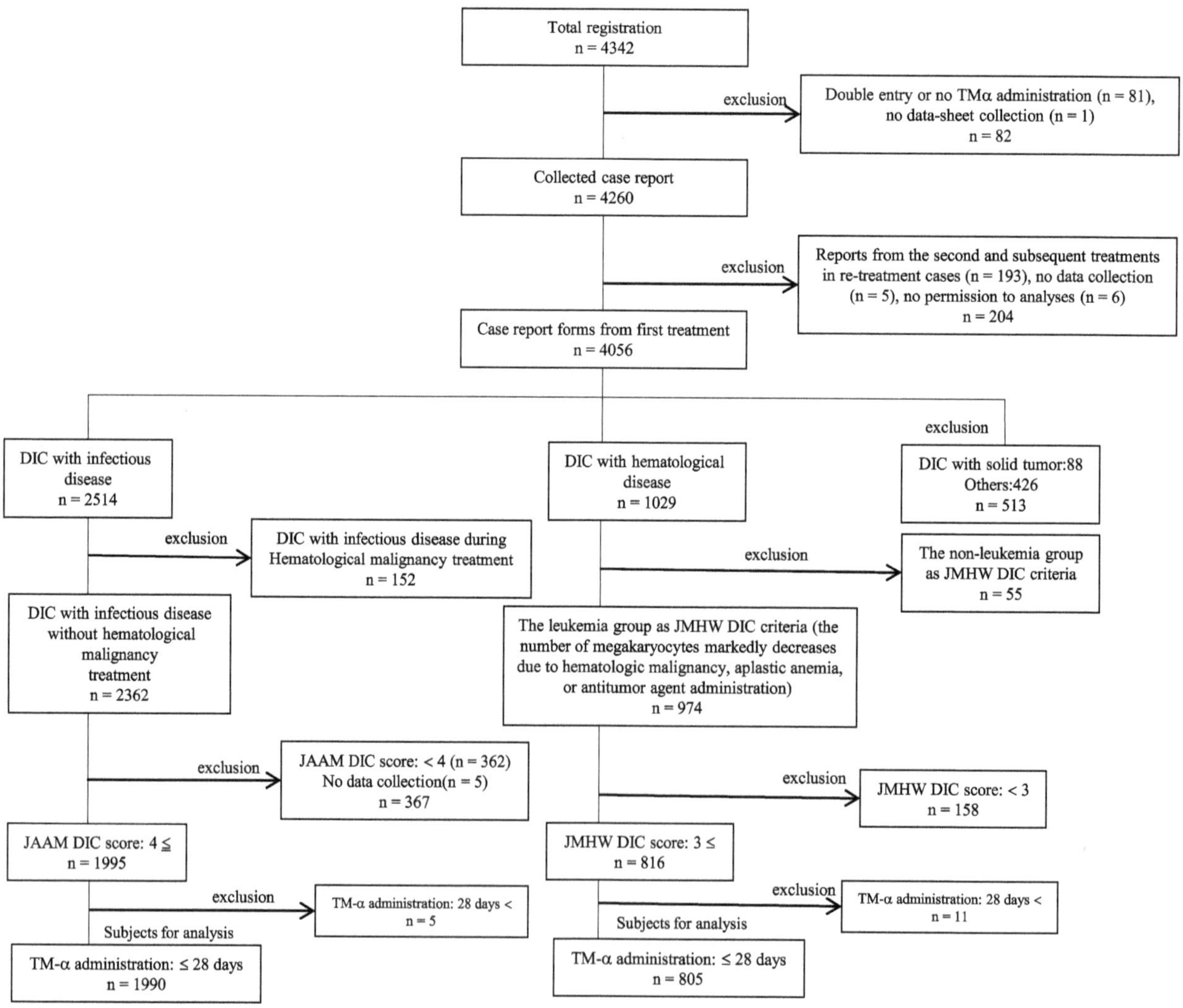

Fig. 1 Subject disposition in this PMS study

perfect. It has been noted that these divisions provide useful benchmarks, though they are arbitrary definitions.

In patients with infectious diseases, SOFA severity was divided into 3 groups (SOFA score 0–6, 7–12, and 13–24), and AT activity was divided into 4 groups (AT activity ≤30, 30% < AT activity ≤50, 50% < AT activity ≤70%, and AT activity > 70%). In patients with hematological diseases, the plasma concentrations of T. Bil and creatinine were divided into 2 groups (< 1.2 mg/dL (normal) and ≥ 1.2 mg/dL), the plasma concentration of LDH was divided into 2 groups (≤ 222 U/L (normal) and > 222 U/L), and underlying diseases were divided into 8 groups (acute myeloid leukemia (AML), acute promyelocytic leukemia (APL), acute lymphoid leukemia (ALL), chronic myeloid leukemia, chronic lymphocytic leukemia, myelodysplastic syndromes, multiple myeloma, and malignant lymphoma (ML)). All analyses were performed using SAS ver. 9.4 (SAS Institute, Co. Ltd., Cary, NC, USA) by EPS Corporation (Tokyo, Japan) according to the statistical analysis plan.

This study was conducted in compliance with the guidelines for Good Post-Marketing Surveillance Practices as required by the Japanese Ministry of Health, Labour, and Welfare, and was performed through a contract agreement with the Japanese Society on Thrombosis and Hemostasis and Asahi Kasei Pharma Corporation.

Results

Backgrounds of DIC patients

The clinical characteristics of the 1990 DIC patients (848 male patients and 1142 female patients) with infectious diseases are summarized in Table 1. The age of the DIC patients with infectious disease at diagnosis ranged from 0 to 102 years (median age, 71 years). The JAAM DIC scores in the DIC patients with infectious diseases ranged from 4 to 8 points (median score, 6). The SOFA score in the DIC patients with infectious diseases ranged from 1 to 23 points (median score, 10). The AT activity in the DIC patients with infectious diseases ranged from 7.5 to 140% (median score, 55).

The clinical characteristics of the 805 DIC patients (318 male patients and 487 female patients) with hematological malignancy are summarized in Table 2. The age of the DIC patients with hematological malignancy at diagnosis ranged from 0 to 92 years (median age, 63 years). Their JMHW DIC scores ranged from 3 to 9 points (median score, 4.0).

Table 1 DIC with infectious disease: Patients' characteristics

Background factor	Median (range) or *n* (%)
Sex, *n* (%)	
Male	848 (42.6)
Female	1142 (57.4)
Age, years	71 (0–102)
DIC duration before TM-α treatment, days	0 (0–51)
Total dose of TM-α, U/kg	369.2 (45–609.5)
Duration of TM-α	6 (1–27)
Prior DIC treatment, n (%)	
+	762 (38.3)
-	1228 (61.7)
Prior heparan sulfate and low molecular weight heparin, *n* (%)	
+	198 (9.9)
-	1792 (90.1)
Prior serine protease inhibitors, *n* (%)	
+	550 (27.6)
-	1440 (72.4)
Prior AT, *n* (%)	
+	375 (18.8)
-	1615 (81.2)
Platelets, 10^4/μL	5.2 (0–126)
PT ratio	1.32 (0.8–19.06)
FDP, μg/mL	29.2 (0.8–3200)
FBG, mg/dL	358.7 (0.1885–1481)
AT, %	55 (7.5–140)
JAAM DIC score	6 (4–8)
SOFA score	10 (1–23)

The plus and minus signs mean with and without treatment, respectively
AT antithrombin, *DIC* disseminated intravascular coagulation, *FBG* fibrinogen, *FDP* fibrin/fibrinogen degradation products, *JAAM* Japanese Association for Acute Medicine, *SOFA* sequential organ failure assessment, *TM-α* recombinant human thrombomodulin

κ coefficient, SOFA severity, and overall survival rate with infectious disease

In DIC with infectious disease, the κ coefficient between DIC resolution and treatment outcome was analyzed according to the SOFA severity group (Table 3). The κ coefficients were 0.234, 0.295, and 0.311 in the SOFA score groups 0–6, 7–12, and 13–24, respectively (Table 3). Thus, DIC resolution and DIC non-resolution accorded with survival and non-survival, respectively, in the groups with higher SOFA severity (Table 3).

In particular, the SOFA score 13–24 group with DIC non-resolution showed poor overall survival (OS) (41.8%). Remarkably, the DIC resolution group achieved higher OS at 28 days than the DIC non-resolution group regardless of the SOFA score group in DIC with infectious disease (Table 3).

κ coefficient, AT activity, and overall survival

Regarding AT activity, the coefficients were not remarkable and were constant among the three groups (Table 3).

The group with AT activity ≤30% with DIC non-resolution showed poor OS (38.8%). Remarkably, the DIC resolution group achieved higher OS at 28 days than the DIC non-resolution group regardless of the AT activity group in DIC with infectious disease (Table 3).

Table 2 DIC with hematological malignancy: Patients' characteristics

Background factor	Median (range) or *n* (%)
Sex, *n* (%)	
Male	318 (39.5)
Female	487 (60.5)
Age, years	63 (0–92)
DIC duration before TM-α treatment, days	0 (−1–31)
Total dose of TM-α, U/kg	380 (100.6–549)
Duration of TM-α	6 (1–28)
Prior DIC treatment, n (%)	
+	197 (24.5)
-	608 (75.5)
Prior heparan sulfate and low molecular weight heparin, n (%)	
+	101 (12.5)
-	70.4 (87.5)
Prior serine protease inhibitors, *n* (%)	
+	92 (11.4)
-	713 (88.6)
Prior AT, *n* (%)	
+	42 (5.2)
-	763 (94.8)
Platelet, 10^4/μL	3.25 (0.2–51.9)
PT ratio	1.26 (0.76–7.55)
FDP, μg/mL	48.4 (0.8–1760)
FBG, mg/dL	204 (14–966.5)
AT, %	86.1 (10.2–150)
JMHW DIC score	4 (3–9)

The plus and minus signs mean with and without treatment, respectively
AT antithrombin, *DIC* disseminated intravascular coagulation, *FBG* fibrinogen, *FDP* fibrin/fibrinogen degradation products, *JMHW* Japanese Ministry of Health and Welfare, *TM-α* recombinant human thrombomodulin

κ coefficients, underlying disease, and overall survival with hematological malignancy

In DIC with hematological malignancy, the κ coefficient of underlying disease was approximately 0.3 regardless of the group of underlying hematological malignancy (AML, APL, ALL, and ML) (Table 4).

The DIC resolution group achieved higher OS at 28 days than the DIC non-resolution group regardless of the underlying disease (except for CML) (Table 4).

κ coefficients, parameters (bilirubin/creatinine/LDH), and overall survival with hematological malignancy

In DIC with hematological malignancy, κ coefficients between DIC resolution and treatment outcome were examined according to T. Bil, creatinine, and LDH levels. The κ coefficients were higher in the abnormal groups of T. Bil and creatinine (Table 5). The κ coefficients for LDH were not remarkable and constant among the groups (Table 5).

The abnormal LDH (> 222 U/L) group with DIC non-resolution showed poor OS (59.5%). The DIC resolution group achieved higher OS at 28 days than the DIC non-resolution group regardless of T. Bil, LDH, and creatinine levels in DIC with hematological malignancy (Table 5).

Discussion

In previous reports dealing with DIC resolution and treatment outcomes of DIC, different and controversial results were shown because of differences in the patients' background characteristics [9, 18]. The present study identified three important points by analyzing DIC resolution and survival or DIC non-resolution and non-survival according to the organ failure, as follows. (i) In DIC with infectious disease, a higher concordance of DIC resolution and survival or DIC non-resolution and non-survival was seen in the severe SOFA score group, but this was not seen for AT activity, for which κ values

Table 3 Relations between DIC resolution and treatment outcome in DIC with infectious disease

Group	OS, % (n)		κ	
	DIC resolution	DIC non-resolution	Point estimate	95% CI, [lower, upper] limits
SOFA score				
0–6	91.3 (84)	69.6 (48)	0.234 ± 0.067	[0.103, 0.366]
7–12	89.7 (201)	58.5 (161)	0.295 ± 0.035	[0.226, 0.365]
13–24	84.5 (71)	41.8 (107)	0.311 ± 0.043	[0.227, 0.394]
AT activity, %				
≤ 30	77.1 (27)	38.8 (38)	0.301 ± 0.074	[0.156, 0.446]
30 < AT activity ≤50	87.3 (117)	51.2 (134)	0.300 ± 0.039	[0.225, 0.376]
50 < AT activity ≤70	90.9 (169)	53.4 (125)	0.353 ± 0.039	[0.277, 0.429]
> 70	89.6 (112)	54.6 (83)	0.333 ± 0.049	[0.238, 0.428]

The degree of accordance between DIC resolution and treatment outcome according to the SOFA score group was analyzed by κ coefficient in DIC with infectious disease
CI confidence interval, *DIC* disseminated intravascular coagulation, *OS* overall survival, *SOFA* sequential organ failure assessment

Table 4 Relations between DIC resolution and treatment outcome in DIC with hematological malignancy

Group	Underlying disease	OS, % (n)		κ	
		DIC resolution	DIC non-resolution	Point estimate	95% CI, [lower, upper] limits
0	AML	88.1 (104)	62.0 (57)	0.276 ± 0.062	[0.155, 0.397]
1	APL	98.6 (73)	67.9 (36)	0.339 ± 0.071	[0.200, 0.478]
2	ALL	95.3 (61)	70.4 (19)	0.301 ± 0.105	[0.096, 0.506]
3	CML	85.7 (6)	100.0 (4)	−0.170 ± 0.149	[− 0.462, 0.122]
4	CLL	66.7 (4)	50.0 (1)	0.147 ± 0.347	[−0.538, 0.823]
5	MDS	82.4 (14)	36.4 (4)	0.467 ± 0.172	[0.131, 0.804]
6	MM	100.0 (6)	55.6 (5)	0.390 ± 0.177	[0.044, 0.736]
7	ML	90.0 (36)	48.0 (24)	0.400 ± 0.086	[0.232, 0.568]

The degree of accordance between DIC resolution and treatment outcome according to the underlying disease of hematological malignancy was analyzed by κ coefficient in DIC with hematological malignancy

ALL acute lymphoid leukemia, *AML* acute myeloid leukemia, *APL* acute promyelocytic leukemia, *CI* confidence interval, *CLL* chronic lymphocytic leukemia, *CML* chronic myeloid leukemia, *DIC* disseminated intravascular coagulation, *MDS* myelodysplastic syndromes, *ML* malignant lymphoma, *MM* multiple myeloma, *OS* overall survival

were constant among the groups. (ii) In DIC with hematological malignancy, a higher concordance of DIC resolution and survival or DIC non-resolution and non-survival was seen in the groups with abnormal T. Bil and creatinine levels, but not for LDH. (iii) The DIC resolution group achieved higher OS at 28 days than the DIC non-resolution group, regardless of any group of background factors (except for the underlying disease of CML).

κ coefficients and SOFA severity and the correlation between SOFA severity and lower overall survival in DIC with infectious disease

Although factors affecting DIC resolution and prognostic factors affecting OS were identified in previous reports [11, 12, 18, 20–26], the clinical significance of the patients' background characteristics according to the organ disfunction for the association between DIC resolution and treatment outcomes in DIC patients with infectious disease has been unclear. Previously, regarding the SOFA in DIC with infectious disease, Yamakawa et al. reported the benefit of anticoagulant therapy in sepsis for SOFA scores of 13–17 in a nationwide multicenter registry in Japan [27]. Moreover, Nakajima et al. reported an association between SOFA scores and mortality in patients with sepsis during the first week in the Japanese Society of Education for Physicians and Trainees in Intensive Care (JSEPTIC) DIC study. In particular, Nakajima et al. reported the importance of the central nervous system category of SOFA (days 1, 3, 7) and the coagulation category of SOFA (day 7) [28]. The present PMS study clearly showed that, in DIC with infectious disease, the κ coefficient was higher in the severe SOFA score group than in the mild group. Thus, a higher concordance of DIC resolution and survival or DIC non-resolution and non-survival was shown in the severe SOFA score group. Consequently, in DIC with infectious disease, treatment with the target of DIC resolution may be essential to improve OS according to the increased severity of organ dysfunction.

Table 5 Relations between DIC resolution and treatment outcome in DIC with hematological malignancy

Group	OS, % (n)		κ	
	DIC resolution	DIC non-resolution	Point estimate	95% CI, [lower, upper] limits
Total bilirubin, mg/L				
< 1.2	92.3 (240)	70.3 (104)	0.251 ± 0.046	[0.162, 0.340]
≥ 1.2	90.3 (56)	43.7 (38)	0.434 ± 0.066	[0.306, 0.563]
Creatinine, mg/dL				
< 1.2	92.1 (268)	66.5 (125)	0.283 ± 0.041	[0.203, 0.364]
≥ 1.2	87.2 (41)	41.7 (25)	0.437 ± 0.080	[0.279, 0.594]
LDH, U/L				
≤ 222	96.2 (51)	69.6 (16)	0.322 ± 0.113	[0.102, 0.543]
> 222	91.4 (255)	59.5 (131)	0.336 ± 0.039	[0.260, 0.412]

The degree of accordance between DIC resolution and treatment outcome according to the total bilirubin level was analyzed by κ coefficient in DIC with hematological malignancy

CI confidence interval, *DIC* disseminated intravascular coagulation, *LDH* lactate dehydrogenase, *OS* overall survival

κ coefficients for T. Bil and creatinine in DIC with hematological malignancy

Previously, regarding prognostic factors for OS in DIC with hematological malignancy [29–31], Bird reported that hyperbilirubinemia (> 18 mg/L) and renal replacement therapy may be prognostic factors affecting OS on univariate analysis in hematological malignancy patients requiring intensive care unit care and intensive care [29]. Moreover, Bird et al. reported mechanical ventilation and ≥ 2 organ failures as prognostic factors for OS on multivariate analysis. The clinical significance of hyperbilirubinemia and abnormal creatinine levels was also reported previously [29]. The cut-offs of hyperbilirubinemia and abnormal creatinine were reported to be > 1.8 mg/dL or > 2.0 mg/dL and > 1.2 mg/dL or > 1.5 mg/dL, respectively [30, 31]. Consistent with these previous reports [29–32], in the present PMS study, the κ coefficients were higher in the groups with abnormal T. Bil and creatinine levels than in those with normal values. Thus, the concordance of DIC resolution and survival or DIC non-resolution and non-survival was greater in the group with abnormal T. Bil and creatinine levels. Consequently, these findings clearly suggest that, in DIC with hematological malignancy, treatment with the target of DIC resolution may be essential to improve OS when there are hyperbilirubinemia and increased creatinine levels.

Importance of DIC resolution for OS regardless of any background factors in infectious disease and hematological malignancy

MDS, MM and ML showed higher tendency of the degree of accordance between DIC resolution and treatment outcome than those of AML, APL and ALL in Table 5.

As for the liver dysfunction in hematological disease with DIC, Chi S et al. reported that Liver dysfunction in ML such as bilirubin, aminotransferases, serum choline esterase, and albumin levels, were worse in patients with DIC than those without DIC, indicating impaired production of coagulation factors [33]. Furthermore, DIC exerts significantly negative impact on prognosis of non-Hodgkin lymphoma [33]. Similarly, in our present study, the liver dysfunction had significant impact for the degree of accordance between DIC resolution and treatment outcome in ML.

As for the renal dysfunction, Radojevic-Skodric S et al. reviewed that acute renal failure originated from cancer infiltration, drug, dehydration, amyloidosis and others represents a severe complication of different malignancies, that causes significant morbidity and mortality [34]. However, little was known and discussed about the affection of DIC for the renal dysfunction in hematological malignancy [34]. In our present study, the renal dysfunction had significant impact for the degree of accordance between DIC resolution and treatment outcome in hematological malignancy.

Consequently, regarding as the organ dysfunction in hematological malignancy, the hematological malignancy may tend to be complicated with organ dysfunction due to the infiltration of tumors or the infection due to immunodeficient state. Thus, to improve the organ dysfunction, the DIC resolution may be the therapeutic target in DIC with hematological malignancy.

The present analyses clearly demonstrated that DIC resolution had a positive impact on OS at 28 days, especially the patients with organ failure showed well concordance between DIC resolution and OS. Thus, treatment with the target of DIC resolution may be essential to improve OS in DIC with infection and hematological malignancy. Further study is needed to elucidate the real clinical impact of DIC resolution on OS at 28 days.

Limitation

The laboratory assays were performed in each site or its contracted laboratory, not one central laboratory, because this study is a post marketing surveillance as a clinical practice. The precision management of laboratory data is well performed in Prefectural Association of Medical Technologists.

Conclusions

In DIC with infectious disease, the κ coefficient in the high SOFA score group may have significance in clinical practice. Similarly, in DIC with hematological malignancy, the strong κ coefficient in the organ failure patients may have an impact in clinical practice. Consequently, the present study clearly demonstrated that the DIC resolution group achieved higher OS at 28 days than the DIC non-resolution group. Finally, DIC resolution can be a possible main target for the treatment of underlying diseases associated with DIC.

Abbreviations

ALL: Acute lymphoid leukemia; AML: Acute myeloid leukemia; APL: Acute promyelocytic leukemia; AT: Antithrombin; CML: Chronic myeloid leukemia; DIC: Disseminated intravascular coagulation; JAAM: Japanese Association for Acute Medicine; JMHW: Japanese Ministry of Health and Welfare; JSEPTIC: Japanese Society of Education for Physicians and Trainees in Intensive Care; LDH: Lactate dehydrogenase; ML: Malignant lymphoma; OS: Overall survival; PMS: Post-marketing surveillance; SOFA: Sequential organ failure assessment; T. Bil: Total bilirubin; TM-α: Thrombomodulin alpha

Acknowledgements

The authors would like to thank all investigators who participated in this PMS study. The authors acknowledge Ms. Hideyo Ohshige for helpful suggestions.

Authors' contributions

NK planed the study design, interpreted the data and wrote the manuscript. HW overviewed and supervised the current study. TU, KK, SM, NI, KS, YS, II and KO advised the study design and interpretation of the data, and reviewed the manuscript. TH planed the study design. All authors read and approved the final manuscript.

Author details
[1]Department of Internal Medicine, Miyazaki Prefectural Miyazaki Hospital, 5-30 Kitatakamatsu, Miyazaki 880-8510, Japan. [2]Department of General Medicine, Mie Prefectural General Medical Center, Mie, Japan. [3]Department of Laboratory Medicine, National Hospital Organization Takasaki General Medical Center, Gunma, Japan. [4]Faculty of Medical Technology, Teikyo University, Tokyo, Japan. [5]Department of Clinical Laboratory Medicine, Tokyo Saiseikai Central Hospital, Tokyo, Japan. [6]Department of Hematology, National Hospital Organization Disaster Medical Center, Tokyo, Japan. [7]Emergency and Critical Care Center, Mie University Hospital, Mie, Japan. [8]Department of Hematology, Uonuma Institute of Community Medicine, Niigata University Medical and Dental Hospital, Niigata, Japan. [9]Department of Hematology, Fukushima Medical University, Fukushima, Japan. [10]Department of Medical Affairs, Asahi Kasei Pharma Corporation, Tokyo, Japan. [11]Department of Surgery, Center for Gastroenterology and Liver Disease, Kitakyushu City Yahata Hospital, Fukuoka, Japan.

References

1. Levi M, Ten Cate H. Disseminated intravascular coagulation. N Engl J Med. 1999;341:586–92.
2. Bick RL. Disseminated intravascular coagulation current concepts of etiology, pathophysiology, diagnosis and treatment. Hematol Oncol Clin North Am. 2003;17:149–76.
3. Gando S, Levi M, Toh CH. Disseminated intravascular coagulation. Nat Rev Dis Primers. 2016;2:16037. https://doi.org/10.1038/nrdp.2016.37.
4. Ikezoe T. Thrombomodulin/activated protein C system in septic disseminated intravascular coagulation. J Intensive Care. 2015;3:1. https://doi.org/10.1186/s40560-014-0050-7.
5. Ikezoe T. Pathogenesis of disseminated intravascular coagulation in patients with acute promyelocytic leukemia, and its treatment using recombinant human soluble thrombomodulin. Int J Hematol. 2014;100:27–37. https://doi.org/10.1007/s12185-013-1463-0.
6. Wada H, Thachil J, Di Nisio M, Kurosawa S, Gando S, Toh CH. The scientific standardization committee on DIC of the international society on thrombosis Haemostasis. Guidance for diagnosis and treatment of DIC from harmonization of the recommendations from three guidelines. J Thromb Haemost. 2013;11:2078–9. https://doi.org/10.1111/jth.12366.
7. Wada H, Asakura H, Okamoto K, et al. Japanese Society of Thrombosis Hemostasis/DIC subcommittee.Expert consensus for the treatment of disseminated intravascular coagulation in Japan. Thromb Res. 2010;125:6–11.
8. Levi M, Toh CH, Thachil J, Watson HG. Guidelines for the diagnosis and management of disseminated intravascular coagulation. British Committee for Standards in Haematology. Br J Haematol. 2009;145:24–33.
9. Saito H, Maruyama I, Shimazaki S, et al. Efficacy and safety of recombinant human soluble thrombomodulin (ART-123) in disseminated intravascular coagulation: results of a phase III, randomized, double-blind clinical trial. J Thromb Haemost. 2007;5:31–41.
10. Mimuro J, Takahashi H, Kitajima I, et al. Impact of recombinant soluble thrombomodulin (thrombomodulin alfa) on disseminated intravascular coagulation. Thromb Res. 2013;131:436–43.
11. Asakura H, Takahashi H, Tsuji H, et al. Post-marketing surveillance of thrombomodulin alfa, a novel treatment of disseminated intravascular coagulation - safety and efficacy in 1,032 patients with hematologic malignancy. Thromb Res. 2014;133:364–70.
12. Eguchi Y, Gando S, Ishikura H, et al. Post-marketing surveillance data of thrombomodulin alfa: sub-analysis in patients with sepsis-induced disseminated intravascular coagulation. J Intensive Care. 2014;2:30.
13. Wada H, Okamoto K, Iba T, et al. Addition of recommendations for the use of recombinant human thrombomodulin to the "expert consensus for the treatment of disseminated intravascular coagulation in Japan". Thromb Res. 2014;134:924–5. https://doi.org/10.1016/j.thromres.2014.07.033.
14. Rhodes A, Evans LE, Alhazzani W, et al. Surviving Sepsis campaign: international guidelines for Management of Sepsis and Septic Shock: 2016. Crit Care Med. 2017;45:486–552. https://doi.org/10.1097/CCM.0000000000002255.
15. Vincent JL, Francois B, Zabolotskikh I, et al. Effect of a recombinant human soluble Thrombomodulin on mortality in patients with Sepsis-associated coagulopathy: the SCARLET randomized clinical trial. JAMA. 2019;321:1993–2002. https://doi.org/10.1001/jama.2019.5358.
16. Yamakawa K, Aihara M, Ogura H, et al. Recombinant human soluble thrombomodulin in severe sepsis: a systematic review and meta-analysis. J Thromb Haemost. 2015;13:508–19. https://doi.org/10.1111/jth.12841.
17. Yamakawa K, Levy JH, Iba T. Recombinant human soluble thrombomodulin in patients with sepsis-associated coagulopathy (SCARLET): an updated meta-analysis. Crit Care. 2019;23:302. https://doi.org/10.1186/s13054-019-2587-2.
18. Okuda A, Ogura T, Imanishi M, et al. Clinical impact of recombinant soluble thrombomodulin for disseminated intravascular coagulation associated with severe acute cholangitis. Gut Liver. 2018;12:471–7. https://doi.org/10.5009/gnl17489.
19. Landis JR, Koch GG. The measurement of observer agreement for categorical data. Biometrics. 1977;33:159–74.
20. Dempfle CE, Wurst M, Smolinski M, et al. Use of soluble fibrin antigen instead of D-dimer as fibrin-related marker may enhance the prognostic power of the ISTH overt DIC score. Thromb Haemost. 2004;91:812–8.
21. Hatada T, Wada H, Nobori T, et al. Plasma concentrations and importance of high mobility group box protein in the prognosis of organ failure in patients with disseminated intravascular coagulation. Thromb Haemost. 2005;94:975–9.
22. Kotajima N, Kanda T, Fukumura Y, et al. Serum thrombomodulin as a prognostic marker of disseminated intravascular coagulation. J Med. 1999;30:19–29.
23. Gando S, Saitoh D, Ogura H, et al. Japanese Association for Acute Medicine Disseminated Intravascular Coagulation (JAAMDIC) study group: natural history of disseminated intravascular coagulation diagnosed based on the newly established diagnostic criteria for critically ill patients: results of a multicenter, prospective survey. Crit Care Med. 2008;36:145–50.
24. Seki Y, Wada H, Kawasugi K, et al. Japanese Society of Thrombosis Hemostasis/DIC subcommittee: a prospective analysis of disseminated intravascular coagulation in patients with infections. Intern Med. 2013;52:1893–8.
25. Kawano N, Tasaki A, Kawano S, et al. Sequential organ failure assessment (SOFA) score as a prognostic factor for disseminated intravascular coagulation patients with infectious disease treated with recombinant human soluble thrombomodulin (rhTM) in clinical practice. Japanese J Transfus Cell Ther. 2017;63:763–79.
26. Ohshiro M, Kuroda J, Kobayashi Y, et al. ADAMTS-13 activity can predict the outcome of disseminated intravascular coagulation in hematologic malignancies treated with recombinant human soluble thrombomodulin. Am J Hematol. 2012;87:116–9.
27. Yamakawa K, Umemura Y, Hayakawa M, et al. Benefit profile of anticoagulant therapy in sepsis: a nationwide multicentre registry in Japan. Crit Care. 2016;20:229.
28. Nakashima T, Miyamoto K, Shimokawa T, et al. The association between sequential organ failure assessment scores and mortality in patients with sepsis during the first week: the JSEPTIC DIC study. J Intensive Care. 2018;1:885066618775959. https://doi.org/10.1177/0885066618775959.
29. Bird GT, Farquhar-Smith P, Wigmore T, Potter M, Gruber PC. Outcomes and prognostic factors in patients with haematological malignancy admitted to a specialist cancer intensive care unit: a 5 yr study. Br J Anaesth. 2012;108:452–9.
30. Benoit DD, Vandewoude KH, Decruyenaere JM, Hoste EA, Colardyn FA. Outcome and early prognostic indicators in patients with a hematologic malignancy admitted to the intensive care unit for a life-threatening complication. Crit Care Med. 2003;31:104–12.
31. Hampshire PA, Welch CA, McCrossan LA, Francis K, Harrison DA. Admission factors associated with hospital mortality in patients with haematological malignancy admitted to UK adult, general critical care units: a secondary analysis of the ICNARC case mix Programme database. Crit Care. 2009;13:R137. https://doi.org/10.1186/cc8016.
32. Cuthbertson BH, Rajalingham Y, Harrison S, McKirdy F. The outcome of haematological malignancy in Scottish intensive care units. J Intensive Care Soc. 2008;9:135–40.

33. Chi S, Ikezoe T. Disseminated intravascular coagulation in non-Hodgkin lymphoma. Int J Hematol. 2015;102:413–9. https://doi.org/10.1007/s12185-015-1854-5.
34. Radojevic-Skodric S, Bogdanovic L, Jovanovic M, Baralic I, Dzamic Z, Gordon R, Ognjanovic S, Basta-Jovanovic G. Acute renal failure in different malignant tumors. Curr Med Chem. 2016;23:2041–6.

5

Increased dose primary thromboprophylaxis in ambulatory patients with advanced pancreatic ductal adenocarcinoma: A single centre cohort study

Anthony Maraveyas[1,2*], Farzana Haque[1,2], Iqtedar Ahmed Muazzam[1], Waqas Ilyas[1] and George Bozas[1]

Abstract

Background: Advanced pancreatic ductal adenocarcinoma (aPDAC) patients have a lifetime all type thromboembolic event (ATTE) rate of 25–35%. Efficacy and safety of increased dose primary thromboprophylaxis (IDPTP) with low molecular heparin (LMWH) given for 3 months has been shown in two prospective randomized trials.

Objectives: To report on efficacy -reduction of all type thromboembolic events (ATTE)-, safety -incidence of Major Bleeding (MB)- and compliance in a single-centre cohort of aPDAC patients receiving first line chemotherapy and LMWH-IDPTP.

Methods: From May 2009 to October 2016, 82 patients received IDPTP –LMWH with dalteparin. Schedule: 55 kg and below: 7500 IU, between 55 and 80 kg: 10,000 IU, above 80 kg: 12,500 IU. MB is reported using the International Society of Thrombosis and Haemostasis (ISTH) criteria. ATTE was defined as any arterial or venous event, incidental or clinically symptomatic, including visceral VTE.

Results: Mean and median time on dalteparin was 10.2 (95%CI 8.1, 12.4) and 8.0 (95%CI 6.2, 9.7) months respectively. ATTE was observed in 7 (8.5%) of patients, with a median time on IDPTP of 6.2 months (95% CI 10.0, 13.2). MB was seen in 10 (12.2%) patients with a median time on IDPTP of 4.5 months (95% CI 1.6, 7.4). Six major bleeds (60%) were the direct or indirect result of aPDAC. Eighty-one patients had died at the time of data collection with a median overall survival time of 8.7 months (95%CI 6.4, 11.0). Thromboembolism and bleeding were late events. No impact of thromboembolism or bleeding on overall survival was observed.

(Continued on next page)

* Correspondence: anthony.maraveyas@hey.nhs.uk
[1]Queen's Centre for Oncology and Haematology, Castle Hill Hospital, Hull University Teachng Hospitals (HUTH), Cottingham HU16 5JQ, UK
[2]Joint Centre for Cancer Studies, Faculty of Health Sciences, The Hull York Medical School, University of Hull, Cottingham HU6 7RX, UK

(Continued from previous page)
Conclusions: IDPTP-dalteparin was associated with lower ATTE occurrence rates than expected and comparable major bleeding rates. ATTE and MB were late events, the majority of MB was from direct or indirect result of locally progressing aPDAC. Since these conditions can frequently arise in aPDAC, IDPTP should be regularly reviewed beyond 3 months.

Keywords: Pancreatic cancer, Thromboprophylaxis, Dalteparin, Ambulatory, Chemotherapy

Introduction

Patients with advanced pancreatic ductal adenocarcinoma (aPDAC) have a seven-fold increased risk of developing arterial and venous thromboembolism as compared to patients with most other solid and haematological malignancies [1, 2]. This translates to about one-third of patients with aPDAC suffering from either a clinically evident or unsuspected (*incidental*) thrombo-embolic event (TE) over short life spans, and the spectrum of these events is broad, ranging from the conventional venous thromboembolism (VTE) of deep vein thrombosis (DVT)-pulmonary embolism (PE) complex, to arterial as well as visceral thromboses [3]. The manifestation of a VTE, apart from the physical and psychological burden inflicted on the cancer patient, generally results in increased health utilization costs of up to 45% per affected cancer patient [4] with VTE incremental costs being the highest in pancreatic cancer patients [5].

The administration of chemotherapy in ambulatory aPDAC patients is an additional risk factor for thromboembolism which further complicates an already challenging management [3]. A modest improvement in survival has been seen with new chemotherapy combinations [6] but they have also resulted in increased VTE risk [7]. The use of currently available anticoagulant agents reduces VTE risk and their efficacy has been proven in randomized trials [8, 9]. Current guidelines from major international bodies have changed recently to suggest that patients with Khorana Score 2 and above should be considered for thromboprophylaxis. This by definition includes patients with pancreatic cancer [10–12]. These guidelines have been updated following the studies AVERT [13, 14] and CASSINI [14, 13] and now include the option of using a direct oral anticoagulant (DOAC). The recent (2018) NICE guidance also suggests that thromboprophylaxis should be considered in aPDAC [13, 15]. Both the FRAGEM study [8] with a 19.6% absolute difference in thromboembolic rates and a number needed to treat (NNT) of 5 and CONKO-04 [9] with an absolute difference of 8.7% in thromboembolic rate and a NNT of 11 provide meaningful clinical benefit.

These studies used increased dose primary thromboprophylaxis (IDPTP) for 3 months. The investigators in CONKO 004 used a two-third -therapeutic dose of enoxaparin (1 mg/kg once daily) for the first 3 months, followed by conventional primary prophylaxis maintenance of 40 mg sc. once daily. The investigators in FRAGEM used full therapeutic dose schedule of dalteparin (200u/kg once daily for 4 weeks reduced to 150u/kg once daily) for the first 3 months at which point thromboprophylaxis was discontinued.

None of the current guidelines offer a clear opinion on LMWH dosing though the ASCO guideline mentions that the studies of LMWH thromboprophylaxis in pancreatic cancer patients were conducted at a 'treatment dose' of LMWH [10] otherwise termed 'supra-prophylactic' [16].

Soon after the LMWH–related data were first published Hull University Teaching Hospitals (HUTH) implemented a dalteparin-based thromboprophyalxis schedule for aPDAC patients using a supra-prophylactic dosing approach. It is this single-centre retrospective real-world cohort series experience in ambulatory aPDAC patients that we present here.

Objectives

To report efficacy, safety and compliance in a real-world cohort of consecutively treated patients with advanced and metastatic pancreatic cancer receiving an increased dose primary thromboprophylaxis (IDPTP) schedule of dalteparin.

Patients and methods

Increased dose primary thromboprophylaxis (IDPTP) schedule

For aPDAC patients deemed fit for systemic treatment we implement a simplified dalteparin increased dose thromboprophylaxis schedule, close to the 75% full therapeutic dose of the CLOT schedule used in the FRAGEM trial and in accordance with the CONKO-04 scheduling of continuing thromboprophylaxis throughout chemotherapy but without mandating a switch to conventional dose or discontinuation.

The IDPTP dalteparin schedule was weight adjusted and ranged as follows. For patients 55 kg and below: 7500 IU, for patients between 55 and 80 kg: 10,000 IU and for patients above 80 kg: 12,500 IU.

Patients

The HPB-MDT database (Taunton & Somerset NHSFT QGCE 1499) was searched for all pancreatic cancer patients that were referred to the cancer centre between May 2009 and October 2016, and then cross referenced with the electronic chemotherapy database (ARIA MedOnc 13.7). Patients were further cross referenced through the electronic medical record system (iSOFT Patient Centre®, CSC™ and Lorenzo®, CSC™) and through the pharmacy dispensary database (Ascribe RX™) for all the outpatient prescriptions for anticoagulants. From these electronic medical records demographic, historic, cancer therapy and cancer specific data were collected for each patient in a retrospective manner. All patients had histological/cytological proof of ductal adenocarcinoma of the pancreas with inoperable locally advanced or metastatic cancer and an multi-disciplinary team (MDT) decision for systemic treatment. Only patients due to receive first line treatment were included. Accordingly, patients had to have haematological and biochemical indices allowing systemic treatment for their cancer to be eligible for treatment with IDPTP. Non-ambulatory patients or patients not considered for systemic cancer treatment either due to performance status or non -alleviated cancer morbidity (e.g. non-relievable obstruction of biliary tree or duodenal outlet) were excluded. Patients with any contraindications for anticoagulation (dalteparin) were excluded. Patients receiving other anticoagulation (e.g. a direct oral anticoagulant or warfarin) and not transferred to IDPTP were excluded.

Outcome measures

Primary aim of the study was to determine the occurrence rate of all type thromboembolic events (ATTE) and rate of major bleeding (MB). ATTE was defined as any TE event -arterial or venous, incidental or clinically symptomatic including visceral VTE and is reported in this manuscript as ATTE or TE. VTE is used for conventional PE/DVT definitions. All recorded ATTE and bleeding events were reviewed by three of the authors (AM, GB & IM) and the nature, severity and classification agreed. Bleeding episodes were classified into major and minor by ISTH criteria. Major bleeding was defined as any fatal bleeding, symptomatic bleeding in a critical area or organ, such as intracranial, intraspinal, intraocular, retroperitoneal, intra-articular or pericardial, or intramuscular with compartment syndrome, and/or bleeding causing a fall in haemoglobin level of $20\,g\,L^{-1}$ ($1.24\,mmol\,L^{-1}$) or more, or leading to transfusion of two or more units of whole blood or red cells [17]. Compliance to anticoagulant treatment was inferred from the time and length of the last recorded outpatient or ward prescription chart.

Approvals & Statistics

The study reported in this manuscript is the result of work that has been classified as an audit. This is a regular undertaking in UK hospitals with the primary goal of maintaining quality standards and identifying areas that need improvement if they fall below accepted standards as set by national or international guidelines. Audits can be retrospective or prospective. As per the NHS Health Research Authority guidelines our study, which can be classified within the audit / service evaluation description, does not require external Research Ethics Committee approval [18]. The NHS Trust governance body which authorises the project is doing so if the study is conducted within the regulatory framework including the Data Protection Act (1998), the Caldicott principles (1997) and the NHS Confidentiality code of practice (2003) [19]. Within this context regulatory approval is sought and obtained based on the quality of the audit and the priority of the area studied and whether it fits within the quality framework of the organization. The endorsement code for our study is 2017.98 issued by the Hull and East Yorkshire Hospitals NHS Trust approved the 1st June 2017.

Patients were registered sequentially on the database. Data were collected in an MS EXCEL®2010 (Microsoft Corp™) spread sheet maintained in a secure virtual hard-drive with restricted access. All analyses were performed with SPSS ver22, IBM Corp®. Descriptive statistics were used to analyse patient characteristics. Data on the TE-free interval (TFI) and bleeding-free interval (BFI) were reported in months from the time of commencing IDPTP up to the index event (last follow-up contact or the date of death). The collected data was also used to calculate the median overall survival of the whole study population and whether TE and bleeding episodes had demonstrated any impact on survival. No patients were lost to follow-up. To correct for competing risk of death the rate of TE or major bleeding for the time intervals was calculated as the number of events divided by the total number of patients at risk (subject months). The Clopper-Pearson method was used for 95% confidence interval. The Kaplan Meier method was utilised to explore the prognostic significance of categorical variables using the log rank test to compare factors.

Results

One hundred and one patients with aPDAC were registered on the database. The search process also identified patients with prior VTE either before cancer diagnosis or during adjuvant treatment for PDAC, patients with concurrent diagnosis of TE and aPDAC or patients on long term anticoagulants due to pre-existing AF (atrial Fibrillation) or unprovoked VTE having been switched to an IDPTP schedule (details in Flow chart Fig. 1).

Demographics and event rates are reported for the whole cohort but the detailed analyses are restricted to

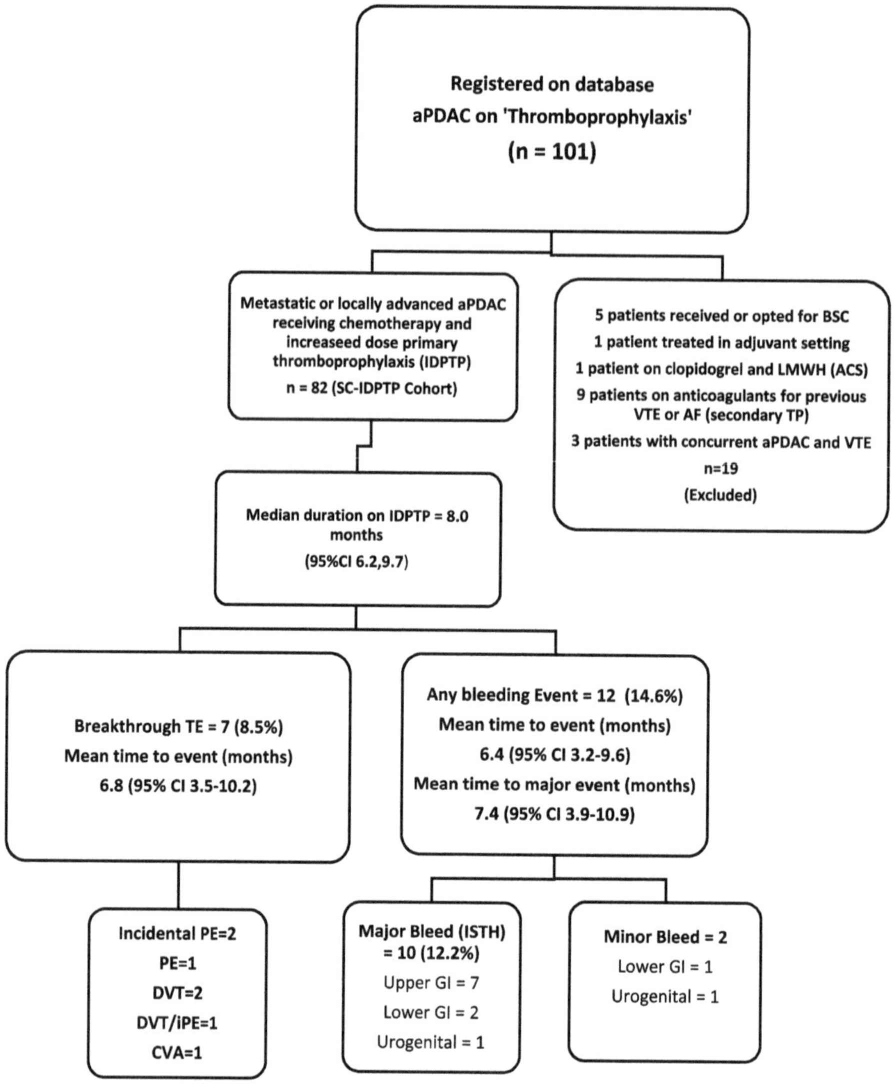

Fig. 1 Study Flow Chart. Patients that were identified using the search strategy described in the methods were entered on to a data base. Their case notes were reviewed and the patients that bore similarities to the experimental cohort of patients treated in the FRAGEM [8] and CONKO-04 [9] studies were selected for further analyses (IDPTP group). The miscellaneous reasons for exclusion are detailed. (ACS = Acute Coronary Syndrome, AF = Atrial Fibrillation, aPDAC = advanced Pancreatic Ductal Adenocarcinoma, BSC = Best Supportive Care, CVA = Cerebrovascular Accident, DVT = Deep Vein Thrombosis, GI = Gastrointestinal, iPE = Incidental Pulmonary Embolism, LMWH = Low Molecular Weight Heparin, PE = Pulmonary Embolism, SC-IDPTP = Study Cohort of Increased Dose Primary Thromboprophylaxis, TE = Thromboembolic Event, TP = Thromboprophylaxis, VTE = Venous Thromboembolism)

the 82 patients that corresponded to an entirely IDPTP cohort.

The median age was 67 years, interquartile range: 61–72 ($n = 82$) and 54% were male. Sixty five per cent of these patients were metastatic and the rest locally advanced. Ninety-six patients were offered first line chemotherapy. Sixty nine per cent of the patients received only one line of treatment, 20% received two lines of treatment and only 2% received a third line of treatment.

Of the IDPTP cohort ($N = 82$) 64 patients (78%) were offered Gemcitabine based first line treatment (48 patients (58%) single agent gemcitabine & 16 (20%) first line gemcitabine-based combination) and 18 patients FOLFIRINOX.

The baseline characteristics of all patients, for the 82 IDPTP patients and the primary endpoint data are shown in Table 1.

In the IDPTP cohort ($n = 82$), patients remained on thromboprophylaxis for a median duration of 8.0 months,95% CI (6.2, 9.7). Seven patients [8.5, 95%CI (3.7,14.6)] developed a TE despite IDPTP and ten patients [12.2, 95%CI (6.1,19.5)] in the IDPTP group had a major bleed and 2 patients [2.4 95%CI (.0,6.1)] had a minor bleed, for a combined event rate of major

Table 1 Patient Characteristics and Events (Major and Minor Bleeding & All-type Thromboembolic Events). Data shown for both the IDPTP Cohort and for thepatients that were excluded

	Miscellaneous TP	IDPTP
Patients	19	82
Age (years)	71 (r41)	67 (r51)
Male	11 (58%)	43 (52%)
Metastatic	10 (53%)	56 (68%)
WHO PS 0–1	10 (53%)	67 (82%)
Chemotherapy	[a]14 (73%)	[b]82 (100%)
FOLFIRINOX	2 (10%)	18 (22%)
Gemcitabine SA	10 (52%)	48 (58%)
Gem-Based Combi	1 (5%)	16 (20%)
No Chemo	5 (27%)	–
Events (Bleeding)	4 (20%)	12 (14.5%)
#Major Bleeding	2 (10%)	10 (12.2%)
Upper GI bleed	2 (100%)	7 (70%)
Lower GI bleed	–	2 (20%)
Genitourinary		1 (10%)
#Minor Bleeding	2 (10%)	[c]2 (2.5%)
Events (AT-TE)	7 (37%)	7 (8.5%)
DVT	2	2
i-PE	2	2
PE	–	1
i-PE & DVT	–	1
CVA	2	1
DVT & Visceral	1	

[a]One patient was treated with adjuvant chemotherapy (5FU/LV)
[b]73% completed one line, 24% two lines & 2.4% three lines of chemotherapy
[c]1 minor bleed was upper GI

bleeding and TE of 18% [$n = 15/82$ as two patients had a TE and a bleed, 95%CI (9.8,26.80)]. The correlation between the IDPTP and the timing of TE and bleeding can be seen in Fig. 2. The cumulative rates of MB at 3 and 12 months of this real-world cohort are compared to the experimental arms of FRAGEM and CONKO-04 where data were available (Table 2). Figure 2a & b present the cumulative hazard of developing a breakthrough TE or bleed. The accelerated phase of the cumulative hazards plot is between 4 and 8 months which is later than the conventional first 3 months seen in patients without thromboprophylaxis, reflecting the median time to TE of 6.2 months (95% CI 10.0, 13.2). This can be seen most clearly in Fig. 2c where the data are corrected for competing risk of death. A similar late pattern is noted for the major bleeding episodes median time to MB 4.5 months (95% CI 1.6, 7.4). The majority of the minor and major bleeding events were upper GI (8/14, 57%) and were directly or indirectly attributable to an aPDAC related cause.

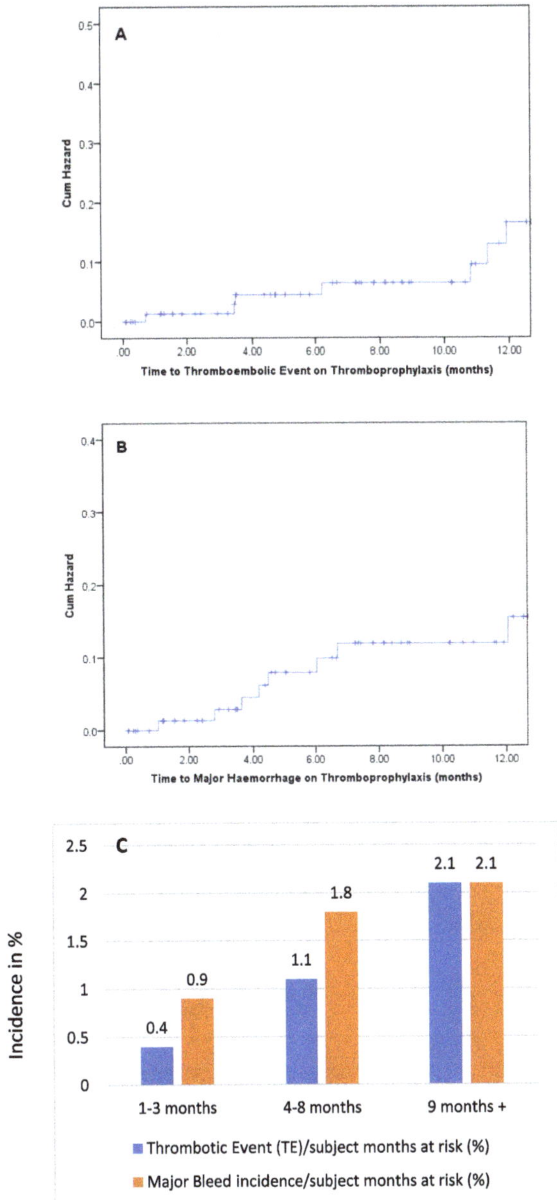

Fig. 2 The Hazard Plots (Kaplan Meier) for: **a** TE events, (**b**) Major Haemorrhage for the first 12 months of follow-up of 82 patients on IDPTP. **c** Incidence of thrombembolic events (TE) and Major Bleeding at different time periods. A lower incidence of TE is seen during the 1st 3 months, 0.4% (95% CI 0.0, 2.4; 1/226 subject months), compared to 2.1% (95% CI 0.4, 5.9; 3/146 subject months) after 8 months (the median duration of thromboprophylaxis in this group). A similar pattern is notable for major bleeding. Incidence was less in the 1st 3 months, 0.9% (95% CI 0.1, 3.1; 2/227 subject months). During the 4–8 months period (280 subject months), 5 patients experienced a major bleeding event that gives an incidence of 1.8% (95% CI 0.6, 4.1). Comparable to that, 3 patients (146 subject months) had major bleed after 8 months with an incidence of 2.1% (95% CI 0.4, 5.9)

Median overall survival time from commencement of IDPTP was 8.7 months, 95%CI, (6.4, 10.9). Neither thrombotic events [12.2 months, 95% CI (10.7, 13.7)

Table 2 Cross-study comparison of three- and 12 month landmark events (Major Bleeding and Thromboembolism)

EVENTS (Cumulative Incidence)	CONKO-04	FRAGEM	SC-IDPTP
Major Bleed (3 Months)	4.5%(7/160)	3%(3/59)	2.4% (2/82)
Major Bleed (12 months)	8.3%(13/160)	Not Reported (N/R)	9.7% (8/82)
Breakthrough PE/DVT (12 months)	5.7%(9/160)	[a]8.4% (5/59)	[b]4.8% (4/82)
Breakthrough All type TE (12 months)	Not Reported (N/R)	12% (7/59)	8.5% (7/82)

[a]Excludes 2 cases of I-PE to be comparable to CONKO-04
[b]Excludes 2 cases of I-PE and the CVA so as to be comparable to CONKO-04

Vs 8.4 months, 95% CI (6.9, 9.9) ($p = 0.79$)] nor bleeding events [median 8.7 months, 95% CI (0.4, 17) Vs 8.4 months, 95% CI (5.8, 11) ($p = 0.94$)] affected survival outcome. Progressive cancer remained the major cause of death (70 out of 82; 85%) as reported in the death certificate and agreed by the investigators. Five of these patients were found to be suffering from MB as a tumour progression-related event during the process of dying from cancer, the context often informing decision making (e.g. restraint on endoscopic or other invasive procedures or reflection on futility of transfusion). For one patient bleeding (uterine) was reported in the death certificate and confirmed by the investigators as the direct cause of death (1.2%). Two patients suffered a fatal thromboembolic event (2.4%). One a CVA and one a pulmonary infarct. Figure 3 charts MB and ATTE in the relevant patients, their timing and their link to the

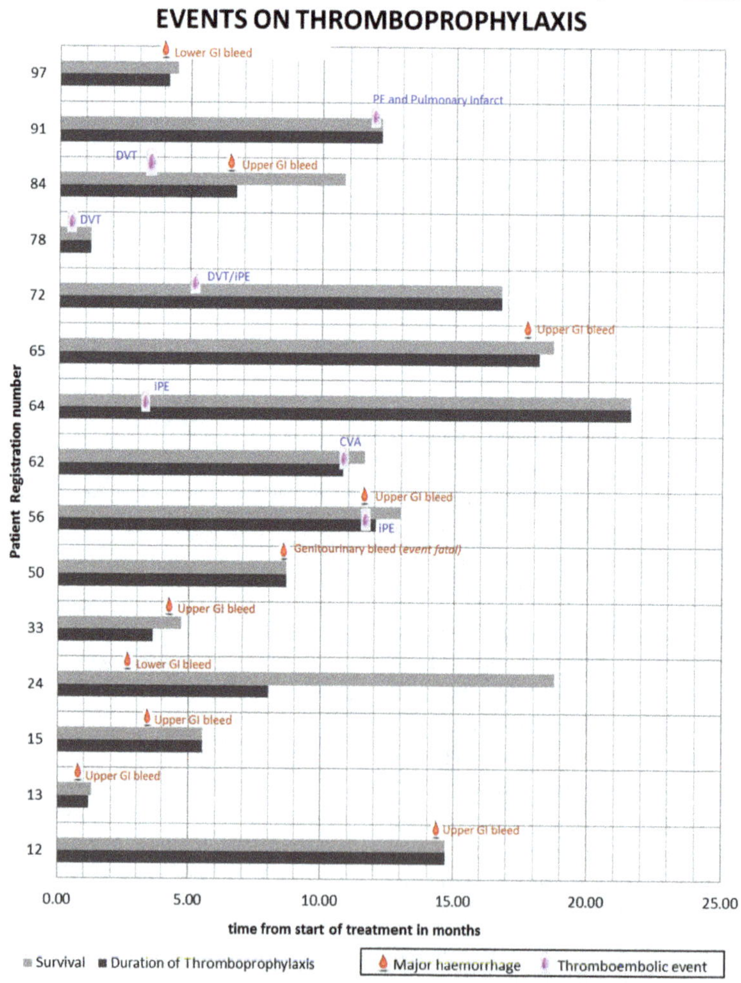

Fig. 3 The 15 patients with events. The sequential database numbers have been retained. (GI = Gastrointestinal, iPE = Incidental Pulmonary Embolism)

presence or absence of IDPTP over the patient lifespan.

Discussion

In this study we show that the 12 month rate of TE (8.5%) is on a par with the TE rate seen in the experimental arm of the FRAGEM study (12%) and much lower than the control arm of FRAGEM (24%). It is also much lower than in historical series (25–35%) [20]. We also show an incidence of clinically apparent VTE rate (PE and DVT) is 4.8%, similar to CONKO-04 (5.7%) and lower than in FRAGEM (8.4%) (Table 2).

The existing RCT evidence pertains to patients receiving first line chemotherapy Gemcitabine either as a single agent or in combination with a platinum [8, 9]. In this study 42% of patients received the more modern combinations of FOLFIRINOX (5-Fluorouracil, Irinotecan, Oxaliplatin) or AG (Abraxane, Gemcitabine). These chemotherapy schedules do not seem to have a diminished thrombosis risk. FOLFIRINOX has a 60% increase in risk of VTE (6.6% compared to gemcitabine alone 4.1%) [21]. In the study HALO-109-202 the rate of VTE in the investigational arm (AG-PEGPH20) was 42% and control arm (Gem Abraxane) was 25% respectively in stage 1 of the trial [22]. 40 mg of enoxaparin saw no VTE reduction in either arm. Only when an enoxaparin-IDPTP schedule was implemented did VTE rate drop to 5 and 6% respectively. A LMWH scale of dose-effectiveness in pancreatic cancer patients has recently been reported and can aid in decision making on IDPTP [7].

Of cancer patients commencing chemotherapy and who experience a VTE, 18.1% will have their first event within the first month, 47% within the first 3 months, and 72.5% within the first 6 months after starting chemotherapy [23].

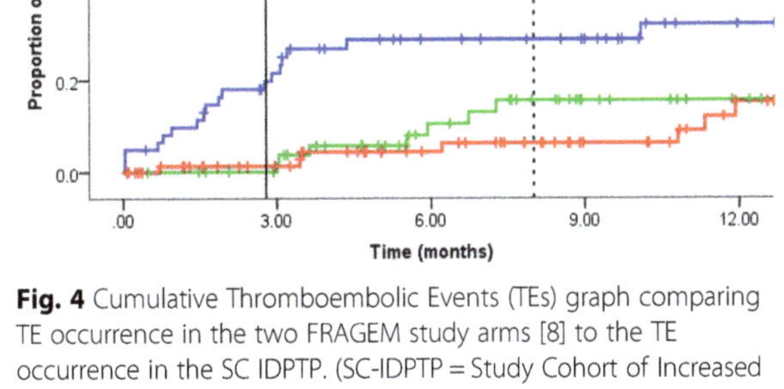

Fig. 4 Cumulative Thromboembolic Events (TEs) graph comparing TE occurrence in the two FRAGEM study arms [8] to the TE occurrence in the SC IDPTP. (SC-IDPTP = Study Cohort of Increased Dose Primary Thromboprophylaxis, TE = Thromboembolic Event)

In our cohort mean time to thrombosis was 6.8 months and TE-rates continue to remain diminished over the first 10 months. This is longer than seen in the FRAGEM-LMWH arm (Fig. 4). The mandatory stop of thomboprophylaxis at 84 days in FRAGEM is likely to have influenced these findings, whereas in this cohort median time on IDPTP was 8.0 months.

Figure 2a demonstrates a very late (around 12 months) manifestation of increasing TE events and may reflect the difficulties and ambivalence of maintaining thromboprophylaxis in patients arriving at the pre-terminal/terminal phase of malignancy. Adherence to anticoagulants may at that point be challenging or deemed futile. Of the 82 patients we report on, 81 had died at the time of data collection. Given the retrospective nature of the data collection for many patients we were not able to establish exact date or cause of LMWH discontinuation.

In this cohort we show that the 12 month MB rate was 12% (10/82). Sixty percent (6/10) of these related to direct (infiltration) or indirect (varices) effects of the malignancy. Of the four remaining major bleeds benign pathology was documented in 2 [uterus (fibroids) and anus (haemorrhoids)] while one upper and one lower GI bleed remained unattributed. Major bleed was an end of life feature in 5 patients but death attributable solely to bleeding was documented in only one case, while in 4 it was part of the end-of-life (EoL) process of pancreatic cancer progression being the underlying cause of death and the management of the bleed tailored to the end of life healthcare context (e.g. restraint on endoscopic or other invasive procedures). There was no survival difference between patients with either TE or bleeding compared to no events.

These data are certainly in keeping with real world data for patients with aPDAC not on thromboprophylaxis [24]. Wang et al. reported on 246 cases of aPDAC, and found an overall bleeding incidence of 13% (32 patients). A multivariate analysis showed that only stage IV disease maintained predictive significance for a bleed [HR 21.94, 95% CI (2.274–211.642)]. Haemorrhage was an end of life event in 7% (18 patients out of 246) [24] with a median time to death of 31 +/− 21 days. No impact of bleeding on survival was seen [HR 1.246,95% CI (0.754–2.060)] [8.7 months,95% CI (0.4, 17) Vs 8.4 months,95% CI (5.8, 11)].

Compliance has been shown to be better with oral agents with large early discontinuation rates for LMWHs [25]. Although injectable anticoagulants are undoubtedly more invasive than a tablet, patient views are nuanced. At least 2 studies [26, 27] have shown that patients will select anticoagulants that are least likely to interfere with their cancer treatment. In the SELECT-D sub-study [27] patients did express a preference for an oral agent, but only if this has equivalent clinical benefit to LMWH. Compliance in this study was very high, with patients staying on dalteparin for a median of 8.0 months. This may also reflect the absence of an alternative oral option in the IDPTP setting.

Very recently an experience of therapeutic dose rivaroxaban (15 mg po bd followed by 20 mg po od) in 28 aPDAC patients receiving Gemcitabine Abraxane showed a 7% major bleeding (2/28) and one breakthrough VTE (asymptomatic) 3.5% [28]. Though promising there are no randomised data in this setting for supraprophylactic DOAC schedules and one has to be aware of the greater propensity of the GI tract cancers to be associated with increased non-major but clinically relevant bleeding as seen in the recent studies of cancer associated thrombosis treatment with rivaroxaban and edoxaban [29, 30].

Two RCTs using conventional thromboprevention schedules of the direct oral anticoagulants rivaroxaban and apixaban in cancer patients have recently published [13, 14].

Patients were risk stratified using the Khorana risk stratification score. In the CASSINI study 54% of patients had either aPDAC (33%) or Gastroesophageal (GO) (21%) while in the AVERT study only 16% belonged to this group (13% aPDAC and 3% GO). Efficacy endpoint subanalysis (prevention of VTE) for the aPDAC patients has been published only for the CASSINI study (HR 0.70, 95%CI 0.344,1.432). The latest iteration of most guidelines (2019) reinforced by the 2 recent randomised studies of DOACS, have formulated thromboprophylaxis recommendations for aPDAC [10, 11]. As the data stand LMWH-IDPTP schedules remain a proven approach for aPDAC while for the DOACs there remains the need for a well-designed single cancer RCT [31].

Our study has shortcomings; the obvious one being that this is a retrospective series and some endpoint comparisons are made to historical data. A further weakness is that we cannot account for selection bias, as we cannot identify and report in retrospect on outcomes of patients that were not offered IDPTP or not switched to IDPTP either due to contraindication or due to omission. Whilst the chemotherapy regimens were mostly gemcitabine based they included patients on different trials with experimental combinations and at least 18% on triple combination FOLFIRINOX. Nevertheless over this period, and contrary to many other cancers, the overall treatment of pancreatic cancer has brought little material change to outcomes. The study is too small to pick up substantial signals of mortality due to adverse events. Our median survivorship for the group is similar with that expected for aPDAC patients as seen in other series and studies. Although the IDPTP-dalteparin guideline adhered to in this study concurs with the CONKO-04 approach of 'lifelong' thromboprophylaxis, the higher doses used throughout in our study compared to CONKO-04 (40 mg enoxaparin) do not allow direct comparisons. We demonstrate VTE control but note that thromboprophylaxis at late stages of advancing pancreatic cancer where bleeding becomes a more common event should be approached cautiously and the use of IDPTP beyond the 3 months used in the major studies should be reviewed on a case by case basis.

Conclusion

In conclusion we demonstrate that, an IDPTP-LMWH with dalteparin reflecting the evidence can be implemented and patients can adhere to it. IDPTP-LMWH was associated with lower ATTE occurrence rates than expected and comparable major bleeding rates. ATTE and MB were late events; the majority of MB was from the GI tract and the direct or indirect result of locally progressing aPDAC. Since these conditions can frequently arise in aPDAC, IDPTP-LMWH beyond 3 months needs to be approached cautiously and should be regularly reviewed due to the increasing risk of GI bleeding in late –progressing- pancreatic cancer.

Abbreviations

AG: Abraxane, Gemcitabine; aPDAC: Advanced Pancreatic Ductal Adenocarcinoma; ATTE: All Type Thromboembolic Events; BFI: Bleeding-free interval; DOAC: Direct Oral Anticoagulant; DVT: Deep Vein Thrombosis; ECOG-PS: Eastern Cooperative Group Performance status; FOLFIRINOX: Folinic Acid, 5-Fluorouracil, Irinotecan, Oxaliplatin; GI: Gastrointestinal; GO: Gastro-oesophageal; IPE: Incidental Pulmonary Embolism; IDPTP: Increased Dose Primary Thromboprophylaxis; LMWH: Low Molecular Weight Heparin; MB: Major Bleeding; NNT: Numbers Needed to Treat; PE: Pulmonary Embolism; PEGPH20: Pegvorhyaluronidase Alfa; TE: Thromboembolic Event; TFI: TE-free Interval; VTE: Venous Thromboembolism

Acknowledgements

We would like to thank Sarah Scargill for pharmacy crosschecking and Anita Gadie for support with the search of the HPB-MDT database.

Authors' contributions

AM Conceived the study, reviewed data, adjudicated/reconciled events and wrote the manuscript. FH Cleaned the data, analysed the data and reviewed the manuscript. IM developed the audit proposal, collected data, cleaned the data and reviewed the manuscript. WI Co-developed the audit proposal, collected data, developed the database, cleaned the data and reviewed the manuscript and submitted final audit report. GB Analysed the data, adjudicated/reconciled events and reviewed the manuscript. All Authors approved the final manuscript.

References

1. Maraveyas A. Pancreatic cancer: a model cancer for the study of the therapeutic effects of anticoagulants. World J Gastroenterol. 2009;1:38–40.
2. Khorana AA, Fine RL. Pancreatic cancer and thromboembolic disease. Lancet Oncol. 2004 Nov;5(11):655–63.
3. Menapace LA, Peterson DR, Berry A, Sousou T, Khorana AA. Symptomatic and incidental thromboembolism are both associated with mortality in pancreatic cancer. Thromb Haemost. 2011;106(2):371–8.
4. Lyman GH, Eckert L, Wang Y, Wang H, Cohen A. Venous thromboembolism risk in patients with cancer receiving chemotherapy: a real-world analysis. Oncologist. 2013;18:1321–9. https://doi.org/10.1634/theoncologist.2013-0226.
5. Khorana AA, Dalal MR, Lin J, Connolly GC. Health care costs associated with venous thromboembolism in selected high-risk ambulatory patients with solid tumors undergoing chemotherapy in the United States. Clinicoecon Outcomes Res. 2013;5:101–8.
6. Hidalgo M, Cascinu S, Kleeff J, Labianca R, Löhr JM, Neoptolemos J, et al. Addressing the challenges of pancreatic cancer: future directions for improving outcomes. Pancreatology. 2015;15(1):8–18.
7. Maraveyas A, Muazzam I, Noble S, Bozas G. Advances in managing and preventing thromboembolic disease in cancer patients. Curr Opin Support Palliat Care. 2017;11(4):347–54.
8. Maraveyas A, Waters J, Roy R, Fyfe D, Propper D, Lofts F, et al. Gemcitabine versus gemcitabine plus dalteparin thromboprophylaxis in pancreatic cancer. Eur J Cancer. 2012;48:1283–92.
9. Pelzer U, Opitz B, Deutschinoff G, Stauch M, Reitzig PC, Hahnfeld S, et al. Efficacy of prophylactic low-molecular weight heparin for ambulatory patients with advanced pancreatic cancer: outcomes from the CONKO-004 trial. J Clin Oncol. 2015;33:2028–34.
10. Lyman GH, Khorana AA, Kuderer NM, Lee AY, Arcelus JI, Balaban EP, et al. Venous thromboembolism prophylaxis and treatment in patients with cancer: American Society of Clinical Oncology clinical practice guideline update. J Clin Oncol. 2013;31:2189–204.
11. National Comprehensive Cancer Network Venous Thromboembolic Disease Clinical Practice Guidelines in Oncology. v.1.2015 edn. www.nccn.org: NCCN, 2015, Guideline.#.
12. Farge D, Debourdeau P, Beckers M, Baglin C, Bauersachs RM, Brenner B, et al. International clinical practice guidelines for the treatment and prophylaxis of venous thromboembolism in patients with cancer. J Thromb Haemost. 2013;11(1):56–70.
13. Khorana AA, Soff GA, Kakkar AK, Vadhan-Raj S, Riess H, Wun T, et al. CASSINI investigators. Rivaroxaban for Thromboprophylaxis in high-risk ambulatory patients with cancer. N Engl J Med. 2019;380(8):720–8.
14. Carrier M, Abou-Nassar K, Mallick R, Tagalakis V, Shivakumar S, Schattner A, et al. Apixaban to prevent venous thromboembolism in patients with cancer. N Engl J Med. 2019;380(8):711–9.
15. Venous thromboembolism in over 16s: reducing the risk of hospital acquired deep vein thrombosis or pulmonary embolism: NICE guidline 21 March 2018. https://www.nice.org.uk/guidance/ng89.
16. Ek L, Gezelius E, Bergman B, Bendahl PO, Anderson H, Sundberg J, et al. Swedish lung cancer study group (SLUSG). Randomized phase III trial of low-molecular-weight heparin enoxaparin in addition to standard treatment in small-cell lung cancer: the RASTEN trial. Ann Oncol. 2018;29(2):398–404.
17. Schulman S, Kearon C. Subcommittee on control of anticoagulation of the scientific and standardization Committee of the International Society on thrombosis and Haemostasis definition of major bleeding in clinical investigations of antihemostatic medicinal products in non-surgical patients. J Thromb Haemost. 2005;3(4):692–4.
18. HRA. Defining research. NHS health Rresearch authority; 2017. http://www.hra-decisiontools.org.uk/research/docs/DefiningResearchTable_Oct2017-1.pdf. Accessed 30 Dec 2017.
19. HQIP. Best practice in clinical audit; 2016. Health quality improvement partnership. http://www.hqip.org.uk/resources/best-practice-in-clinical-audit-hqip-guide/. Accessed 30 Dec 2017.
20. Campello E, Ilich A, Simioni P, et al. The relationship between pancreatic cancer and hypercoagulability: a comprehensive review on epidemiological and biological issues. Br J Cancer. 2019;121:359–71. https://doi.org/10.1038/s41416-019-0510-x.
21. Conroy T, Desseigne F, Ychou M, Bouché O, Guimbaud R, Bécouarn Y, et al. FOLFIRINOX versus gemcitabine for metastatic pancreatic cancer. N Engl J Med. 2011;364:1817–25 10.1056.
22. Hingorani SR, Zheng L, Bullock AJ, Seery TE, Harris WP, Sigal DS, et al. HALO 202: randomized phase II study of PEGPH20 plus nab-paclitaxel/gemcitabine versus nab-paclitaxel/gemcitabine in patients with untreated, metastatic pancreatic ductal adenocarcinoma. J Clin Oncol. 2018;36(4):359–66.
23. Khorana AA, Dalal M, Lin J, Connolly GC. Incidence and predictors of venous thromboembolism (VTE) among ambulatory high-risk cancer patients undergoing chemotherapy in the United States. Cancer. 2013;119(3):648.
24. Wang YU, Yuan C, Liu X. Characteristics of gastrointestinal hemorrhage associated with pancreatic cancer: a retrospective review of 246 cases. Mol Clin Oncol. 2015;3(4):902–9.
25. Khorana AA, McCrae KR, Milentijevic D, Fortier J, Nelson WW. Laliberté F, et al current practice patterns and patient persistence with anticoagulant treatments for cancer-associated thrombosis. Res Pract Thromb Haemost. 2017;1(1):14–22.
26. Noble S, Matzdorff A, Maraveyas A, Holm MV, Pisa G. Assessing patients' anticoagulation preferences for the treatment of cancer-associated thrombosis using conjoint methodology. Haematologica. 2015;100(11):1486–92.
27. Hutchinson A, Rees S, Young A, Maraveyas A, Date K, Johnson MJ. Oral anticoagulation is preferable to injected, but only if it is safe and effective: an interview study of patient and carer experience of oral and injected anticoagulant therapy for cancer-associated thrombosis in the select-d trial. Pal Med (Manuscript accepted for publication). 2019;33(5):510.
28. Yu KH, Mantha S, Tjan C, Kaufmann ES, Brenner R, Lowery MA. Pilot study of gemcitabine, nab-paclitaxel, PEGPH20, and rivaroxaban for advanced pancreatic adenocarcinoma: an interim analysis. J Clin Oncol. 2018;36(4_suppl):405.
29. Young AM, Marshall A, Thirlwall J, Chapman O, Lokare A, Hill C, et al. Comparison of an Oral factor Xa inhibitor with low molecular weight heparin in patients with cancer with venous thromboembolism: results of a randomized trial (SELECT-D). J Clin Oncol. 2018;36(20):2017–23.
30. Raskob GE, van Es N, Verhamme P, Carrier M, Di Nisio M, Garcia D, et al. Edoxaban for the treatment of cancer-associated venous thromboembolism. N Engl J Med. 2018;378(7):615–24.
31. Agnelli G. Direct Oral anticoagulants for Thromboprophylaxis in ambulatory patients with cancer. N Engl J Med. 2019;380(8):781–3.

6

Blue toe syndrome caused by emboli from anomalous left atrial septal pouch thrombus

Snehasis Pradhan*, Kciku Gresa, Jan-Peter Röing genannt Nölke and Hans-Joachim Trappe

Abstract

Background: Left atrial septal pouches (LASPs) are a relatively newly described but common anatomical cardiac variant thought to be associated with atrial fibrillation (AF) and cardio-embolic stroke. Blue toe syndrome (BTS) describes ischemic changes in the toes due to microembolisation of the digital arteries. Establishing the etiology of BTS is vital so that the underlying cause can be treated. Here we describe the first case of BTS arising due to emboli from LASP thrombus arising on a background of new-onset AF.

Case presentation: A 65-year-old man presented with a two-day history of progressive painful swelling and bluish-purple discoloration of the second and fourth toes of his left foot and new-onset AF. Tests for hypercoagulability disorders were negative. Duplex ultrasound and CT angiography excluded deep venous thrombosis and an absence of embolus, thrombus, or occlusion in the arterial tree in the lower extremities bilaterally, so BTS was diagnosed. While transthoracic echocardiography and chest CT initially showed no cardiac abnormalities or mural thrombus, subsequent transesophageal echocardiography revealed a LASP with an associated pedunculated thrombus. The affected toes were amputated due to wet gangrene, but the patient recovered well with thrombus resolution after anticoagulation.

Conclusion: The presence of a LASP in the absence of any other identifiable cause of BTS should trigger careful investigation of the interatrial septum, preferably using a multimodality imaging approach. The possibility that LASPs may not merely be an innocent bystander but a causative mechanism for peripheral ischemia must be considered.

Keywords: Blue toe syndrome (BTS), Left atrial septal pouch (LASP), Interatrial septum, Atrial fibrillation, CT angiography, Transesophageal echocardiography, Case report

Background

"Blue toe syndrome" (BTS; also known as "trashfoot") was first reported in 1976 and describes the sudden onset of acute pain and cyanosis in one or more toes due to microembolization of the digital arteries from a proximal source via a patent arterial tree in an otherwise well-perfused foot [1]. In the original case series describing BTS [2], all 31 patients had direct angiographic evidence of a proximal source of emboli in the vascular tree. The etiologies of BTS can be divided into three main groups: (i) emboli from the cardiac and arterial system; (ii) acquired hypercoagulability disorders; and (iii) syndromes leading to peripheral vascular pathology (Table 1) [1]. These causes are not mutually exclusive; for example, cryoglobulinemia can induce vasculitis and subsequent digital arteriolar and capillary thrombosis. Therefore, definitive subclassification of BTS can be difficult in some cases.

* Correspondence: infosnehasis@gmail.com
Department of Cardiology and Angiology, Marien Hospital Herne, Ruhr-University of Bochum, Hoelkeskampring 40, 44625 Herne, Germany

Table 1 Differential diagnosis of blue toe syndrome

Emboli from cardiac and arterial system	Hypercoagulability disorders	Peripheral vascular pathology
Left atrial septal pouch Atrial fibrillation Left ventricular aneurysm Cholesterol crystal Penetrating ulcers or Aneurysms of the aorto-iliac-femoral arterial system Cardiac/vascular tumors (myxoma) Valvular heart disease Endocarditis Iatrogenic injury	Antiphospholipid syndrome Malignancy Thrombocytopenic purpura Disseminated intravascular coagulation Warfarin skin necrosis Cryoglobulinemia Myeloproliferative disorders	Raynaud's phenomenon Perniosis Frostbite Infectious and noninfectious inflammation Medication-induced vasoconstriction

In clinical practice, obliteration of the digital arterial vascular network can be devastating. In most instances, the cessation of arterial blood flow and consequent ischemic necrosis requires amputation of the affected toes or foot and may be life-threatening. Arterial embolisms are a primary cause of acute lower limb ischemia, with the majority of emboli, particularly in BTS, originating from the left heart as a result of intracavitary thrombus fragmentation [3]. Atrial fibrillation, left ventricular aneurysm, bacterial endocarditis, valvular heart disease, vascular tumors, and penetrating ulcerated atherosclerotic plaques in the aorto-iliac-femoral arterial system with or without aneurysm are the major sources of microembolisation [4]. A minority are paradoxical venous emboli that pass through an intracardiac shunt [5].

Atrial septal pouches are a relatively newly described but common anatomical cardiac variant [6]. An atrial septal pouch is a kangaroo pocket-like structure in the interatrial septum, without an interatrial shunt. They develop from postnatal partial fusion of the septum primum and the septum secundum, either opening into the left atrium or, less frequently, into the right atrium. Left atrial septal pouch (LASP) is very common, with an overall prevalence of 40.8%; right septal pouches and double pouches are present in 5.1 and 3.7% of individuals, respectively [7]. Although right septal pouches seem to have no clinical significance, the high prevalence of LASPs and evidence of a possible association between LASP and various cardio-embolic consequences, especially atrial fibrillation and cardio-embolic stroke [8], suggest that LASP may not be entirely benign. Therefore, LASP could also be a source of cardio-embolism in BTS, although this phenomenon has yet to be described in the literature.

Here we report the first case of thrombus formation within a LASP in the absence of a patent foramen ovale giving rise to BTS. This case illustrates the previously unrecognized potential of LASP thrombus giving rise to BTS, which in this case was complicated by digital necrosis, gangrene, and the eventual need for toe amputation.

Case presentation

A 65-year-old man presented to the emergency department with two-day history of progressive painful swelling and bluish-purple discoloration of his second and fourth toes of his left foot (Fig. 1) and history of palpitations for a week. There was no recent history of any trauma, fever, or excessive sensitivity or exposure to the cold. He had a past medical history of hypertension, for which he was prescribed ramipril, metoprolol, and amlodipine, and he was overweight.

On arrival, he was tachycardic at 188 bpm and his blood pressure was 110/70 mmHg. Clinical examination of his extremities revealed that the left second and fourth toes were tender, slightly edematous, and purple-blue, with a sharp demarcation between discolored and normal skin at the base of both digits. There were no signs of livedo reticularis or superficial lymphangitic streaking in the feet or lower extremities. BTS was suspected because peripheral pulses were palpable in both legs and there was no clinical evidence of ischemia in the contralateral limb.

His CRP was elevated at 10 mg/L (normal < 0.5 mg/L) and D-dimers were positive. Brain natriuretic peptide was elevated, but renal and liver function parameters were well within normal limits. Diagnostic screening for diabetes was negative. Laboratory tests for thrombophilia, which included disorders of plasminogen activation,

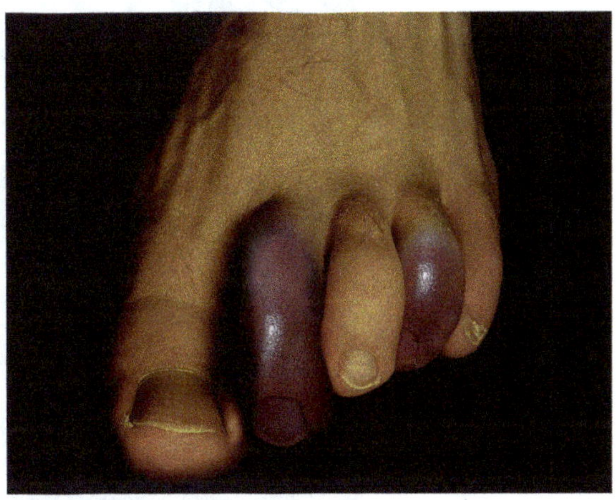

Fig. 1 Bluish purple ischemic discoloration of the second and fourth toes of the left foot consistent with blue toe syndrome

antithrombin III deficiency, protein C and protein S deficiency, and serum homocysteine levels, were negative.

Left foot radiographs showed mild swelling but no evidence of osteomyelitis or fracture. Duplex ultrasound excluded deep venous thrombosis and confirmed patency of the outflow vessels without any embolus, thrombus, or occlusion in the arterial tree in the lower extremities bilaterally. A bedside transthoracic echocardiography (TTE) showed normal left ventricular systolic function without any segmental wall motion abnormality. The interatrial septum was thickened but there was no visible mobile echogenic mass in the atria. Signs of pulmonary hypertension were absent. An electrocardiogram showed tachycardia with new-onset irregularly irregular atrial fibrillation.

The patient underwent immediate thoracic, abdominal, and lower limb contrast-enhanced computed tomographic angiography (CTA), which excluded any proximal thromboembolic source or occlusions in the arterial tree (Fig. 2). There was no evidence of abdominal aortic aneurysm or severe atheromatous plaque. Chest CT incidentally revealed a focal left atrial cavity filling defect adjacent to the atrial septum. This defect was initially suspected to be a cardiac CT artifact in light of the irregular tachyarrhythmia at the time of acquisition, which can generate slow-moving, turbulent blood due to atrial hypokinesia (Fig. 3).

Initially, it was decided to treat the case conservatively with anticoagulant (heparin drip) and anti-aggregate medications (intravenous PGE1). He was admitted to vascular surgery with a diagnosis of acute limb ischemia. He was placed on an empiric antibiotic regimen (ciprofloxacin), received morphine for pain, and was sent to the telemetry unit.

However, the patient experienced extreme worsening of progressive pain overnight that prevented sleep and further bluish discoloration of both toes. Since he had palpable posterior tibial and dorsalis pedis pulses and a normal ankle-brachial index (1.2; normal > 1), the vascular surgery team decided against toe amputation. An emergency percutaneous vascular angiography showed

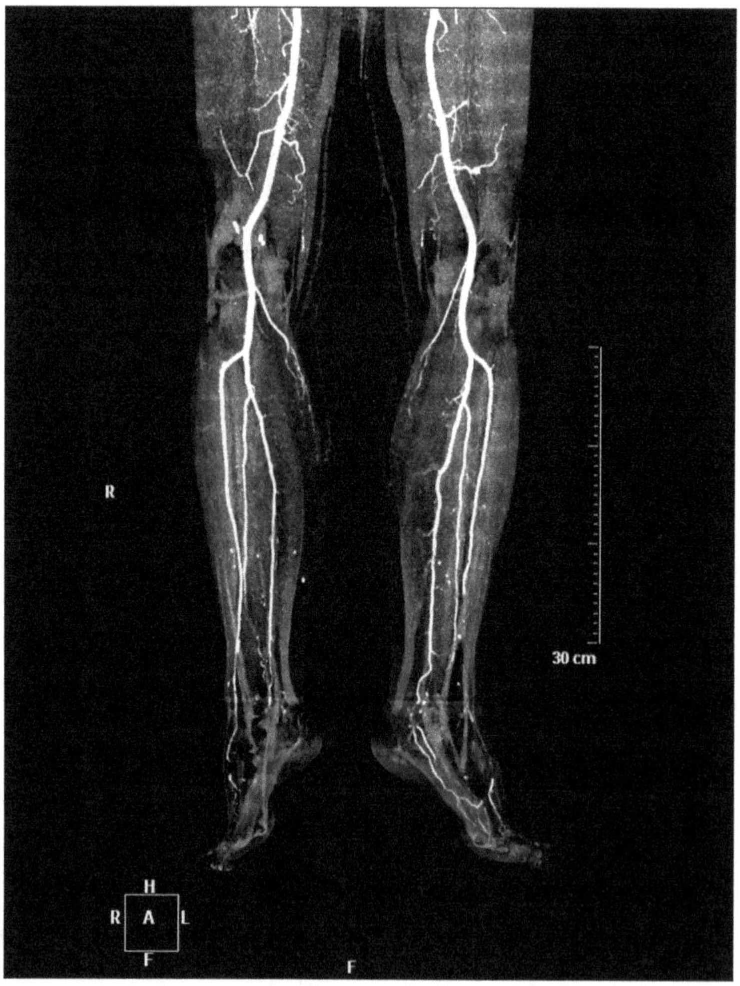

Fig. 2 Coronal maximum intensity projection (MIP) of lower extremity CT angiography ruling out any significant arterial stenosis

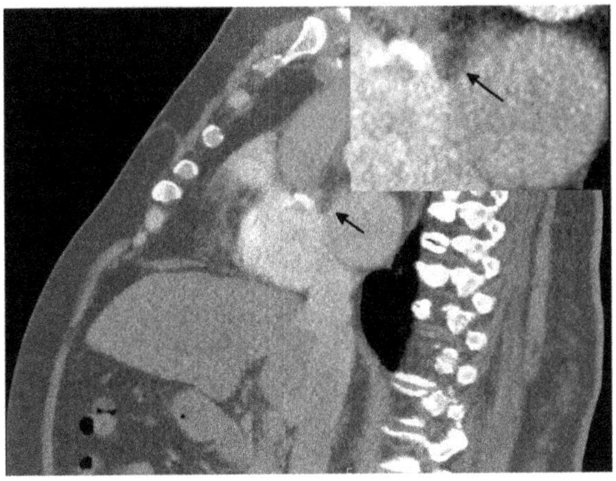

Fig. 3 Chest computed tomographic image demonstrating a filling defect adjacent to the interatrial septum (arrow)

no further evidence of any major arterial occlusion of the arterial tree of the left limb.

Given the high index of suspicion of an intracavitary thrombus in the setting of new-onset AF and progression of the cyanotic toes, the patient was referred to cardiology, where transesophageal echocardiography (TEE) revealed an anomalous LASP and a 7.7 × 3 mm mobile ovoid echodense mass with a narrow stalk attached to the pouch, suggestive of thrombus (Fig. 4, Supplemental Video 1). There was no evidence of an atrial septal defect on color Doppler flow imaging (Fig. 5), although it showed mild mitral regurgitation. A further confirmatory agitated saline test to exclude atrial septal defect was not performed to avoid dislodging the thrombus. No mass was seen in the left atrial appendage (LAA; peak emptying velocity 63 cm/s). Three-dimensional (3D) TEE visualized the LASP containing a mobile thrombus with attachment of its stalk to the interatrial septum (Fig. 6 and Supplemental Video 2). It is noteworthy that the exact shape and volume of the mass were revealed by the 3D acquisition.

The AF was managed with metoprolol, to which the patient responded adequately. Taking the small size of thrombus into consideration, absence of direct evidence of transit of thrombus in the right atrium, and following multidisciplinary case discussions involving cardiology, vascular surgery, and cardiac surgery, the decision was made to continue therapeutic anticoagulation and observe. Despite the intravenous administration of ciprofloxacin and later tazobactam, after microbiological wound swab culture of normal flora (*Staphylococcus aureus* and *Proteus species*), the patient developed wet gangrene and demarcated necrosis in his second and fourth toes. Antibiotic therapy was stopped given the lack of apparent benefit. The patient was offered amputation, to which he agreed due to the likelihood of faster healing and rehabilitation. Histopathological examination of the resected specimen confirmed tissue necrosis without signs of malignancy.

His incisions healed 2 weeks later and, after a short period of physiotherapy, he was able to ambulate. No further infection was detected. The patient was transitioned to an oral anticoagulant (apixaban) and he was discharged from the hospital. He remained well 3 months after discharge, with follow-up TEE showing no evidence of thrombus (Fig. 7). On further imaging with agitated saline contrast, there was no evidence of an atrial septal defect (Fig. 8).

Extensive tests for autoimmune disorders potentially responsible for thrombotic events, including anti-lupus erythematosus, antinuclear, antimitochondrial, and antiphospholipid antibodies, were all negative. Therefore, the LASP, together with new-onset AF, were likely to have resulted in thrombus formation within the septal pouch with subsequent embolization to the digital artery of the left foot to cause BTS.

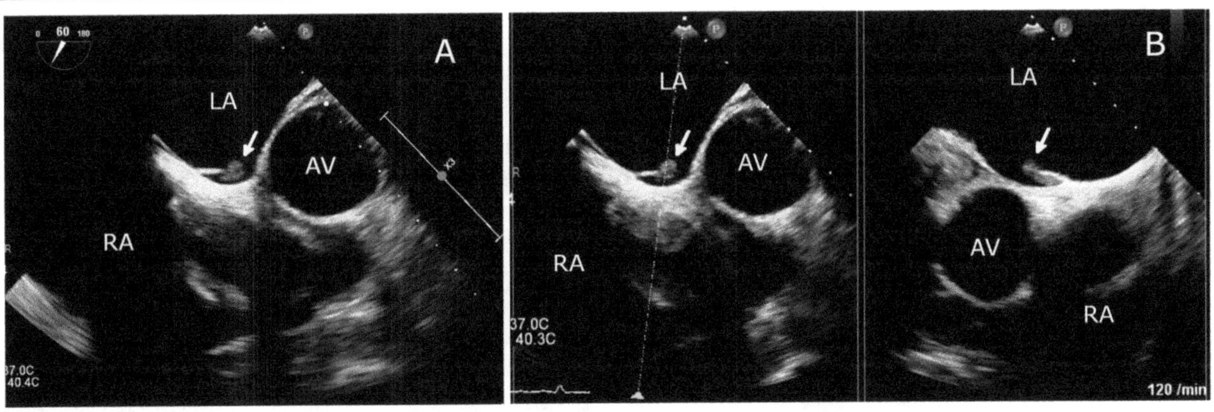

Fig. 4 Transesophageal echocardiogram (TEE) showing LASP opening into the left atrial (LA) cavity and a 7.7 × 3 mm thrombus (arrow), with the site of attachment corresponding to the LASP (2-dimensional TEE images at 60° [**a**] and biplane TEE image [**b**] of the interatrial septum). AV, aortic valve; LA, left atrium; RA, right atrium

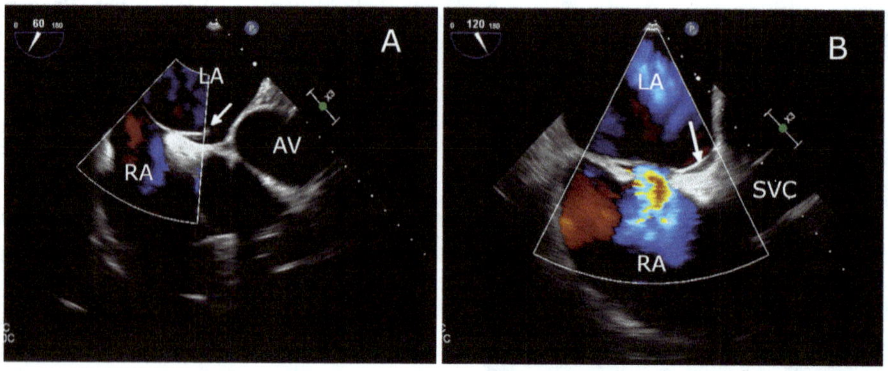

Fig. 5 Left atrial septal pouch thrombus seen by color Doppler imaging during TEE at 60° [**a**] and 120° [**b**] and simultaneously excluding any probable atrial septal defect

Discussion and conclusions

Here we describe a unique case of LASP thrombosis causing catastrophic BTS necessitating toe amputation. Careful history taking and physical examination will often suggest the etiology of BTS, but in this case, the etiology was more difficult to establish. The spontaneous onset of painful bluish discoloration of the toes in BTS usually signifies an arterial embolic disorder, and emboli can be dislodged thrombi, fragments of atheromatous plaque, or rarely, tumor cells or other foreign bodies [4, 9]. Therefore, thrombogenic and atherogenic sources of emboli need to be considered in the diagnostic work-up. Establishing the underlying cause of BTS is important since, if untreated, BTS arising from embolization can frequently be complicated by further emboli that not only result in the loss of digits but also the forefoot and limb, sometimes causing death [10].

In our case, the patient presented with acute onset painful and cyanotic second and fourth toes of the left foot. A thrombogenic etiology initially seemed more likely in this patient given the negative history of trauma, new-onset tachycardia and AF, and distal ischemia appearing simultaneously in multiple toes. Non-invasive vascular assessment such as Duplex ultrasound (DUS) is the first-line imaging modality in any form of acute limb ischemia (ALI) due to its diagnostic accuracy, wide availability, low cost, rapidity, and non-invasiveness. DUS can accurately determine the presence of an aneurysm and the anatomic location and degree of thromboembolic obstruction, hence should be performed in all patients [11]. Neither aneurysm nor thromboembolic obstruction were apparent on DUS of the arteries proximal to the ischemic digits, contralateral arteries, and the venous system in our patient at the time of presentation. Alternatively, CT angiography (CTA) is a rapid and accurate imaging modality that can diagnose and grade the extent of vascular disease in ALI and is the preferred imaging modality for establishing the source of the embolism [12]. In our patient, CTA showed no evidence of abdominal aortic aneurysm, atheromatous plaque, or any source of emboli proximal to the digital ischemia. Percutaneous vascular angiography, while considered the "gold standard" test [13], is mostly reserved for cases in which non-invasive evaluation is uninformative and in our case was inconclusive for any immediate thromboembolic event. Of note, invasive angiography should not be used as the first diagnostic modality and should not replace non-invasive imaging for the definitive diagnosis of BTS.

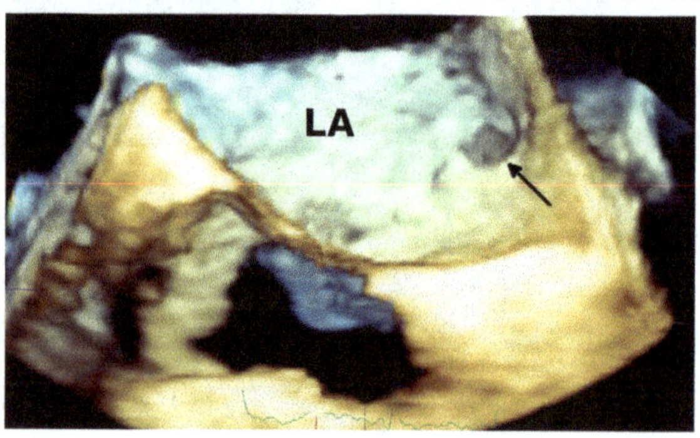

Fig. 6 3D reconstruction of the TEE image of the left side of the interatrial septum to visualize the sessile thrombus (arrows) adherent to the septum and arising from the left atrial septal pouch

TTE is a useful screening tool for detecting a potential cardiac source of peripheral embolism [14]. In our patient, TTE revealed no visible signs of ventricular aneurysm, vegetations, intracavitary masses, or thrombus. A failure to visualize the source of an embolus on TTE does not rule out its existence, so routine TTE can be omitted when there is a clear cardiac source of embolism clinically.

Other possible differential diagnoses of the etiology of toe ischemia included hyperviscosity disorders such as cryoglobulinemia, antiphospholipid syndrome, thrombotic thrombocytopenic purpura, and disseminated intravascular coagulation [15]. Cryoglobulinemia may cause purpuric or

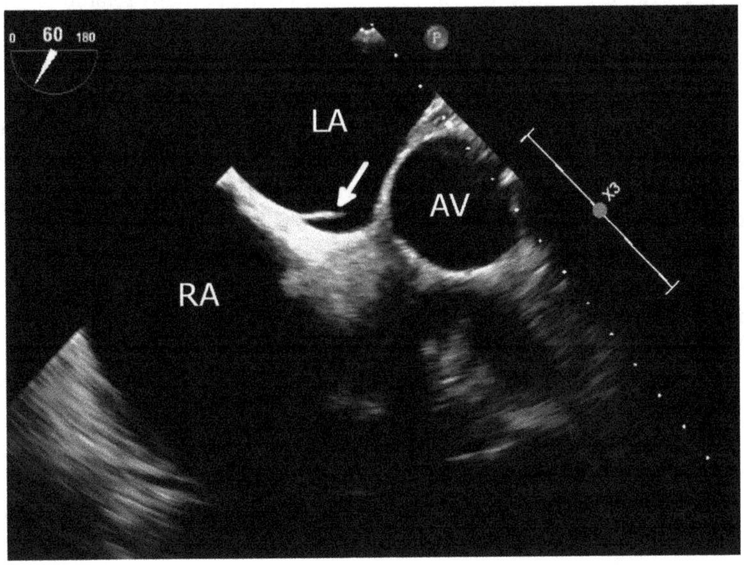

Fig. 7 Follow-up TEE showing LASP opening into the left atrium with complete resolution of thrombus after 3 months of oral anticoagulation

gangrenous lesions in the distal extremities, and its incidence is increasing due to its association with hepatitis C infection [16]. However, in cryoglobulinemia, skin lesions would typically be more widespread and unlikely to affect, as in our case, only two toes. Antiphospholipid syndrome is an autoimmune disease causing thrombotic events in both the arterial and venous systems. Thrombotic thrombocytopenic purpura causes BTS through occlusion of arterioles with platelet plugs. Disseminated intravascular coagulation typically arises secondary to another condition such as sepsis or malignancy, causing widespread activation of the clotting cascade generating excess thrombin and resulting in vascular occlusion. Extensive laboratory and clinical investigations for these conditions were all normal or negative in our case.

Several vasoconstrictive disorders such as Raynaud's phenomenon, perniosis, and frostbite may cause BTS through exposure to cold or even emotional stress [17]. In perniosis, digital cyanosis results from exposure to low temperatures (above freezing) and damp and is often accompanied by blisters, erosions, and eventually ulcers. In BTS secondary to vasculitides, inflammatory white blood cells adhere to and damage vessel walls, leading to narrowing, occlusion, and possible rupture of the small- and medium-sized vessels. Reynaud's phenomenon was unlikely in our patient given the limited involvement of

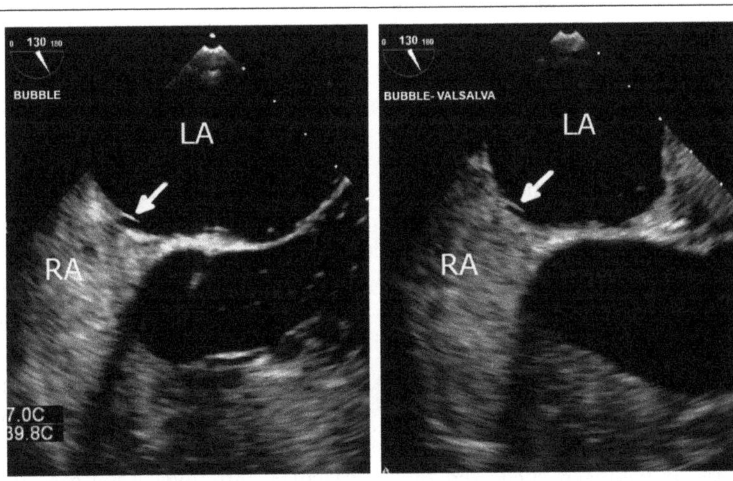

Fig. 8 Agitated saline contrast imaging in a follow-up TEE with the bicaval view showing a LASP (arrow) and absence of an atrial septal defect both at rest (**a**) and with Valsalva maneuver (**b**)

two toes without any visible cyanosis in the upper extremities. Vasculitis was improbable, considering the sharply demarcated border between the affected and unaffected skin, the absence of fever, and lack of response to antibiotic therapy. Taking the negative history of cold exposure into consideration, perniosis or frostbite were unlikely etiologies.

Many routine medications and recreational drugs such as cocaine and cannabis have been reported to precipitate BTS. Prescribed medications reported to cause BTS include systemic vasopressors (epinephrine, dopamine), imipramine, amphotericin B, and warfarin, which may trigger transient deficiency of protein C [18]. Physicians should therefore be aware of the side effects of these drugs and medications. In our case, the patient's prescribed medications had no such thrombogenic potential.

TEE identified a LASP with an ovoid thrombus projecting into the left atrium with its peduncle entrapped in the septal pouch. No other embolic source was identified. Moreover, the possibility of a septic emboli in the absence of endocardial infect focus was rather unlikely. Of note, infective endocarditis is the most common embolic source in septic emboli.

A 3D TEE provided further comprehensive evaluation of the interatrial septum with further confirmation of the presence of thrombus. 2D and 3D TEE are superior to TTE for identifying anatomic anomalies of the interatrial septum, since TEE may not adequately visualize the anatomy [19]. Atrial septal pouches can also be identified by CT imaging, and physician must be able to recognize it. However, CT is not required in patients who have already undergone TEE and is not the diagnostic modality of choice. Indeed, in our case, the LASP was initially considered to be an artifact on CT, with a later comparison of the filling defect adjacent to the atrial septum in CT images with TEE evaluation of the thrombus showing an identical morphology and securing the diagnosis. TEE seems to be superior to CT for the clinical identification and comprehensive evaluation of atrial septal pouches. Moreover, TEE should be the preferred imaging technique due to its ability to unambiguously detect patent foramen ovale.

The high prevalence of atrial septal pouches (over 50% of the healthy population), especially LASPs in ~ 40% of healthy individuals, means that the pouch is not a pathology rather a normal human anatomical variant (7). LASPs can promote blood stasis and are a potential site for thrombus formation according to Virchow's triad. The most frequent cause of atrial thrombus formation is AF, which generally occurs in mitral valve lesions of rheumatic origin, hyperthyroidism, acute myocardial infarction, and cardiosclerosis. It has been suggested that LASPs are associated with a higher prevalence of AF, perhaps as a consequence of the developmental fusion of the PFO or perhaps related to the presence of scar tissue in the septal pouch apex, which may be proarrhythmogenic [20]. The hemodynamic changes caused by AF, causing the formation of mural thrombi within LASPs, are also a critical source of emboli. Apart from AF, several other reported risk factors for pouch thrombogenesis include elevated left ventricular filling pressure in heart failure, possible local endothelial injury, and a hypercoagulable state.

The relevant discussion point in our case was to establish the relationship between LASP, AF, and BTS. In reality, it is difficult to know whether the AF precipitated the thrombus, whether the LASP triggered the AF, or whether the thrombus formed in the LASP and the new-onset AF was incidental - but the important educational point is that these were the undeniable possibilities and LASPs have been reported in association with AF, transient ischemic attacks and thrombi along the left atrial septum [6–8]. However, there is much to be learned before one can place such a discovery in perspective.

Through multimodal imaging, our case confirmed thrombus formation inside a LASP in the setting of new-onset AF. In non-valvular AF, left atrial thrombi almost exclusively arise in the LAA due to decreased velocity (< 20–30 cm/s). However, in our case, no thrombi were detected in the LAA. Taking this into account and extensive exclusion of other possible etiologies, we suggest that the LASP was the most likely source of emboli causing lower extremity digital ischemia.

As noted above, numerous conditions can lead to toe necrosis, so management is entirely dependent on the cause. Although we had no definite etiology at the time of establishing treatment, we took appropriate and safe action to minimize ischemic insult and prevent progression by prescribing anticoagulation and a vasodilator (PGE1). In the clinical setting of critical ischemia of uncertain etiology, a cocktail protocol including anticoagulation, vasodilators, antiplatelet agents, and eventually pulsed glucocorticoids with or without vasodilatory prostaglandins (PG) should be considered as a salvage treatment until a definite diagnosis is made. Dry gangrene is not an emergency and can be treated conservatively before any final surgical intervention. However, wet gangrene is a semi-emergency because it may cause cellulitis, sepsis, and severe necrosis, and it often warrants immediate surgical attention. For better functional outcomes and faster rehabilitation, as in our case, amputation may be the best approach.

To conclude, BTS is an acute limb ischemia that is frequently misdiagnosed on initial presentation. Since pedal pulses are often palpable, the physician may discount vascular pathology. Our case demonstrates that LASPs may serve as a nidus for thrombus formation, particularly in patients with impaired cardiac function and stagnant atrial flow. LASPs can therefore be a source of embolic complications leading to BTS. We suggest that the presence of a LASP in the absence of any other risk factor should trigger careful investigation of the interatrial septum, preferably using a multimodality imaging approach for the diagnosis of LASP masses. The possibility that LASPs may not simply be an innocent bystander but a causative mechanism for peripheral ischemia must be considered.

Abbreviations

AF: Atrial fibrillation; ALI: Acute limb ischemia; BTS: Blue toe syndrome; CRP: C-reactive protein; CTA: Computed tomography angiography; DUS: Duplex ultrasound; LAA: Left atrial appendage; LASP: left atrial septal pouch; PGE: Prostaglandin E; TEE: Transesophageal echocardiography; TTE: Transthoracic echocardiography

Acknowledgements

We acknowledge support by the DFG Open Access Publication Funds of the Ruhr-Universität Bochum.

Note: CARE guidelines/methodology have been thoroughly adhered while preparing this case report.

Authors' contributions

SP analyzed and interpreted the patient data and wrote the manuscript; GK, JN and HJT collected all the relevant materials, including the figures and medical records. All authors have read and approved the final submitted version.

References

1. Blue Toe Syndrome. https://www.escardio.org/Journals/E-Journal-of-Cardiology-Practice/Volume-2/Blue-Toe-Syndrome-Title-Blue-Toe-Syndrome. Accessed 24 Mar 2020.
2. Karmody AM. "Blue toe" syndrome: an indication for limb salvage Surgery. Arch Surg. 1976;111:1263.
3. Becquemin J-P, Kovarsky S. Arterial emboli of the lower limbs: analysis of risk factors for mortality and amputation. Ann Vasc Surg. 1995;9:S32–8.
4. Making the diagnosis when the patient has "blue toes". - PubMed - NCBI. https://www.ncbi.nlm.nih.gov/pubmed/7982584. Accessed 30 Mar 2020.
5. Greenberg JW, Goff ZD, Mooser AC, Wittgen CM, Smeds MR. Acute Limb Ischemia Secondary to Patent Foramen Ovale–Mediated Paradoxical Embolism: A Case Report and Systematic Review of the Literature. Ann Vasc Surg. 2020;:S0890509619310659.
6. Krishnan SC, Salazar M. Septal pouch in the left atrium. JACC Cardiovasc Interv. 2010;3:98–104.
7. Mazur M, Jasinska KA, Walocha JA. The morphology, clinical significance and imaging methods of the atrial septal pouch: a critical review. Transl Res Anat. 2018;13:7–11.
8. Gurudevan SV, Shah H, Tolstrup K, Siegel R, Krishnan SC. Septal Thrombus in the left atrium: is the left atrial Septal pouch the culprit? JACC Cardiovasc Imaging. 2010;3:1284–6.
9. Sidawy AN, Perler BA. Rutherford's Vascular Surgery and Endovascular Therapy. 9th ed. New York: Elsevier; 2018. p. 2832. In: Rutherford's vascular Surgery and endovascular Therap. 9th ed. New York: Elsevier.
10. Brewer ML, Kinnison ML, Perler BA, White RI. Blue toe syndrome: treatment with anticoagulants and delayed percutaneous transluminal angioplasty. Radiology. 1988;166:31–6.
11. Olinic D-M, Stanek A, Tătaru D-A, Homorodean C, Olinic M. Acute limb ischemia: an update on diagnosis and management. J Clin Med. 2019;8. https://doi.org/10.3390/jcm8081215.
12. Oweis Y, Viets Z, Shetty AS. Role of lower extremity run-off CT angiography in the evaluation of acute vascular disease. Abdom Radiol. 2017;42:1028–45.
13. Singh H, Cardella JF, Cole PE, Grassi CJ, McCowan TC, Swan TL, et al. Quality improvement guidelines for diagnostic arteriography. J Vasc Interv Radiol. 2002;13:1–6.
14. Gossage JA, Ali T, Chambers J, Burnand KG. Peripheral arterial embolism: prevalence, outcome, and the role of echocardiography in management. Vasc Endovasc Surg. 2006;40:280–6.
15. Hirschmann JV, Raugi GJ. Blue (or purple) toe syndrome. J Am Acad Dermatol. 2009;60:1–20.
16. Hyperviscosity Syndrome in Cryoglobulinemia: Clinical Aspects and Therapeutic Considerations. Semin Thromb Hemost. 2003;29:473–8.
17. Brown PJ, Zirwas MJ, English JC. The purple digit: an algorithmic approach to diagnosis. Am J Clin Dermatol. 2010;11:103–16.
18. Aboud AA, Abrams M, Mancini AJ. Blue toes after stimulant therapy for pediatric attention deficit hyperactivity disorder. J Am Acad Dermatol. 2011; 64:1218–9.
19. Zisa D, Faletra FF, Wessler BS, Halin NJ, Reddy P, Patel AR, et al. Ridges and pouches: a case series of anomalous atrial Septal fusion. CASE Cardiovasc Imaging Case Rep. 2019;4:7–17.
20. Hołda MK, Koziej M, Wszołek K, et al. Left atrial accessory appendages, diverticula, and left-sided septal pouch in multi-slice computed tomography. Association with atrial fibrillation and cerebrovascular accidents. Int. J. Cardiol. 2017;244:163–8.

7

International longitudinal registry of patients with atrial fibrillation and treated with rivaroxaban: RIVaroxaban Evaluation in Real life setting (RIVER)

Jan Beyer-Westendorf[1,2*], A. John Camm[3], Keith A. A. Fox[4], Jean-Yves Le Heuzey[5], Sylvia Haas[6], Alexander G. G. Turpie[7], Saverio Virdone[8], Ajay K. Kakkar[8,9] and for the RIVER Registry Investigators

Abstract

Background: Real-world data on non-vitamin K oral anticoagulants (NOACs) are essential in determining whether evidence from randomised controlled clinical trials translate into meaningful clinical benefits for patients in everyday practice. RIVER (RIVaroxaban Evaluation in Real life setting) is an ongoing international, prospective registry of patients with newly diagnosed non-valvular atrial fibrillation (NVAF) and at least one investigator-determined risk factor for stroke who received rivaroxaban as an initial treatment for the prevention of thromboembolic stroke. The aim of this paper is to describe the design of the RIVER registry and baseline characteristics of patients with newly diagnosed NVAF who received rivaroxaban as an initial treatment.

Methods and results: Between January 2014 and June 2017, RIVER investigators recruited 5072 patients at 309 centres in 17 countries. The aim was to enroll consecutive patients at sites where rivaroxaban was already routinely prescribed for stroke prevention. Each patient is being followed up prospectively for a minimum of 2-years. The registry will capture data on the rate and nature of all thromboembolic events (stroke / systemic embolism), bleeding complications, all-cause mortality and other major cardiovascular events as they occur. Data quality is assured through a combination of remote electronic monitoring and onsite monitoring (including source data verification in 10% of cases). Patients were mostly enrolled by cardiologists ($n = 3776$, 74.6%), by internal medicine specialists 14.2% ($n = 718$) and by primary care/general practice physicians 8.2% ($n = 417$). The mean (SD) age of the population was 69.5 (11.0) years, 44.3% were women. Mean (SD) $CHADS_2$ score was 1.9 (1.2) and CHA_2DS_2-VASc scores was 3.2 (1.6). Almost all patients (98.5%) were prescribed with once daily dose of rivaroxaban, most commonly 20 mg (76.5%) and 15 mg (20.0%) as their initial treatment; 17.9% of patients received concomitant antiplatelet therapy. Most patients enrolled in RIVER met the recommended threshold for AC therapy (86.6% for 2012 ESC Guidelines, and 79.8% of patients according to 2016 ESC Guidelines).

Conclusions: The RIVER prospective registry will expand our knowledge of how rivaroxaban is prescribed in everyday practice and whether evidence from clinical trials can be translated to the broader cross-section of patients in the real world.

(Continued on next page)

* Correspondence: Jan.Beyer@uniklinikum-dresden.de
[1]Thrombosis Research Unit, Department of Medicine I, Division Haematology, University Hospital "Carl Gustav Carus" Dresden, Dresden, Germany
[2]Kings Thrombosis Service, Department of Haematology, Kings College London, London, UK
Full list of author information is available at the end of the article

(Continued from previous page)
Trial registration: Unique identifier: NCT02444221. Registerd 14 May 2015; Retrospectively Registered.

Keywords: Anticoagulant, Antithrombotic, Atrial fibrillation, Outcomes, Registry, Rivaroxaban

Introduction

Atrial fibrillation (AF) is the most common sustained arrhythmia reported in adult patients [1] and is associated with an at least a five-fold increase in the risk of stroke [2]. Vitamin K antagonists (VKAs) formed the cornerstone of thromboembolic prophylaxis in patients with AF for many years. The introduction of non-vitamin K anticoagulants (NOACs), dabigatran, rivaroxaban, apixaban and edoxaban, provided physicians with agents with comparative efficacy and reduced potential for bleeding compared with vitamin K antagonists, while removing the need for dose titration, periodic laboratory testing and dietary restrictions that are necessary with VKAs [3–5]. Evidence from phase III studies showed that rivaroxaban, one of several oral direct Factor Xa inhibitors, is noninferior to warfarin for the reduction of stroke or systemic embolism in patients with AF [6] and significantly reduces rates of intracranial and fatal haemorrhages, but not rates of bleeding overall. Based on these results, rivaroxaban is now licensed in more than 130 countries worldwide for stroke prevention in patients with AF.

Real-world studies (such as Global Anticoagulant Registry in the FIELD-AF [GARFIELD-AF] [7], ORBIT-AF I & II) [8] and GLORIA-AF [9, 10] have demonstrated that NOAC use is increasing while post-marketing surveillance (such as Xarelto® for Prevention of Stroke in Patients with Atrial Fibrillation [XANTUS]) [11] and national studies (such as the EXPAND) [12, 13] have provided data on the safety and efficacy of rivaroxaban. However, most of these registries are disease-specific and evaluate treatment patterns and outcomes across different treatments. Therefore, additional drug-specific observational studies are needed to assess the clinical and health economic risk-benefits of specific NOACs and in particular, the factors that could potentially influence outcomes (e.g. compliance with prescribing guidelines, persistence and comorbidities) for each treatment alternative separately. This paper describes the design of the RIVaroxaban Evaluation in Real life setting (RIVER) registry and baseline characteristics of patients who received rivaroxaban as part of routine care. This paper also includes an initial analysis of the doses of rivaroxaban used in everyday practice as well as a description of the use of rivaroxaban in combination with antiplatelets in different clinical scenarios, such as patients with renal impairment.

Methods

Registry design

RIVER (NCT02444221) is an ongoing international, multicentre, prospective registry of patients with newly diagnosed non-valvular AF and at least one investigator-determined risk factor for stroke. All patients in this study are initiated on an anticoagulant treatment with rivaroxaban at the start of the study, as part of routine care, for the prevention of thromboembolic stroke.

The goal of the RIVER registry is to provide insights into the clinical management and related outcomes of rivaroxaban-treated AF patients across a number of regions and across the spectrum of healthcare systems, in routine clinical management. Follow-up will be for a minimum of 2 years (and up to 3.5 years) after enrolment (Fig. 1). By capturing data from unselected patients treated in everyday practice, the registry has the potential to identify best practices as well as deficiencies in the use of rivaroxaban as an initial treatment for the prevention of thromboembolic events in patients with newly diagnosed AF. The decision to initiate treatment with rivaroxaban was based on clinicians' judgement. The study was designed such that the collection of data would not interfere with the clinical management of patients or with the prescribing behaviours of attending physicians. No specific treatments, tests, or procedures were mandated or withheld from patients, and patients were free to withdraw from the registry at any time. The registry will not only assess the rates of clinical outcomes such as stroke and systemic embolization (SE), all-cause mortality, other major cardiovascular events (Table 1) and bleeding complications, but will also evaluate therapy persistence (including discontinuation, interruption and changes to the treatment regimen). Patients' experiences using antithrombotic treatment will also be evaluated by the Anti-Clot Treatment Scale (ACTS) questionnaire.

Independent ethics committee and hospital-based institutional review board approvals were obtained, as necessary, for the registry protocol. The registry is being conducted in accordance with the Declaration of Helsinki, local regulatory requirements, and the International Conference on Harmonisation–Good Pharmacoepidemiological and Clinical Practice guidelines.

RIVER is an independent academic research initiative, sponsored by the Thrombosis Research Institute, London, UK and funded by an unrestricted research grant from Bayer AG, Berlin, Germany. The authors are responsible for the design and conduct of this study, data analyses and the final content of this paper.

Registry population and site selection

The study population consists of consecutive patients enrolled prospectively at sites where rivaroxaban was

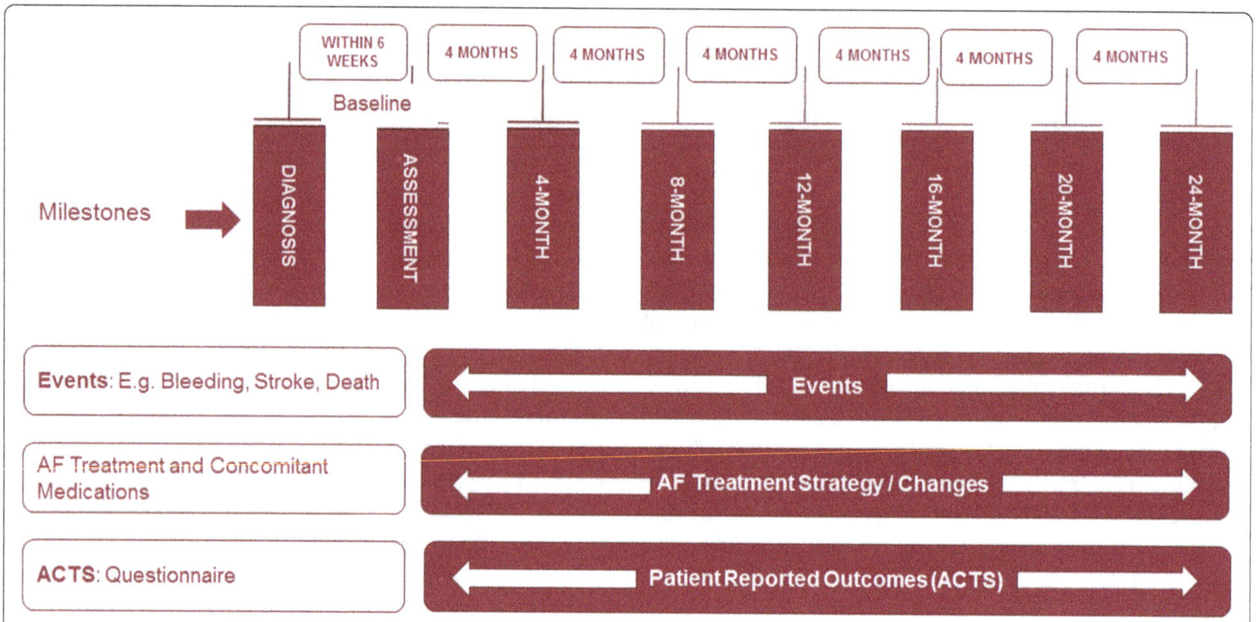

Fig. 1 Cohort design and data collection

already routinely prescribed for stroke prevention before the start of the study. Patients, receiving rivaroxaban as part of usual care, were eligible for enrolment in RIVER if they were 18 years of age or older, had a new diagnosis of non-valvular AF within the past 6 weeks and at least one additional risk factor for stroke as identified by the investigator. These risk factors were identified by clinicians and were not restricted to those in stroke prevention guidelines. Patients with a transient reversible cause of AF, those for whom follow-up is not foreseen or possible and all patients participating in interventional studies were excluded. All patients provided written informed consent to participate. Sites were chosen to be representative of AF care settings and included patients from multiple settings: office-based specialists, hospital departments (neurology, cardiology, geriatrics, internal medicine, and emergency room), anticoagulant clinics and general or family practice settings.

Data capture

Patient visits are not mandated, but data collection using the electronic case report form (eCRF) occurs at 4-monthly intervals and captures all relevant data from the patients' medical notes (Fig. 1).

A summary of the evaluations performed at baseline and at follow-up visits are outlined in Additional file 1: Table S1. At entry into the study, data are collected on the care setting, patients' demographics, vital signs, medical history, the nature of AF (paroxysmal versus persistent versus permanent), symptoms of AF, relevant medications, including antiplatelet and other concomitant therapy and investigator-identified risk factors for stroke. The results from investigations at diagnosis (including ECG morphology, left ventricular ejection fraction [LVEF] measurement, full blood count, haemoglobin, platelet count and creatinine) are recorded as well as the treatment strategy for AF (rate or rhythm control) and the date and outcome following cardioversion. Information is collected on initial rivaroxaban dosing regimen, start and stop dates, changes in therapy, compliance concerns and the reason for suspending or terminating therapy sooner than intended (such as bleeding, patient decision, and/or physician decision). Patients' experiences using antithrombotic treatment are recorded on the Anti-Clot Treatment Scale (ACTS) questionnaire at 4, 12 and 24 months.

At 4-monthly intervals, all routinely performed tests (including INR, haemoglobin, platelet count and creatinine), vital signs as well as all major events (stroke/transient ischaemic attack [TIA], bleeding / site of bleeding, death, myocardial infarction [MI], acute coronary syndrome [ACS], peripheral embolism) and hospitalizations/medical consultations are documented. For patients who switch to VKAs, data are also collected on INR, INR frequency and outcome related to INR fluctuation.

Based on the data collected from the eCRF, healthcare resource consumption will be captured so that the economic burden of AF can be computed both overall and per patient per year from the perspective of the payer, e.g. national health service, public/private/statutory insurance etc.

Data management

Data are captured using an electronic case report form (eCRF) designed by eClinicalHealth Services, Stirling, UK and submitted electronically via a secure website to

Table 1 Main Outcome Measures in the RIVER Registry

Outcome	Category
Clinical events	Stroke (ischaemic and haemorrhagic)
	Transient ischaemic attack
	Peripheral/non-central nervous system embolism
	Pulmonary embolism
	Heart failure
	Myocardial infarction
	Hospitalization
	Sudden cardiac death
	Non-cardiovascular death
Bleeding events	
Severity	Major
	Non-major clinically relevant
	Minor
Location	Intracranial
	Ears, nose, throat
	Gastrointestinal
	Genitourinary
	Vascular access site
Outcome	Recovered
	Permanently disabled
	Fatal
Healthcare utilization used for bleeding event	Hospitalization
	Emergency room visit
	Surgery for bleeding
	Transfusion
	Physician consultation
Therapy persistence	Rate of discontinuation
	Duration of time on therapy
	Reasons for discontinuation
Hospitalization for any event	
Any other hospital visits	Inpatient, outpatient, and emergency room
Patients treated with vitamin K antagonists	Frequency and timing of monitoring
	INR recordings in relation to therapeutic range
	Location of testing (self-monitoring, general practitioner clinic, anticoagulant clinic, etc.)
	Dose adjustments
	Use of bridging anticoagulation necessitated by interruption of vitamin-K antagonist
	Outcomes in relation to INR fluctuation
Patients treated with additional antithrombotic therapy	Therapy changes (discontinuation, temporary interruptions and use of bridging therapy)
	Reasons of therapy changes (if applicable)
	Patient treatment satisfaction using the Anti-Clot Treatment Scale (ACTS), depending on the cohort and country (at 4, 12, 24 months)

INR International normalise ratio

the registry-coordinating centre at the Thrombosis Research Institute who were responsible for checking the completeness and accuracy of data collected from medical records. All patients are assigned a unique identifier and personal identifiable data will be removed at the hospital source, ensuring anonymity and protecting confidentiality.

Data quality is assured through a combination of remote electronic monitoring and more conventional onsite monitoring (including source data verification in 10% of all cases), based on the validated quality assurance programmes developed in GARFIELD-AF [14]. The milestones for study eCRFs will be examined by the registry-coordinating centre at TRI, to ascertain completeness, accuracy and data queries sent to participating sites.

Statistical analysis

Continuous variables are expressed as mean ± standard deviation (SD). Categorical variables are expressed as frequencies and percentages. Statistical analysis was performed using SAS software version 9.4 (SAS Institute Inc., Cary, NC, USA).

Results

Baseline characteristics

A total of 5282 patients were screened for inclusion into RIVER between January 2014 and June 2017 and 5072 eligible patients were included in the analysis of baseline characteristics. Reasons for the exclusion of 210 patients after screening were: age ≤ 18 years (n = 4), alternative AC treatment before rivaroxaban after AF diagnosis (n = 68), no risk factor for stroke (n = 22), patients having transient AF secondary to a reversible cause (n = 10), difficulty in long-term follow-up (n = 24), patients participating in other interventional studies (n = 9) and, finally, patients whose enrolment was not confirmed due to either patient' decision (n = 27), death before consent (n = 1) or other reason (n = 77). Patients were recruited from 309 centres in 17 countries, including patients from Europe (67.5%) (Denmark, France, Germany, Italy, Netherlands, Poland, Russia, Spain, Sweden and UK), the Middle East (17.0%) (Egypt, Saudi Arabia and UAE), Australia (5.8%), Canada (3.0%), Brazil (5.6%) and Mexico (1.1%) (Fig. 2). Overall, 74.6% (n = 3779) of

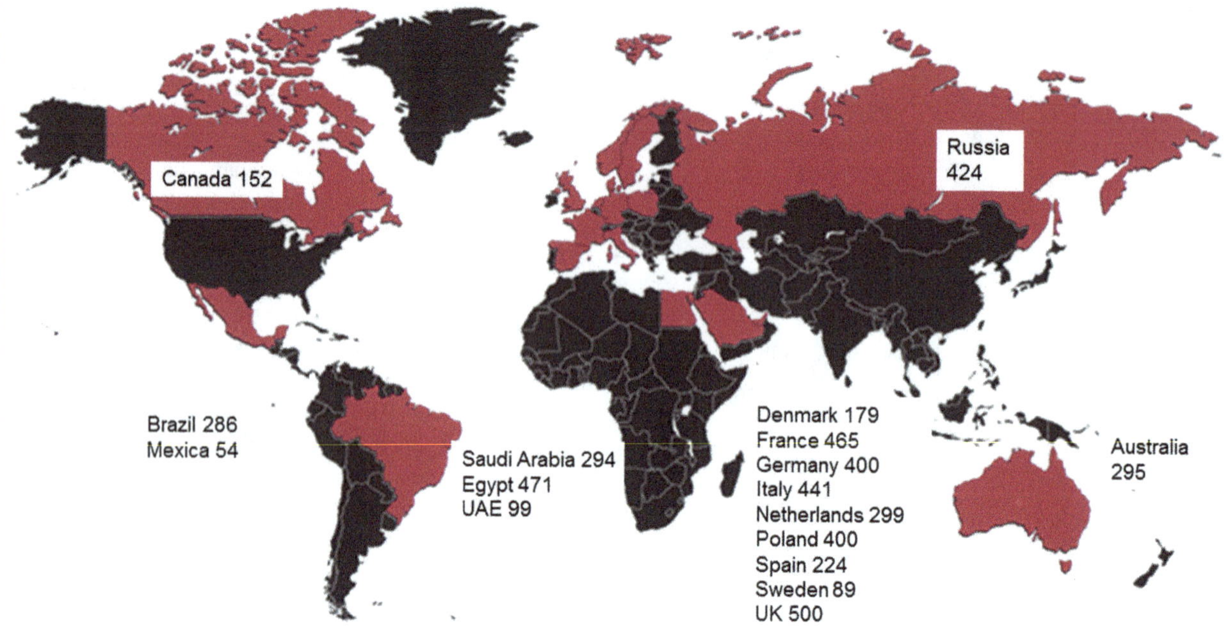

Fig. 2 Location of enrolling centres of the RIVER Registry. Australia, Brazil, Canada, Denmark, Egypt, France, Germany, Italy, Mexico, Netherlands, Poland, Russia, Saudi Arabia, Spain, Sweden, UAE, UK

patients were enrolled by cardiologists, 14.2% (n = 718) by internal medicine specialists and 8.2% (n = 417) by primary care/general practice physicians.

Approximately one-third of patients (36.3%) in RIVER had asymptomatic AF. Among patients with symptomatic AF, the most common symptoms identified were: palpitations (38.4%), shortness of breath (31.8%) and tiredness (14.8%). At the time of enrolment, 34.2% of patients had new or unclassified AF, 34.2% paroxysmal AF, 15.9% persistent AF and 15.7% permanent AF.

Baseline patient characteristics are described in Table 2. Caucasian patients represented the largest proportion of patients in RIVER (80.3%), followed by Asians (5.6%) and hispanic/latino (3.7%). Beyond the factors defined by CHA_2DS_2-VASc, physicians also considered the following as risk factors for stroke: smoking (12.8%), obesity (19.5%), renal impairment (5.9%), proteinuria (1.9%), enlarged left atrium (16.4%), thyroid disease (5.9%), thrombus in atrium (0.4%) and Chagas disease (0.3%) amongst others (5.3%).

Figure 3 describes the prescribing of rivaroxaban according to the post-hoc calculation of $CHADS_2$ and CHA_2DS_2-VASc risk scores. The mean (SD) $CHADS_2$ score for patients in the RIVER was 1.9 (1.2) and 58.2% of patients had a score of ≥2. Mean (SD) CHA_2DS_2-VASc score was 3.2 (1.6). Most patients met the recommended threshold for AC therapy (86.6% for 2012 ESC Guidelines, and 79.8% of patients according to 2016 ESC Guidelines). The mean (SD) HAS-BLED score was 1.1 (0.8) and the most commonly observed bleeding risk factors were: hypertension (77.0%), stroke (7.2%) and moderate-to-severe chronic renal disease (4.6%).

Rivaroxaban dosing

Almost all patients (98.5%) were prescribed a single daily dose of rivaroxaban, most commonly 20 mg (76.5%) and 15 mg (20.0%), while10 mg dose was prescribed in only 2.1% of patients. Creatinine clearance (CrCl) information at enrolment, estimated using the Cockcroft-Gault equation, was available for 3135 of 5072 (61.8%) patients. The mean (SD) CrCl was 83.7 (35.6) mL/min and 86.1% patients had CrCl ≥50 mL/min. Among patients with normal or mild renal function (CrCl ≥50 mL/min), 2246 (83.2%) were prescribed the recommended rivaroxaban dose of 20 mg once daily at baseline. Of patients with moderate or severe renal impairment (CrCl 15–49 mL/min), 246 (56.6%) patients received rivaroxaban 15 mg once daily, 173 (39.8%) 20 mg once daily and 16 (3.7%) 10 mg once daily as their initial treatment.

Concomitant antiplatelet therapy

Just under one in five (17.9%, n = 904) patients who were initiated on rivaroxaban at the time of diagnosis of AF also received concomitant antiplatelet therapy. Use of antiplatelets was higher in patients with a CHA_2DS_2-VASc score of 2–9 (19.4%) than in those with a score of 0 or 1 (9.2%) (Table 3); 229 (25.3%) had a coronary artery disease (CAD), peripheral artery disease (PAD) or carotid disease. The most frequently used antiplatelet was aspirin in 55.4%

Table 2 Patient baseline characteristics

Variable	All patients (*N* = 5072)
Age[a], mean ± SD, years	69.5 ± 11.0
Age group[a], n (%)	
< 65	1444 (28.5)
65–74	1861 (36.8)
≥ 75	1758 (34.7)
Women, n (%)	2248 (44.3)
Body mass index[b], mean ± SD, kg/m^2	29.2 ± 5.7
Smoking status (current/previous)[c], n (%)	1599 (35.3)
Pulse[d], mean ± SD, bpm	85.1 ± 23.8
Medical history, n (%)	
Hypertension[e]	3899 (77.0)
Hypercholesterolemia[f]	2208 (43.9)
Diabetes mellitus[g]	1227 (24.3)
Congestive heart failure[h]	1086 (21.5)
Family history of cardiac disease[i]	618 (19.5)
Coronary artery disease[j]	967 (19.2)
Current/History of myocardial infarction or unstable angina[k]	718 (14.2)
Chronic renal disease[l]	
Mild renal dysfunction (stage: I, II)	629 (13.5)
Moderate renal dysfunction (stage: III)	186 (4.0)
Severe renal dysfunction (stage: IV, V)	26 (0.6)
Left ventricular ejection fraction < 40%[m]	305 (9.5)
Stroke[n]	365 (7.2)
Transient ischaemic attack[o]	331 (6.5)
Aortic or peripheral artery disease[p]	252 (5.0)
Carotid disease[q]	212 (4.2)
History of bleeding[r]	118 (2.3)
Heavy alcohol consumption[s]	83 (2.1)
Venous thromboembolism[t]	103 (2.0)
Systemic embolization[u]	56 (1.1)
Cirrhosis[v]	20 (0.4)

[a]Data not available for 9 patients
[b]Data not available for 586 patients
[c]Data not available for 541 patients
[d]Data not available for 236 patients
[e]Data not available for 9 patients
[f]Data not available for 44 patients
[g]Data not available for 12 patients
[h]Data not available for 10 patients
[i]Data not available for 1906 patients
[j]Data not available for 28 patients
[k]Data not available for 21 patients
[l]Data not available for 395 patients
[m]Data not available for 1870 patients
[n]Data not available for 11 patients
[o]Data not available for 15 patients
[p]Data not available for 29 patients
[q]Data not available for 53 patients
[r]Data not available for 12 patients
[s]Data not available for 1087 patients
[t]Data not available for 15 patients
[u]Data not available for 17 patients
[v]Data not available for 24 patients

(501 of 904) of cases. Overall, 76 (1.5%) patients received dual AP therapy (ADP receptor/P2Y12 and aspirin).

Discussion

Observational or non-interventional studies play an important role in understanding and confirming the clinical effectiveness of interventions across the spectrum patients treated in routine practice.

RIVER is an independent treatment registry of patients from multiple care settings and patients were recruited from 17 countries. The RIVER registry provides an opportunity to examine the impact of different recruitment methodologies on the characteristics and treatment between the registries. RIVER will document the regional heterogeneity in the clinical presentation and assess the rate of stroke and systemic embolism and other outcomes with specific reference to the incidences of bleeding and therapy persistence. In addition to reporting on pre-specified clinical and economic outcomes, analyses from RIVER will be hypothesis generating, allowing the exploration of some aspects of the real world situation of AF treated with rivaroxaban and outcomes.

The degree of the patient involvement in the generation of data is expected to provide unique insights into the patients-reported outcomes. Patients' experiences using antithrombotic treatment are recorded on the Anti-Clot Treatment Scale (ACTS) questionnaire. It is important to recognise that registries differ in their design, recruitment strategies, care settings, geographical spread and duration of follow-up. Compared to other ongoing prospective registries, the RIVER registry has the potential to capture the medicosocial burden due to AF in large-scale populations by employing broad inclusion and exclusion criteria in widely representative populations of patients with AF who are treated with rivaroxaban, across a range of clinical settings and to capture long-term follow-up data. The value of RIVER registry is enhanced by a quality assurance programme and the supervision of an independent Audit Committee, which oversees site-dependent verification, remote site monitoring and electronic database monitoring to ensure data quality.

There are several phase IV studies (such as XANTUS [15], XAPASS [16], XASSURE (China; NCT02784717) and registries such as EXPAND [12]) are ongoing to investigate rivaroxaban in patients with NVAF in the real world clinical settings. The XANTUS program comprises three studies: the XANTUS (Europe, Israel plus enrollment in Canada; NCT01606995), XANTUS-EL (Eastern Europe, Eastern Mediterranean, Middle East, Latin America; NCT01800006) and XANAP (Asia-Pacific; NCT01750788) [15].

Other ongoing non-interventional registries also provide real-world data on the effectiveness and safety of

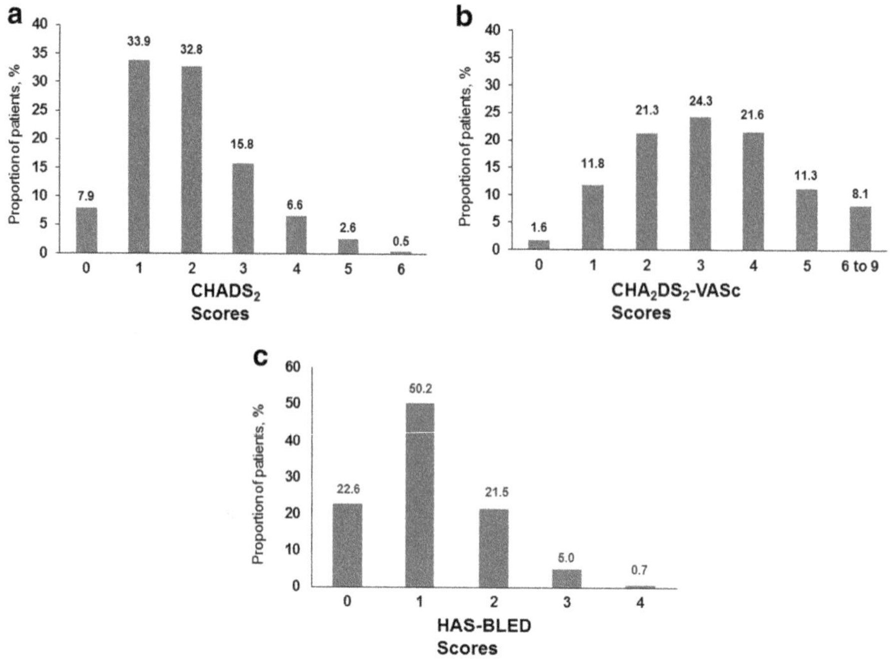

Fig. 3 **a** Distribution of $CHADS_2$ **b** CHA_2DS_2-VASc and **c** HAS-BLED scores in RIVER registry

rivaroxaban, including the Global Anticoagulant Registry in the FIELD (GARFIELD)-AF [17], Outcomes Registry for Better Informed Treatment of Atrial Fibrillation (ORBIT)-AF I and ORBIT-AF II [18–21], the Global Registry on Long-Term Oral Anthithrombotic Treatment in pateints with Atrial Fribrillation (GLORIA-AF) [22–26] and the Dresden NOAC Registry [4, 5, 27] (Additional file 1: Table S2).

In the presence of all these observational studies, it is important to highlight that there are a number of important differences in the design (Additional file 1: Table S2) and recruitment of patients (Table 4) in RIVER compared with these registries. RIVER is a drug-specific registry which allows studying of drug utilisation patterns including off-label use whereas other registries like GARFIELD-AF, ORBIT-AF I and II, GLORIA-AF and Dresden NOAC are disease-specific registries which allows studying of burden, management and adverse event profile for different treatments related to a specific disease. Patients in XANTUS study were enrolled from Europe, Israel and Canada whereas EXPAND and XAPASS are Japanese registries and ORBIT-AF registries are restricted to USA [19]. By contrast, RIVER has recruited patients more broadly from 17 countries (outside Asia) including Europe, the Middle East, Latin America, Canada and Australia. RIVER provides long-term follow-up data of 2 years where as XANTUS had 1-year follow-up data. RIVER is a more contemporary population in which physicians' now have experience using the NOACs. It is a known phenomenon that, after approval of new treatments, physicians start to apply these first in well-defined, carefully selected populations and expand it to other populations only after a period of time, which is reflecting learning curves based on increased comfort levels, communication of real world evidence, newly reported and previously unknown side effects or changes in guideline recommendations [28–30] . As a consequence, treatment patterns and outcomes may shift over time, which is an additional justification for more

Table 3 Use of antithrombotic drugs at diagnosis, overall and according to CHA_2DS_2-VASc scores of 0 or 1 and 2–9

Antiplatelet drug[a]	All patients ($N = 5072$)	Patients with CHA_2DS_2-VASc score of 0 or 1 ($N = 665$)	Patients with CHA_2DS_2-VASc score 2–9 ($N = 4289$)
Acetylsalicylic acid (Aspirin/ASA)	501 (9.9)	28 (4.2)	465 (10.8)
ADP receptor/P2Y12 inhibitor	248 (4.9)	13 (2.0)	231 (5.4)
Glycoprotein IIb/IIIa inhibitor	9 (0.2)	2 (0.3)	6 (0.1)
Prostaglandin analogue	71 (1.4)	5 (0.8)	66 (1.5)
Other antiplatelet	8 (0.2)	2 (0.3)	6 (0.1)

[a]Categories are not mutually exclusive

Table 4 Comparison of adjusted demographic and baseline clinical characteristics of patients from the international observational studies (RIVER and XANTUS), studies from Japan (EXPAND and XAPASS)

	RIVER	XANTUS	EXPAND	XAPASS
Patients, n	5072	6784	7141	11,308
Patient characteristics				
Sex, male (%)	55.7	59.2	67.7	61.9
Age, mean (SD)	69.5 (11.0)	71.5 (10.0)	71.6 (9.4)	73.1 (9.9)
BMI (Kg/m^2), mean (SD)	29.2 (5.7)	28.3 (5.0)	NA	23.7 (3.8)
Creatinine clearance (mL/min), mean (SD)	83.7 (35.6)	NA	69.7 (26.2)	67.7 (28.9)
< 15 mL/min (%)	0.02	0.3	NA	0.03
≥ 15 to < 30 mL/min (%)	0.7	1.1	1.9	2.8
≥ 30 to < 50 mL/min (%)	7.9	8.0	18.9	21.1
≥ 50 mL/min (%)	53.4	56.2	74.6	68.0
Missing (%)	38.0	34.4	4.7	8.2
Rivaroxaban dose/dosing frequency				
10 mg once daily	105 (2.1)	NA	NA	NA
15 mg once daily	1008 (20.0)	1410 (20.8)	NA	NA
20 mg once daily	3867 (76.5)	5336 (78.7)	NA	NA
10 mg twice daily	10 (0.2)	NA	NA	NA
15 mg twice daily	29 (0.6)	NA	NA	NA
20 mg twice daily	22 (0.4)	NA	NA	NA
10/15/20 mg other dosing Regimen	12 (0.2)	35 (0.5)	NA	NA
$CHADS_2$ score, mean (SD)	1.9 (1.2)	2.0 (1.3)	2.1 (1.3)	2.2 (1.3)
CHA_2DS_2-VASc score, mean (SD)	3.2 (1.6)	3.4 (1.7)	3.4 (1.7)	NA
Comorbidity/medical history				
Congestive heart failure (%)	21.5	18.6	26.1	25.0
Hypertension (%)	77.0	74.7	70.9	74.3
Diabetes mellitus (%)	24.3	19.6	24.3	22.3
History of stroke/systemic embolism/TIA (%)	13.4	19.0	24.1	23.7
Type of AF				
Paroxysmal (%)	34.2	40.7	44.8	NA
Non-paroxysmal (persistent/permanent) (%)	31.6	40.8	55.2	NA
First diagnosed (%)	34.2	18.5	NA	NA

contemporary registries in SPAF. All of these factors may have contributed to the observed differences in patients characteristics, compared to other observational studies. Patients in RIVER are younger (mean age: 69.5 years) than patients in EXPAND (72 yrs), XANTUS (71.5 yrs), and XAPASS (73.1 yrs). The mean BMI of patients in RIVER (29.2 ± 5.7) is higher than the Japanese cohort in XAPASS (23.7 ± 3.8) (Table 4). Patients studied in RIVER also had a lower risk of stroke, with a mean $CHADS_2$ scores of 1.9 and 13.4% had a prior episode of stroke/TIA/SE when compared to XANTUS (2.0, 19%) and XAPASS (2.2; 23.7%) (Additional file 1: Table S3) and other NOAC trials, such as RE-LY (2.1; 20.0%) and ARISTOTLE (2.1; 19.4%) [31, 32]. Nevertheless, patients in RIVER had high rates of comorbid conditions, including hypertension, hypercholesterolemia, diabetes mellitus, heart failure and coronary artery disease. However, the proportion of patients (8.6%) with renal dysfunction (CrCl < 50 mL/min) in RIVER was similar to XANTUS (9.4%) and lower than Japanese cohorts in EXPAND (20.8%) and XAPASS (23.9%); although this observation may be due to the higher rates of missing CrCl measurements in both RIVER (38.0%) and XANTUS (34.4%). Patients in RIVER and XANTUS received higher doses of rivaroxaban (20 mg/day [76.5 and 78.7% patients, respectively], 15 mg/day [20.0 and 20.8% patients, respectively]) than those in EXPAND (15 mg/day [56.6% patients], 10 mg/day [43.4% patients]), suggesting differences in the guideline recommendations, which are based on patient characteristics and ethnicities. In Japan, the regular dosage of rivaroxaban is 15 mg od, which is lower than the global recommended dosage of 20 mg od [33].

Limitations

Centres for recruitment of patients in RIVER are limited to the care settings where rivaroxaban is the standard of care and includes sites beyond those included in the clinical trials. Comparison of different treatments are not possible, as there is no comparator drug, while larger prospective disease registries such as GARFIELD-AF and drug registries such as GLORIA-AF and ORBIT-II have the potential to conduct comparative effectiveness studies between different treatment approaches in newly diagnosed patients with AF. Due to the observational design of the study, laboratory and other investigations could not mandated.

Strengths

RIVER is an independent academic research initiative. This prospective registry was designed with a novel approach to outcomes research, including the following features: sites representation of national AF care settings, randomised site selection, unselected eligible patients enrolled consecutively with a follow-up period of minimum 2 years and extensive monitoring and audit including source data verification of 10% of all eCRFs. Like other prospective registries, such as GARFIELD-AF [34], RIVER aims to overcome the inherent bias of retrospective cohort studies from national registries and claims databases [35–39].

Conclusion

The RIVER prospective registry will expand our knowledge of how rivaroxaban is prescribed in everyday practice and whether evidence from clinical trials can be translated to the broader cross-section of patients treated in the real world.

Abbreviations

AC: Anticoagulant; ACTS: Anti-Clot Treatment Scale; AF: Atrial fibrillation; AP: Antiplatelet; CAD: Coronary artery disease; CHF: Congestive heart failure; CrCl: Creatinine clearance; LVEF: Left ventricular ejection fraction; MI: Myocardial infarction; NOACs: Non-vitamin K anticoagulants; NVAF: Non-valvular atrial fibrillation; PAD: Peripheral artery disease; SE: Systemic embolism; TIA: Transient ischaemic attack; VKA: Vitamin K antagonist

Acknowledgments

We thank the physicians, nurses, and patients involved in the RIVER Registry. SAS programming support was provided by Madhusudana Rao (Thrombosis Research Institute, London, UK). Editorial support was provided by Surekha Damineni (Thrombosis Research Institute, London, UK).

RIVER Registry Investigators

Global Steering Committee

Ajay K. Kakkar (UK) (Chair), A. John Camm (UK), Jean-Yves Le Heuzey (France), Keith A.A. Fox (UK), Jan Beyer-Westendorf (Germany), Sylvia Haas (Germany), Alexander G.G. Turpie (Canada).

Publications Committee

A. John Camm (UK) (Chair), Ajay K. Kakkar (UK), Karen S. Pieper (USA), Gloria Kayani (UK).

Audit Committee

Keith A.A. Fox (UK), Bernard J. Gersh (USA).

RIVER Principal Investigators

Europe

Denmark: P. Hildebrandt, H. Dominguez, W. Comuth, L. Frost, D.S. Moller, H. Christensen, L.M. Bruun.

France: A. Milhem, J. Gauthier, C. Mielot, S. Chanseaume, S. Chopra, A. Amlaiky, O. Tricot, V. Sierra, A. Dompnier, N. Zannad, A. Pinzani, A. Quatre, J. Mansourati, L. Fauchier, N. Badenco, E. Gandjbakhch, K.F. Chachoua, V. Malquarti, F. Pierron, F. Sacher, J. Taieb, J.M. Davy, E. Marijon, N. Lellouche, A. Leenhardt, A. Salem, I. Lesto, J.J. Muller, R. Garcia, J.P. Neau, J.B. Berneau.

Germany: N. Schön, D. Gulba, K.F. Appel, J. Merke, J. Dshabrailov, C. Bauknecht, O. Scheuermann, T. Schröder, W. Jung, A. Kopf, J. Brachmann, M. Leschke, J. Taggeselle, M. Seige, T. Läßig, S. Appel, M. Schmiedl, K. Müller, G.U. Heinz, C. Axthelm, K. Eberhard, B. Hügl, T. Schwarz, U. Sechtem.

Italy: A. Falanga, V. Rubino, L. Calo, W. Ageno, F. Massari, D. Imberti, L. Di Gennaro, F. Gaita, A. Margonato, G. Cannava, F. Capasso, I. Diemberger, F. Pelliccia, A. Cafolla, S. Bardari, L. Mattei, L. Ruocco, G. Boriani, D. Poli, S. Testa, C. Indolfi, R. Quintavalla, L. Mos.

Netherlands: G. Ladyjanskaia, I. Aksoy, M. Van De Wetering, L. Theunissen, F. Den Hartog, R. Nijmeijer, R. Van De Wal, S. Reinders, M. Patterson, E.D. Melker, R. Troquay.

Poland: J. Korecki, A. Szyszka, F. Diks, J. Sumis, J. Cygler, B. Miklaszewicz, E. Litwiejko-Pietrynczak, P. Napora, G. Drelich, T. Kawka-Urbanek, J.K. Wranicz, M. Mierzejewski, A. Drzewiecka, D. Wronska, I. Fares, J. Baska, K. Stania, W. Krzyzanowski, P. Miekus, M. Tyminski.

Russia: D. Dronov, S. Zenin, E. Isaeva, A. Lopukhov, V. Yakusevich, D. Kuznetsov, T. Kameneva, E. Pokushalov, V. Karetnikova, I. Dik, I. Karpushina, D. Nikolin, A. Doletsky, A. Ardashev, A. Timofeeva, O. Miller, N. Lyamina, Y. Shubik, S. Boldueva.

Spain: J.L. Blanco Coronado, C. Gonzalez Juanatey, E. Otero, D. Alonso, J. Torres Llergo, J. Gonzalez Lama, J.A.V. De Prada Tiffe, F.J. Garcia Seara, J.J. Gomez Doblas, J.A. Riancho, J.L. Clua-Espuny, J. Motero, V.I. Arrarte, D. Martin Raymondi, G. Isasti Aizpurua, F. Marin, J.A. Nieto, J. Fernandez Portales, P. Alvarez Garcia.

Sweden: I. Torstensson, B. Cederin, T. Kalm, U. Rosenqvist, J. Thulin, A. Hajimirsadeghi, M. Crisby.

UK: A. Manoj, A. Bakhai, A. Mistri, M. Krishnan, S. Kumar, S. Kirubakaran, H. Thomas, J. Camm, F. Ahmed, A.M. Ross, K. Barry, R. Stockwell, A. Broadley, M. Mamun, K. Chatterjee, J. Cooke, J. McCready, D. Dutta, K. John, P. Pandya, R. Howlett, P.Vinson, Lim, P. Foley, D. Bruce, A. Dixit, D. Broughton, J. Taylor, R. Schilling.

Middle East

Egypt: K. Leon, K. Saeed, S. Shaheen, M. Tawfik, A. Mortadda, M. Seleem, M.S.I Aly, G. Kazamel, M. Elbadry, S. Kamal, M. Hassan, M. Mostafa, M.E.S. Medhat, Ekhlas, R. Ghaleb, M.O. Taha, I. Daoud, H. Al Din, A.M. Imam, M.A. El Hameed, Helmy.

Saudi Arabia: M. Al-Murayeh, N. Akhtar, B.M. Matto, M.A. Ghani, O.A. Amoudi, M.M. Morsy, A.A.F Bashir, Y. M. Al Hossni, B. Al Ghamdi, Zia-ul-Sabah, S. Mir, D. Dardir, A. Masswary, A.R. Al Shehri, A. Masswary, J. Iqbal, M.A.J. Almansori.

UAE: C.G Venkitachalam, J. Kurian, J. Rao, A. Aisheh, A.A. Albawab, B. Subbaraman, A. Amanat, K.J. Esfehani, R. Lochan, A. Bin Brek, B. Mittal, Y. Ghazi, M. Krishna, S.B. Tabatabaei, P.S. Thoppil, S. Nasim, S. El Khider Nour.

Latin America

Brazil: P. Barros, A.P. Almeida, M. Andrade, B. Garbelini Jr., T.L. Silvestrini, Á.R. Alves Junior, C.E.B. de Lima, A. Kormann, G.G. de Lima, C. Halperin, D. Salvadori Junior, A.F. Freitas Jr., J.R. Gemelli, C.E. Ornelas, J.M.M. Dantas, J.L. Aziz, L.M. Backes, W.S. Barroso, M.S. Paiva, J. A. de Figueiredo Neto, F.R. dos Santos, J.A. de Lima Neto, R. Bergo, P.R. Salvador Junior.

Mexico: A.G. López, J.C.P Alva, M.A.A. Gamba, F.G. Padilla-Padilla, A.E.B. Ruiz.

Other countries:

Canada: J. Berlingieri, A. Bakbak, M. Gupta, K. Saunders, A. Costa-Vitali, P.R. Beaudry, R. Bhargava, Y. Khaykin, J.S. Healey, E. Crystal, Nadeau, Dhillon.

Australia: Alistair Begg, C. Anderson, S. Baveja, D. Cross, A. Catanchin, D. Brieger, K.T. Lim, P. Davidson, R. Tan, R. Bhindi, J. Hickey, J. Layland, M. Bloch, C. Itty, B. Singh, P. Carroll, A. Lee, G. Starmer, R. Lehman.

Author details

[1]Thrombosis Research Unit, Department of Medicine I, Division Haematology, University Hospital "Carl Gustav Carus" Dresden, Dresden, Germany. [2]Kings Thrombosis Service, Department of Haematology, Kings College London, London, UK. [3]Cardiology Clinical Academic Group, Molecular and Clinical Sciences Institute, St. George's University of London, London, UK. [4]Centre for Cardiovascular Science, University of Edinburgh, Edinburgh, UK. [5]Georges Pompidou Hospital, René Descartes University, Paris, France. [6]Formerly Klinikum rechts der Isar, Technical University of Munich, Munich, Germany. [7]McMaster University, Hamilton, Canada. [8]Thrombosis Research Institute, London, UK. [9]University College of London, London, UK.

References

1. Franken RA, Rosa RF, Santos SC. Atrial fibrillation in the elderly. J Geriatr Cardiol. 2012;9(2):91–100.
2. Ezekowitz MD. Atrial fibrillation: the epidemic of the new millennium. Ann Intern Med. 1999;131(7):537–8.
3. Ruff CT, Giugliano RP, Braunwald E, Hoffman EB, Deenadayalu N, Ezekowitz MD, et al. Comparison of the efficacy and safety of new oral anticoagulants with warfarin in patients with atrial fibrillation: a meta-analysis of randomised trials. Lancet. 2014;383(9921):955–62.
4. Beyer-Westendorf J, Ebertz F, Forster K, Gelbricht V, Michalski F, Kohler C, et al. Effectiveness and safety of dabigatran therapy in daily-care patients with atrial fibrillation. Results from the Dresden NOAC registry. Thromb Haemost. 2015;113(6):1247–57.
5. Helmert S, Marten S, Mizera H, Reitter A, Sahin K, Tittl L, et al. Effectiveness and safety of apixaban therapy in daily-care patients with atrial fibrillation: results from the Dresden NOAC registry. J Thromb Thrombolysis. 2017;44(2):169–78.
6. Patel P, Pandya J, Goldberg M. NOACs vs. Warfarin for stroke prevention in nonvalvular atrial fibrillation. Cureus. 2017;9(6):e1395.
7. Camm AJ, Accetta G, Ambrosio G, Atar D, Bassand JP, Berge E, et al. Evolving antithrombotic treatment patterns for patients with newly diagnosed atrial fibrillation. Heart. 2017;103(4):307–14.
8. Steinberg BA, Gao H, Shrader P, Pieper K, Thomas L, Camm AJ, et al. International trends in clinical characteristics and oral anticoagulation treatment for patients with atrial fibrillation: results from the GARFIELD-AF, ORBIT-AF I, and ORBIT-AF II registries. Am Heart J. 2017;194:132–40.
9. Huisman MV, Rothman KJ, Paquette M, Teutsch C, Diener HC, Dubner SJ, et al. The changing landscape for stroke prevention in AF: findings from the GLORIA-AF registry phase 2. J Am Coll Cardiol. 2017;69(7):777–85.
10. Huisman MV, Lip GY, Diener HC, Dubner SJ, Halperin JL, Ma CS, et al. Design and rationale of global registry on long-term Oral antithrombotic treatment in patients with atrial fibrillation: a global registry program on long-term oral antithrombotic treatment in patients with atrial fibrillation. Am Heart J. 2014;167(3):329–34.
11. Camm AJ, Amarenco P, Haas S, Hess S, Kirchhof P, Kuhls S, et al. XANTUS: a real-world, prospective, observational study of patients treated with rivaroxaban for stroke prevention in atrial fibrillation. Eur Heart J. 2016; 37(14):1145–53.
12. Shimokawa H, Yamashita T, Uchiyama S, Kitazono T, Shimizu W, Ikeda T, et al. The EXPAND study: efficacy and safety of rivaroxaban in Japanese patients with non-valvular atrial fibrillation. Int J Cardiol. 2018;258:126–32.
13. Ikeda T, Atarashi H, Inoue H, Uchiyama S, Kitazono T, Yamashita T, et al. Study design and baseline characteristics of the EXPAND study: evaluation of effectiveness and safety of Xa inhibitor, rivaroxaban for the prevention of stroke and systemic embolism in a Nationwide cohort of Japanese patients diagnosed as non-Valvular atrial fibrillation. Tohoku J Exp Med. 2016;240(4):259–68.
14. Fox KAA, Gersh BJ, Traore S, Camm AJ, Kayani G, Krogh A, et al. Evolving quality standards for large-scale registries: the GARFIELD-AF experience. Eur Heart J Qual Care Clin Outcomes. 2017;3:114–22.
15. Camm AJ, Amarenco P, Haas S, Hess S, Kirchhof P, van Eickels M, et al. XANTUS: rationale and design of a noninterventional study of rivaroxaban for the prevention of stroke in patients with atrial fibrillation. Vasc Health Risk Manag. 2014;10:425–34.
16. Ogawa S, Minematsu K, Ikeda T, Kitazono T, Nakagawara J, Miyamoto S, et al. Design and baseline characteristics of the xarelto post-authorization safety & effectiveness study in Japanese patients with atrial fibrillation (XAPASS). J Arrhythm. 2018;34(2):167–175.
17. Kakkar AK, Mueller I, Bassand JP, Fitzmaurice DA, Goldhaber SZ, Goto S, et al. International longitudinal registry of patients with atrial fibrillation at risk of stroke: global anticoagulant registry in the FIELD (GARFIELD). Am Heart J. 2012;163(1):13–9.e1.
18. Piccini JP, Fraulo ES, Ansell JE, Fonarow GC, Gersh BJ, Go AS, et al. Outcomes registry for better informed treatment of atrial fibrillation: rationale and design of ORBIT-AF. Am Heart J. 2011;162(4):606–12.e1.
19. Steinberg BA, Blanco RG, Ollis D, Kim S, Holmes DN, Kowey PR, et al. Outcomes registry for better informed treatment of atrial fibrillation II: rationale and design of the ORBIT-AF II registry. Am Heart J. 2014;168(2):160–7.
20. Steinberg BA, Shrader P, Thomas L, Ansell J, Fonarow GC, Gersh BJ, et al. Off-label dosing of non-vitamin K antagonist Oral anticoagulants and adverse outcomes: the ORBIT-AF II registry. J Am Coll Cardiol. 2016;68(24):2597–604.
21. Kaufman BG, Kim S, Pieper K, Allen LA, Gersh BJ, Naccarelli GV, et al. Disease understanding in patients newly diagnosed with atrial fibrillation. Heart. 2018;104(6):494–501.
22. Mazurek M, Huisman MV, Rothman KJ, Paquette M, Teutsch C, Diener HC, et al. Gender differences in antithrombotic treatment for newly diagnosed atrial fibrillation: the GLORIA-AF registry program. Am J Med. 2018;131(8): 945–55.e3.
23. Huisman MV, Rothman KJ, Paquette M, Teutsch C, Diener HC, Dubner SJ, et al. Two-year follow-up of patients treated with dabigatran for stroke prevention in atrial fibrillation: global registry on long-term antithrombotic treatment in patients with atrial fibrillation (GLORIA-AF) registry. Am Heart J. 2018;198:55–63.
24. McIntyre WF, Conen D, Olshansky B, Halperin JL, Hayek E, Huisman MV, et al. Stroke-prevention strategies in north American patients with atrial fibrillation: the GLORIA-AF registry program. Clin Cardiol. 2018;41(6):744–51.
25. Huisman MV, Ma CS, Diener HC, Dubner SJ, Halperin JL, Rothman KJ, et al. Antithrombotic therapy use in patients with atrial fibrillation before the era of non-vitamin K antagonist Oral anticoagulants: the global registry on long-term oral antithrombotic treatment in patients with atrial fibrillation (GLORIA-AF) phase I cohort. Europace. 2016;18(9):1308–18.
26. Mazurek M, Huisman MV, Rothman KJ, Paquette M, Teutsch C, Diener HC, et al. Regional differences in antithrombotic treatment for atrial fibrillation: insights from the GLORIA-AF phase II registry. Thromb Haemost. 2017; 117(12):2376–88.
27. Hecker J, Marten S, Keller L, Helmert S, Michalski F, Werth S, et al. Effectiveness and safety of rivaroxaban therapy in daily-care patients with atrial fibrillation. Results from the Dresden NOAC registry. Thromb Haemost. 2016;115(5):939–49.
28. Anderson TS, Lo-Ciganic WH, Gellad WF, Zhang R, Huskamp HA, Choudhry NK, et al. Patterns and predictors of physician adoption of new cardiovascular drugs. Healthc (Amst). 2018;6(1):33–40.
29. Loo SY, Dell'Aniello S, Huiart L, Renoux C. Trends in the prescription of novel oral anticoagulants in UK primary care. Br J Clin Pharmacol. 2017; 83(9):2096–106.
30. Lubloy A. Factors affecting the uptake of new medicines: a systematic literature review. BMC Health Serv Res. 2014;14:469.
31. Connolly SJ, Ezekowitz MD, Yusuf S, Eikelboom J, Oldgren J, Parekh A, et al. Dabigatran versus warfarin in patients with atrial fibrillation. N Engl J Med. 2009;361(12):1139–51.
32. Granger CB, Alexander JH, McMurray JJ, Lopes RD, Hylek EM, Hanna M, et al. Apixaban versus warfarin in patients with atrial fibrillation. N Engl J Med. 2011;365(11):981–92.
33. JCS Joint Working Group. Guidelines for pharmacotherapy of atrial fibrillation (JCS 2013). Circ J. 2014;78(8):1997–2021.
34. Fox KAA, Accetta G, Pieper KS, Bassand JP, Camm AJ, Fitzmaurice DA, et al. Why are outcomes different for registry patients enrolled prospectively and retrospectively? Insights from the global anticoagulant registry in the FIELD-atrial fibrillation (GARFIELD-AF). Eur Heart J Qual Care Clin Outcomes. 2018; 4(1):27–35.
35. Chan YH, See LC, Tu HT, Yeh YH, Chang SH, Wu LS, et al. Efficacy and safety of apixaban, dabigatran, rivaroxaban, and warfarin in Asians with nonvalvular atrial fibrillation. J Am Heart Assoc. 2018;7(8). https://doi.org/10.1161/JAHA.117.008150.
36. Lip GYH, Pan X, Kamble S, Kawabata H, Mardekian J, Masseria C, et al. Discontinuation risk comparison among 'real-world' newly anticoagulated atrial fibrillation patients: Apixaban, warfarin, dabigatran, or rivaroxaban. PLoS One. 2018;13(4):e0195950.
37. Lip GY, Pan X, Kamble S, Kawabata H, Mardekian J, Masseria C, et al. Major

bleeding risk among non-valvular atrial fibrillation patients initiated on apixaban, dabigatran, rivaroxaban or warfarin: a "real-world" observational study in the United States. Int J Clin Pract. 2016;70(9):752–63.

38. Nielsen PB, Skjoth F, Sogaard M, Kjaeldgaard JN, Lip GY, Larsen TB. Effectiveness and safety of reduced dose non-vitamin K antagonist oral anticoagulants and warfarin in patients with atrial fibrillation: propensity weighted nationwide cohort study. BMJ. 2017;356:j510.
39. Beyer-Westendorf J, Gelbricht V, Forster K, Ebertz F, Kohler C, Werth S, et al. Peri-interventional management of novel oral anticoagulants in daily care: results from the prospective Dresden NOAC registry. Eur Heart J. 2014; 35(28):1888–96.

8

A brief review on resistance to $P2Y_{12}$ receptor antagonism in coronary artery disease

Ellen M. K. Warlo[1,2,3*], Harald Arnesen[1,2,3] and Ingebjørg Seljeflot[1,2,3]

Abstract

Background: Platelet inhibition is important for patients with coronary artery disease. When dual antiplatelet therapy (DAPT) is required, a $P2Y_{12}$-antagonist is usually recommended in addition to standard aspirin therapy. The most used $P2Y_{12}$-antagonists are clopidogrel, prasugrel and ticagrelor. Despite DAPT, some patients experience adverse cardiovascular events, and insufficient platelet inhibition has been suggested as a possible cause. In the present review we have performed a literature search on prevalence, mechanisms and clinical implications of resistance to $P2Y_{12}$ inhibitors.

Methods: The PubMed database was searched for relevant papers and 11 meta-analyses were included. $P2Y_{12}$ resistance is measured by stimulating platelets with ADP ex vivo and the most used assays are vasodilator stimulated phosphoprotein (VASP), Multiplate, VerifyNow (VN) and light transmission aggregometry (LTA).

Discussion/conclusion: The frequency of high platelet reactivity (HPR) during clopidogrel therapy is predicted to be 30%. Genetic polymorphisms and drug-drug interactions are discussed to explain a significant part of this inter-individual variation. HPR during prasugrel and ticagrelor treatment is estimated to be 3–15% and 0–3%, respectively. This lower frequency is explained by less complicated and more efficient generation of the active metabolite compared to clopidogrel. Meta-analyses do show a positive effect of adjusting standard clopidogrel treatment based on platelet function testing. Despite this, personalized therapy is not recommended because no large-scale RCT have shown any clinical benefit. For patients on prasugrel and ticagrelor, platelet function testing is not recommended due to low occurrence of HPR.

Keywords: P2Y12 receptor antagonists, Residual platelet reactivity, Clopidogrel, Prasugrel, Ticagrelor

Background

Platelet inhibition is pivotal to reduce cardiovascular events (CVE) in patients with coronary artery disease (CAD). The cornerstone in such treatment is aspirin, but when dual antiplatelet treatment (DAPT) is required, adding a $P2Y_{12}$ inhibitor is usually recommended. The most used $P2Y_{12}$ inhibitors are clopidogrel, prasugrel and ticagrelor. Their different properties are shown in Fig. 1 and Table 1.

Clopidogrel has been the most used $P2Y_{12}$ inhibitor in routine clinical practice for years and has been the subject of a considerable amount of research. DAPT with aspirin and clopidogrel was previously the preferred combination, but this changed after prasugrel and ticagrelor were introduced. Prasugrel has replaced clopidogrel in patients with ST-elevation myocardial infarction (STEMI) after percutaneous coronary intervention (PCI), and ticagrelor is preferred in patients with non-ST elevation myocardial infarction (NSTEMI) after PCI [2, 4]. Prasugrel and ticagrelor reduce new cardiovascular events more efficiently in these patient populations, but on the other hand more bleeding complications are reported [5, 6]. After elective PCI in patients with stable CAD, clopidogrel is still the first choice [7].

Despite DAPT, some patients still experience recurrent cardiovascular events. This may be due to many reasons, but insufficient platelet inhibition has been suggested a possible cause, and inter-individual differences in response are well known. A challenge in antiplatelet

* Correspondence: emkw@warlo.no
[1]Center for Clinical Heart Research, Department of Cardiology, Oslo University Hospital, Pb 4956 Nydalen, 0424 Oslo, Norway
[2]Faculty of Medicine, University of Oslo, Oslo, Norway
Full list of author information is available at the end of the article

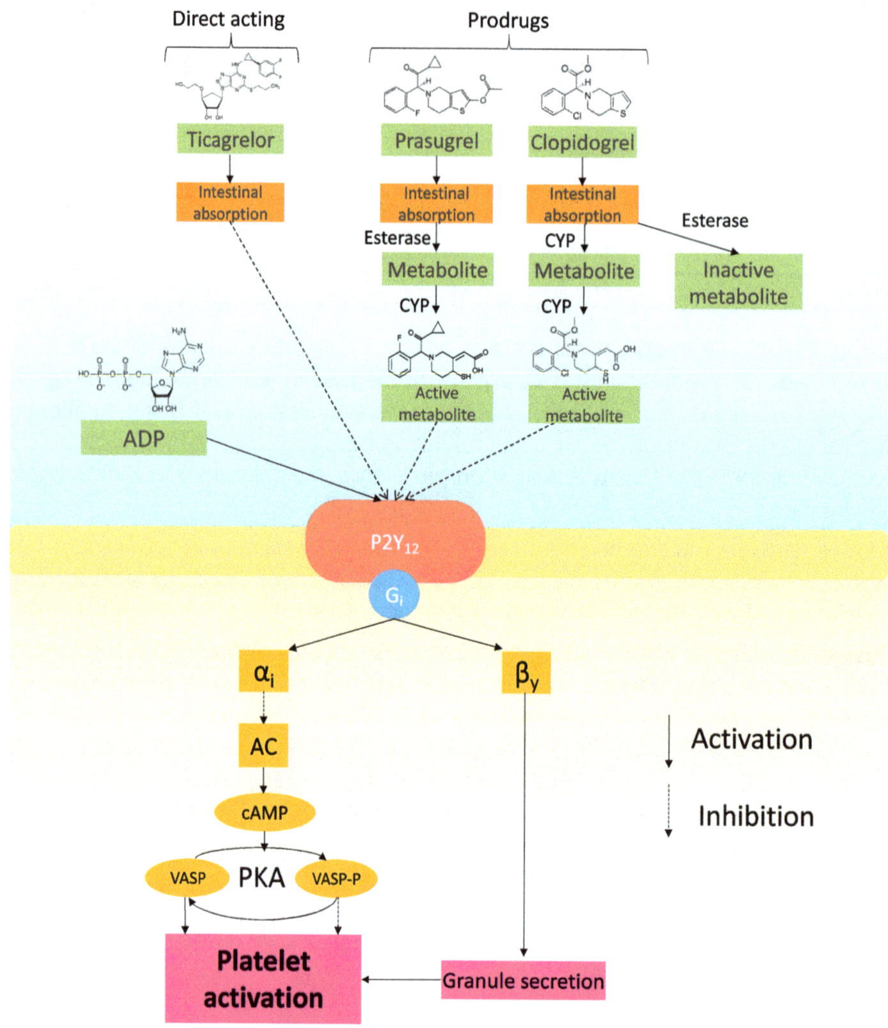

Fig. 1 The role of the $P2Y_{12}$ receptors in ADP stimulated platelet activation. Adapted from [1]

therapy is the lack of a standardized way to titrate the drug dose to achieve sufficient platelet inhibition and personalize treatment, like we can do with lipid-lowering and blood pressure medication [8].

Lack of response to antiplatelet therapy, termed resistance, non-responsiveness or high platelet reactivity (HPR) despite use of platelet inhibitors, has been widely studied. It has been distinguished between clinical and laboratory non-responsiveness. Clinical non-responsiveness is discussed when platelet-inhibited patients experience cardiovascular events. Laboratory non-responsiveness is defined when platelets still are active ex vivo despite treatment. These phenomena have only to some degree been shown to overlap [8].

Also, non-compliance i.e. patients not taken their medication, has to be considered when discussing the responsiveness/resistance phenomenon in clinical practice. This is, however, not discussed in the present review.

Studies on platelet non-responsiveness were initially focused on aspirin which has been extensively studied. When clopidogrel was introduced, this phenomenon was early addressed, and has later been studied also with regard to other $P2Y_{12}$ inhibitors. The interindividual response variability to clopidogrel is well established [9, 10]. Response variability to ticagrelor and prasugrel, on the other hand, is less known.

Table 1 Properties of the different P2Y12 inhibitors. Adapted and modified from [2, 3]

	Clopidogrel	Prasugrel	Ticagrelor
Chemical class	Thienopyridine	Thienopyridine	Cyclopentyl-triazolopyrimidine
	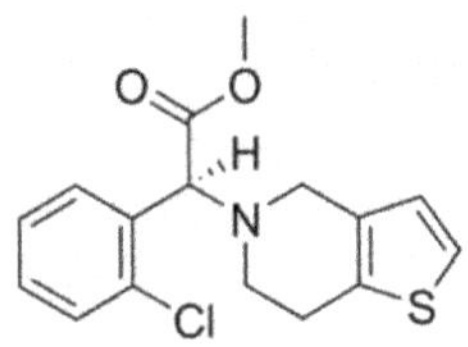	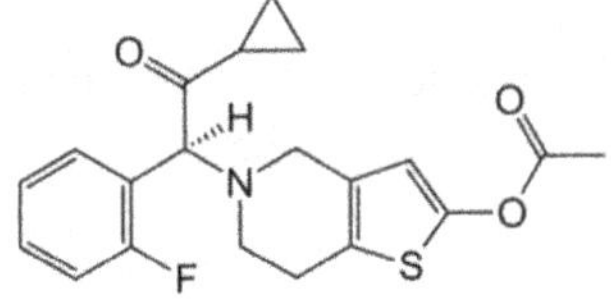	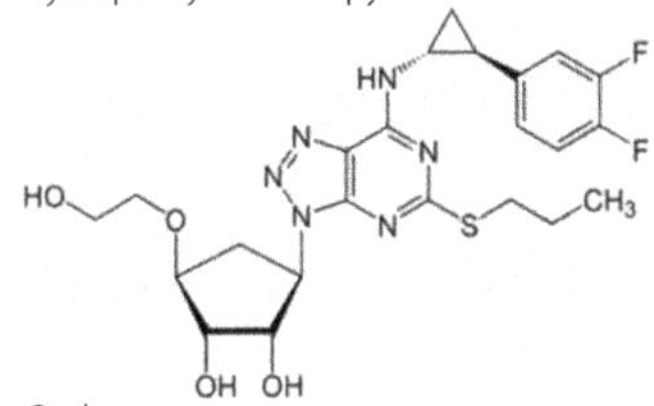
Administration	Oral	Oral	Oral
Dose	300–600 mg orally then 75 mg a day	60 mg orally then 10 mg a day	180 mg orally then 90 mg twice a day
Binding reversibility	Irreversible	Irreversible	Reversible
Binding site	ADP-binding site	ADP-binding site	Allosteric binding site
Activation	Prodrug, with variable liver metabolism	Prodrug, with predictable liver metabolism	Active drug, with additional active metabolite
Onset of loading dose effect	2–6 h	30 min	30 min
Duration of effect	3–10 days	7–10 days	3–5 days
Plasma half-life of active $P2Y_{12}$ inhibitor	30–60 min	30–60 min	6–12 h
Inhibition of adenosine reuptake	No	No	Yes

The aim of the present work was to summarize the literature on prevalence, mechanisms and clinical implications of resistance to $P2Y_{12}$ receptor inhibitors and give a conclusion based on the reports available.

Methods

ESC Guidelines on "Ischaemic Heart Disease and Acute Cardiac Care" [2, 11, 12] were used to discuss clinical guidelines for the different states of coronary artery disease.

Search strategy

Studies until the 11th of December 2017 were included in the literature search. The PubMed database was used. Phrases or synonyms for "$P2Y_{12}$ receptor antagonists" and "drug resistance" (shown below), were used identifying 1228 papers. When limiting the search to English language and last 5 years, in which the novel $P2Y_{12}$ inhibitors have been incorporated into clinical practice, the number of papers was reduced to 540.

Our search strategy was as following:

("Purinergic P2Y Receptor Antagonists" [mesh] OR ((ADP[Title] OR P2Y12[Title]) AND (Antagonist[Title] OR blocker*[Title])) OR clopidogrel[Title] OR prasugrel[Title] OR ticagrelor[title]) AND ("Drug Resistance"[Mesh] OR "Pharmacogenetics"[Mesh] OR resistance[Title] OR respons*[Title] OR respond*[Title] OR toleran*[Title] OR nonrespon*[Title] OR reactiv*[Title]) AND "last 5 years"[PDat] AND English[lang]*

Further focus on systematic reviews by adding "systematic[sb]" to the search strategy, identified 26 papers. To discover any potential Cochrane reviews, we added "Cochrane Database Syst Rev"[Journal] to the search, but 0 papers were found.

Of the 26 systematic reviews, we excluded 8 due to lack of power, studying non-CAD population or not being meta-analyses.

Results on genetic aspects (7 papers) were excluded from this review due to the complexity without obvious relevance for functionality, other than one specific single nucleotide polymorphisms (SNPs)'s influence on clopidogrel function. The topic is to some degree featured in the discussion. Thus, 11 meta-analyses are included.

Methods to determine $P2Y_{12}$ resistance/non-responsiveness

$P2Y_{12}$resistance is measured by stimulating platelets with ADP ex vivo. There are different assays for this purpose and the most used are measure of vasodilator stimulated phosphoprotein (VASP), Multiplate, VerifyNow (VN) and light transmission aggregometry (LTA). Platelet aggregometry induced by ADP is a functional test with a more global aggregation measure than e.g. VASP, which is more specific to drug action at subcellular levels. Aggregometry is the basic principle for VerifyNow, Multiplate and LTA.

Determination of drug response by all these methods have shown to predict clinical outcome in a significant number of patients after PCI [13]. Nevertheless, the expert consensus guidelines do not recommend LTA unless none of the other assays are available. This is due to lack of standardization of this method [13]. The currently recommended assays are therefore the VerifyNow, the Multiplate assay and the VASP assay. However, in clinical practice VerifyNow and Multiplate are preferred due to their standardized and user-friendly set up.

Another issue is determination of cut-off values for the definition of "laboratory non-responsiveness". The optimal threshold is still being investigated and may vary depending on the clinical situation. The current recommendation is 208 PRU with VerifyNow, 46 AU with the Multiplate assay and 50% with the VASP assay [13].

Discussion

Meta-analyses on laboratory non-responsiveness to P2Y12 antagonism

The number of patients included in the analyses investigating laboratory non-responsiveness range from 445 to 5395. This variation may be explained by different inclusion criteria, the number of drugs included and the type and number of laboratory methods used. A summary of the meta-analyses on laboratory non-responsiveness are shown in Table 2.

Meta-analyses on clinical outcome of non-responsiveness to P2Y12 antagonism

In the analyses investigating clinical outcome the number of patients varies from 605 to 28,178. This wide range may also be explained by different inclusion criteria, the number of drugs included, different study design and follow-up time, in addition to the laboratory methods used. A summary of the meta-analyses on clinical outcome are shown in Table 3.

Prevalence and mechanisms of high platelet reactivity (HPR) in $P2Y_{12}$-antagonists

Clopidogrel

The prevalence of high platelet reactivity (HPR) during clopidogrel treatment is high. However, the estimates have been inconsistent and dependent on the laboratory methods and cut off values used. From the expert consensus guidelines from 2014, the prevalence is predicted to be approximately 30% [13], which also fits with the meta-analysis by D'Ascenzo, F. et al. (Table 3).

Which factors that cause this huge variation in clopidogrel response is not fully resolved, but the most important factors seem to be genetic polymorphisms and drug-drug interactions [25].

Table 2 Platelet function testing on different antiplatelet therapies and regimens

Authors (year)	Study design	Population	No. studies (no. patients)	Drug and/or intervention	Lab method	Laboratory outcome Data are presented mainly as mean difference in PR with 95% confidence interval (CI) or frequency (%) of HPR
Zhang, H. et al. (2017) [14]	Meta-analysis of RCTs [12] and registry studies [6]	Patients with CAD	16 (2187)	Prasugrel vs. ticagrelor	VN and/or VASP	For the LD, the difference in PR between the prasugrel and ticagrelor groups was [10.80 (− 9.81, 31.40), $p = 0.30$] using the VN test and [− 2.87 (− 6.35, 0.60), $p = 0.10$] using the VASP test. For the MD, the PR was lower in the ticagrelor group than in the prasugrel group, [− 43.37 (− 60.53, − 26.21), $p < 0.01$] using the VN test and [− 9.23 (− 15.82, − 2.64), p < 0.01], using the VASP test.
Lhermusier, T. et al. (2015) [15]	Meta-analysis	Patients with CAD	29 (5395)	Ticagrelor vs. prasugrel vs. clopidogrel	VASP, VN, LTA	Compared with clopidogrel 75 mg, both prasugrel 10 mg and ticagrelor 90 mg × 2 were associated with lower PRU [− 117 (− 134.1, − 100.5)] and [− 159.7 (− 182.6, − 136.6)], respectively), lower PRI [− 24.2 (− 28.2, − 20.3) and [− 33.6 (− 39.9, − 27.6)], respectively), and lower MPA [− 11.8 (− 17, − 6.3) and [− 20.7 (− 28.5, − 12.8)], respectively). Similar results were obtained comparing clopidogrel 75 mg with 150 mg. with prasugrel 10 mg, ticagrelor 90 mg × 2 was associated with lower PRU [− 42.5 (− 62.9, − 21.9)], lower PRI [− 9.3 (− 15.6, − 3.5)], and lower MPA [− 8.9 (− 16.4, − 1.2)].
Lemesle, G. et al. (2015) [16]	Meta-analysis of RCTs [1] and registry studies [6]	Patients with CAD	14 (1822)	Prasugrel vs. ticagrelor	VASP, VN	The frequency of HPR was significantly lower in the ticagrelor group: 1.5% vs. 9.8% ($p < 0.0001$). In studies testing impact of LD, the frequency of HPR was 4.5% (ticagrelor) vs. 13.2% (prasugrel) ($p = 0.07$). In studies testing impact of MD, the frequency was 0.6% (ticagrelor) vs. 7.8% (prasugrel) (p < 0.0001).
Alexopoulos, D. et al. (2014) [17]	Meta-analysis	Patients with CAD	8 (445)	Ticagrelor	VN	Distribution of PR during ticagrelor MD was highly skewed toward lower values. No case of HPR (cut-off ≥230 PRU) was observed. Age and BMI positively affected PR, while current smoking lowered PR.

Table 3 Clinical outcome with different antiplatelet therapies and regimens

Authors (year)	Study design	Population	No. studies (no. patients)	Drug and/or intervention	Lab method	Clinical outcome Risk Ratio (RR) or Odds ratio (OR) with 95% CI are mainly given
Zhou, Y. et al. (2017) [18]	Meta-analysis of RCTs	Patients with CAD undergoing PCI	13 (7290)	CAT vs. IAT based on platelet function testing	VASP, VN, LTA, Multiplate	Testing-guided IAT was associated with a significant reduction in MACE [RR: 0.55 (0.36, 0.84), $p = 0.005$], CV death [RR: 0.60 (0.38–0.96), p = 0.03], ST [RR: 0.58 (0.36, 0.93), $p = 0.02$] and TVR [RR: 0.33 (0.14–0.76), $p = 0.009$] compared to CAT. No significant difference in rate of bleeding events.
Xu, L. et al. (2016) [19]	Meta-analysis of RCTs	Patients with CAD undergoing PCI	13 (5111)	CAT vs. IAT based on platelet function testing	VASP, VN, LTA, Multiplate, TEG	The incidences of CV death, nonfatal MI, and stent thrombosis were significantly lower in the IAT group than in the CAT group [RR: 0.45, (0.36, 0.57), $p < 0.00001$], whereas bleeding was similar between the two groups [RR: 1.05 (0.86, 1.27), $p = 0.65$].
Reny, J. et al. (2016) [20]	Meta-analysis of prospective cohorts and RCTs	Patients with symptomatic atherothrombosis	13 (6478)	Clopidogrel	LTA	The strength of the association between PR and the risk of MACE increased significantly ($p = 0.04$) with the number of risk factors present (age > 75 years, ACS at inclusion, diabetes, and hypertension). No association was detected in patients with no risk factor ($p = 0.48$).
Ma, W. et al. (2015) [21]	Meta-analysis of RCTs	Patients undergoing PCI	17 (4822)	CAT vs. IAT with and without platelet function testing.	VASP, VN, LTA, Multiplate	IAT was generally associated with a significant reduction in the risk of MACE [OR: 0.52 (0.39, 0.71), $p < 0.0001$]. The subgroup with HPR did also benefit from IAT compared to CAT [OR: 0.54 (0.38, 0.77), $p =$ 0.0007]. The observed benefits were mainly attributed to treatment-associated reduction in ST [OR: 0.43 (0.23, 0.78), $p = 0.006$] and TVR [OR: 0.38 (0.20, 0.74), $p = 0.004$]. No difference in the rate of major/minor bleeding event between IAT or CAT [OR: 0.80 (0.56, 1.13), $p = 0.21$].
Lin, L. et al. (2015) [22]	Meta-analysis of RCTs	Patients undergoing PCI	8 (3865)	CAT vs. IAT in patients with HPR	VASP, VN, LTA, Multiplate	In patients with HPR, IAT significantly reduced the risk of MACE/MACCE [RR: 0.59 (0.39, 0.88), $p = 0.01$], CV death [RR: 0.33, (0.12, 0.97), $p = 0.04$], ST [RR: 0.43 (0.20, 0.92), $p = 0.03$], and TVR [RR 0.31 (0.10, 0.93), $p = 0.04$], without increasing major bleeding [RR 0.75 (0.43, 1.31), $p = 0.31$] compared with CAT.
D'Ascenzo, F. et al. (2014) [23]	Meta-analysis	Patients with CAD	26 (28178)	Aspirin vs. clopidogrel	VN, LTA, Multiplate, TEG,	HPR was reported in 29% of patients on clopidogrel. HPR was not an independent prognostic indicator of adverse cardiac events in patients with either stable and unstable coronary disease for adverse cardiac events.
Chen, J. et al. (2013) [24]	Meta-analysis	Population with ACS	8 (605)	Clopidogrel with and without PPI	VASP, VN, Multiplate	Compared to clopidogrel treatment alone, patients who received both a PPI and clopidogrel had less of a decrease in the PRI [WMD: 8.18 (6.81, 9.56), $p <$ 0.00001], less ADP-induced platelet aggregation inhibition [WMD: 7.28 (2.44, 12.11), $p = 0.003$], higher PRU [WMD: 40.58 (19.31, 61.86), $p = 0.0002$], and higher risks of clopidogrel resistance [OR: 2.49 (1.49, 4.14), $p = 0.0005$]. However, no significant differences for the incidences of MACE were found.

BMI body mass index, *CAT* conventional antiplatelet therapy, *CV* cardiovascular, *HPR* high platelet reactivity, *IAT* intensified antiplatelet therapy, *LD* loading dose, *LPR* low platelet reactivity, *LTA* light transmission aggregometry, *MACCE* major adverse cardiac and cerebrovascular events, *MACE* major adverse cardiovascular events, *MI* myocardial infarction, *MD* maintenance dose, *MPA* maximal platelet aggregation, *OR* odds ratio, *PPI* protein pump inhibitors, *PR* platelet reactivity, *PRI* platelet reactivity index, *PRU* platelet reactivity units, *RCTs* randomized controlled trials, *RR* relative risk, *SD* standard dose, *ST* stent thrombosis, *TEG* thrombelastography, *TVR* target vessel revascularization, *VASP* Vasodilator stimulated phosphoprotein, *VN* VerifyNow-P2Y$_{12}$, *WMD* weighted mean differenc

Hepatic activation of clopidogrel and conversion into an active metabolite is essential for the inhibition of the P2Y$_{12}$ receptor [26, 27]. This metabolization is dependent of the cytochrome P450 isoenzymes (CYPs) [28]. The isoenzymes CYP2C19 is shown to be of particular interest and is said to explain 12–15% of the variable response to clopidogrel [10]. About 25 SNPs coding for CYP2C19 have been described in which CYP2C19*2 seems to be of most importance, i.e. shown to reduce serum concentration of the active metabolite and also to reduce inhibition of platelet aggregation [29, 30].

Reduced function of CYP2C19 has been reported to increase the risk for MACE [31, 32].

Drug interactions can also affect clopidogrel response. Rifampicin induces several CYPs, including CYP2C19, and leads to higher levels of active clopidogrel with subsequent greater $P2Y_{12}$ receptor blockade [33]. Ketoconazole on the other hand inhibits CYP3A4 and leads to reduced clopidogrel activation [34]. Proton pump inhibitors (PPI) depend on CYP2C19 metabolism like clopidogrel. Chen et al. have reported that combining these drugs increase the risk of clopidogrel resistance, but may be clinically unimportant, as no significant difference in major adverse cardiac events were observed [24]. Treatment with statins which are metabolized by CYP3A4 has shown not or only slightly to reduce platelet reactivity, but not to affect clinical outcome [35, 36].

Other factors that are discussed to contribute to low clopidogrel response are poor absorption, P2Y12 receptor polymorphisms, increased platelet turnover, different clinical factors like sex, diabetes, kidney disease, obesity, hypercholesterolemia [23, 25, 37].

Prasugrel and ticagrelor

There is broad scientific consensus that patients on prasugrel or ticagrelor are less susceptible to HPR than patients on clopidogrel, as also shown from the results in Table 2. Like the estimates for clopidogrel resistance, there has also been discrepancy between the reported prevalence of resistance to both prasugrel and ticagrelor.

The variation in the reported prevalence's may partly be due to lack of methodological standardization. Difference in the HPR definition across the studies is one limitation [16], but it also seems like PR varies depending on loading sequence, pre-treatment with clopidogrel, time point of testing, switching strategy, and patient population included [38].

Lemesle et al. have published a meta-analysis (Table 2) and included studies looking at the rate of HPR in the acute phase during loading dose (LD), but also during maintenance dose (MD) [16]. When isolating studies that tested PR after loading dose, no significant differences between the ticagrelor and prasugrel group were found. Nevertheless, when testing the impact of the maintenance dose, the rate of HPR was significantly lower in the ticagrelor group. The overall rate of HPR was significantly lower in the ticagrelor vs. prasugrel group [15]. Also the meta-analysis by Zhang et al. describe PR to be similar between the ticagrelor and prasugrel group after loading dose, but lower in the ticagrelor group during maintenance dose [14]. The meta-analysis by Lhermusier et al., though only including studies during maintenance dose, supports this observation [15].

The rate of HPR on prasugrel and ticagrelor treatment has not been established, but it is on maintenance dose estimated to be 3–15% for patients on prasugrel and 0–3% for ticagrelor treated patients [25]. Despite the low PR for both drugs, comparisons have shown that ticagrelor is the most potent platelet inhibitor and has the lowest prevalence of HPR [15, 16, 25].

The differences in HPR between clopidogrel, prasugrel and ticagrelor can partly be explained by the differences in their pharmacokinetics. Prasugrel has more efficient generation of active metabolite compared to clopidogrel [39, 40]. It is less dependent of CYP2C19 metabolism, and therefore not as affected by genetic variants of this enzyme [41, 42]. The most potent agent, ticagrelor, is an active drug and is not dependent on enzyme activation, i.e. is less susceptible to drug-drug interactions or pharmacogenetic influences [43]. Nevertheless, it has been shown that levels of active ticagrelor are affected by genetic variants of SLCO1B1 (solute carrier organic anion transporter family member 1B1) and UGT2B7 (UDP glucuronosyltransferase family 2 member B7). These gene variants have, however, not been shown to have any clinical implication [41]. PR on ticagrelor was affected by age, BMI and smoking status i.e. patients with increasing age and BMI have higher PR, and smokers lower PR [17]. Nevertheless, the PR on ticagrelor was generally very low and the rate of non-responders was 0% in this meta-analysis.

HPR as a predictor of clinical outcome and personalized antiplatelet therapy

Multiple studies have shown that patients with HPR during clopidogrel treatment are at greater risk for MACE [10]. Because of this, individualization of antiplatelet therapy based on platelet function testing has been studied in several RCTs. The principle in these trials has mainly been to compare the effect of intensified antiplatelet therapy (IAT) against conventional antiplatelet therapy (CAT) on clinical outcome in patients with HPR. The IAT protocols differ in the studies and is either increasing the clopidogrel dose or changing to prasugrel or ticagrelor. The results from these studies are diverging.

A meta-analysis performed by Zhou et al. (Table 3) found that patients undergoing PCI treated with IAT based on platelet function testing had reduced risk of MACE, CV death, stent thrombosis and target vessel revascularization, without any increase in the risk of bleeding [18]. Xu et al. found similar results in their meta-analysis with significantly reduced risk of CV death, nonfatal MI and stent thrombosis in the IAT group [19]. Ma et al. also found that patients with HPR did benefit from IAT compared to conventional antiplatelet therapy (CAT), where the observed benefits were mainly attributed to treatment-associated reduction in stent thrombosis and target vessel revascularization [21].

Even though these meta-analyses reach the same conclusion, they are similar and with some exceptions based on the same studies.

Despite similar results from these three meta-analysis, no large-scale randomized clinical trial has demonstrated any benefit of personalized antiplatelet therapy [37]. The GRAVITAS trial found no difference in clinical endpoints when comparing high dose vs. low dose clopidogrel among patients with HPR undergoing PCI. The number of clinical endpoints in this study was, however, very low, and less than half of the estimated number in the power calculations (5%). In addition, the platelet function testing was undertaken 12–24 h after the PCI, which may be have been too late to affect the outcome [44]. The TRIGGER-PCI study found that switching from clopidogrel to prasugrel in patients with HPR lead to a reduction in platelet reactivity, but no improvement in clinical outcome were observed. However, the trial was stopped prematurely after 6 months due to a lower endpoint rate than expected, and the study did therefore not achieve the desired power. And also in this study, platelet function testing with subsequent adjustment was not done before the morning after PCI [45]. The ARCTIC trial randomly assigned patients to a strategy with platelet function monitoring and treatment adjustment in non-responders, or to standard therapy without monitoring. Of the patients with HPR, about 80% received an increased clopidogrel dose, while only approximately 3% were started on prasugrel. The study showed no significant improvement in clinical endpoints with platelet function testing and subsequent drug adjustment as compared with the conventional strategy [46]. The ANTARCTIC trial randomized patients with acute coronary syndrome (ACS) above 75 years to prasugrel with or without platelet function monitoring with drug adjustment when indicated. They observed no differences in clinical outcome between the two groups [47].

In the meta-analysis by Reny et al. it was reported that the association between the risk of MACE and HPR significantly increases with the number of risk factors [20]. They suggest that the association between MACE and PR is dependent of the patient's cardiovascular profile. The risk factors that are thought to increase the risk of PR and MACE are among others age > 75, ACS at inclusion, diabetes and hypertension. This is supported by another meta-analysis where HPR did not increase the risk of adverse events after adjusting for risk factors [23]. Lack of multivariate analysis may have confounded the evaluation of the independent risk of HPR and may be the reason why all RCTs have failed when trying to show a beneficial effect of individualized antiplatelet therapy based on platelet function testing. The conflicting results between the meta-analyses and the large RCTs may also be due to publication bias.

Antiplatelet therapy and platelet function testing in clinical practice

The current guidelines for DAPT is to combine aspirin with a $P2Y_{12}$ blocker. Which $P2Y_{12}$ blocker depends on the clinical situation. For stable CAD patients after elective PCI, DAPT with clopidogrel is recommended, but for patients presenting with ACS prasugrel or ticagrelor are preferred [12].

The ESC guidelines do not recommend platelet function testing in routine clinical practice before or after elective stenting [12]. This is because no large-scale RCT has demonstrated any beneficial effect of adjusting therapy based on platelet function testing during clopidogrel treatment. With regards to prasugrel and ticagrelor, tailored therapy based on platelet function has not been that widely investigated, as HPR on these drugs is rare. Thus, platelet function testing is not recommended in these patients either [13].

The "ACCF/AHA/SCAI Guideline for PCI" also states that platelet function testing should not be used in routine clinical practice. Nevertheless, they say that testing may be considered in patients at high risk for MACE and that alternative agents such as prasugrel and ticagrelor might be considered in clopidogrel-treated patients with HPR [48]. However, these guidelines are from 2011 and are not based on the results from more recent RCTs.

Conclusion

The prevalence of HPR is greater in patients treated with clopidogrel (approximately 30%) compared to patients on the more novel antiplatelet agents prasugrel (3–15%) and ticagrelor (0–3%). These differences are likely due to different drug pharmacokinetics where prasugrel and ticagrelor have more efficient generation of active metabolite compared to clopidogrel.

Although meta-analyses show an effect of adjusting standard clopidogrel treatment based on platelet function testing, personalized therapy is not recommended because no large-scale RCT have shown any clinical benefit. Nevertheless, it should be noticed that the performed RCTs were underpowered to show any clinical effect. Personalized therapy is neither recommended for patients on prasugrel nor ticagrelor due to low occurrence of HPR on these respective drugs.

Abbreviations

ACS: acute coronary syndrome; AU: aggregation units; BMI: body mass index; CAD: coronary artery disease; CAT: conventional antiplatelet therapy; CV: cardiovascular; CVE: cardiovascular events; CYPs: cytochrome P450 isoenzymes; DAPT: dual antiplatelet therapy; ESC: European Society of Cardiology; HPR: high platelet reactivity; IAT: intensified antiplatelet therapy; LD: loading dose; LPR: low platelet reactivity; LTA: light transmission aggregometry; MACCE: major adverse cardiac and cerebrovascular events; MACE: major adverse cardiovascular events; MD: maintenance dose; MI: myocardial infarction; MPA: maximal platelet aggregation; NSTEMI: non-ST elevation myocardial infarction; OR: odds ratio; PCI: percutaneous coronary intervention; PPI: protein pump inhibitors; PR: platelet reactivity;

PRI: platelet reactivity index; PRU: platelet reactivity units; RCTs: randomized controlled trials; RR: relative risk; SD: standard dose; ST: stent thrombosis; STEMI: ST-elevation myocardial infarction; TEG: thrombelastography; TVR: target vessel revascularization; VASP: Vasodilator stimulated phosphoprotein; VN: VerifyNow-P2Y$_{12}$; WMD: weighted mean difference

Acknowledgments
The authors would like to thank librarian Maria S. Isachsen for excellent assistance with the literature search.

Author details
[1]Center for Clinical Heart Research, Department of Cardiology, Oslo University Hospital, Pb 4956 Nydalen, 0424 Oslo, Norway. [2]Faculty of Medicine, University of Oslo, Oslo, Norway. [3]Center for Heart Failure Research, University of Oslo, Oslo, Norway.

References

1. Angiolillo DJ, Ueno M. Optimizing platelet inhibition in clopidogrel poor metabolizers: therapeutic options and practical considerations. JACC Cardiovasc Interv. 2011;4(4):411–4. https://doi.org/10.1016/j.jcin.2011.03.001.
2. Roffi M, Patrono C, Collet J-P, Mueller C, Valgimigli M, Andreotti F, et al. 2015 ESC guidelines for the management of acute coronary syndromes in patients presenting without persistent ST-segment elevationTask force for the Management of Acute Coronary Syndromes in patients presenting without persistent ST-segment elevation of the European Society of Cardiology (ESC). Eur Heart J. 2016;37(3):267–315. https://doi.org/10.1093/eurheartj/ehv320.
3. Angiolillo DJ, Rollini F, Storey RF, Bhatt DL, James S, Schneider DJ, et al. International expert consensus on switching platelet P2Y12 receptor-inhibiting therapies. Circulation. 2017;136(20):1955–75. https://doi.org/10.1161/circulationaha.117.031164.
4. Steg PG, James SK, Atar D, Badano LP, Blomstrom-Lundqvist C, Borger MA, et al. ESC guidelines for the management of acute myocardial infarction in patients presenting with ST-segment elevation. Eur Heart J. 2012;33(20):2569–619. https://doi.org/10.1093/eurheartj/ehs215.
5. Wiviott SD, Braunwald E, McCabe CH, Montalescot G, Ruzyllo W, Gottlieb S, et al. Prasugrel versus clopidogrel in patients with acute coronary syndromes. N Engl J Med. 2007;357(20):2001–15. https://doi.org/10.1056/NEJMoa0706482.
6. Wallentin L, Becker RC, Budaj A, Cannon CP, Emanuelsson H, Held C, et al. Ticagrelor versus clopidogrel in patients with acute coronary syndromes. N Engl J Med. 2009;361(11):1045–57. https://doi.org/10.1056/NEJMoa0904327.
7. Montalescot G, Sechtem U, Achenbach S, Andreotti F, Arden C, Budaj A, et al. 2013 ESC guidelines on the management of stable coronary artery disease: the task force on the management of stable coronary artery disease of the European Society of Cardiology. Eur Heart J. 2013;34(38):2949–3003. https://doi.org/10.1093/eurheartj/eht296.
8. Kuliczkowski W, Witkowski A, Polonski L, Watala C, Filipiak K, Budaj A, et al. Interindividual variability in the response to oral antiplatelet drugs: a position paper of the working group on antiplatelet drugs resistance appointed by the section of cardiovascular interventions of the polish cardiac society, endorsed by the working group on thrombosis of the European society of cardiology. Eur Heart J. 2009;30(4):426–35. https://doi.org/10.1093/eurheartj/ehn562.
9. Parodi G, Marcucci R, Valenti R, Gori AM, Migliorini A, Giusti B, et al. High residual platelet reactivity after clopidogrel loading and long-term cardiovascular events among patients with acute coronary syndromes undergoing PCI. JAMA. 2011;306(11):1215–23. https://doi.org/10.1001/jama.2011.1332.
10. Oliphant CS, Trevarrow BJ, Dobesh PP. Clopidogrel response variability: review of the literature and practical considerations. J Pharm Pract. 2016;29(1):26–34. https://doi.org/10.1177/0897190015615900
11. Ibanez B, James S, Agewall S, Antunes MJ, Bucciarelli-Ducci C, Bueno H, et al. 2017 ESC guidelines for the management of acute myocardial infarction in patients presenting with ST-segment elevationThe task force for the management of acute myocardial infarction in patients presenting with ST-segment elevation of the European Society of Cardiology (ESC). Eur Heart J. 2018;39(2):119–77. https://doi.org/10.1093/eurheartj/ehx393.
12. Valgimigli M, Bueno H, Byrne RA, Collet J-P, Costa F, Jeppsson A, et al. 2017 ESC focused update on dual antiplatelet therapy in coronary artery disease developed in collaboration with EACTSThe task force for dual antiplatelet therapy in coronary artery disease of the European Society of Cardiology (ESC) and of the European Association for Cardio-Thoracic Surgery (EACTS). Eur Heart J. 2018;39(3):213–60. https://doi.org/10.1093/eurheartj/ehx419.
13. Aradi D, Storey RF, Komocsi A, Trenk D, Gulba D, Kiss RG, et al. Expert position paper on the role of platelet function testing in patients undergoing percutaneous coronary intervention. Eur Heart J. 2014;35(4):209–15. https://doi.org/10.1093/eurheartj/eht375.
14. Zhang H, Zhang P, Dong P, Yang X, Wang Y, Zhang H, et al. Effect of ticagrelor versus prasugrel on platelet reactivity: a meta-analysis. Coron Artery Dis. 2017;28(7):597–604. https://doi.org/10.1097/mca.0000000000000541.
15. Lhermusier T, Lipinski MJ, Tantry US, Escarcega RO, Baker N, Bliden KP, et al. Meta-analysis of direct and indirect comparison of ticagrelor and prasugrel effects on platelet reactivity. Am J Cardiol. 2015;115(6):716–23. https://doi.org/10.1016/j.amjcard.2014.12.029.
16. Lemesle G, Schurtz G, Bauters C, Hamon M. High on-treatment platelet reactivity with ticagrelor versus prasugrel: a systematic review and meta-analysis. J Thromb Haemost. 2015;13(6):931–42. https://doi.org/10.1111/jth.12907.
17. Alexopoulos D, Xanthopoulou I, Storey RF, Bliden KP, Tantry US, Angiolillo DJ, et al. Platelet reactivity during ticagrelor maintenance therapy: a patient-level data meta-analysis. Am Heart J. 2014;168(4):530–6. https://doi.org/10.1016/j.ahj.2014.06.026.
18. Zhou Y, Wang Y, Wu Y, Huang C, Yan H, Zhu W, et al. Individualized dual antiplatelet therapy based on platelet function testing in patients undergoing percutaneous coronary intervention: a meta-analysis of randomized controlled trials. BMC Cardiovasc Disord. 2017;17(1):157. https://doi.org/10.1186/s12872-017-0582-6.
19. Xu L, Hu XW, Zhang SH, Li JM, Zhu H, Xu K, et al. Intensified antiplatelet treatment reduces major cardiac events in patients with Clopidogrel low response: a meta-analysis of randomized controlled trials. Chin Med J. 2016;129(8):984–91. https://doi.org/10.4103/0366-6999.179786.
20. Reny JL, Fontana P, Hochholzer W, Neumann FJ, Ten Berg J, Janssen PW, et al. Vascular risk levels affect the predictive value of platelet reactivity for the occurrence of MACE in patients on clopidogrel. Systematic review and meta-analysis of individual patient data. Thromb Haemost. 2016;115(4):844–55. https://doi.org/10.1160/th15-09-0742.
21. Ma W, Liang Y, Zhu J, Wang Y, Wang X. Meta-analysis appraising high maintenance dose clopidogrel in patients who underwent percutaneous coronary intervention with and without high on-clopidogrel platelet reactivity. Am J Cardiol. 2015;115(5):592–601. https://doi.org/10.1016/j.amjcard.2014.12.013.
22. Lin L, Wang H, Chen YF, Lin WW, Wang CL, Lin CH. High maintenance dose of clopidogrel in patients with high on-treatment platelet reactivity after a percutaneous coronary intervention: a meta-analysis. Coron Artery Dis. 2015;26(5):386–95. https://doi.org/10.1097/mca.0000000000000246.
23. D'Ascenzo F, Barbero U, Bisi M, Moretti C, Omede P, Cerrato E, et al. The prognostic impact of high on-treatment platelet reactivity with aspirin or ADP receptor antagonists: systematic review and meta-analysis. Biomed Res Int. 2014;2014:610296. https://doi.org/10.1155/2014/610296.
24. Chen J, Chen SY, Lian JJ, Zeng XQ, Luo TC. Pharmacodynamic impacts of proton pump inhibitors on the efficacy of clopidogrel in vivo--a systematic review. Clin Cardiol. 2013;36(4):184–9. https://doi.org/10.1002/clc.22094.
25. Thomas MR, Storey RF. Clinical significance of residual platelet reactivity in patients treated with platelet P2Y12 inhibitors. Vasc Pharmacol. 2016;84:25–7. https://doi.org/10.1016/j.vph.2016.05.010

26. Zhang Y, Zhang S, Ding Z. Role of P2Y12 receptor in thrombosis. Adv Exp Med Biol. 2017;906:307–24. https://doi.org/10.1007/5584_2016_123.
27. Thomas MR, Storey RF. Optimal management of antiplatelet therapy and proton pump inhibition following percutaneous coronary intervention. Curr Treat Options Cardiovasc Med. 2012;14(1):24–38. https://doi.org/10.1007/s11936-011-0157-2.
28. Siller-Matula JM, Trenk D, Schror K, Gawaz M, Kristensen SD, Storey RF, et al. Response variability to P2Y12 receptor inhibitors: expectations and reality. JACC Cardiovasc Interv. 2013;6(11):1111–28. https://doi.org/10.1016/j.jcin.2013.06.011.
29. Mao L, Jian C, Changzhi L, Dan H, Suihua H, Wenyi T, et al. Cytochrome CYP2C19 polymorphism and risk of adverse clinical events in clopidogrel-treated patients: a meta-analysis based on 23,035 subjects. Arch Cardiovasc Dis. 2013;106(10):517–27. https://doi.org/10.1016/j.acvd.2013.06.055.
30. Pettersen AA, Arnesen H, Opstad TB, Seljeflot I. The influence of CYP 2C19*2 polymorphism on platelet function testing during single antiplatelet treatment with clopidogrel. Thromb J. 2011;9:4. https://doi.org/10.1186/1477-9560-9-4.
31. Mega JL, Close SL, Wiviott SD, Shen L, Hockett RD, Brandt JT, et al. Cytochrome p-450 polymorphisms and response to clopidogrel. N Engl J Med. 2009;360(4):354–62. https://doi.org/10.1056/NEJMoa0809171.
32. Simon T, Verstuyft C, Mary-Krause M, Quteineh L, Drouet E, Meneveau N, et al. Genetic determinants of response to clopidogrel and cardiovascular events. N Engl J Med. 2009;360(4):363–75. https://doi.org/10.1056/NEJMoa0808227.
33. Judge HM, Patil SB, Buckland RJ, Jakubowski JA, Storey RF. Potentiation of clopidogrel active metabolite formation by rifampicin leads to greater P2Y12 receptor blockade and inhibition of platelet aggregation after clopidogrel. J Thromb Haemost. 2010;8(8):1820–7. https://doi.org/10.1111/j.1538-7836.2010.03925.x.
34. Farid NA, Payne CD, Small DS, Winters KJ, Ernest CS 2nd, Brandt JT, et al. Cytochrome P450 3A inhibition by ketoconazole affects prasugrel and clopidogrel pharmacokinetics and pharmacodynamics differently. Clin Pharmacol Ther. 2007;81(5):735–41. https://doi.org/10.1038/sj.clpt.6100139.
35. Park JS, Cha KS, Lee HW, Oh JH, Choi JH, Lee HC, et al. Platelet reactivity and clinical outcomes in patients using CYP3A4-metabolized statins with clopidogrel in percutaneous coronary intervention. Heart Vessel. 2017;32(6):690–9. https://doi.org/10.1007/s00380-016-0927-6.
36. Trenk D, Hochholzer W, Frundi D, Stratz C, Valina CM, Bestehorn HP, et al. Impact of cytochrome P450 3A4-metabolized statins on the antiplatelet effect of a 600-mg loading dose clopidogrel and on clinical outcome in patients undergoing elective coronary stent placement. Thromb Haemost. 2008;99(1):174–81. https://doi.org/10.1160/th07-08-0503.
37. Franchi F, Angiolillo DJ. Novel antiplatelet agents in acute coronary syndrome. Nat Rev Cardiol. 2015;12(1):30–47. https://doi.org/10.1038/nrcardio.2014.156.
38. Siller-Matula JM, Hintermeier A, Kastner J, Kreiner G, Maurer G, Kratochwil C, et al. Distribution of clinical events across platelet aggregation values in all-comers treated with prasugrel and ticagrelor. Vasc Pharmacol. 2016;79:6–10. https://doi.org/10.1016/j.vph.2016.01.003.
39. Wiviott SD, Trenk D, Frelinger AL, O'Donoghue M, Neumann FJ, Michelson AD, et al. Prasugrel compared with high loading- and maintenance-dose clopidogrel in patients with planned percutaneous coronary intervention: the Prasugrel in comparison to Clopidogrel for inhibition of platelet activation and aggregation-thrombolysis in myocardial infarction 44 trial. Circulation. 2007;116(25):2923–32. https://doi.org/10.1161/circulationaha.107.740324.
40. Wallentin L, Varenhorst C, James S, Erlinge D, Braun OO, Jakubowski JA, et al. Prasugrel achieves greater and faster P2Y12receptor-mediated platelet inhibition than clopidogrel due to more efficient generation of its active metabolite in aspirin-treated patients with coronary artery disease. Eur Heart J. 2008;29(1):21–30. https://doi.org/10.1093/eurheartj/ehm545.
41. Friede K, Li J, Voora D. Use of Pharmacogenetic information in the treatment of cardiovascular disease. Clin Chem. 2017;63(1):177–85. https://doi.org/10.1373/clinchem.2016.255232.
42. Mega JL, Close SL, Wiviott SD, Shen L, Hockett RD, Brandt JT, et al. Cytochrome P450 genetic polymorphisms and the response to prasugrel: relationship to pharmacokinetic, pharmacodynamic, and clinical outcomes. Circulation. 2009;119(19):2553–60. https://doi.org/10.1161/circulationaha.109.851949.
43. Birkeland K, Parra D, Rosenstein R. Antiplatelet therapy in acute coronary syndromes: focus on ticagrelor. J Blood Med. 2010;1:197–219. https://doi.org/10.2147/jbm.s9650.
44. Price MJ, Berger PB, Teirstein PS, Tanguay JF, Angiolillo DJ, Spriggs D, et al. Standard- vs high-dose clopidogrel based on platelet function testing after percutaneous coronary intervention: the GRAVITAS randomized trial. JAMA. 2011;305(11):1097–105. https://doi.org/10.1001/jama.2011.290.
45. Trenk D, Stone GW, Gawaz M, Kastrati A, Angiolillo DJ, Muller U, et al. A randomized trial of prasugrel versus clopidogrel in patients with high platelet reactivity on clopidogrel after elective percutaneous coronary intervention with implantation of drug-eluting stents: results of the TRIGGER-PCI (testing platelet reactivity in patients undergoing elective stent placement on Clopidogrel to guide alternative therapy with Prasugrel) study. J Am Coll Cardiol. 2012;59(24):2159–64. https://doi.org/10.1016/j.jacc.2012.02.026.
46. Collet JP, Cuisset T, Range G, Cayla G, Elhadad S, Pouillot C, et al. Bedside monitoring to adjust antiplatelet therapy for coronary stenting. N Engl J Med. 2012;367(22):2100–9. https://doi.org/10.1056/NEJMoa1209979.
47. Cayla G, Cuisset T, Silvain J, Leclercq F, Manzo-Silberman S, Saint-Etienne C, et al. Platelet function monitoring to adjust antiplatelet therapy in elderly patients stented for an acute coronary syndrome (ANTARCTIC): an open-label, blinded-endpoint, randomised controlled superiority trial. Lancet. 2016;388(10055):2015–22. https://doi.org/10.1016/s0140-6736(16)31323-x.
48. Levine GN, Bates ER, Blankenship JC, Bailey SR, Bittl JA, Cercek B, et al. 2011 ACCF/AHA/SCAI guideline for percutaneous coronary intervention: a report of the American College of Cardiology Foundation/American Heart Association task force on practice guidelines and the Society for Cardiovascular Angiography and Interventions. Circulation. 2011;124(23):e574–651. https://doi.org/10.1161/CIR.0b013e31823ba622.

9

Non-vitamin K antagonist oral anticoagulants vs. vitamin-K antagonists in patients with atrial fibrillation and chronic kidney disease: A nationwide cohort study

Emma Kirstine Laugesen[1*], Laila Staerk[1,2], Nicholas Carlson[3,4], Anne-Lise Kamper[4], Jonas Bjerring Olesen[1], Christian Torp-Pedersen[5], Gunnar Gislason[1,2,3,6] and Anders Nissen Bonde[1,2]

Abstract

Background: We aimed to compare effectiveness and safety of non-vitamin K antagonist oral anticoagulants (NOACs) versus vitamin-K antagonists (VKA) in atrial fibrillation (AF) patients with chronic kidney disease (CKD) not receiving dialysis.

Methods: By using personal identification numbers, we cross-linked individual-level data from Danish administrative registries. We identified every citizen with a prior diagnosis of AF and CKD who initiated NOAC or VKA (2011–2017). An external analysis of 727 AF patients with CKD (no dialysis) was performed to demonstrate level of kidney function in a comparable population. Study outcomes included incidents of stroke/thromboembolisms (TEs), major bleedings, myocardial infarctions (MIs), and all-cause mortality. We used Cox proportional hazards models to determine associations between oral anticoagulant treatment and outcomes.

Results: Of 1560 patients included, 1008 (64.6%) initiated VKA and 552 (35.4%) initiated NOAC. In a comparable population we found that 95.3% of the patients had an estimated glomerular filtration rate (eGFR) < 59 mL/min. Patients treated with NOAC had a significantly decreased risk of major bleeding (hazard ratio (HR): 0.47, 95% confidence interval (CI): 0.26–0.84) compared to VKA. There was not found a significant association between type of anticoagulant and risk of stroke/TE (HR: 0.83, 95% CI: 0.39–1.78), MI (HR: 0.45, 95% CI: 0.18–1.11), or all-cause mortality (HR: 0.99, 95% CI: 0.77–1.26).

Conclusion: NOAC was associated with a lower risk of major bleeding in patients with AF and CKD compared to VKA. No difference was found in risk of stroke/TE, MI, and all-cause mortality.

Keywords: Atrial fibrillation, Chronic kidney disease, NOAC, VKA, Oral anticoagulation

Background

Atrial fibrillation (AF) and chronic kidney disease (CKD) often coexist [1], and the number of patients suffering from both conditions is rising globally as a result of an aging population.

The risk of stroke, systemic thromboembolism (TE) and myocardial infarction (MI) is higher among such patients than in AF patients with no renal disease [2, 3]. A Danish study found that AF patients with non-end-stage CKD or end-stage CKD had an increased risk of stroke/TE of 50 and 83%, respectively, compared to patients without renal disease [2]. Furthermore, the presence of both disorders increases the risk of bleeding [2], causing treatment for such patients to be complicated. Oral anticoagulation (OAC) reduces risk of stroke/TE and all-cause mortality in the general AF population [4], and AF patients at high stroke risk are recommended lifelong therapy with a vitamin-K antagonist (VKA) or non-VKA OAC (NOAC). NOACs have proven superior or noninferior to VKA in AF patients in regards of stroke and TE

* Correspondence: emma.laugesen@hotmail.com
[1]Department of Cardiology, Copenhagen University Hospital Herlev and Gentofte, Kildegaardsvej 28, 2900 Hellerup, Denmark
Full list of author information is available at the end of the article

prevention, they are prescribed in fixed dosages, and do not require international normalized ratio (INR) monitoring [5]. Both have also proven effective in reducing the risk of stroke/TE in AF patients with mild to moderate CKD [3, 6], but patients with severe CKD have been excluded from randomized clinical trials of NOAC vs. VKA [5, 7]. All NOACs are dependent on renal clearance [6], which is why there have been concerns regarding efficacy and safety in patients with more advanced CKD.

The aim of this study was to determine whether patients diagnosed with both AF and CKD can benefit from NOAC treatment the same way as patients without CKD can.

Methods

Data sources

Every Danish citizen has a unique civil registration number, which allows cross-linkage of several nationwide registries on an individual basis. All discharge diagnoses from Danish hospitals are registered in the nationwide Danish National Patient Registry [8]. Every drug dispensed from a Danish pharmacy is registered in the Danish Prescription Registry [9]. Civil status, yearly income, and cause of death is registered in the Danish Civil Registration System [10].

Study population

We identified all Danish citizens with a prior diagnosis of AF and CKD who initiated OAC between 22nd August 2011 (the day that the first NOAC, dabigatran, was approved in Denmark) and 30th June 2017 (data in our registries were available until this date). We excluded patients with valvular AF, defined as mechanical heart valve or rheumatic heart disease [11]. Additionally, we excluded patients on dialysis where use of all NOACs are off-label, and patients initiated on edoxaban because of the low number of patients, and due to its short period of time available on the market.

External analysis of CKD diagnosis

Since plasma creatinine measurements were not available after 2011, we had no data on kidney function in the study population. We estimated kidney function level by studying a comparable CKD population. Thus, we created a dataset of 727 AF patients with CKD (no dialysis) diagnosed between 1997 and 2011who were first-time initiators of OAC. All patients in the dataset had one plasma creatinine measurement within 90 days before initiating OAC. Plasma creatinine values were registered in databases from either a general practitioner or from a hospital clinic. eGFR was calculated using the CKD-EPI equation [12].

Baseline characteristics

Co-medication and co-morbidity were defined as in earlier studies [13, 14]. In brief, co-medication was identified as prescriptions claimed during the last 180 days before inclusion in the study, and co-morbidity was identified from hospital diagnoses during the last 5 years before inclusion. All ICD and ATC codes used to define our dataset are listed in Additional file 1.

Outcomes and follow-up

Patients were followed after initiation of OAC until whichever of the following occurred first: emigration, death from any cause, discontinuation/switch of OAC, 30th June 2017, or 1 year after initiation of OAC. We investigated the following events: i) major bleeding, defined as gastrointestinal, urogenital, airway, intraocular or intracranial bleeding causing hospitalization [15], ii) stroke/thromboembolism defined as ischemic or unclassified stroke or systemic arterial thromboembolism [16], iii) myocardial infarction, and iv) all-cause mortality.

Statistics

Continuous covariates are presented as medians with interquartile range (IQR), and categorical covariates are presented as frequencies with percentages. *P*-value for differences in baseline characteristics were calculated using Wilcoxon-Mann Whitney or Chi square test for continuous or categorical variables, respectively. *P*-value for trend in initiation patterns over time was calculated using the Cochran-Armitage test. Cumulative incidence of event according to OAC was calculated using the Aalen-Johansen estimator, with death considered as competing risk. P-value for difference in cumulative incidence was calculated using Grey's test. Cox proportional hazards models were used to examine the association between use of OAC and outcome. Models with bleeding as outcome were adjusted for age, sex, prior bleeding, prior stroke, liver disease, hypertension, and use of non-steroid anti-inflammatory drugs, acetylsalicylic acid, adenosine phosphate receptor inhibitors (ADPi), and income during the previous year. Models with stroke/thromboembolism or myocardial infarction as outcome were adjusted for age, sex, prior stroke, heart failure, hypertension, coronary heart disease, diabetes, and use of acetylsalicylic acid, ADPi and income during the previous year. Models with all-cause mortality as outcome were adjusted for age, sex, prior stroke, prior bleeding, coronary heart disease, heart failure, hypertension, diabetes, liver disease, use of non-steroid anti-inflammatory drugs, acetylsalicylic acid, ADPi, and income during the previous year. A Kolmogorov type test was used to examine the proportional hazards assumption, and we tested for interaction between OAC and relevant variables, including age, sex, income, and treatment with ADPi or acetylsalicylic acid in all models. G-formula based on the Cox models were used to calculate standardized absolute risk according to VKA or NOAC treatment each month during follow-up, and 1000 bootstraps calculated confidence intervals. As sensitivity, we calculated hazard ratios (HRs) according to time-varying OAC instead of OAC at baseline, e.g., patients could switch exposure status (switch between VKA and NOAC treatment or vice versa) during follow up. HRs were also calculated according to OAC among AF patients with CKD when not censoring at shift or discontinuation. All analyses were performed with SAS version 9.4, and R version 3.4.1 [17].

Results

Study population

A study population of 1560 patients with non-valvular AF and CKD was identified during our study period from 2011 to 2017 (Fig. 1). From this population, 1008 (64.6%) patients initiated VKA treatment while 552 (35.4%) initiated NOAC treatment. Of these 552 patients, 302 (54.7%) received apixaban, 170 (30.8%) received rivaroxaban, and 80 (14.5%) received dabigatran. In the NOAC population, 428 patients (77.5%) received reduced dose. The patients who initiated a NOAC were more likely to be older and more often women compared to the VKA patients (Table 1). No significant differences were found between the two groups of patients in terms of other co-morbidities or co-medications. Other baseline characteristics are listed in Table 1.

External analysis of kidney function in a comparable population

In a comparable population of 727 patients with a diagnosis code of AF and CKD, 3 (0.4%) of included patients had an eGFR > 90 mL/min/1.73m^2, 31 (4.3%) an eGFR 60–90 mL/min/1.73m^2, 312 (42.9%) an eGFR 30–59 mL/min/1.73m^2, 319 (43.9%) an eGFR 15-29 mL/min/1.73m^2, and 62 (8.5%) an eGFR < 15 mL/min/1.73m^2 (Additional file 2).

In aggregation, 95.3% had an eGFR < 59 mL/min/1.73m^2 while 52.4% had an eGFR < 29 mL/min/1.73m^2.

Risk of stroke/TE, major bleeding, myocardial infarction, and all-cause mortality

Between 2011 and 2017, a significant increase in initiation of NOAC among AF patients with CKD was observed (p-value for trend < 0.001) (Fig. 2). In 2011, 15% initiated a NOAC compared to 60% in late 2017.

Table 2 shows adjusted HRs and 1-year standardized absolute risks of stroke/TE, major bleeding, MI, and all-cause mortality according to type of OAC.

The 1-year standardized absolute risk of stroke/TE in AF patients with CKD treated with NOAC was 2.0% (95% confidence interval (CI): 0.8–3.3%) and for the group treated with VKA 2.4% (95% CI: 1.4–3.5%). There was no significant difference between the risk of stroke/TE among NOAC patients compared to VKA patients (HR: 0.83, 95% CI: 0.39–1.78).

The 1-year standardized absolute risk of major bleeding in the study population was 2.8% (95% CI: 1.5–4.3%) among NOAC patients and 5.9% (95% CI: 4.4–7.5%) among VKA patients. There was a significant association between major bleeding and type of OAC. Patients receiving a NOAC had a significant lower risk of major bleeding compared to patients on VKAs (HR: 0.47, 95% CI: 0.26–0.84). We found no significant association

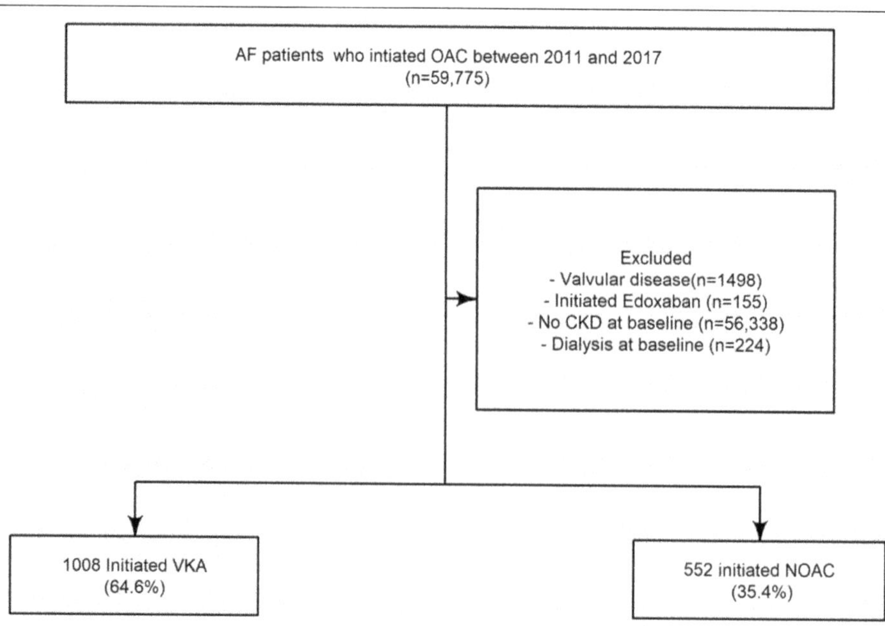

Fig. 1 Flowchart of the selection of the study population

Table 1 Baseline characteristics of the study population

	Patients receiving VKA (*n* = 1008)	Patients receiving NOAC (*n* = 552)	*P* value
Median age (IQR)	78.00 (71.00–84.00)	80.00 (72.00–86.00)	< 0.001
Male (%)	645 (64.0)	314 (56.9)	0.007
Oral anticoagulation			< 0.001
Apixaban (%)		302 (54.7)	
Dabigatran (%)		80 (14.5)	
Rivaroxaban (%)		170 (30.8)	
VKA (%)	1008 (100.0)		
Income group (quartiles)			< 0.001
1st (lowest)	382 (37.9)	143 (25.9)	
2nd	297 (29.5)	165 (29.9)	
3rd	208 (20.6)	159 (28.8)	
4th (highest)	121 (12.0)	85 (15.4)	
Comorbidity			
Hypertension	914 (90.7)	494 (89.5)	0.454
Previous stroke (%)	175 (17.4)	102 (18.5)	0.629
Previous bleeding (%)	256 (25.4)	122 (22.1)	0.164
Heart failure (%)	395 (39.2)	195 (35.3)	0.147
Ischemic heart disease (%)	440 (43.7)	231 (42.9)	0.492
Peripheral artery disease (%)	101 (10.0)	63 (11.4)	0.391
Diabetes (%)	260 (25.8)	141 (25.5)	0.962
Liver disease (%)	28 (2.8)	19 (3.4)	0.563
Alcohol abuse (%)	38 (3.8)	31 (5.6)	0.117
Comedication			
ADPi (%)	164 (16.3)	107 (19.4)	0.138
Aspirin (%)	537 (53.3)	277 (50.2)	0.264
Statin (%)	503 (49.9)	275 (49.8)	1.000
Beta-blocker (%)	582 (57.7)	303 (54.9)	0.302
RASi (%)	524 (52.0)	294 (53.3)	0.667
NSAID (%)	116 (11.5)	69 (12.5)	0.619

Abbreviations: *IQR* Interquartile Range, *ADPi* Adenosine diphosphate inhibitor, *VKA* Vitamin K antagonist, *RASi* Renin angiotensin system inhibitor, *NOAC* Nonvitamin K oral anticoagulants, *NSAID* Non-steroid anti-inflammatory drugs

between use of NOAC and risk of MI (HR: 0.50, 95% CI: 0.21–1.19). The 1-year standardized absolute risk of MI was 1.1% (95% CI: 0.4–2.2%) in the NOAC treated population and 2.5% (95% CI: 1.5–3.7%) in the VKA treated population. The 1-year standardized absolute risk of all-cause mortality was 23.2% (95% CI: 19.4–27.0%) in AF patients with CKD on NOAC and 23.3% (95% CI, 20.2–26.3%) for the ones on VKA. No differences were found in all-cause mortality between the two groups of patients (HR: 0.99, 95% CI: 0.77–1.26). Figure 3 illustrates the standardized absolute risks of stroke/TE, major bleeding, MI, and all-cause mortality, described above, in the first year following drug initiation among patients on NOAC or VKA, respectively. Cumulative incidences of the study outcomes yielded similar results (Additional file 3).

Sensitivity analyses

Analyses of the study outcomes were repeated with time-varying OAC, allowing patients to change treatment group after inclusion. These results were similar to main analyses (Additional file 4). When not censoring at shift or discontinuation of OAC, outcomes were similar to main results as well (Additional file 5).

Discussion

This nationwide study examined the risk of stroke/TE, major bleeding, MI, and all-cause mortality in AF patients with CKD, comparing patients treated with NOAC to VKA. We had the following important findings: 1) there was a significant gradual increase in the use of NOACs during our study period, 2) NOACs were associated with

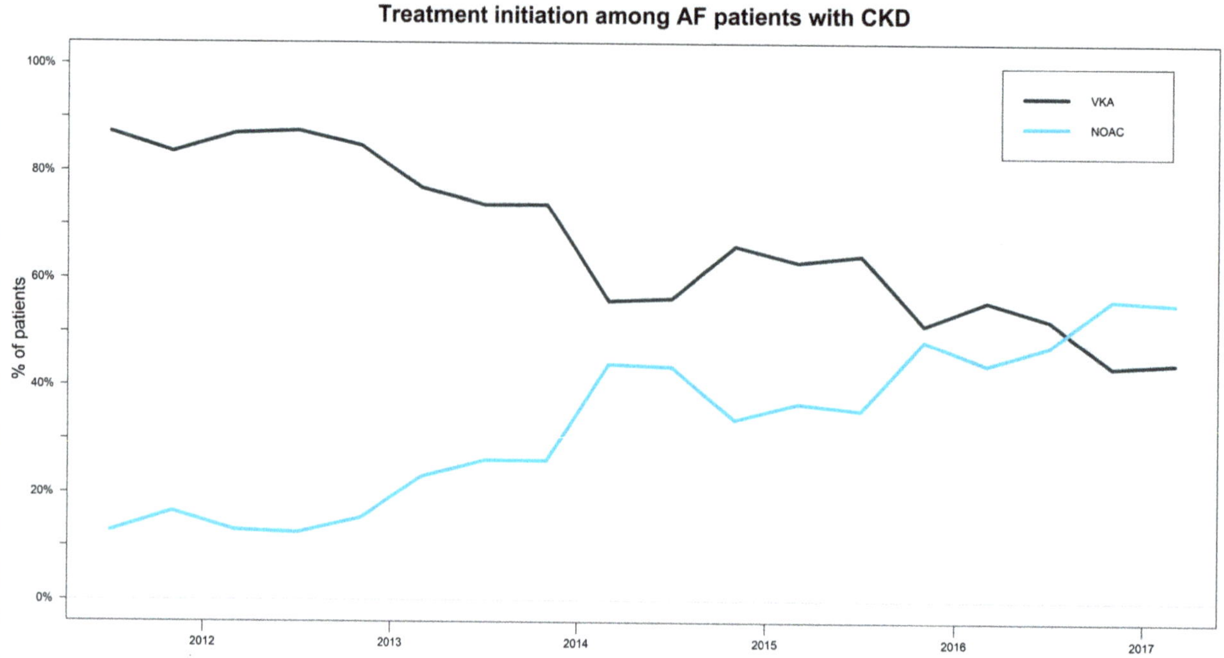

Fig. 2 Treatment initiation patterns of OAC among AF patients with CKD between 2011 and 2017

a significantly lower risk of major bleeding compared to VKA, 3) there were no significant differences in occurrence of MI, stroke/TE or all cause-mortality with NOAC or VKA, and 4) the risk of death of any cause was almost one in four following the first year after treatment initiation among patients with AF and CKD.

Efficacy and safety of NOAC vs. VKA

VKAs such as warfarin are cleared mainly by the liver [6], but there are no randomized studies comparing warfarin with placebo among patients with moderate or severe CKD. Warfarin-related hemorrhages increase as creatinine clearance decreases [6], and time in therapeutic range with warfarin is often low among patients with CKD, which indicates a need for careful monitoring of these patients or a different treatment strategy. Our study investigated three types of alternatives to warfarin for stroke prevention among AF patients with CKD, specifically apixaban, rivaroxaban, and dabigatran. The renal clearance of these NOACs ranges between 27 and 80% [6], and there has been concern that reduced drug-elimination due to impaired GFR may lead to an increased risk of severe bleedings. Our findings do not support this concern, as NOAC treatment was associated with lower risk of major bleeding compared to VKA in our real-world population.

When looking at NOACs as a group for patients with non-end-stage CKD, our results seem to be in accordance with the existing literature. Effect of renal function on efficacy and safety of NOACs vs. warfarin among AF patients has been studied in the pivotal phase 3 NOAC trials, both among patients with reduced renal function at baseline and among patients with deteriorating renal function after randomization. A review collected data from randomized trials to compare safety and efficacy of NOACs vs. warfarin. They included patients with AF and an eGFR between 15 and 60 mL/min (CKD stages 3 and 4) and concluded that NOACs were just as effective in stroke prevention without causing an increased risk of bleeding, primarily reflecting AF patients with CKD stage 3 [18]. Apixaban reduced risk of bleeding compared to warfarin

Table 2 Risk of events according to OAC in AF and CKD patients

	Number of events	Hazard ratio (95%CI)	Standardized absolute risk (95%CI)
Stroke/thromboembolism			
VKA	21	1.00 (reference)	2.4% (1.4–3.5%)
NOAC	11	0.83 (0.39–1.78)	2.0% (0.8–3.3%)
Major bleeding			
VKA	55	1.00 (reference)	5.9% (4.4–7.5%)
NOAC	15	0.47 (0.26–0.84)	2.8% (1.5–4.3%)
Myocardial infarction			
VKA	22	1.00 (reference)	2.5% (1.5–3.7%)
NOAC	7	0.45 (0.18–1.11)	1.1% (0.4–2.2%)
All-cause mortality			
VKA	183	1.00 (reference)	23.3% (20.2–26.3%)
NOAC	106	0.99 (0.77–1.26)	23.2% (19.4–27.0%)

Abbreviations: *CI* Confidence interval, *OAC* Oral anticoagulation, *VKA* Vitamin-K antagonist, *NOAC* Nonvitamin K oral anticoagulant

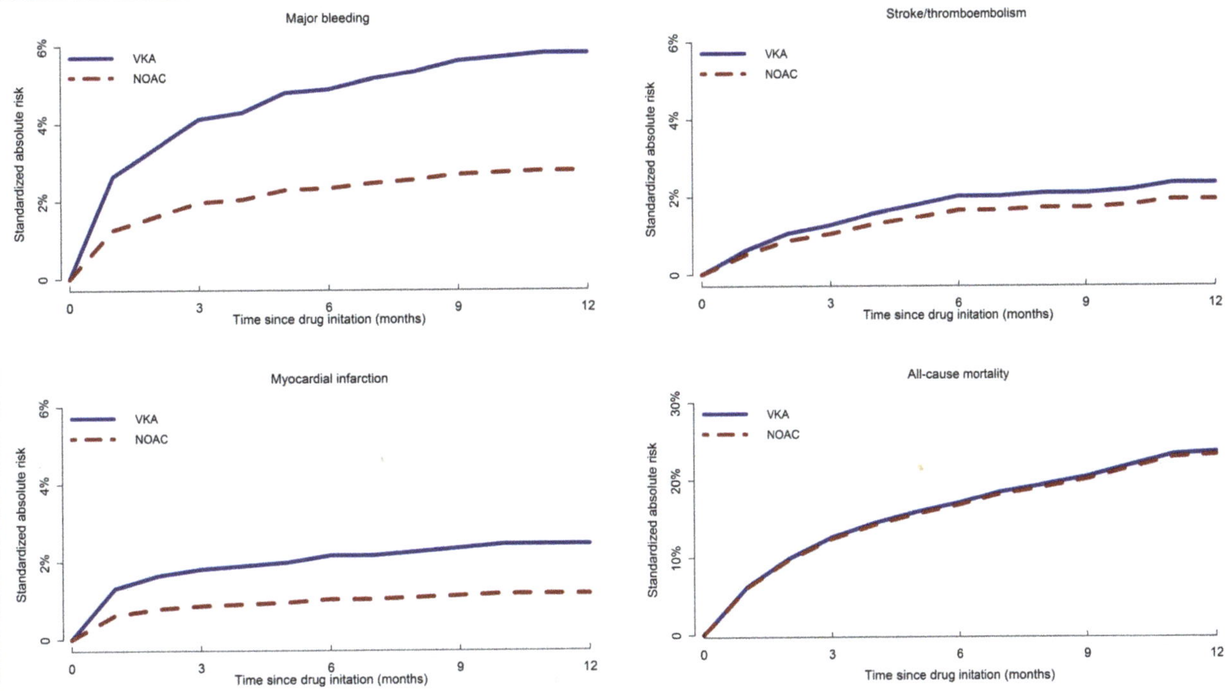

Fig. 3 Standardized absolute risk of event according to type of OAC among AF patients with CKD

to a higher degree among patients with renal impairment than among patients without renal impairment [7], whereas dabigatran showed a significant trend towards higher risk of major bleeding compared to warfarin among patients with low renal function compared to patients with normal renal function [19]. The majority of the NOAC-treated study population (54.7%) received apixaban and a minority (14.5%) received dabigatran. Our results could reflect the beneficial safety of apixaban among patients with non-end-stage CKD found in trials, and extent the findings from randomized trials to a real-world population.

Combining results from clinical trials and real-world studies, NOACs have been shown to be as effective and safe as VKA in AF patients with non-end-stage CKD, but all pivotal phase 3 randomized trials of warfarin vs. NOACs excluded patients with creatinine clearance (CrCl) < 25 mL/min [6], and it is still not known whether NOACs are safe and effective among these patients. Renal function data was not available in our study cohort, but by looking at the kidney function distribution in the comparable population, it seems reasonable to hypothesize that a significant number of patients in our study also had an eGFR< 30 mL/min/1.73m^2.

Our findings are also consistent with a recent retrospective cohort study that investigated the effects of apixaban vs. warfarin in AF patients on dialysis [20]. They found comparable efficacy with apixaban compared to warfarin, but a significant 28% lower HR of major bleeding. The results regarding use of NOACs in end-stage CKD are still hypothesis-generating and ongoing randomized controlled trials will most likely settle these questions.

Mortality risk among patients with AF and CKD

In our study the risk of dying of any cause was almost one in four following the first year of treatment with NOAC or VKA treatment (23.2% vs. 23.3%). Albeit we found a significant lower risk of major bleeding and a non-significant lower risk of stroke/TE and MI associated with NOAC compared with VKA treatment, we did not find any signals of a difference in the mortality risk between NOAC vs. VKA treatment. Regarding the outcomes major bleeding, stroke/TE, and MI, we adjusted the statistical analyses for competing risk of death, but it was evident that our study population comprising of vulnerable and frail individuals with AF and CKD had a high risk of death, independent of NOAC or VKA treatment. In routine clinical practice, the high risk of death must be taken into account when managing patients with AF and CKD. Also, even though there was no difference in mortality risk between NOAC and VKA treatment, a major bleeding event can be serious or fatal, and the different bleeding risk with NOAC and VKA treatment (2.8% vs. 5.9%) may be considered when initiating OAC treatment among these frail patients.

Strengths and limitations

In this nationwide study, we included all Danish citizens on OAC with a diagnosis of both AF and CKD. We had practically no loss to follow-up and, the majority of diagnoses used, such as CKD, AF, bleedings, and strokes have been validated [8]. Our study also had limitations. Since creatinine values were not available after 2011, we could not determine exact renal function for each patient in our study, which is a major limitation. For the same reason, we were not able to determine if there was a difference in renal function between the NOAC and VKA population. This limitation makes it difficult to draw definite conclusions from our study and the limitation should be kept in mind when interpreting the results. Further research should be done to establish the effects of NOAC in patients with AF and CKD. The number of study outcomes for stroke/TE and MI are limited, making it uncertain to draw conclusions for those outcomes. The limited number of study outcomes also prevented us from analyzing differences among the individual types of NOAC. Some comorbidities such as hypertension and diabetes were identified using data on prescribed medicine, meaning that patients treated solely with lifestyle changes could not be identified. Also, it is possible that some patients were prescribed this medication for other unknown purposes. The hypertension definition, however, has previously been validated with a positive predictive value of 80% [14]. Finally, as in most observational studies, unmeasured confounding or residual confounding affecting our results cannot be ruled out.

Conclusion

In patients with AF and CKD, treatment with NOAC was associated with a lower risk of major bleeding, but was not found to either decrease or increase the risk of stroke/TE, MI or all-cause mortality compared to VKA.

Abbreviations

ADPi: Adenosine phosphate receptor inhibitors; AF: Atrial fibrillation; ATC: Anatomical therapeutic chemical classification; CI: Confidence interval; CKD: Chronic kidney disease; CrCl: Creatinine clearance; eGFR: Estimated glomerular filtration rate; GFR: Glomerular filtration rate; HR: Hazard ratio; ICD: International classification of disease; INR: International normalized ratio; IQR: Interquartile range; MI: Myocardial infarction; NOAC: Non-vitamin K antagonist oral anticoagulant; NSAID: Non-steroid anti-inflammatory drugs; OAC: Oral anticoagulation; RASi: Renin angiotensin system inhibitor; TE: Thromboembolism; VKA: Vitamin-K antagonists

Acknowledgements

Not applicable.

Authors' contributions

All authors contributed substantially to the conception and design, acquisition of data, or analysis, and interpretation of data. EKL drafted the article, and LS, NC, A-LK, JBO, CTP, GG, and ANB revised it critically for important intellectual content. All authors gave final approval of the version to be submitted.

Author details

[1]Department of Cardiology, Copenhagen University Hospital Herlev and Gentofte, Kildegaardsvej 28, 2900 Hellerup, Denmark. [2]Department of Clinical Medicine, Faculty of Health and Medical Sciences, University of Copenhagen, Blegdamsvej 3B, 2200 Copenhagen, Denmark. [3]The Danish Heart Foundation, Vognmagergade 7, 1120 Copenhagen, Denmark. [4]Department of Nephrology, Copenhagen University Hospital Rigshospitalet, Blegdamsvej 9, 2100 Copenhagen, Denmark. [5]Department of Health Science and Technology, Aalborg University, Fredrik Bajers vej 5, 9100 Aalborg, Denmark. [6]The National Institute of Public Health, University of Southern Denmark, Studiestræde 6, 1455 Copenhagen, Denmark.

References

1. Alonso A, Lopez FL, Matsushita K, et al. Chronic kidney disease is associated with the incidence of atrial fibrillation: the atherosclerosis risk in communities (ARIC) study. Circulation. 2011;123(25):2946–53. https://doi.org/10.1161/CIRCULATIONAHA.111.020982.
2. Olesen JB, Lip GYH, Kamper A-L, et al. Stroke and bleeding in atrial fibrillation with chronic kidney disease. N Engl J Med. 2012;367(7):625–35. https://doi.org/10.1056/NEJMoa1105594.
3. Bonde AN, Lip GYH, Kamper AL, et al. Net clinical benefit of antithrombotic therapy in patients with atrial fibrillation and chronic kidney disease: a nationwide observational cohort study. J Am Coll Cardiol. 2014;64(23):2471–82. https://doi.org/10.1016/j.jacc.2014.09.051.
4. Hart RG, Pearce LA, Aguilar MI. Meta-analysis : antithrombotic therapy to prevent stroke in patients who have Nonvalvular atrial fibrillation. Ann Intern Med. 2007;146:857–67. https://doi.org/10.7326/0003-4819-146-12-200706190-00007.
5. Ruff CT, Giugliano RP, Braunwald E, et al. Comparison of the efficacy and safety of new oral anticoagulants with warfarin in patients with atrial fibrillation: a meta-analysis of randomised trials. Lancet. 2014;383(9921):955–62. https://doi.org/10.1016/S0140-6736(13)62343-0.
6. Chan KE, Giugliano RP, Patel MR, et al. Nonvitamin K anticoagulant agents in patients with advanced chronic kidney disease or on dialysis with AF. J Am Coll Cardiol. 2016;67(24):2888–99. https://doi.org/10.1016/j.jacc.2016.02.082.
7. Lanas F, Hanna M, Wallentin L, et al. Efficacy of apixaban when compared with warfarin in relation to renal function in patients with atrial fibrillation: insights from the ARISTOTLE trial. Eur Heart J. 2012;33(22):2821–30. https://doi.org/10.1093/eurheartj/ehs274.
8. Schmidt M, Schmidt SAJ, Sandegaard JL, Ehrenstein V, Pedersen L, Sørensen HT. The Danish national patient registry: a review of content, data quality, and research potential. Clin Epidemiol. 2015;7:449–90. https://doi.org/10.2147/CLEP.S91125.
9. Kildemoes HW, Sorensen HT, Hallas J. The Danish National Prescription Registry. Scand J Public Health. 2011;39(7 Suppl):38–41. https://doi.org/10.1177/1403494810394717.
10. Pedersen CB. The Danish civil registration system. Scand J Public Health. 2011;39(September 2010):22–5. https://doi.org/10.1177/1403494810387965.
11. De CR, Camm AJ. What is "valvular" atrial fibrillation? A reappraisal. Eur Heart J. 2014;35(47):3328–35. https://doi.org/10.1093/eurheartj/ehu352.
12. Levey AS, Stevens LA, Schmid CH, et al. A new equation to estimate glomerular filtration rate. Ann Intern Med. 2009;150(9):604–12. https://doi.org/10.7326/0003-4819-150-9-200905050-00006.
13. Staerk L, Lip GY, Olesen JB, et al. Stroke and recurrent haemorrhage associated with antithrombotic treatment after gastrointestinal bleeding in patients with atrial fibrillation: nationwide cohort study. Bmj. 2015;351:h5876. https://doi.org/10.1136/bmj.h5876.
14. Olesen JB, Lip GY, Hansen ML, et al. Validation of risk stratification schemes for predicting stroke and thromboembolism in patients with atrial fibrillation: nationwide cohort study. Bmj. 2011;342:d124. https://doi.org/10.1136/bmj.d124.
15. Cunningham A, Stein CM, Chung CP, Daugherty JR, Smalley WE, Ray WA. An automated database case definition for serious bleeding related to oral anticoagulant use. Pharmacoepidemiol Drug Saf. 2011;20(6):560–6. https://doi.org/10.1002/pds.2109.
16. Krarup LH, Boysen G, Janjua H, Prescott E, Truelsen T. Validity of stroke diagnoses in a National Register of patients. Neuroepidemiology. 2007;28(3):150–4. https://doi.org/10.1159/000102143.

17. R Core Team. R: a language and environment for statistical computingR Found Stat Comput. URL http://www.R-project.org/. 2017. Vienna: R Foundation for Statistical Computing; 2017.
18. Kimachi M, Furukawa TA, Kimachi K, Goto Y, Fukuma S, Fukuhara S. Direct oral anticoagulants versus warfarin for preventing stroke and systemic embolic events among atrial fibrillation patients with chronic kidney disease. Cochrane Database Syst Rev. 2017;11. https://doi.org/10.1002/14651858.CD011373.
19. Hohnloser SH, Eikelboom JW, Ezekowitz MD, et al. Efficacy and safety of dabigatran compared with warfarin in relation to baseline renal function in patients with atrial fibrillation. Circulation. 2013;129(9):961–70. https://doi.org/10.1161/circulationaha.113.003628.
20. Siontis KC, Zhang X, Eckard A, et al. Outcomes associated with Apixaban use in end-stage kidney disease patients with atrial fibrillation in the United States. Circulation. 2018;138(15):1519–29. https://doi.org/10.1161/CIRCULATIONAHA.118.035418.

10

Multiple trauma including pelvic fracture with multiple arterial embolization

Takahito Miyake[1*], Hideshi Okada[1], Norihide Kanda[1], Fuminori Yamaji[1], Haruka Okamoto[1], Hiroaki Ushikoshi[1], Kei Noguchi[2], Hiroyuki Tomita[2], Shozo Yoshida[1] and Shinji Ogura[1]

Abstract

Background: Pelvic fracture with high energy trauma has a high mortality rate, especially in men. In addition, severe multiple trauma, major hemorrhage, and administration of red blood cells predict mortality in elderly patients with pelvic fracture. We herein report a rare case in which multiple arterial embolization occurred after pelvic fracture.

Case presentation: An 83-year-old male cyclist was transported to our hospital after being struck by a car. On arrival, he was diagnosed with multiple trauma, including rib fractures with hemothorax, lumbar fractures of the transverse process, and injuries in the right acetabulum, left adrenal gland, and liver. He underwent massive transfusion and transcatheter arterial embolization due to extravasation from the right superior gluteal artery and left adrenal gland. On the second day, owing to right lower leg ischemia, serum creatinine kinase and myoglobin levels were markedly elevated from the reference value; hence, a right above-knee amputation was performed 12 h after the accident. However, both protein levels remained high after amputation, resulting in acute renal injury, which was treated via hemodiafiltration on hospital day 3. In addition, sustained low efficiency hemodialysis and plasma exchange were performed on hospital day 4. Despite these treatments, the patient's hemodynamics did not improve, and he died on hospital day 8. The autopsy revealed necropsy of the iliopsoas muscles and the digestive tract.

Conclusions: The causes of the patient's death were considered to be persistent rhabdomyolysis and severe hypotension due to iliopsoas necrosis and peritonitis due to digestive tract necrosis. Multiple arterial embolization caused by consumption coagulopathy associated with multiple trauma may account for severe outcomes in this case.

Keywords: Pelvic fracture, Arterial embolization, Coagulopathy, Fibrinolysis

Background

Pelvic fracture with high energy trauma has a high mortality rate, especially in men [1]. It can lead to severe multiple trauma or major hemorrhage and may require administration of red blood cells [1, 2]. Its complications, especially in elderly patients, include diabetes, cardiovascular disease, and chronic kidney disease. Chronic diseases sometimes influence the treatment outcomes of patients with pelvic fracture.

We report a case in which multiple arterial embolization occurred after pelvic fracture.

Case presentation

An 83-year-old male cyclist was transported to our hospital 3 h after being struck by a car. He had a medical history of acute coronary syndrome, for which he had received aspirin. His vital signs were as follows: respiratory rate, 22 breaths per minute; oxygen saturation on room air, 96%; systolic blood pressure, 125/67 mmHg;

* Correspondence: tmiyake@gifu-u.ac.jp
[1]Advanced Critical Care Center, Gifu University Hospital, 1-1 Yanagido, Gifu 501-1194, Japan
Full list of author information is available at the end of the article

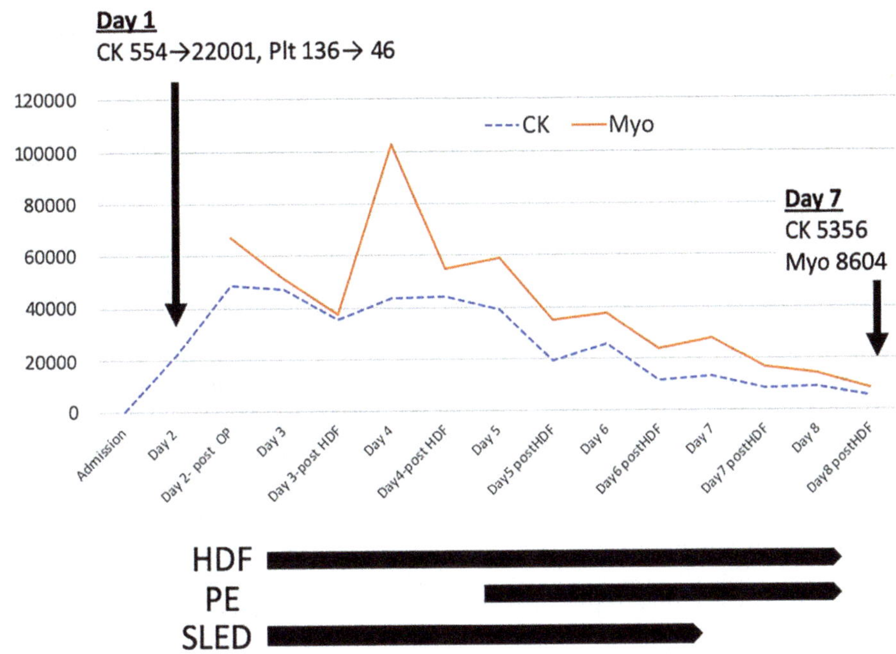

Fig. 1 Time course of treatment. CK: creatinine kinase, Plt: platelets, Myo, myoglobin, OP: operation, HDF: hemodiafiltration, PE: plasma exchange, SLED: sustained low efficiency dialysis

pulse rate, 88 beats per minute; and body temperature, 36.1 °C. On arrival, the patient was diagnosed with multiple trauma including multiple rib fractures with hemothorax, lumbar fractures of the transverse process, and injuries to the right acetabulum, left adrenal gland, left ureter, and liver. His injury severity score was 22. We also found extravasation of the right superior gluteal artery and left adrenal gland. He was treated with a transfusion of red blood cells, plasma, and platelets, and the ratio was 12:26:30. He was administered 1 g tranexamic acid when he presented in this very acute phase. He also required intubation and transcatheter arterial embolization. There was no evidence of stray embolic materials in the external iliac arteries.

Laboratory test findings (normal ranges in parentheses) on hospital day 1 were as follows: white blood cells, 22.08 × 10^3/μL (3.3–8.6 × 10^3/μL); serum hemoglobin, 11.7 mg/dL (13.7–16.8 mg/dL); platelets, 136 × 10^3/μL (158–348 × 10^3/μL); serum creatinine kinase, 554 U/L (59–248 U/L); partial thromboplastin time, 37 s (25–38 s); international normalized ratio, 1.31; prothrombin time, 15.8 s (9.8–12.1 s); serum fibrinogen, 107 mg/dL (200–400 mg/dL); serum D-dimer, 155.7 μg/mL (< 1.0 μg/mL); and serum fibrinogen degradation products (FDPs), 305.5 μg/mL (< 5.0 μg/mL). Arterial blood gas analysis revealed mixed acidosis with an elevated serum lactate level (FiO_2, 0.50; pH, 7.186; pCO_2, 49.4 Torr; pO_2 89.1 Torr; HCO_3, 18 mmol/L; base excess −9.9; and lactate, 44 mg/dL).

On hospital day 2, his serum creatine kinase was markedly elevated to 22,001 U/L. At this point, his right lower leg showed signs of ischemia including lack of a pulse in the right dorsalis pedis artery and cold leg. These findings suggested arterial embolization in the right thigh, and 12 h after the injury, a right above-knee amputation was performed. However, the serum creatine kinase, 44,229 U/L and myoglobin levels, 67,000 ng/mL (< 110 ng/mL), were markedly high, even after the amputation, resulting in acute renal injury, which was treated via hemodiafiltration on hospital day 3. In addition, sustained low efficiency hemodialysis and plasma exchange were performed on hospital day 4. The patient's creatine kinase and myoglobin levels are plotted against his time course of treatment in Fig. 1. The changes in hemostatic markers are shown in Table 1. Despite these treatments, the patient's hemodynamic status, which included high catecholamine levels, did not improve, and he died on hospital day 8.

Table 1 The Changes of Hemostatic Markers

		Day1	Day2	Day 3	Day 4	Day 5
Platelet	X10^3/μL	136	46	50	65	74
PT time	seconds	37	34.7	32	32.2	68.6
PT-INR		1.31	1.12	1	1.06	1.07
Fibrinogen	mg/dL	107	199	351	553	692
FDP	μg/mL	305.5	116.9	70.6	26.5	17.7
D-dimer	μg/mL	155.7	51.3	28.7	10.2	5.7

PT: Prothrombin, PT-INR: Prothrombin-International normalized ratio, FDP: Fibrin degradation product

We performed an autopsy 50 h after his death. A general postmortem examination revealed pulmonary effusion and alveolar hemorrhage. Liver congestion and fatty liver were observed, as were tubular necrosis and nephrosclerosis in the kidney. There were also signs of arteriosclerosis in the aorta, common iliac arteries, and coronary artery. Necrosis was more severe in the left iliopsoas muscle than in the right. Although there were no signs of initial gastrointestinal tract injuries, irreversible ischemic colitis from the descending colon to the rectum was revealed. In addition, thrombus becoming fibrinized was detected in multiple organs in the autopsy specimen. (Fig. 2). This may explain the peritonitis and sepsis, which resulted in continued shock and ultimately death.

Discussion and conclusions

The causes of the patient's death were considered to be persistent rhabdomyolysis due to iliopsoas necrosis and severe hypotension due to digestive tract necrosis and consequent septic peritonitis. Multiple arterial embolization caused by the consumption coagulopathy arising from multiple trauma may be responsible for the acute embolization in multiple organs (Fig. 3). This was a very rare case of pelvic fracture with multiple organ thrombosis.

Acute traumatic coagulopathy develops in approximately 25% of severely injured individuals. It increases the risk of multisystem organ failure substantially and the risk of death by 4-fold [3]. "Trauma-induced coagulopathy" is the generally accepted term for coagulopathy in patients with massive hemorrhage and multiple trauma. It is an endogenous hypocoagulable state unrelated to iatrogenic events. Several distinct but highly integrated mechanisms have been implicated in trauma-induced coagulopathy; these include those resulting in protein C activation, platelet deficit and dysfunction, endothelial involvement, and microparticle formation. The protein C pathway is considered to be a key mediator of traumatic-induced coagulopathy owing to its downstream effects (e.g., thrombin diversion, deactivation of coagulation factors, and derepression of fibrinolysis) [4].

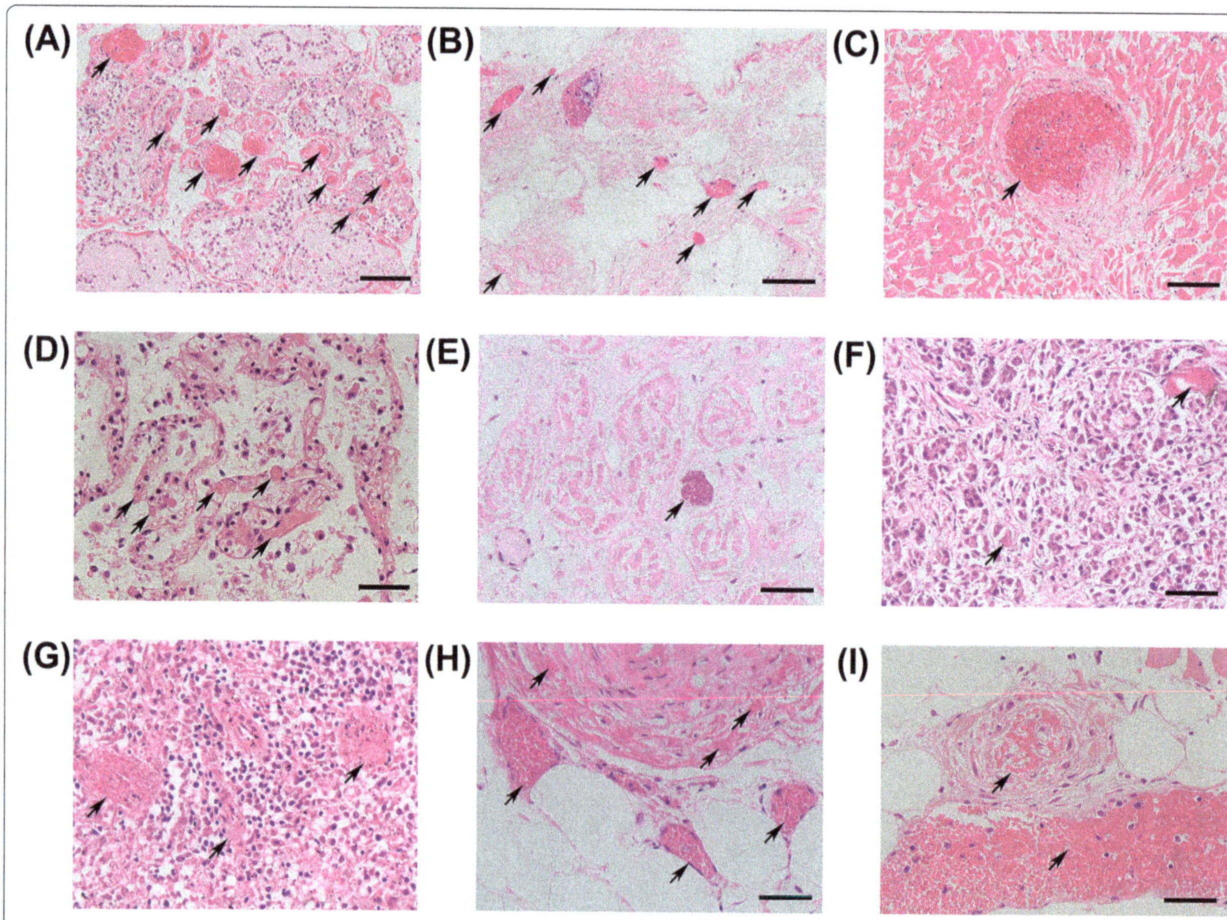

Fig. 2 Multiple organ embolization in autopsy specimen. **a** Thyroid gland, (**b**) Left ventricle with atrophic change, (**c**) Left ventricle with non-atrophic change, (**d**) Lung, (**e**) Kidney, (**f**) Pancreas, (**g**) Spleen, (**h**) Transversus Colon, (**i**) Iliopsoas muscle. Arrows indicate thrombus becoming fibrinized. Bars: 50 μm

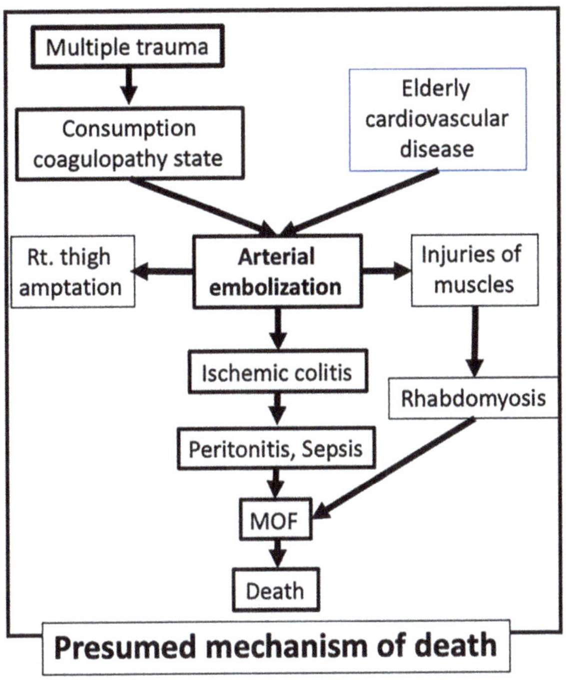

Fig. 3 Presumed mechanism of death. Rt: right, MOF: multiple organ failure

The most common coagulation disorder after injury is hypercoagulability. Hypofibrinolysis or fibrinolysis shutdown is an integral component of postinjury hypercoagulopathy [3] and is an independent predictor of adverse outcomes after injury, including death [4]. In a previous report, persistent fibrinolysis shutdown independently predicted mortality with an odds ratio of 8.48 ($p = 0.022$) [3]. In reports in which fibrinolysis phenotypes were stratified via thromboelastography (TEG), fibrinolysis shutdown, physiologic fibrinolysis, and hyperfibrinolysis were defined as ≤0.8%, 0.8–3, and > 3% of clot lysis at 30 min after maximum clot strength, respectively [5, 6]. Recent prospective cohort data suggest that severely injured patients more often present with fibrinolysis shutdown than with hyperfibrinolysis or physiologic fibrinolysis [4].

Although we did not perform TEG in this case, we presume that fibrinolysis shutdown was the key mechanism underlying the multiple embolisms in our case. Initial D-dimer and FDP levels were significantly higher than normal, which suggests that the hypercoagulopathy in the very acute phase.

It is very rare that multiple arterial embolization occurs after multiple trauma, while acute traumatic coagulopathy is often caused by severe trauma. This might explain why severely injured patients could die from hypercoagulation during the acute phase of trauma and may account for the paucity of pelvic fractures associated with multiple arterial emboli in the literature.

In this case, the aspirin the patient had received for acute coronary syndrome may have affected the hypercoagulation status, allowing him to survive for several days after his injury. Several studies have investigated the effect of antiplatelet therapy in critically ill patients [7–11]. In the meta-analysis by Du et al., aspirin administration correlated with low hospital mortality rates in critically ill patients [7]. Several observational studies have shown that antiplatelet drugs reduce the levels of biomarkers such as C-reactive protein, soluble CD62P, CD54, and pro-inflammatory cytokines [8, 9]. Other studies, however, found that aspirin did not lower mortality [10] or improve acute lung injuries [11]. Although the role of aspirin in patients in intensive care units is still controversial, aspirin may have some benefits in acute critical situations [7–11].

In addition, this patient's serum lactate was classified as mixed acidosis. It was thought that the mechanism of this acidosis was mainly from a diabetic origin such as microangiopathy.

In conclusion, severe inflammation caused by severe trauma might cause multiple emboli by inducing hypercoagulation and can sometimes leads to death. Our study highlights the confirmation of hypercoagulation in autopsy. In such situations, anticoagulation therapy may prolong survival.

Abbreviations

FDP: fibrinogen degradation products; TEG: thromboelastography

Acknowledgements

The authors would like to thank the paramedical crews for the data.

Authors' contributions

TM, NK, SY, FY, HO, HU, and SO treated the patient. KN and HT performed pathological autopsy. TM wrote the manuscript. H. Okada revised and edited the manuscript. All authors have read and approved the final manuscript.

Author details

[1]Advanced Critical Care Center, Gifu University Hospital, 1-1 Yanagido, Gifu 501-1194, Japan. [2]Department of Tumor Pathology, Gifu University School of Medicine, 1-1 Yanagido, Gifu 501-1194, Japan.

References

1. Holstein JH, Culemann U, Pohlemann T, working group mortality in pelvic fracture patients. What are predictors of mortality in patients with pelvic fractures? Clin Orthop Relat Res. 2012;470:2090–7.

2. Ojodu I, Pohlemann T, Hopp S, Rollmann MF, Holstein JH, Herath SC. Predictors of mortality for complex fractures of the pelvic ring in the elderly: a twelve-year review from a German level I trauma center. Injury. 2015;46: 1996–8.
3. Meizoso JP, Karcutskie CA, Ray JJ, Namias N, Schulman CI, Proctor KG. Persistent fibrinolysis shutdown is associated with increased mortality in severely injured trauma patients. J Am Coll Surg. 2017;224:575–82.
4. Cohen MJ, Christie SA. Coagulopathy of trauma. Crit Care Clin. 2017;33(1): 101–18.
5. Moore HB, Moore EE, Gonzalez E, Chapman MP, Chin TL, Silliman CC, et al. Hyperfibrinolysis, physiologic fibrinolysis, and fibrinolysis shutdown: the spectrum of postinjury fibrinolysis and relevance to antifibrinolytic therapy. J Trauma Acute Care Surg. 2014;77:811–7 discussion, 817.
6. Moore HB, Moore EE, Liras IN, Gonzalez E, Harvin JA, Holcomb JB, et al. Acute fibrinolysis shutdown after injury occurs frequently and increases mortality: a multicenter evaluation of 2,540 severely injured patients. J Am Coll Surg. 2016;222:347–55.
7. Du F, Jiang P, He S, Song D, Xu F. Antiplatelet therapy for critically ill patients: a pairwise and Bayesian network meta-analysis. Shock. 2018;49: 616–24.
8. Al Harbi SA, Tamim HM, Al-Dorzi HM, Sadat M, Arabi YM. Association between aspirin therapy and the outcome in critically ill patients: a nested cohort study. BMC Pharmacol Toxicol. 2016;17:5.
9. Muhlestein JB. Effect of antiplatelet therapy on inflammatory markers in atherothrombotic patients. Thromb Haemost. 2010;103:71–82.
10. Weber M, Bhatt DL, Brennan DM, Hankey GJ, Steinhubl SR, Johnston SC, et al. High-sensitivity C-reactive protein and clopidogrel treatment in patients at high risk of cardiovascular events: a substudy from the CHARISMA trial. Heart. 2011;97:626–31.
11. Kor DJ, Erlich J, Gong MN, Malinchoc M, Carter RE, Gajic O, et al. Lung injury prevention study I: association of prehospitalization aspirin therapy and acute lung injury: results of a multicenter international observational study of at-risk patients. Crit Care Med. 2011;39:2393–400.

Changes in anticoagulant prescription in Dutch patients with recent-onset atrial fibrillation: Observations from the GARFIELD-AF registry

J. Seelig[1,16*], F. W. A. Verheugt[2], M. E. W. Hemels[1,3], L. Illingworth[4], A. Lucassen[5], H. Adriaansen[6], M. C. M. Bongaerts[7], M. Pieterse[8], J. P. R. Herrman[9], P. Hoogslag[10], W. Hermans[11], B. E. Groenemeijer[12], L. V. A. Boersma[13,14], K. Pieper[4,15], H. ten Cate[16,17] and on behalf of the GARFIELD-AF Investigators

Abstract

Background: For the improvement of AF care, it is important to gain insight into current anticoagulation prescription practices and guideline adherence. This report focuses on the largest Dutch subset of AF-patients, derived from the GARFIELD-AF registry.

Methods: Across 35 countries worldwide, patients with newly diagnosed 'non-valvular' atrial fibrillation (AF) with at least one additional risk factor for stroke were included. Dutch patients were enrolled in five, independent, consecutive cohorts from 2010 until 2016.

Results: In the Netherlands, 1189 AF-patients were enrolled. The prescription of non-vitamin K antagonist oral anticoagulants (NOAC) has increased sharply, and as per 2016, more patients were initiated on NOACs instead of vitamin K antagonists (VKA). In patients with a class I recommendation for anticoagulation, only 7.5% compared to 30.0% globally received no anticoagulation. Reasons for withholding anticoagulation in these patients were unfortunately often unclear.

Conclusions: The data from the GARFIELD-AF registry shows the rapidly changing anticoagulation preference of Dutch physicians in newly diagnosed AF. Adherence to European AF guidelines in terms of anticoagulant regimen would appear to be appropriate. In absence of structured follow up of AF patients on NOAC, the impact of these rapid practice changes in anticoagulation prescription in the Netherlands remains to be established.

Keywords: Atrial fibrillation, Anticoagulants, Inappropriate prescribing, Guideline adherence

Introduction

In the Netherlands, AF patients on vitamin K antagonist (VKA) therapy are routinely managed by specialized anticoagulation clinics. Back in 2012, a report from the health council of the Netherlands endorsed the careful introduction of NOACs, given the lack of real-world data, absence of specific antidotes, and a substantial risk of non-compliance due to a lack of monitoring [1]. These factors resulted in a slower uptake of a NOAC-based approach in comparison to other countries [2]. However, based on a decision-related Markov model, it was recently calculated that an increase in NOAC prescription in the Netherlands would result in higher quality of life [3, 4]. Moreover, given the increasing real-world data on NOACs versus VKAs, uncertainties about

* Correspondence: jseelig@rijnstate.nl
[1]Department of Cardiology, Rijnstate, Wagnerlaan 55, 6815 AD Arnhem, the Netherlands
[16]Department of Internal Medicine, Cardiovascular Research Institute Maastricht, Maastricht, the Netherlands
Full list of author information is available at the end of the article

the safety of these drugs have diminished. It is therefore important to monitor anticoagulation prescription trends for AF in the Netherlands, which are currently unknown. This will give insights in how to further improve our AF care.

Moreover, insight in adherence to AF-guidelines could also help to improve AF care. In the Netherlands, it is estimated that the prevalence of AF is around 2.0% in 2020, expected to increase to 3.2% by 2050 [5]. In parallel, in subjects with AF the ischemic stroke rate will rise, primarily due to ageing and an increase in patients with multiple morbidities [5–7]. This increases health-care related costs and reduces quality of life. To minimise these aspects, it is important that AF guidelines are adhered to, as non-adherence is associated with increased ischemic stroke and mortality rates [8, 9].

This report expands on previously published Dutch GARFIELD-AF data, and demonstrates changes in antithrombotic treatment initiation in newly diagnosed AF in the Netherlands [2]. We compare the results with the global GARFIELD-AF cohort, and with recommendations of the most recent European AF-guidelines [10].

Methods

Design

GARFIELD-AF was a multicentre, prospective registry of patients with recent onset non-valvular AF from over a 1000 centres in 35 countries worldwide. Globally, the recruitment of patients started in December 2009 and was completed in August 2016. In the Netherlands, patients were included as of November 2010. Patients were enrolled in five independent, consecutive cohorts 1) 2009–2011, 2) 2011–2012, 3) 2013–2014, 4) 2014–2015, and 5) 2015–2016. Data used was from the October 2017 dataset.

Population

Patients diagnosed with 'non-valvular' AF within the previous 6 weeks, aged ≥18 years, and with at least one investigator-determined risk factor for stroke were considered eligible for inclusion. Patients were excluded if; 1) follow-up with a physician was considered unlikely or impossible, 2) there was a potentially reversible, transient cause for AF, or 3) they were enrolled in a controlled clinical trial. For each country, a sufficient number of investigator sites from different care settings were identified.

Data collection

All data were made anonymous and were imported to a secured, electronic case report form (eCRF), which was designed by Dendrite Clinical Systems Ltd. (Henley-on-Thames, UK). Oversight of operations and data management were done by the Thrombosis Research Institute [TRI] (London, UK), which is the sponsor and coordinating centre. A detailed description of the methods can be found elsewhere [11]. The study is registered at ClinicalTrials.gov (unique identifier: NCT01090362).

At inclusion, patient characteristics such as demographics, medical history, vital signs, and type and dose of antithrombotic therapy were recorded. Amongst others, the components of the CHA_2DS_2-VASc stroke risk and HAS-BLED bleeding risk scores were collected [12, 13]. Vascular disease was defined as the combination of a history of acute coronary syndrome with peripheral and/or coronary artery disease. Chronic kidney disease was defined according to the National Kidney Foundation guidelines [14].

Analysis

Continuous variables are expressed as means with standard deviation, and categorical variables as frequencies with percentages. Data from patients with missing values were not removed from the analyses. Follow-up data was not analysed due to a lack of power. Similarly, no *p*-values were calculated. Data analysis was performed with SAS Enterprise Guide, version 7.1 (SAS Institute Inc., Cary, NC, USA).

Results

Population

In the Netherlands, 1189 out of 52,014 patients (2.3%) were enrolled across 16 sites. Across the different Dutch cohorts were 199 (1; 2009–2011), 410 (2; 2011–2012), 357 (3; 2013–2014), 155 (4; 2014–2015), and 161 (5; 2015–2016) AF patients enrolled. In the Netherlands and worldwide, the mean age was 70.7 and 69.7 years, respectively, and 42.4% compared to 44.2% of patients were female. At baseline, hypertension (65.5%), hypercholesterolemia (36.0%), diabetes mellitus (20.0%), and coronary artery disease (18.7%) were the most common comorbidities in the Dutch cohorts. The mean CHA_2DS_2-VASc (3.1 vs. 3.2) and HAS-BLED (1.4 vs. 1.4) scores were comparable between the Dutch and overall cohort, respectively. Compared to the worldwide cohort, more patients were enrolled in cardiology departments (90.2% vs. 65.7%) in the Dutch subset. Further baseline characteristics are described in Table 1.

Changes in antithrombotic therapy

Of all 35 participating countries, the percentage of patients on oral anticoagulation at AF diagnosis was on average highest in the Netherlands (89.9%). A comparison of anticoagulation treatments (with or without concomitant antiplatelet therapy) between the five different cohorts, demonstrates a rise in the prescription of NOACs from 0.0 to 60.9% over the years (Fig. 1). Conversely, a decrease in VKA prescription from 88.9 to

Table 1 Baseline characteristics of Dutch and all included patients

	Netherlands (*N* = 1189)	World (*N* = 52,014)
Female sex, n (%)	504 (42.4)	22,987 (44.2)
Age, mean (sd)	70.7 (9.9)	69.7 (11.5)
< 65, n (%)	311 (26.2)	15,693 (30.2)
65–74, n (%)	426 (35.8)	16,948 (32.6)
≥ 75, n (%)	452 (38.0)	19,373 (37.2)
BMI (kg/m^2), mean (sd)	28.5 (5.3)	27.8 (5.7)
Congestive Heart Failure, n (%)	82 (6.9)	10,397 (20.0)
Hypertension, n (%)	775 (65.5)	39,585 (76.3)
Diabetes Mellitus, n (%)	238 (20.0)	11,540 (22.2)
Stroke/TIA, n (%)	137 (11.5)	5954 (11.4)
PE or DVT, n (%)	22 (1.9)	1356 (2.6)
Coronary artery disease, n (%)	222 (18.7)	11,232 (21.6)
Acute Coronary Syndrome, n (%)	166 (14.0)	4895 (9.5)
Chronic Kidney Disease, n (%)		
None	377 (31.7)	23,919 (46.0)
Stages 1 to 2	629 (52.9)	16,508 (31.7)
Stages 3 to 5	118 (9.9)	5373 (10.3)
History of Bleeding, n (%)	25 (2.1)	1317 (2.5)
Hypercholesterolemia, n (%)	422 (36.0)	20,940 (41.6)
CHA_2DS_2-VASc	3.1 (1.5)	3.2 (1.6)
HAS-BLED	1.4 (0.9)	1.4 (0.9)
Care Setting Speciality at Diagnosis, n (%)		
Cardiology	1097 (92.3)	34,165 (65.7)
Other Hospital Departments	30 (2.5)	10,434 (20.1)
Primary Care / General Practice	62 (5.2)	7410 (14.2)

BMI Body mass index, *VKA* Vitamin K antagonist, *NOAC* Non-vitamin K antagonist oral anticoagulant, *TIA* Transient ischaemic attack, *PE* Pulmonary embolism, *DVT* Deep venous thrombosis

34.8% was observed. The proportion of patients on antiplatelet monotherapy decreased from 6.1 to 2.5%. The proportion of patients not treated with antithrombotics reduced from 5.1 to 1.9%. In the most recent cohort, the proportion of patients on antiplatelet drug therapy (2.5%) or no antithrombotic therapy (1.9%) were both the lowest of all participating countries.

Guideline adherence and reasons of not prescribing anticoagulation

Within the Dutch cohorts, 79.4% of patients had a class I recommendation for anticoagulation for stroke prevention in AF (i.e. males CHA_2DS_2-VASc ≥2, and females CHA_2DS_2-VASc ≥3), according to ESC guidelines [10]. Of these patients, 92.5% were treated with oral anticoagulants, 4.8% with antiplatelet monotherapy, and 2.7% with no antithrombotic therapy (Fig. 2). In patients with a class IIa recommendation for stroke prevention in AF (i.e. males CHA_2DS_2-VASc = 1, and females CHA_2DS_2-VASc = 2; 16.6% of patients), 82.6% of patients were treated with oral anticoagulants, 6.0% with antiplatelet monotherapy, and 11.4% with no antithrombotic therapy. In patients with no increased stroke risk according to the CHA_2DS_2-VASc score (i.e. males CHA_2DS_2-VASc = 0, and females CHA_2DS_2-VASc = 1; 4.0% of patients), 66.7% were treated with oral anticoagulants, 4.4% with antiplatelet monotherapy, and 28.9% with no antithrombotic therapy.

Unfortunately, in the Netherlands and worldwide, reasons for not prescribing anticoagulants in males with CHA_2DS_2-VASc ≥2, and females with CHA_2DS_2-VASc ≥3 were often recorded as 'unknown' (28.8% versus 39.4%) or 'other' (40.9 versus 22.4%). Excluding these options, the most frequently reported reasons in the Netherlands were 'low stroke risk' (12.1%) and 'bleeding risk' (7.6%) (Table 2). In the worldwide cohort, excluding Dutch patients, the main reasons for not prescribing anticoagulants were 'patient refusal' (7.8%), 'bleeding risk' (7.2%), 'low risk of stroke' (5.8%) and 'already taking antiplatelet drugs for other medical condition' (5.4%).

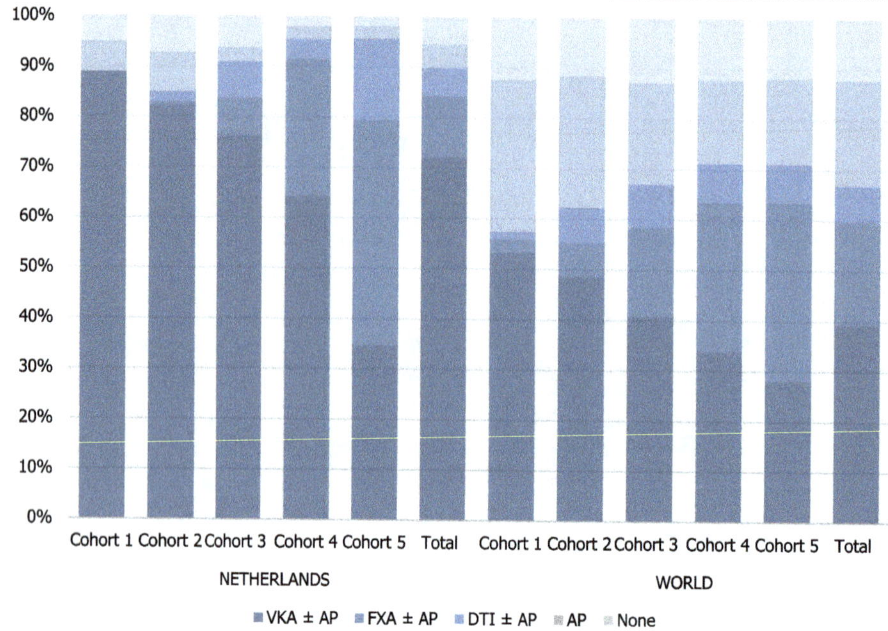

Fig. 1 Treatment at diagnosis by cohort (*VKA* Vitamin K Antagonist, *AP* Antiplatelet Drug, *FXa* Factor Xa inhibitor, *DTI* Direct Thrombin Inhibitor)

Discussion

GARFIELD-AF was the largest, worldwide, prospective registry of newly diagnosed AF patients. In the Netherlands, 1189 patients were enrolled, making it the largest Dutch AF-cohort available to date. This manuscript provides a unique insight in the rapid changes in anticoagulation management of novel AF, which had not been described since the introduction of the NOACs in the Netherlands. The comparison between NOAC uptake rates in the Netherlands vs other countries is important, as this could have influenced the quality of Dutch AF care. Future studies will have to analyze how these differences have impacted the safety and efficacy of AF care. Moreover, this is the first report describing nationwide adherence to AF-guidelines in the Netherlands and explores reasons for withholding oral

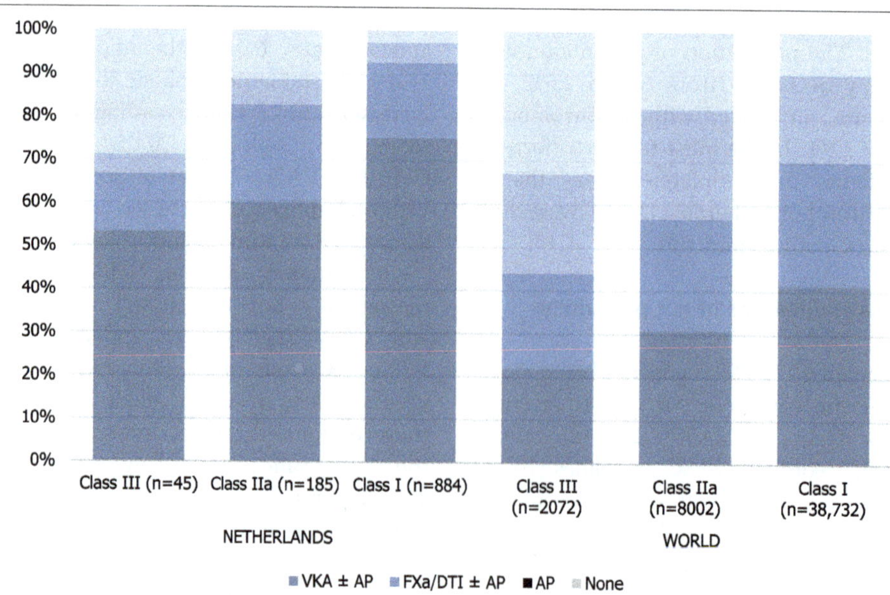

Fig. 2 Treatment at diagnosis by Class of Recommendation according to the 2016 ESC AF-guidelines (*VKA* Vitamin K Antagonist, *AP* Antiplatelet Drug, *FXa* Factor Xa inhibitor, *DTI* Direct Thrombin Inhibitor)

Table 2 Main reasons anticoagulant not used in males with CHA_2DS_2-VASc ≥2, and females with CHA_2DS_2-VASc ≥3

	Netherlands (*N* = 66) n (%)	World (*N* = 11,630) n (%)
Alcohol abuse	0 (0.0)	48 (0.4)
Already taking AP for other medical condition	3 (4.5)	628 (5.4)
Patient refusal	1 (1.5)	911 (7.8)
Previous bleeding event	2 (3.0)	211 (1.8)
Taking medication contraindicated or cautioned for use with OAC	1 (1.5)	78 (0.7)
Other	12 (18.2)	1682 (14.5)
Unknown	19 (28.8)	4588 (39.4)
Physician's choice	28 (42.4)	3484 (30.0)
Bleeding risk	5 (7.6)	836 (7.2)
Concern over patient compliance	0 (0.0)	412 (3.5)
Guideline recommendation	0 (0.0)	237 (2.0)
Fall risk	0 (0.0)	401 (3.4)
Low risk of stroke	8 (12.1)	677 (5.8)
Other	15 (22.7)	921 (7.9)

AP Antiplatelet drug, *OAC* Oral anticoagulation

anticoagulation in AF, which gives insight in how to further improve our AF care. Also, this country-specific evaluation may also be of help in improving care when comparisons are made with anticoagulant management in other countries.

In the Netherlands, there was initially a slow shift to more NOAC prescription, compared to the rest of the world. However, as of 2014–2015, the anticoagulation landscape has changed rapidly, resulting in more newly diagnosed AF patients treated with NOACs than VKA as of 2016. Our findings were comparable to a recent analysis of anticoagulant pharmaceutical dispensing data of naïve oral anticoagulation starters for any indication in the Netherlands [15]. A possible explanation for this initial slow shift could be that there is a well-organized system of specialized anticoagulation clinics in the Netherlands. In these clinics, the monitoring of compliance and complications of VKA treatment through regularly scheduled follow-up checks is aimed at minimising risks accompanying VKA treatment. Although NOACs have been repeatedly shown to be at least as effective and safe as VKAs in both randomized controlled trials and real-world data, a lack of monitoring could have contributed to a hesitation to shift to a more NOAC based approach. This is not unreasonable, as without a regular check of factors such as renal function, weight or age, patients are often (± 10% in two recent Dutch AF-studies), treated with a too high or too low NOAC dose [16, 17]. Moreover, early discontinuation of (N)OAC treatment can be as high as 50% at 6 months in certain patient groups [18, 19]. Frequently mentioned reasons for early discontinuation are (minor) bleeding, other anticoagulant-related side-effects, and a lack of the perceived need for anticoagulation [20, 21]. Therefore, international guidelines recommend structured follow up of patients on NOACs (ESC) including assessment of adherence to medication, complications, interactions and regular (at least annual, but more often on indication) check on renal and liver functions [22]. For the Netherlands, much of this burden will come down on the shoulders of prescribers (mainly cardiologists) and for the long term on general practitioners. It is imperative that, based on national guidance documents such as the "Landelijke Standaard Ketenzorg Antistolling" (LSKA) 2.0 and the updated "Landelijke Transmurale Afspraak antistolling" (manuscript in preparation), the chronic care for patients on NOACs becomes well organized [23].

In GARFIELD-AF, the Netherlands had the highest proportion of patients on oral anticoagulation at diagnosis (89.9%). In the most recent cohort, Dutch patients had the lowest proportions of antiplatelet monotherapy (2.5%) or no antithrombotic therapy (1.9%). For patients with a class I recommendation for anticoagulation, 7.5% of patients were undertreated according to the ESC guidelines [10]. Compared to the worldwide cohort (30.0%), this proportion is relatively low. In patients with a class III recommendation for anticoagulation (i.e. CHA_2DS_2-VASc 0 in males, CHA_2DS_2-VASc 1 in females), the proportion of patients on anticoagulation is high (66.6%) [10]. Although there is no chronic indication for anticoagulation in these patients, the guideline recommends at least 3 weeks of pre-treatment with oral anticoagulation in late cardioversions [10]. The ACWAS trial showed that in patients with recent-onset (< 36 h) AF, a delayed cardioversion strategy led to spontaneous conversion within 48 h in 69% of patients [24]. In a post-hoc analysis of the ACUTE trial, nearly 50% of patients with pre-existing AF of ≤1 week had a spontaneous cardioversion [25]. It is likely that patients with recent-onset, newly diagnosed AF without risk factors for stroke are often 'overtreated' with anticoagulation, given the high rate of spontaneous conversion. It is therefore worth researching if there are possibilities to safely limit the prescription of anticoagulants in these patients.

Although the proportion of undertreated patients in the Netherlands was relatively low, there is still room for improvement. In GARFIELD-AF, main reasons for not prescribing anticoagulants in patients with a class I recommendation for anticoagulation for stroke prevention in AF were often not clear. In patients with a clear recorded reason for withholding anticoagulation, a 'low risk of stroke' (12.1%) and 'bleeding risk' (7.6%) were the most common reasons in the Dutch cohort. Depicting patients with 2 or more non-sex related stroke risk factors as having a 'low risk of stroke' is contradictory, and the precise reasoning behind it is unknown. It would be valuable to gather more information on reasons for withholding anticoagulation, and to evaluate if withholding anticoagulation in these groups is a safe approach.

This study has several limitations. As described before, the high proportion of patients included in Dutch cardiology departments limits the external validity of this study to nationwide clinical practice. Moreover, the number of patients was too low, and the mean follow-up was too short, to relate major adverse events to CHA_2DS_2-VASc scores or changes in anticoagulant treatment practices. Moreover, reasons for not prescribing anticoagulants were extracted from the medical records and were not confirmed by the prescribing physician, and a large proportion of reasons could not be recorded and were classified as 'other'. Further research without these limitations is necessary. DUTCH-AF (Dutch trial register number: NL7464) is a largescale registration of newly diagnosed AF-patients in the Netherlands, which does not have these limitations and could provide further answers [26].

Conclusion

The data from the GARFIELD-AF registry shows the rapidly changing anticoagulation preference of Dutch physicians in newly diagnosed AF. Adherence to European AF guidelines in terms of anticoagulant regimen would appear to be appropriate. In absence of structured follow up of AF patients on NOAC, the impact of these rapid practice changes in anticoagulation prescription in the Netherlands and in relation to other countries remains to be established.

Acknowledgements

The authors of this paper extend their gratitude to the following 17 Dutch medical centres whose participation in GARFIELD-AF has made the results presented in this paper possible: Diaconessenhuis Meppel; Gelre Ziekenhuizen, Apeldoorn; Huisartsenpraktijk Buiks; Medisch Centrum Alkmaar; St. Jans Gasthuis; Onze Lieve Vrouwe Gasthuis; Sint Franciscus Gasthuis; Slingeland Ziekenhuis; St. Antonius Ziekenhuis; St. Elisabeth Ziekenhuis; Stichting Cardiologie Amsterdam; Stichting Trombosedienst Maastricht; Trombosedienst Apeldoorn-Zutphen; Trombosedienst Ziekenhuis Rivierenland; WECOR; Ziekenhuis Bethesda. Our special thanks go out to the following individuals, who supervised patient enrolment in their capacity as principal investigators: Dr. Henk Adriaansen (Research Interne Trombosedienst Geneeskunde Apeldoorn-Zutphen), Dr. L.V.A. Boersma (St. Antonius Ziekenhuis), Dr. Maarten Bongaerts (Trombosedienst Ziekenhuis Rivierenland), Dr. Paul Buiks (Huisartsenpraktijk Buiks), Dr. B.E. Groenemeijer (Research Interne Gelre Ziekenhuizen, Apeldoorn), Dr. Coen Guldener (WECOR), Dr. Walter Hermans (St. Elisabeth Ziekenhuis), Dr. Jean-Paul Herrman (Onze Lieve Vrouwe Gasthuis), Dr. P.A.M. Hoogslag (Diaconessenhuis Meppel), Dr. Andreas Lucassen (St. Jans Gasthuis), Dr. P.R. Nierop (Sint Fransiscus Gasthuis), Dr. M.G.C. Pieterse (Stichting Cardiologie Amsterdam), Dr. W. Terpstra (Slingeland Ziekenhuis), and Dr. S.H.K. The (Treant Zorggroep – locatie Bethesda).

Author's contributions

JS, MEWH and HTC drafted the report. HTC was the national study coordinator. KP and LI analysed the data. All authors contributed, read and approved the final manuscript.

Author details

[1]Department of Cardiology, Rijnstate, Wagnerlaan 55, 6815 AD Arnhem, the Netherlands. [2]Department of Cardiology, Onze Lieve Vrouwe Gasthuis (OLVG), Amsterdam, the Netherlands. [3]Radboud University Medical Centre, Nijmegen, the Netherlands. [4]Thrombosis Research Institute, London, UK. [5]Department of Cardiology, St. Jans Gasthuis, Weert, the Netherlands. [6]Anticoagulation Clinic, Gelre Ziekenhuizen, Apeldoorn-Zutphen, the Netherlands. [7]Anticoagulation Clinic, Ziekenhuis Rivierenland, Tiel, the Netherlands. [8]Stichting Cardiologie Amsterdam, Amsterdam, the Netherlands. [9]Department of Cardiology, Onze Lieve Vrouwe Gasthuis, Amsterdam, the Netherlands. [10]Department of Cardiology, Isala Diaconessenhuis, Meppel, the Netherlands. [11]Department of Cardiology, Elisabeth-TweeSteden Ziekenhuis, Tilburg, the Netherlands. [12]Department of Cardiology, Gelre Ziekenhuizen, Apeldoorn, the Netherlands. [13]Department of Cardiology, St. Antonius Ziekenhuis, Nieuwegein, the Netherlands. [14]Department of Cardiology, Amsterdam UMC, Amsterdam, the Netherlands. [15]Duke Clinical Research Institute, Durham, USA. [16]Department of Internal Medicine, Cardiovascular Research Institute Maastricht, Maastricht, the Netherlands. [17]Anticoagulation Clinic Maastricht, Maastricht, the Netherlands.

References

1. New anticoagulants - a well-dosed introduction: health council of the Netherlands. 2012. [27-Jan-2020]. Available from: https://www.healthcouncil.nl/documents/advisory-reports/2012/05/15/new-anticoagulants-a-well-dosed-introduction.
2. Ten Cate V, Ten Cate H, Verheugt FW. The Global Anticoagulant Registry in the FIELD-Atrial Fibrillation (GARFIELD-AF): exploring the changes in anticoagulant practice in patients with non-valvular atrial fibrillation in the Netherlands. Neth Hear J. 2016;24(10):574–80.
3. Verhoef TI, Redekop WK, Hasrat F, de Boer A, Maitland-van der Zee AH. Cost effectiveness of new oral anticoagulants for stroke prevention in patients with atrial fibrillation in two different European healthcare settings. Am J Cardiovasc Drugs. 2014;14(6):451–62.
4. de Jong LA, Koops M, Gout-Zwart JJ, et al. Trends in direct oral anticoagulant (DOAC) use: health benefits and patient preference. Neth J Med. 2018;76(10):426–30.

5. Krijthe BP, Kunst A, Benjamin EJ, et al. Projections on the number of individuals with atrial fibrillation in the European Union, from 2000 to 2060. Eur Heart J. 2013;34(35):2746–51.
6. Son MK, Lim NK, Park HY. Trend of prevalence of atrial fibrillation and use of oral anticoagulation therapy in patients with atrial fibrillation in South Korea (2002-2013). J Epidemiol. 2018;28(2):81–7.
7. Proietti M, Laroche C, Nieuwlaat R, et al. Increased burden of comorbidities and risk of cardiovascular death in atrial fibrillation patients in Europe over ten years: a comparison between EORP-AF pilot and EHS-AF registries. Eur J Intern Med. 2018;55:28–34.
8. Kirchhof P, Ammentorp B, Darius H, et al. Management of atrial fibrillation in seven European countries after the publication of the 2010 ESC guidelines on atrial fibrillation: primary results of the PREvention oF thromboemolic events--European Registry in Atrial Fibrillation (PREFER in AF). Europace. 2014;16(1):6–14.
9. Gorin L, Fauchier L, Nonin E, et al. Prognosis and guideline-adherent antithrombotic treatment in patients with atrial fibrillation and atrial flutter: implications of undertreatment and overtreatment in real-life clinical practice; the Loire Valley atrial fibrillation project. Chest. 2011;140(4):911–7.
10. Kirchhof P, Benussi S, Kotecha D, et al. 2016 ESC guidelines for the management of atrial fibrillation developed in collaboration with EACTS. Eur Heart J. 2016;37(38):2893–962.
11. Kakkar AK, Mueller I, Bassand JP, et al. International longitudinal registry of patients with atrial fibrillation at risk of stroke: Global Anticoagulant Registry in the FIELD (GARFIELD). Am Heart J. 2012;163(1):13–9 e1.
12. Lip GY, Nieuwlaat R, Pisters R, Lane DA, Crijns HJ. Refining clinical risk stratification for predicting stroke and thromboembolism in atrial fibrillation using a novel risk factor-based approach: the euro heart survey on atrial fibrillation. Chest. 2010;137(2):263–72.
13. Pisters R, Lane DA, Nieuwlaat R, et al. A novel user-friendly score (HAS-BLED) to assess 1-year risk of major bleeding in patients with atrial fibrillation: the euro heart survey. Chest. 2010;138(5):1093–100.
14. International Society of Nephrology. Chapter 2: definition, identification, and prediction of CKD progression. Kidney Int Suppl. 2013;3(1):63–72.
15. van den Heuvel JM, Hovels AM, Buller HR, et al. NOACs replace VKA as preferred oral anticoagulant among new patients: a drug utilization study in 560 pharmacies in the Netherlands. Thromb J. 2018;16:7.
16. Jacobs MS, van Hulst M, Campmans Z, Tieleman RG. Inappropriate non-vitamin K antagonist oral anticoagulants prescriptions: be cautious with dose reductions. Neth Hear J. 2019;27(7–8):371–7.
17. Pisters R, van Vugt SPG, Brouwer MA, et al. Real-life use of rivaroxaban in the Netherlands: data from the Xarelto for prevention of stroke in patients with atrial fibrillation (XANTUS) registry. Neth Hear J. 2017;25(10):551–8.
18. Harper P, Pollock D, Stephens M. Dabigatran persistence and adherence in New Zealand: a nationwide retrospective observational study. BMJ Open. 2018;8(4):e020212.
19. Beyer-Westendorf J, Ehlken B, Evers T. Real-world persistence and adherence to oral anticoagulation for stroke risk reduction in patients with atrial fibrillation. Europace. 2016;18(8):1150–7.
20. Lip GYH, Pan X, Kamble S, et al. Discontinuation risk comparison among 'real-world' newly anticoagulated atrial fibrillation patients: Apixaban, warfarin, dabigatran, or rivaroxaban. PLoS One. 2018;13(4):e0195950.
21. McHorney CA, Spain CV. Frequency of and reasons for medication non-fulfillment and non-persistence among American adults with chronic disease in 2008. Health Expect. 2011;14(3):307–20.
22. Steffel J, Verhamme P, Potpara TS, et al. The 2018 European heart rhythm association practical guide on the use of non-vitamin K antagonist oral anticoagulants in patients with atrial fibrillation. Eur Heart J. 2018;39(16): 1330–93.
23. Landelijke Standaard Keten Antistolling 2.0 (LSKA): KNMP. Available from: https://www.knmp.nl/patientenzorg/samenwerking/landelijke-standaard-keten-antistolling-2-0-lska. Accessed 27 Jan 2020.
24. Pluymaekers N, Dudink E, Luermans J, et al. Early or delayed cardioversion in recent-onset atrial fibrillation. N Engl J Med. 2019;380(16):1499–508.
25. Tejan-Sie SA, Murray RD, Black IW, et al. Spontaneous conversion of patients with atrial fibrillation scheduled for electrical cardioversion: an ACUTE trial ancillary study. J Am Coll Cardiol. 2003;42(9):1638–43.
26. DUTCH-AF Registry. Prospective evaluation of dosing and adherence of anticoagulant treatment and the risk for bleeding in atrial fibrillation: Nederlands Trial Register. 2018. [12-07-2019]. Available from: https://www.trialregister.nl/trial/7464.

12

Therapeutic duplication of anticoagulants: A retrospective study of frequency and consequences in a tertiary referral hospital

Ramin Rahmanzade[1,2], Francisco Cabrera Diaz[3], Claudia Zaugg[3], Philipp Schuetz[1,4] and Ali Reza Salili[4,5*]

Abstract

Background: Anticoagulants are commonly prescribed in medical practices and could be of significant harm in the case of medication errors. We conducted a retrospective observational study to determine the frequency and consequences of the therapeutic duplication of anticoagulants (TDA). As a secondary objective, we aimed to determine the characteristics of the population in which TDA occurs.

Methods: We conducted a retrospective observational study among admitted patients who concomitantly received at least two anticoagulants from August 2017 to August 2018.

Results: A total of 107 patients with TDA are included in the research. The patients with TDA have a mean age of 73. The TDA population has a high rate of associated comorbidities with 69% of patients having arterial hypertonia, 40% with chronic kidney disease, 26% with a history of malignancy, and 20.5% with a history of stroke. More than 65% of patients were under anticoagulation before admission, mostly due to atrial fibrillation. The TDA occurred in more than 95% of cases in the first week or the last week of hospitalization. Patients had a high risk of bleeding prior to the TDA-event with about 62.5% of TDA patients having a HAS-BLED score at least 3. A total of 8 patients showed a significant Hemoglobin (Hb)-drop of at least 10 g/L within 24 h after TDA-event. Two patients had a new or worsened hematuria following TDA-event.

Conclusion: TDA occurred in 0.8% of patients who were under anticoagulation and in 6.7% of patients who received direct oral anticoagulants (DOACs). TDA led in about 7.4% of cases to hemoglobin-relevant bleeding. The old patients with significant comorbidities and a high HAS-BLED score were mainly affected. The female gender and presence of anemia independently predicted the occurrence of bleeding following TDA.

Keywords: Anticoagulants, Therapeutic duplication, Medication error, Direct Oral anticoagulants, HAS-BLED score

Background

According to the Institute for Safe Medication Practice (ISMP), anticoagulants (AC) are among the high-alert-medications leading to serious complications in the case of medication errors [1]. Anticoagulants are involved in about 8.3% of all medication errors in admitted patients [2]. In a 5-year retrospective study, the investigators showed that medication errors are the cause by about 48.8% of all AC-associated adverse drug events [3]. The AC-related medication errors have been mainly classified according to the phase of medication process in which the error occurred into the following types: prescribing, transcribing, dispensing, administering, and monitoring errors. A cross-sectional study on the nation-wide drug safety system of the Netherland showed that the

* Correspondence: Salili@hin.ch
[4]Department of Internal Medicine, Kantonsspital Aarau, 5000 Aarau, Switzerland
[5]Department of Internal Medicine, Clinical Pharmacology, Kantonsspital Aarau, Tellstrasse 51, CH-5001 Aarau, Switzerland
Full list of author information is available at the end of the article

anticoagulant-related medication errors occurred mainly in the prescribing phase and in about 56% of errors low-molecular-weight heparins (LMWH) were involved [2, 4]. Consistently, a similar descriptive study on the Danish patient safety database reported that the adverse medication incidents occurred mainly in the prescribing phase, and in about 80% of incidents the errors occurred at the time of sector changes (at the admission, at the discharge or ward changes) [5]. Among different error subtypes occurring in the prescribing phase, the wrong dosage and drug omission have been repeatedly found as the most common subtype [3, 6, 7]. The affected patients are usually polymorbid patients in advanced ages having already high risk of bleeding [8]. Accordingly, the AC-related errors occur in about 78% of cases in adults older than 60 and in about one third in patients older than 80 years old [9]. The studies reporting the frequency and complications of AK-related medication errors have been mostly done based on data reported to in-house or nation-wide drug safety systems [2, 3, 5–10]. As many of these errors could not be detected or have not been reported by medical staff, the studies based on them could not reliably show the incidence of errors, the affected population and the rate of consequences. For different categories of ACs, the rates of adverse complications, especially the minor and major bleeding rates, have been well studied and reported [11–15].

Therapeutic duplication is defined a concomitant prescription of two medications of the same class. Except for a few indications, the therapeutic duplication of anticoagulants is contraindicated. The therapeutic duplication of anticoagulants (TDA) as a medication error occurring in the prescribing phase was rarely studied [15–17]. As the primary objective, we aimed to comprehensively investigate the incidence, affected population, and complications of TDA in patients admitted to our tertiary referral hospital from 21.08.2017 to 20.08.2018.

Material & Methods

Patient population

This study is a retrospective observational study conducted at Kantonsspital Aarau (KSA), a 450-bed tertiary care facility providing medical and surgical care in all general and specialized medical disciplines. The hospital utilizes an electronic health care system (KISIM) gathering all medical and personal data of patients who received medical care, either in the in-patient or out-patient settings. For the present study, we retrospectively collected data of admitted patients who received AC, using a computerized export from KISIM. From this population, all patients admitted to our in-patient wards, from 21.08.2017–20.08.2018, who concomitantly received at least two ACs were included in the study. Subsequently, the following patients have been excluded: [1] patients who received the anticoagulants separately, no overlapping prescription [2] patients treated with the combination of heparin/ heparin analogs and vitamin-K antagonists (VKA) [3] patients receiving heparin/ heparin analogs based on a weekly schedule such as patients with hemodialysis [4] patients receiving low-dose alteplase (= < 4 mg) as a catheter locking solution [5] patients treated with the combination of VKA and direct oral anticoagulants (DOACs) during bridging from DOACs to VKA [6] patients who explicitly refused participation in research.

Data collection and data analysis

An export (in CSV format) from the data storage system of our hospital (KISIM) has been done. The mentioned export includes all patients who received at least one anticoagulant during their admission from 21.08.2017–20.08.2018 and contains the following patient data: Patient number, admission number, sex, date of birth, dates of admission and discharge, the name of prescribed anticoagulants and the dates of prescriptions.

According to the inclusion and exclusion criteria, the patients with a contraindicated combination of at least two AK have finally been separated. Then, the corresponding demographic, biochemical and clinical parameters, listed ahead were extracted from KISIM: The para-clinical data including pre-error and post-error Hb (Hemoglobin of patient before and after TDA), pre-error and post-error urine analysis, the indication of anticoagulant therapy, the presence of underlying diseases predisposing, the patients to develop complications under TDA: arterial hypertension, renal insufficiency and its stage according to glomerular filtration rate (GFR), hepatic failure, active gastrointestinal ulcers, concomitant therapy with NSAIDs or thrombocyte-aggregation inhibitors, documented bleeding diathesis (abnormal post-operative bleeding, hematoma or bleeding under anticoagulant therapy in therapeutic dosage, known hemorrhagic disorders, etc.), prior anticoagulant therapy, history of recent trauma or upcoming surgery and finally the HAS-BLED score prior to the error. Subsequently, all demographic, biochemical and, clinical data of patients included in the study have been analyzed statistically. The variables have consequently been analyzed descriptively. A Hemoglobin-drop of more than 10 g/L within 24 h has been considered as significant Hb-drop. The patients have been divided into two groups with and without significant Hb-drop. Then, unpaired T-test and chi-square have been done for continuous and binary variables, respectively, to study whether there is a significant difference in variables between two patients groups. Finally, to determine the effect of different risk factors on the bleeding following

TDA, we performed univariate and multivariate logistic regression.

Results

Among 14,812 patients who received at least one anticoagulant, a total of 1663 patients met the inclusion criteria and have been treated with at least two anticoagulants during the one-year period. Among them, a total of 1542 patients met the first five exclusion criteria. Then, a total of 121 patients with a contraindicated combinative therapy of at least two anticoagulants (TDA) have been found. Out of them, 14 patients refused to participate and the rest 107 patients were included in the study (Fig. 1).

Characteristics of the patient population

Age and sex

From 107 patients included in the study, 72 patients (67.3%) were male and 35 patients (32.7%) were female. The minimum and maximum of age were 35 and 97 years, respectively. The mean age of patients was 73.67 (95% CI, 71.2–76.1) and 78.5% of the population was older than 65 years.

Duration of hospitalization

The minimum and maximum of the hospitalization duration were 2 and 72, respectively. The mean duration of hospitalization was 11.52 (CI 95%: 9.69–13.3).

HAS-BLED score related comorbidities

Table 1 shows the HAS-BLED score related parameters and comorbidities. Out of 107 patients, 74 patients (69.1%) had long-term arterial hypertension. Among them, 38 patients had uncontrolled arterial hypertension with systolic blood pressure higher than 160 mmHg. A high proportion of patients (40.1%) had a known chronic kidney disease prior to admission. Among them, 69.7% had stage 3 or higher. Abnormal liver function has been found in 7 patients; 2 with liver cirrhosis and 5 others with hepathopaties. 20.5% of patients had a history of stroke. A relevant co-medication with other drugs (aspirin, clopidogrel, NSAIDs etc.) or alcohol consumption predisposing patients to bleed has been found in 28 Out of 107 patients (26%). Seventeen patients were known to have a bleeding diathesis or a history of major bleeding (See Table 1).

HAS-BLED score of the patient population

The minimum and maximum values of the HAS-BLED scores were 0 and 7, respectively. The mean HAS-BLED score was 2.37 (CI95% 2.14–2.60) with a median value of 3.

Further bleeding-related parameters and comorbidities

Table 2 shows further bleeding-related comorbidities, which have not been considered by the HAS-BLED score. Most of the patients with TDA (66.3%) were on

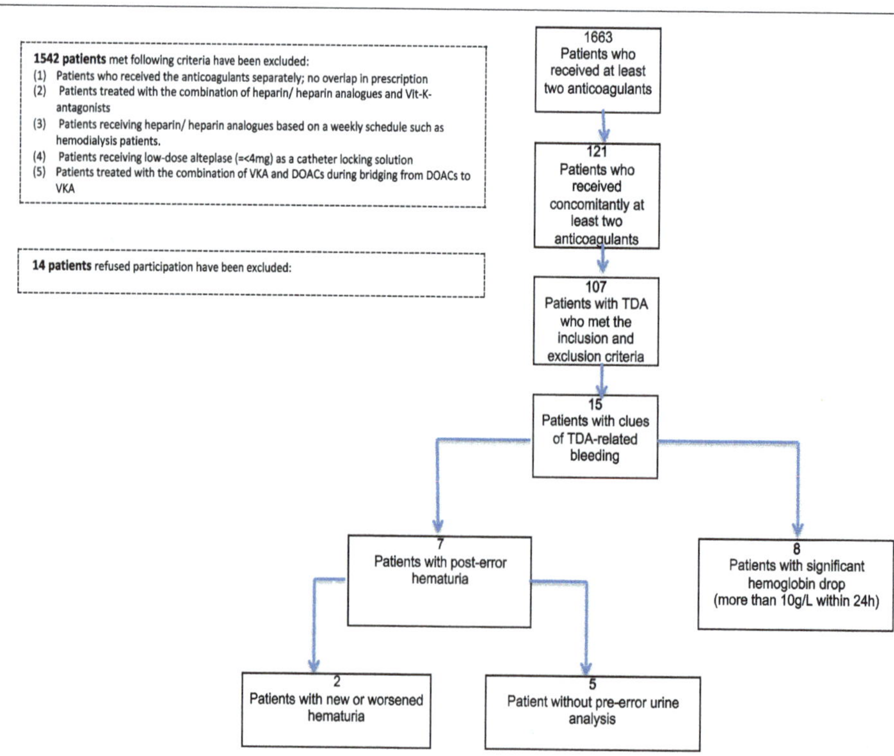

Fig. 1 Diagram of patient recruitment

Table 1 A summary of the HAS-BLED score, the modified score and the related parameters

Characteristics	TDA-patients (whole population)	Patients with Hb-relevant bleeding	Patients without Hb-relevant bleeding
Number	107	8	99
Hypertension	38	4	34
Abnormal liver function	7	1	6
Abnormal renal function	43	3	40
Stroke	22	3	19
Labile INR	15	2	13
Bleeding diathesis	17	2	15
Age > 65 years	84	7	77
Drug/alcohol concomitantly	28	2	26
Prior anticoagulation	71	7	64
HAS-BLED score	2.37 (CI95% 2.14–2.60)	3.00 (CI95% 2.26–3.74)	2.32 (CI95% 2.08–2.55)
Modified HAS-BLED score	3.04 (CI95% 2.8–3.28)	3.87 (CI95% 3.09–4.66)	2.97 (CI95% 2.72–3.22)

anticoagulation prior to admission. Diabetes mellitus has been found in 24 patients (22.4%). A high proportion of patients had a history of malignancy (26.1%). Out of 107 included patients, 22 (20.5%) and 13 (12.1%) patients had ischemic and congestive heart disease, respectively. A fresh fracture or traumatic brain injury has been found in 12 patients (11.2%) at admission. A total of 13 patients had a documented thrombocytopenia prior to the medication error and 8 patients had a known falling tendency. The higher proportion of patients with falling tendency, ischemic heart disease or with a history of a recent fracture or brain injury in the group of patients with Hb-relevant bleeding may put this population at increased risk of complications following TDA. Although it was suggested that anemia might contribute to the major bleeding under rivaroxaban therapy [18], our study showed a lower proportion of anemia in patients with Hb-relevant bleeding. However, the risk of bleeding may also be affected by other contributing factors. It has been demonstrated that AC therapy in patients with deep vein thrombosis leads more frequently to bleeding complications compared to those with atrial fibrillation [19, 20]. While 50% of patients with Hb-relevant bleeding were under anticoagulation due to AF, this was 68% for patients without Hb-relevant bleeding.

Modified HAS-BLED score

Prior anticoagulation is not considered in the HAS-BLED score as a parameter. Indeed, the HAS-BLED score is employed to assess the risk of bleeding following initiation of an anticoagulative therapy in patients who were not being anticoagulated before. As taking anticoagulants before admission might have predictive value for hemorrhagic complications, we defined and applied a modified score by considering the prior anticoagulation as a criterion as well. Another rationale for employing the modified value is the above-mentioned finding that

Table 2 Comparison of bleeding-relevant characteristics of TDA-patients with and without significant Hb-drop

Parameters	TDA-patients (whole population)	Patients with Hb-relevant bleeding	Patients without Hb-relevant bleeding
Pre-error anemia	63.5%	37.5%	65.6%
Pre-error thrombocytopenia	12.1%	12.5%	12.1%
History of malignancy	26.1%	0%	26.1%
Diabetes mellitus	22.4%	0%	22.4%
Ischemic/ congestive heart disease	IHD 20.5% CHF 12.1%	IHD 37.5% CHF 12.5%	IHD 19.1% CHF 12.1%
Recent fracture or traumatic brain injury	11.2%	37.5%	0.9%
Falling tendency	7.5%	12.5%	0.7%
Death in 2 years	18.6%	25%	18.1%

more than 66% of TDA patients were on anticoagulation therapy before admission. The modified HAS-BLED score was calculated by adding one score for prior anticoagulation to the HAS-BLED score. Using the modified HAS-BLED score, the TDA population had a mean value of 3.04 (CI95% 2.8–3.28) and about 67.2% of TDA patients had a score of at least 3.

Rate of anemia prior to medication error of TDA

For 8 patients, no hemoglobin value has been found within 72 h prior to TDA. The rest population with available pre-error hemoglobin had a mean hemoglobin value of 120.7 (95%CI: 116–125). 47 out of 72 male patients (65.2%) had a hemoglobin value lower than 140 g/L prior to error. From 35 females with TDA, 21 patients (60%) had a hemoglobin value lower than 120 g/L prior to error. Altogether, 63.5% of TDA patients had anemia prior to medication error.

TDA-related parameters

Duration of TDA

TDA lasted, before detection and stop, for a minimum and maximum of 1 and 7 days, respectively. The mean duration of TDA was 1.68 (95%CI: 1.42–1.94).

Time interval of the error to admission or discharge

TDA events began in a short interval to the admission or discharge with a mean interval of 2.15 (95%CI: 1.67–2.63). In 21.4% of patients the TDA began on the day of admission or discharge. In 86 patients (83.3%) the error began in within 72 h after admission or before discharge (Table 3).

Table 3 Comparison of characteristics of TDA between the whole population and the patients with Hb-relevant bleeding

Parameters	TDA-patients (whole population)	Patients with Hb-relevant bleeding
Duration of TDA	1.68 (SD 1.35)	1.75 (SD 1.38)
Time interval to admission or discharge	2.15 (SD 2.55)	1.125 (SD 0.99)
Ward	Medical 38.3% Surgical 61.6%	Medical 62.5% Surgical 37.5%
Setting	27.1% post-op 8.4% pre-op	25% post-op 12.5% pre-op
Involved anticoagulants	LMWH+DOAC (88.7%) Heparin+DOAC (7.5%) LMWH+Heparin (0.9%) DOAC+DOAC (0.9%) VKA + Heparin+LMWH (0.9%)	LMWH+DOAC (87.5%) LMWH+Heparin (12.5%)

Involved anticoagulants

Most of TDA patients (88.7%) had a combination of LMWH and DOAC. Among them, 43 patients (40.1%) have been treated with the combination of rivaroxaban and dalteparin and 40 patients (37.3%) with the combination of apixaban and dalteparin as the most common combination of anticoagulants (Table 4).

In total, 1737 patients received at least one DOAC during the study period. Among them, 104 patients who were included in the study (See Table 4) and all of the excluded TDA patients i.e. fourteen patients were under DOAC at the time of TDA. Thus, TDA occurred in 6.7% of patients who were on DOAC.

Significant Hb-drop following TDA

To evaluate the bleeding events following TDA, we considered the last hemoglobin value reported within 72 h prior to TDA and the fist hemoglobin value reported within 72 h after TDA. A significant Hb-drop, which could be attributed to the bleeding, has been defined as a drop of more than 10 g/L within 24 h. The patients with Hb-drops, which could be attributed to other etiologies rather than TDA such as infusion therapies, operations, or known bleeding sources prior to TDA have been excluded. A total of 8 patients showed significant Hb-drop, which could not be attributed to the above-mentioned etiologies; please see the diagram of patients recruitment. Among patients with significant Hb-drop, two patients were taking platelet-aggregation inhibitors before admission; one was on acetylsalicylic acid 100 mg daily and the other one was on clopidogrel 75 mg daily (See case 2 in the part *case study*). Both drugs were stopped upon admission i.e. none of the patients with significant Hb-drop was receiving any platelet-aggregation inhibitors at the time of TDA events.

Table 4 The anticoagulants involved in the TDA events of 107 patients

Involved ACs	Nr. of patients	Proportion
Rivaroxaban+Dalteparin	43	40.1%
Apixaban+ Dalteparin	40	37.3%
Dabigataran+ Dalteparin	7	6.5%
Edoxaban+ Dalteparin	5	4.6%
Liquemin+ Rivaroxaban	3	2.8%
Liquemin+Apixaban	2	1.8%
Liquemin+ Dalteparin	2	1.8%
Edoxaban+Heparin	1	0.9%
Liquemin+Dabigatran	1	0.9%
Apixaban+rivaroxaban	1	0.9%
apixaban+Arixtra	1	0.9%
Marcoumar+Liquemin+Dalteparin	1	0.9%

Bleeding risk factors in TDA patients

Unpaired T-test and Chi-square have been done for all continuous and binary variables, respectively, to study whether there is a significant difference of variables between two patients groups; with and without Hb-drop. A p-value under 0.05 was defined as statistically significant. The relative risks for binary variables are also reported in Table 4. The modified HAS-BLED score was the only variable that reached the statistical significance in the univariate analysis ($p = 0.028$). HAS-BLED score did not reach the significance ($p = 0.064$). Although the TDA patients with a history of prior anticoagulation may have been at increased risk of bleeding (RR 3.55, CI95% 0.45–27.75), this risk was not translated into statistically significant TDA events ($p = 0.22$) (Table 5).

Legend. The odds ratio and relative risks of bleeding risk factors have been reported. As mentioned above, the prior anticoagulation carries the highest relative risk between 3 and 4.

Multivariate logistic regression

To determine the effect of different risk factors on the bleeding following TDA, we performed a multivariate logistic regression. Only two of 15 independent variables did reach the significance. The female sex (odds ratio .075 and a coefficient – 2.58) and anemia (odds ratio .022 and a coefficient – 3.80) showed a p-value lower than 0.05. As the coefficient of these variables is negative, they negatively correlate the risk of bleeding secondary to TDA. The HAS-BLED score ($p = 0.97$) and the modified HAS-BLED score ($p = 0.13$) did not reach the statistical significance in the multivariate regression model.

Discussion

Anticoagulants are among the most common medications prescribed in clinical practice. Although they are given in order to prevent or treat life-threatening thromboembolic events such as stroke or pulmonary embolism, they could be dangerous if prescribed inappropriately [21]. Owing to the wider therapeutic window and the lower dietary and drug interactions of the newer ACs, which are recently introduced to the pharmaceutical marketing, their use in clinical practice have been surpassed that of vitamin-K-antagonists or heparin. However, lack of specific blood tests with wide availability for monitoring the blood level and therapeutic effects of the newer ACs increases the risk of drug bioaccumulation and, subsequently, the risk of bleeding complications [19]. Owing to being critically harmful in the case of medication errors, the AC- related errors have been largely studied. The adverse drug events commonly occur in clinical practice, with a frequency of 6.5 pro 100 admissions [22]. It has been shown that ACs-related medication errors comprise a significant proportion of medication errors in in-patient setting (7.2–8.3%) and the subsequent adverse events could be prevented in about 70% of cases [3, 22]. Pizza et al. retrospectively analyzed the AC-associated adverse drug events reported to the drug safety system at their tertiary hospital and showed that AC-associated medication errors comprise about half of the adverse drug events owing to AC prescription [3]. In addition, these errors have been significantly correlated with the 30-days mortality rate and higher hospitalization costs [3, 23]. Despite similarities of different reports in showing the high frequency of AC-associated medication errors, there are discrepancies regarding involved AKs, the proportion of error types, and the time of medication process in which errors occur. It seems that the following parameters play a role in the above-mentioned discrepancies: time of study according to the first appearance of the newer AK agents, pharmaceutical market varieties among countries in which the studies have been done, and the difference in study population [2, 24].

Table 5 Univariate Analysis results

Risk factor	Odds ratio	Relative risk
Anemia	0.17 (CI 95% 0.03–0.87)	0.19
Abnormal renal function	0.78 (CI 95% 0.18–3.45)	0.8
Uncontrolled hypertention	1.91(CI 95% 0.45–8.12)	1.82
Stroke	2.53 (CI 95% 0.55–11.51)	2.32
Age > 65	2.0 (CI 95% 0.23–17.14)	1.92
Prior anticoagulation	3.83 (CI 95% 0.45–32.39)	3.55
Abnormal liver function	2.21 (CI 95% 0.23–21.05)	2.04
Drug/ alcohol concomitantly	0.94 (CI 95% 0.18–4.93)	0.94
Bleeding diathesis	1.87 (CI 95% 0.34–10.14)	1.76

Although the therapeutic duplication of ACs (TDA) has been written in medication safety alerts of national and international healthcare authorities [25, 26], there is not any study investigating the frequency and the rate of bleeding complications secondary to TDA. From a total of 1663 patients who have received at least two ACs during a 1-year period, 121 TDA-events have occurred. Out of them, 14 patients have been excluded due to the refusal of participation. The TDA-events have been mainly occurred, in about 88.7% of cases, due to the concomitant prescription of DOACs and LMWHs in prophylactic doses. Consistently, Andreica I. et al. reported a few similar cases in which a new LMWH has been prescribed while the previous DOAC prophylaxis was not discontinued [7]. Among 105 patients treated with heparins, either LMWH or unfractionated heparin, 85 patients received the heparin in prophylactic dose (8 patients with low-dose and other with a prophylactic dose higher than 3400 U/d) and 20 patients in therapeutic dose. Leonardi MJ, et al. in a systematic review of 33 randomized control trial evaluating the bleeding complications of patients under pharmacologic deep vein thrombosis (DVT) prophylaxis showed that patients receiving a high prophylactic dose of LMWH carry a risk of 5.8, 1 and 0.3% for hematuria, gastrointestinal (GI) bleeding and retroperitoneal bleeding, respectively. In comparison, the patients receiving a low dose of LMWH develop in < 0.6% of cases the above-mentioned bleeding complications [27].

We found that the risk for TDA was higher for Rivaroxaban and apixaban compared to other DOACs. Although we do not know the specific reason for this difference, these drugs are routinely prescribed for patients with AF, which may also have a higher likelihood for receiving other anticoagulants. Thus, it is rather the patient population and underlying indication that puts them at risk than the drugs itself.

Although different clinical trials have already investigated the bleeding complications due to each AC class and compared them together, there is not any study reporting the bleeding rate following the concomitant prescription of two different classes i.e. the TDA [9, 11–14]. It has been shown that DOACs, although are not superior to the VKA in the prevention of stroke and systemic embolism, are associated with a significantly lower rate of intracranial bleeding [28–30].

According to the ROCKET-AF study, clinically relevant bleeding occurs in about 14.9% of atrial fibrillation (AF) patients (per year) under rivaroxaban therapy [31]. The ARISTOTLE study assessed the risk of major bleeding complications in AF patients receiving apixaban as prophylaxis for stroke. The study reported a bleeding rate of 0.33, 0.13, 0.68, and 0.01% for intracranial hemorrhage, hematuria, GI bleeding, and retroperitoneal bleeding, respectively [14]. Notably, about 85% of TDA patients received either apixaban or rivaroxaban, mostly as prophylaxis for AF.

Herein, we showed that TDA leads in about 7.5% of cases to Hb-relevant bleeding, defined as a drop more than 10 g/L in hemoglobin value within 24 h, in the first 3 days following the medication error. Pre-existing anemia has been attributed to an increased risk of bleeding secondary to AC therapy [31, 32]. From a clinical standpoint, a bleeding event in anemic patients, especially those with ischemic heart disease, would be more devastating. In our study, 63.5% of patients had anemia and 20.5% suffered from ischemic heart disease (IHD) before a TDA occurred. Furthermore, anemia and IHD were documented in 3 patients with TDA-associated bleeding. It has been demonstrated that AC therapy in patients with deep vein thrombosis leads more frequently to bleeding complications compared to those with atrial fibrillation [19, 20]. It has been attributed to the concomitant polymorbidity and polypharmacy of patients with venous thromboembolic disorders [33, 34]. Among patients with TDA, more than 70% have been subjected to AC-therapy due to atrial fibrillation. Therefore, in a population with a higher proportion of patients having been anticoagulated due to venous thromboembolism (VTE) rather than AF, TDA may lead more frequently to bleeding events compared with the present study.

To assess the risk of bleeding prior to AC therapy, different risk stratification tools have been proposed. Among them, HAS-BLED score, HEMORR2HAGES, and ATRIA are most commonly used in daily clinical practice. Despite having a similar accuracy, the HAS-BLED score has the best predictive performance [35, 36]. These scores have been mainly validated for VKA therapy in patients with atrial fibrillation and, therefore, modified HAS-BLED scores have been proposed to assess the risk of bleeding in other circumstances such as anticoagulation with DOACs [18, 37]. Except for one patient who has been found to concomitantly receive VKA, LMWH, and heparin and one other who received the combination of LMWH and heparin, all other TDA patients have been anticoagulated by DOACs. In addition, 66.3% of TDA patients have been received anticoagulants prior to admission. Prior anticoagulation, though, increases the bleeding risk, has not been considered in the above-mentioned risk stratification tools. Therefore, we measured in addition to the currently used HAS-BLED score, a modified score considering the prior anticoagulant therapy similar to the other medications such as NSAIDs and thrombocyte-aggregation inhibitors, as well.

Interestingly, TDA patients have a mean HAS-BLED score of more than 2 and a mean modified score of more than 3, indicating that the affected population carries a high bleeding risk before TDA-event takes place. In addition to the components of the HAS-BLED score, other clinical parameters predisposing the AK-receiving patients to bleed such as the history of congestive heart failure, cerebrovascular disease, and anemia have also been taken into account. Tamaya et al. in a case-control study of 3252 patients with atrial fibrillation under rivaroxaban-therapy showed that the mentioned parameters, which have been ignored by HAS-BLED score, are the strongest predictive factors for bleeding complications [18]. Consistently, more than 20% of patients with TDA had a positive history of stroke or TIA. In addition, about 18.6% of TDA patients were died at the time of this study i.e. within a maximum of 2 years after discharge from hospital. Altogether, the above-mentioned findings indicate that the involved population carries a high risk of bleeding.

Case study

Herein, we discuss two cases with TDA-related bleeding.

1. An 80 years-old hypertensive patient presented with frequent episodes of syncope to the emergency department. The clinical examination revealed a hematoma in the right flank and also multiple ecchymosis in different locations. The falling episodes have been attributed to the side effects of Ropinirole, a dopamine agonist prescribed due to Parkinson's disease. Furthermore, atrial fibrillation has been detected at the time of admission and, therefore, anticoagulation with apixaban (5 mg daily) has begun. As admitted, the patient received also dalteparin in prophylactic dose (5000 IU daily). Just after 1 day of combinative anticoagulation with apixaban and dalteparin the hemoglobin value dropped from 140 g/L to the 117 g/L. The occurrence of gastrointestinal bleeding presenting with melena has been documented during admission. One day later the TDA has been detected and stopped.
2. A 98 years-old hypertensive patient under long-term anticoagulation with rivaroxaban (20 mg daily) due to atrial fibrillation, presented with bilateral subarachnoidal hemorrhage following falling. The patient has been admitted in neurosurgical ward and her medical situation improved without surgical intervention. She received clopidogrel (75 mg daily) following percutaneous transluminal angioplasty, since 1 month ago. Both medications have been stopped at admission. Instead, she received dalteparin in prophylactic dose (2500 IE daily), which has been continued until 1 day before discharge. Clopidogrel and rivaroxaban have been re- started from 4 and 3 days before discharge, respectively. Thus, the patient received for 2 days the combination of Clopidogrel, rivaroxaban and dalteparin, which resulted in Hb-drop from 86 g/L to the 66 g/L (= the Hb-value at first day of TDA) within 48 h. The Hb-value after the second day of TDA is not available. The patient was still disoriented at discharge and has been referred to the nursing house. Finally, she died 3 days after discharge. However, the reason of death is not documented, whether it occurred due to the deterioration of the pre-existing intracranial hemorrhage or/ and new bleeding events or not.

Conclusion

TDA is a known medication error of anticoagulants with unknown frequency and clinical importance. We retrospectively analyzed the frequency and complications of TDA in all patients admitted to our 450-bed referral hospital in 1 year. TDA occurred in 0.8% of patients who were under anticoagulation and in 6.7% of patients who received DOACs. TDA led in about 7.4% of cases to hemoglobin-relevant bleeding. The old patients with significant comorbidities and a high HAS-BLED score were mainly affected. The female gender and presence of anemia independently predicted the occurrence of bleeding following TDA. Moreover, we introduced a new risk stratification tool by modifying the HAS-BLED score to assess the risk of bleeding in patients who are already on anticoagulants. Although the significance of the proposed tool is the matter of further investigations with larger sample sizes, our analyses suggest the consideration of prior anticoagulation as a parameter in addition to the HAS-BLED score lead to the prediction of the bleeding risk in TDA patients at a higher level of significance.

Abbreviations

TDA: therapeutic duplication of anticoagulants; AC: anticoagulants,; LMWH: low-molecular weight heparins; DOAC: direct oral anticoagulants; VKA: vitamin-K antagonists; DVT: deep vein thrombosis; IHD: ischemic heart disease; VTE: venous thromboembolism; AF: atrial fibrillation

Authors' contributions

RR and ARS and PS devised the project, the main conceptual idea and proof outlines. RR and FC and CZ gathered the data. RR analyzed the data. RR took the lead in writing the manuscript and wrote it in consultation with other authors. RR and PS and ARS revised and proofed the final version of the manuscript. The author(s) read and approved the final manuscript.

Author details

Faculty of Medicine, University of Basel, Basel, Switzerland. [2]Biomedical Research and Training, University Hospital Basel, Basel, Switzerland. [3]Hospital Pharmacy, Kantonsspital Aarau, Aarau, Switzerland. [4]Department of Internal Medicine, Kantonsspital Aarau, 5000 Aarau, Switzerland. [5]Department of Internal Medicine, Clinical Pharmacology, Kantonsspital Aarau, Tellstrasse 51, CH-5001 Aarau, Switzerland.

References

1. Institute for Safe Medication Practices Canada. ISMP List of High-Alert Medications in Acute Care Settings. Inst Safe Medicat Pract. 2014.
2. Dreijer AR, Diepstraten J, Bukkems VE, Mol PGM, Leebeek FWG, Kruip MJHA, et al. Anticoagulant medication errors in hospitals and primary care: a cross-sectional study. Int J Qual Heal Care. 2019;31(5):346–52.
3. Piazza G, Nguyen TN, Cios D, Labreche M, Hohlfelder B, Fanikos J, et al. Anticoagulation-associated adverse drug events. Am J Med. 2011;124(12): 1136–42.
4. Barr D, Epps QJ. Direct oral anticoagulants: a review of common medication errors. J Thromb Thrombolysis. 2019;47(1):146–54.

5. Henriksen JN, Nielsen LP, Hellebek A, Poulsen BK. Medication errors involving anticoagulants: data from the Danish patient safety database. Pharmacol Res Perspect. 2017;5(3):e00307.
6. Tran E, Duckett A, Fisher S, Bohm N. Appropriateness of direct oral anticoagulant dosing for venous thromboembolism treatment. J Thromb Thrombolysis. 2017;43(4):505–13.
7. Andreica, Ivyruth; Grissinger M. Oral Anticoagulants: A Review of Common Errors and Risk Reduction Strategies. Pennsylvania Patient Saf Auth. 2015; 12(2):54–61.
8. Grissinger M, Gaunt MJ, Rich DS. Avoiding medication errors: reducing harm in residents using oral anticoagulants. Consult Pharm. 2016;31(6):294–303.
9. Desai RJ, Williams CE, Greene SB, Pierson S, Hansen RA. Anticoagulant medication errors in nursing homes: characteristics, causes, outcomes, and association with patient harm. J Healthc Risk Manag. 2013;33(1):33–43.
10. Uppuluri EM, McComb MN, Shapiro NL. Implementation of a direct Oral anticoagulation screening Service at a Large Academic Medical Center Provided by a pharmacist-managed Antithrombosis clinic as a method to expand antithrombotic stewardship efforts. J Pharm Pract. 2020;33(3):271–5.
11. Eriksson BI, Borris LC, Friedman RJ, Haas S, Huisman M V., Kakkar AK, et al. Rivaroxaban versus enoxaparin for thromboprophylaxis after hip arthroplasty. N Engl J Med. 2008;358:2765–75.
12. Landman GW, Gans ROB. Oral rivaroxaban for symptomatic venous thromboembolism. New England Journal of Medicine. 2011;364(12):1178.
13. Agnelli G, Buller HR, Cohen A, Curto M, Gallus AS, Johnson M, et al. Oral apixaban for the treatment of acute venous thromboembolism. N Engl J Med. 2013;369(9):799–808.
14. Granger CB, Alexander JH, McMurray JJV, Lopes RD, Hylek EM, Hanna M, et al. Apixaban versus warfarin in patients with atrial fibrillation. N Engl J Med. 2011;365:981–92.
15. Narum S, Solhaug V, Myhr K, Johansen PW, Brørs O, Kringen MK. Warfarin-associated bleeding events and concomitant use of potentially interacting medicines reported to the Norwegian spontaneous reporting system. Br J Clin Pharmacol. 2011;71(2):254–62.
16. Teklay G, Shiferaw N, Legesse B, Bekele ML. Drug-drug interactions and risk of bleeding among inpatients on warfarin therapy: a prospective observational study. Thromb J. 2014;12:20.
17. Fusco JA, Paulus EJ, Shubat AR, Miah S. Warfarin and Rivaroxaban Duplication: A Case Report and Medication Error Analysis. Drug Saf - Case Reports. 2015;2(1):5.
18. Tamayo SG, Simeone JC, Nordstrom BL, Patel MR, Yuan Z, Sicignano NM, et al. Risk Factors for Major Bleeding in Rivaroxaban Users With Atrial Fibrillation. Journal of the American College of Cardiology. 2016;68(10): 1144–6.
19. Crowther MA, Warkentin TE. Bleeding risk and the management of bleeding complications in patients undergoing anticoagulant therapy: focus on new anticoagulant agents. Blood. 2008;111(10):4871–9.
20. Shoeb M, Fang MC. Assessing bleeding risk in patients taking anticoagulants. In: J Thrombosis Thrombolysis. 2013;35(3):312–9.
21. Landefeld CS, Beyth RJ. Anticoagulant-related bleeding: clinical epidemiology, prediction, and prevention. Am J Med. 1993;95(3):315–28.
22. Poon EG, Cina JL, Churchill W, Patel N, Featherstone E, Rothschild JM, et al. Medication dispensing errors and potential adverse drug events before and after implementing bar code technology in the pharmacy. Ann Intern Med. 2006;145(6):426–34.
23. Bates DW, Spell N, Cullen DJ, Burdick E, Laird N, Petersen LA, et al. The costs of adverse drug events in hospitalized patients. J Am Med Assoc. 1997; 277(17):1351.
24. Budnitz DS, Lovegrove MC, Shehab N, Richards CL. Emergency hospitalizations for adverse drug events in older Americans. N Engl J Med. 2011;365:2002–12.
25. Quick-Alert 41, Stiftung für Patientensicherheit, Schweiz.
26. Commission TJ. National Patient Safety Goals Effective January 1 , 2015. Hosp Natl Patient Saf Goals. 2015.
27. Leonardi MJ, McGory ML, Ko CY. The rate of bleeding complications after pharmacologic deep venous thrombosis prophylaxis: a systematic review of 33 randomized controlled trials. Arch Surg. 2006;141(8):790–7.
28. Ruff CT, Giugliano RP, Braunwald E, Hoffman EB, Deenadayalu N, Ezekowitz MD, et al. Comparison of the efficacy and safety of new oral anticoagulants with warfarin in patients with atrial fibrillation: a meta-analysis of randomised trials. Lancet. 2014;383(9921):955–62.
29. Wang YP, Kehar R, Iansavitchene A, Lazo-Langner A. Bleeding risk in non-Valvular atrial fibrillation patients receiving direct Oral anticoagulants and warfarin: a systematic review and meta-analysis of observational studies. Blood. 2019;134(Supplement_1):3672.
30. Lobraico-Fernandez J, Baksh S, Nemec E. Elderly bleeding risk of direct Oral anticoagulants in Nonvalvular atrial fibrillation: a systematic review and meta-analysis of cohort studies. Drugs in R and D. 2019;19(3):235–45.
31. Patel MR, Mahaffey KW, Garg J, Pan G, Singer DE, Hacke W, et al. Rivaroxaban versus warfarin in nonvalvular atrial fibrillation. N Engl J Med. 2011;365:883–91.
32. Beyth RJ, Quinn LM, Landefeld CS. Prospective evaluation of an index for predicting the risk of major bleeding in outpatients treated with warfarin. Am J Med. 1998;105(2):91–9.
33. Levine MN, Raskob G, Landefeld S, Kearon C. Hemorrhagic complications of anticoagulant treatment. Chest. 2001;119(1 Suppl):108S–121S.
34. Lassen MR, Gallus A, Raskob GE, Pineo G, Chen D, Ramirez LM. Apixaban versus enoxaparin for thromboprophylaxis after hip replacement. N Engl J Med. 2010;363(26):2487–98.
35. Hellenbart EL, Faulkenberg KD, Finks SW. Evaluation of bleeding in patients receiving direct oral anticoagulants. Vasc Health Risk Manag. 2017;13:325–42.
36. Lip GYH, Nieuwlaat R, Pisters R, Lane DA, Crijns HJGM, Andresen D, et al. Refining clinical risk stratification for predicting stroke and thromboembolism in atrial fibrillation using a novel risk factor-based approach: the euro heart survey on atrial fibrillation. Chest. 2010;137(2):263–72.
37. Tsu LV, Berry A, Wald E, Ehrlich C. Modified HAS-BLED score and risk of major bleeding in patients receiving dabigatran and rivaroxaban: a retrospective, case-control study. Consult Pharm. 2015;30(7):395–402.

13

Circulating histone H3 levels in septic patients are associated with coagulopathy, multiple organ failure and death

Yayoi Yokoyama[1], Takashi Ito[1,2*], Tomotsugu Yasuda[1], Hiroaki Furubeppu[1], Chinatsu Kamikokuryo[1], Shingo Yamada[3], Ikuro Maruyama[2] and Yasuyuki Kakihana[1]

Abstract

Background: Nuclear histone proteins are released into the extracellular space, and act as major mediators of coagulopathy and remote organ failure in septic animals. However, the circulating histone levels in septic patients have not been precisely quantified.

Methods: Using a novel enzyme-linked immunosorbent assay for histone H3 detection, we measured the serum histone H3 levels in 85 patients admitted to the intensive care unit because of infectious diseases. We then evaluated the associations of circulating histone H3 levels with organ failure, coagulopathy, and mortality.

Results: Circulating histone H3 levels were significantly higher in patients with coagulopathy, and were positively correlated with numbers of organ failures. Circulating histone H3 levels were also associated with fatal outcome. Receiver-operating characteristic analyses revealed that the predictive performance of circulating histone H3 levels for mortality was higher than that of conventional inflammatory markers, including white blood cell count, C-reactive protein, and cell-free DNA.

Conclusions: Circulating histone H3 levels are associated with coagulopathy, multiple organ failure, and death in patients requiring intensive care because of infectious diseases.

Keywords: Histone, Sepsis, Coagulopathy, Organ failure

Background

Sepsis is defined as life-threatening organ dysfunction caused by a dysregulated host response to infection [1]. Although the dysregulated host response has not been clarified in detail, an overwhelming inflammatory response is considered to be involved. Nevertheless, anti-inflammatory agents, such as tumor necrosis factor antagonists and interleukin-1 receptor antagonists, did not improve the survival of patients with sepsis [2].

Recent studies have suggested that extracellular histones are potential mediators of death in sepsis [3, 4]. Histones, the most abundant proteins in the nucleus, comprise linker histone H1 and core histones H2A, H2B, H3, and H4. In animal models of sepsis, histones are released into the extracellular space [3, 5] where they induce platelet aggregation [6, 7], endothelial damage [3, 8], and cardiac injury [5, 9]. These findings have led to increased demand for convenient and reproducible methods to measure circulating histone levels. However, there are no reliable methods for quantitative analysis of circulating histone levels other than semiquantitative western blot analysis [3, 5, 10]. In the present study, we quantified serum histone H3 levels using a newly developed enzyme-linked immunosorbent assay (ELISA) and found that circulating histone H3 levels

* Correspondence: takashi@m3.kufm.kagoshima-u.ac.jp
[1]Department of Emergency and Intensive Care Medicine, Kagoshima University Graduate School of Medical and Dental Sciences, 8-35-1 Sakuragaoka, Kagoshima 890-8544, Japan
[2]Department of Systems Biology in Thromboregulation, Kagoshima University Graduate School of Medical and Dental Sciences, Kagoshima, Japan
Full list of author information is available at the end of the article

are associated with coagulopathy, multiple organ failure, and death in septic patients.

Materials and methods

Patients

This was a single-center observational study. Between July 2015 and June 2016, 85 patients with suspected infections were admitted to the intensive care unit (ICU) at Kagoshima University. Written informed consent was obtained from all patients for using residual blood samples of routine blood tests for research. Residual serum samples from the patients were anonymized and used for measurements of histone H3 and cell-free DNA. Clinical data, including Systemic Inflammatory Response Syndrome (SIRS) score, Acute Physiology and Chronic Health Evaluation II (APACHE II) score without the Glasgow Coma Scale component, and Sequential Organ Failure Assessment (SOFA) score without the Glasgow Coma Scale component, were also anonymized and analyzed. Organ failure was defined as a score of ≥3 for each component of the SOFA score [11]. Shock was defined as a score of ≥2 for the cardiovascular component of the SOFA score combined with a lactate level of > 2 mmol/L. Diagnosis of disseminated intravascular coagulation (DIC) was made according to the criteria established by the Japanese Association for Acute Medicine [12]. The study was compliant with the Declaration of Helsinki and approved by the Ethics Committee of Kagoshima University.

Measurement of histone H3 levels

Serum histone H3 levels were measured as recently described [9]. Briefly, polystyrene microtiter plates (Nunc, Roskilde, Denmark) were coated with 100 μL of 1 mg/L anti-histone H3 antibody (Shino-Test Corporation, Sagamihara, Japan). After washing and blocking, the plates were incubated with 100 μL of diluted calibrator and serum samples for 24 h at room temperature. After washing, the plates were incubated with 100 μL of anti-histone H3 peroxidase-conjugated antibody (Shino-Test Corporation) for 2 h at room temperature. The plates were washed again, and incubated with the chromogenic substrate 3,3′,5,5′-tetra-methylbenzidine (Dojindo Laboratories, Kumamoto, Japan). The reaction was terminated with 0.35 mol/L Na_2SO_4, and the absorbance at 450 nm was measured with a microplate reader (Model 680; Bio-Rad, Hercules, CA). A standard curve was obtained with purified calf thymus histone H3 (Roche, Stockholm, Sweden). The lower detection limit of the ELISA was 2 ng/mL, and linearity was observed in the range up to 250 ng/mL. The ELISA specifically detected histone H3, and did not react with other histone family proteins, including histone H2A, H2B and H4 [9].

Measurement of serum cell-free DNA levels

Cell-free DNA levels in serum samples were quantified as described [13]. Briefly, diluted serum samples (1:10 dilution in PBS) were mixed with SytoxGreen (2 μM; Molecular Probes, Eugene, OR). Fluorescence was recorded by an Infinite M200 fluorometer (Tecan, Männedorf, Switzerland) with 485 nm excitation and 538 nm emission. Autofluorescence was determined in samples without SytoxGreen and defined as background fluorescence. DNA concentrations were calculated based on a standard curve of known DNA concentrations.

Measurement of blood cell counts and C-reactive protein levels

Complete blood counts were performed using an automated counting device (XE-5000; Sysmex Corporation, Kobe, Japan). Serum C-reactive protein (CRP) levels were examined with a BioMajesty JCA-BM6070 (Jeol Ltd., Tokyo, Japan).

Statistical analysis

Circulating histone H3 levels are shown as median (lower quartile–upper quartile) or box plot with lower extreme, lower quartile, median, upper quartile, and upper extreme values, as well as outliers. Circulating histone H3 levels in patients with multiple organ failure were compared with those in patients without organ failure by the Shirley–Williams test. Circulating histone H3 levels, cell-free DNA levels, CRP levels, and white blood cell counts in 28-day survivors were compared with those in non-survivors by the Mann–Whitney *U* test. Receiver-operating characteristics (ROC) curve analysis with area under the curve (AUC) calculation was used to quantify the predictive performance of each parameter for 28-day mortality. A *p*-value of less than 0.05 was considered statistically significant.

Results

Baseline characteristics of the patients

The baseline characteristics of the 85 patients are shown in Table 1. All 85 patients had a SOFA score of ≥2, and 10 patients (15.7%) were SIRS-negative. Forty patients (47.1%) were associated with shock and 12 patients (14.1%) had died by day 28.

Circulating histone H3 levels are associated with multiple organ failure

Using a newly developed ELISA for histone H3 detection, we measured the circulating histone H3 levels in the 85 patients with infectious diseases. Serum histone H3 levels in patients without organ failure were 1.0 (0.2–2.7) ng/mL, while those in patients with one, two, and three or more organ failures were 3.0 (1.1–5.3), 4.9 (2.5–9.2), and 5.5 (2.2–36.1) ng/mL, respectively (Fig. 1).

Table 1 Basic characteristics of the study patients

	All	Survivors ($n = 73$)	Non-survivors ($n = 12$)	p
Age	71 (24–92)	71 ($n = 73$)	72.5 ($n = 12$)	0.64
APACHE II	19 (7–33)	18.5 ($n = 72$)	22 ($n = 12$)	0.09
SOFA score	8 (3–17)	7 ($n = 68$)	11 ($n = 11$)	< 0.01
SIRS score	3 (0–4)	3 ($n = 72$)	3 ($n = 12$)	0.72
DIC score	4 (1–8)	4 ($n = 65$)	6 (n = 12)	0.06
Shock (%)	47.1	42.5	75	0.06

APACHE II Acute Physiology and Chronic Health Evaluation II without the Glasgow Coma Scale component, *SOFA* Sequential Organ Failure Assessment without the Glasgow Coma Scale component, *SIRS* Systemic Inflammatory Response Syndrome, *DIC* Disseminated Intravascular Coagulation;
Data for all subjects are shown as median (minimum–maximum), and data for survivors and non-survivors are shown as median (number of patients analyzed)
Differences between survivors and non-survivors were analyzed by the Mann–Whitney *U* test, except for shock (%), which was analyzed by the chi-square test

Serum histone H3 levels in patients with multiple organ failure were significantly higher than those in patients without organ failure ($p < 0.05$).

Circulating histone H3 levels are associated with coagulation failure

We evaluated the serum histone H3 levels in patients with each organ failure in the SOFA score, including respiratory, cardiovascular, coagulation, liver, and renal failures (Fig. 2a). Serum histone H3 levels in patients with coagulation failure (11.1 [2.2–48.4] ng/mL) were significantly higher than those in patients without coagulation failure (3.5 [1.5–5.9] ng/mL). Serum histone H3 levels did not differ significantly between patients with (4.5 [1.6–9.2] ng/mL) and without (3.4 [1.6–5.6] ng/mL) cardiovascular failure. However, extremely high histone H3 levels, including those exceeding 20 ng/mL, were associated with cardiovascular failure. It remained unclear whether serum histone H3 levels in patients with liver failure were higher than those in patients without liver failure because only two patients had liver failure on day 1. Serum histone H3 levels did not differ significantly between patients with and without respiratory failure or renal failure. These results indicate that circulating histone H3 levels are associated with organ failure, especially coagulation failure.

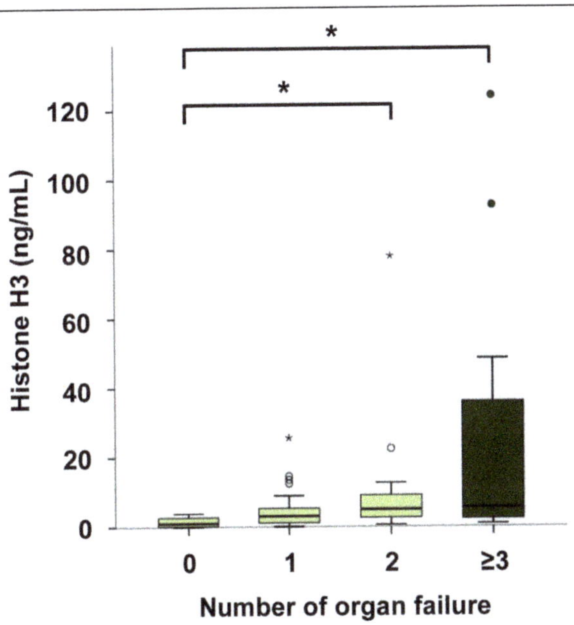

Fig. 1 Circulating histone H3 levels are associated with multiple organ failure. Serum histone H3 levels within 24 h of ICU admission (day 1) were analyzed in septic patients. The septic patients were divided into four groups: those without organ failure ($n = 5$), those with one organ failure ($n = 38$), those with two organ failures ($n = 27$), and those with three or more organ failures ($n = 14$ for three organ failures; $n = 1$ for four organ failures). Data for four patients were missing because serum samples were not obtained within 24 h of ICU admission. Differences between serum histone H3 levels in patients with and without organ failure were analyzed by the Shirley–Williams test. *$p < 0.05$ versus patients without organ failure

In the assessment of SOFA, coagulation failure is defined as the platelet count of less than 50,000/μL [11]. Since thrombocytopenia represents only a part of coagulation failure, we then assessed the status of DIC. Among 77 patients whose DIC scores on Day 1 were available, 43 patients were defined as a state of DIC. Serum histone H3 levels in patients with DIC (4.8 [2.2–9.9] ng/mL) were significantly higher than those in patients without DIC (2.3 [0.9–5.2] ng/mL). In particular, patients with a DIC score of ≥6 showed significantly high serum histone H3 levels (Fig. 2b). These findings indicate that circulating histone H3 levels are associated with a state of DIC.

Circulating histone H3 levels are associated with fatal outcome

The association between serum histone H3 levels and fatal outcome was evaluated (Fig. 3). Serum histone H3 levels in non-survivors (9.1 [5.1–13.0] ng/mL) were significantly higher than those in survivors (3.4 [1.5–6.1] ng/mL) at day 1. Serum histone H3 levels also tended to be higher in non-survivors on days 3, 5, and 7, but the differences did not reach statistical significance. Other conventional inflammatory markers, including serum cell-free DNA, CRP, and white blood cell counts, did not differ between survivors and non-survivors. To quantify

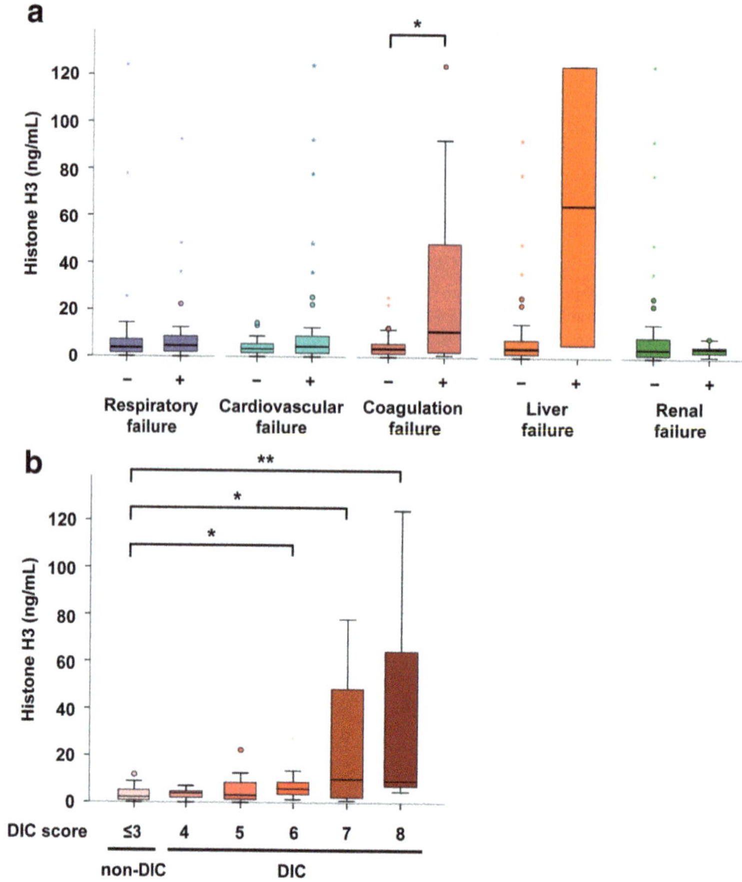

Fig. 2 Circulating histone H3 levels are associated with coagulation failure. **a** Serum histone H3 levels within 24 h of ICU admission (day 1) were analyzed in septic patients with and without respiratory failure ($n = 47$ and $n = 31$, respectively), cardiovascular failure ($n = 51$ and $n = 30$, respectively), coagulation failure ($n = 14$ and $n = 67$, respectively), liver failure ($n = 2$ and $n = 79$, respectively), and renal failure ($n = 19$ and $n = 62$, respectively). Data for four patients were missing because serum samples were not obtained within 24 h of ICU admission. Differences between serum histone H3 levels in patients with and without each organ failure were analyzed by the Mann–Whitney U test. $^*p < 0.05$ versus patients without organ failure. **b** Serum histone H3 levels within 24 h of ICU admission (day 1) were analyzed in septic patients. The septic patients were divided into six groups: those without DIC ($n = 32$), those with DIC score of 4 ($n = 9$), DIC score of 5 ($n = 11$), DIC score of 6 ($n = 9$), DIC score of 7 ($n = 6$), and DIC score of 8 ($n = 7$). Data for 11 patients were missing because serum samples were not obtained within 24 h of ICU admission or DIC scores were not available. Differences of serum histone H3 levels between patients with and without DIC were analyzed by the Shirley–Williams test. $^{**}p < 0.01$ and $^*p < 0.05$ versus patients without DIC

the predictive performance of serum histone H3 levels for 28-day mortality, we analyzed the ROC curves. As shown in Fig. 4, the predictive performance of serum histone H3 levels was higher than that of the other conventional inflammatory markers. These results indicate that increased histone H3 levels are associated with fatal outcome.

Discussion

In this study, we showed that circulating histone H3 levels were associated with coagulopathy, multiple organ failure, and death in septic patients, and could be a useful marker for recognizing the severity of sepsis. Although this study did not demonstrate a cause–effect relationship between histone H3 elevation and organ failure, previous studies indicated that histone H3 exacerbated organ failure in non-clinical settings. Circulating histones were toxic toward vascular endothelial cells [3, 8] and cardiomyocytes [5], and caused microvascular thrombosis and myocardial dysfunction [4, 9]. Histones induced platelet aggregation in vitro and thrombocytopenia in vivo [6, 7]. Furthermore, neutralizing antibodies against histones were able to rescue mice from lethal sepsis [3]. These findings indicate that high concentrations of circulating histones can be a direct inducer of organ failure, rather than a consequence of organ failure. Thus, evaluation of histone H3 levels is important for not only recognizing disease severity, but also developing strategies for histone-targeted therapy.

Circulating histone H3 levels showed a strong positive correlation with cell-free DNA levels, a modest but significant positive correlation with white blood cell counts,

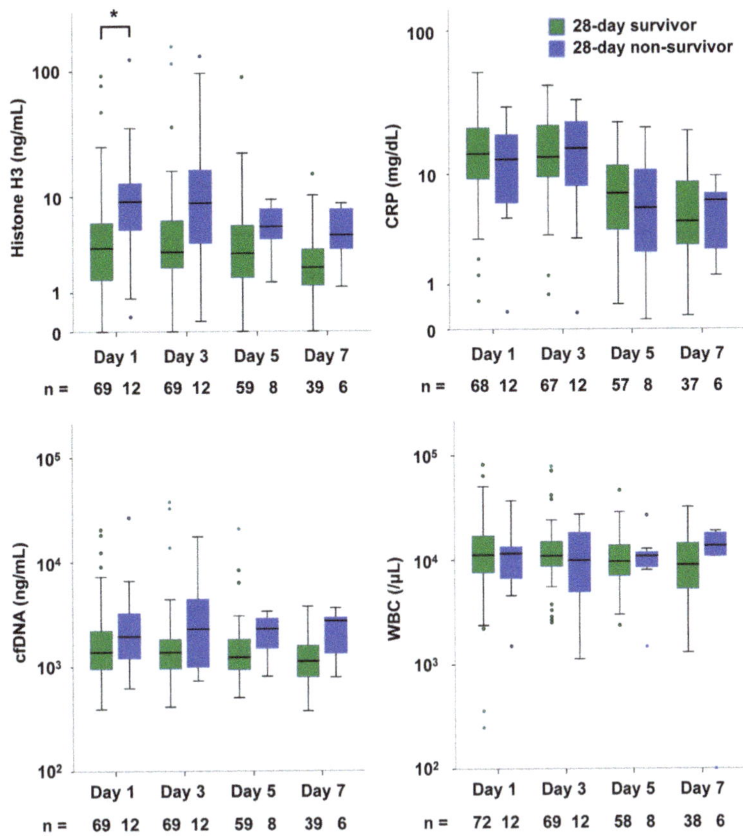

Fig. 3 Circulating histone H3 levels are associated with fatal outcome. Serum histone H3 levels, cell-free DNA (cfDNA) levels, CRP levels, and white blood cell (WBC) counts on days 1, 3, 5, and 7 were analyzed. Patients who left the ICU were not followed up even if they survived. Differences between 28-day survivors and non-survivors were analyzed by the Mann–Whitney *U* test. $^{*}p < 0.05$

and no correlation with CRP levels. Circulating histone H3 levels also showed a significant positive correlation with SOFA score. The predictive performance of circulating histone H3 levels for mortality was higher than that of white blood cell counts, CRP, and cell-free DNA. However, it remains inconclusive whether circulating histone H3 levels are a better prognostic marker than cell-free DNA. Although cell-free DNA can be measured in several ways, including ultraviolet absorbance [14], DNA dyes [13, 15], and polymerase chain reaction [14, 16], the methods are not standardized. So, further studies are needed for standardization of cell-free DNA measurement and evaluation of good prognostic markers in septic conditions.

Limitations of the study

The present study has several limitations. First, the diagnosis of sepsis was not confirmed in all patients in this retrospective study. The inclusion criteria were patients admitted to the ICU because of infectious disease. All subjects had a SOFA score of ≥2. Thus, if their SOFA scores before ICU admission (baseline SOFA scores) were assumed to be zero, all of them met the criteria for sepsis-3 [1]. However, the baseline SOFA scores were not available in this study. Second, patients who left ICU were not followed up. As a result, 73 survivors and 12 non-survivors were analyzed on day 1, while only 39 survivors and 6 non-survivors were analyzed on day 7. This may distort the composition of patients, because mild cases in the survivor group and severe cases in the non-survivor group are predisposed to leave the ICU. Third, the number of patients enrolled in this study was relatively low and did not allow us to conduct multivariate analysis. For larger scale clinical studies, the large-scale preparation of reagents, including anti-histone H3 antibodies, is required. Fourth, the specificity of circulating histone H3 levels for sepsis was yet to be investigated. It is conceivable that circulating histone H3 levels are increased in non-septic diseases, such as trauma and pancreatitis [8]. Another limitation of this study is that our ELISA did not measure other members of the histone family, including histone H4. Previous studies have shown that histone H4 is the most potent stimulator of platelets and endothelial cells, while histone H3 is the second most potent stimulator [3, 6]. The reason why we decided to measure histone H3 rather than histone H4 is the

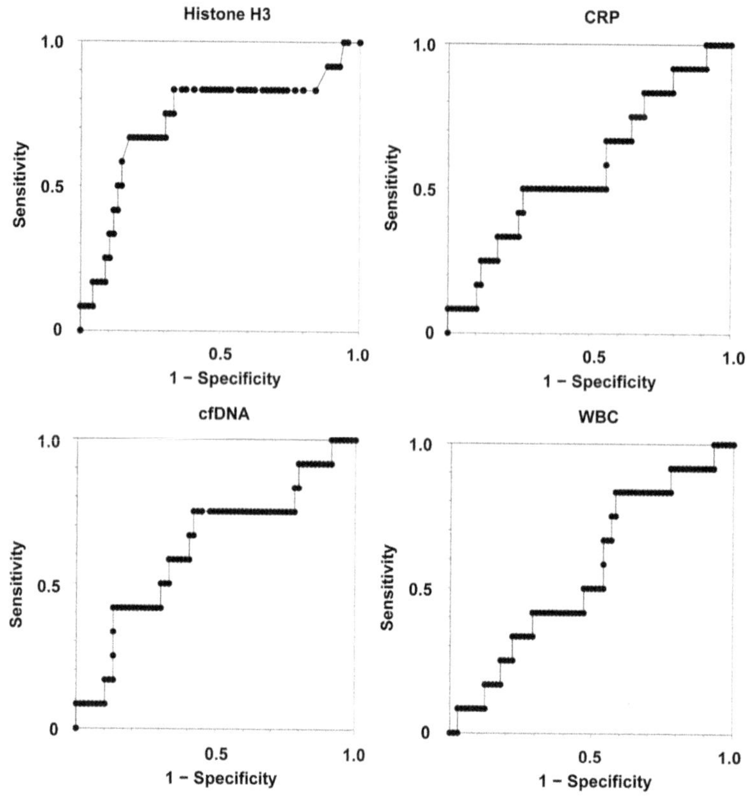

Fig. 4 Predictive performance of circulating histone H3 levels for mortality is higher than that of conventional inflammatory markers. ROC curves for serum histone H3 levels, cell-free DNA (cfDNA) levels, CRP levels, and white blood cell (WBC) counts on day 1 are shown. The AUCs for histone H3 levels, cfDNA levels, CRP levels, and WBC counts were 0.73, 0.63, 0.58, and 0.56, respectively

availability of good purified histone H3 as a reference standard. Further analyses of histone H4 levels and histone H3 levels may improve our understanding of the pathogenesis of sepsis.

Conclusions

Using a newly developed ELISA for histone H3 detection, we measured the circulating histone H3 levels in patients requiring intensive care because of infectious diseases. Circulating histone H3 levels were associated with coagulopathy and multiple organ failure in these patients.

Abbreviations

APACHE: Acute Physiology and Chronic Health Evaluation; AUC: Area under the curve; cfDNA: Cell-free DNA; CRP: C-reactive protein; ELISA: Enzyme-linked immunosorbent assay; ICU: Intensive care unit; PBS: Phosphate-buffered saline; ROC: Receiver-operating characteristic; SIRS: Systemic inflammatory response syndrome; SOFA: Sequential organ failure assessment; WBC: White blood cell

Acknowledgements

The authors thank Miho Yamada, Ryo Sugimoto, and all members of the clinical laboratory in Kagoshima University Hospital for their excellent technical assistance. The authors also thank Alison Sherwin, PhD, from Edanz Group (www.edanzediting.com/ac) for editing a draft of this manuscript.

Author details

[1]Department of Emergency and Intensive Care Medicine, Kagoshima University Graduate School of Medical and Dental Sciences, 8-35-1 Sakuragaoka, Kagoshima 890-8544, Japan. [2]Department of Systems Biology in Thromboregulation, Kagoshima University Graduate School of Medical and Dental Sciences, Kagoshima, Japan. [3]R&D Center, Shino-Test Corporation, Sagamihara, Japan.

References

1. Singer M, Deutschman CS, Seymour CW, Shankar-Hari M, Annane D, Bauer M, Bellomo R, Bernard GR, Chiche JD, Coopersmith CM, et al. The third international consensus definitions for Sepsis and septic shock (Sepsis-3). JAMA. 2016;315(8):801–10.
2. Hotchkiss RS, Karl IE. The pathophysiology and treatment of sepsis. N Engl J Med. 2003;348(2):138–50.
3. Xu J, Zhang X, Pelayo R, Monestier M, Ammollo CT, Semeraro F, Taylor FB, Esmon NL, Lupu F, Esmon CT. Extracellular histones are major mediators of death in sepsis. Nat Med. 2009;15(11):1318–21.
4. Ito T. PAMPs and DAMPs as triggers for DIC. J Intensive Care. 2014;2(1):67.
5. Alhamdi Y, Abrams ST, Cheng Z, Jing S, Su D, Liu Z, Lane S, Welters I, Wang G, Toh CH. Circulating histones are major mediators of cardiac injury in patients with Sepsis. Crit Care Med. 2015;43(10):2094–103.
6. Fuchs TA, Bhandari AA, Wagner DD. Histones induce rapid and profound thrombocytopenia in mice. Blood. 2011;118(13):3708–14.
7. Nakahara M, Ito T, Kawahara KI, Yamamoto M, Nagasato T, Shrestha B, Yamada S, Miyauchi T, Higuchi K, Takenaka T, et al. Recombinant Thrombomodulin protects mice against histone-induced lethal thromboembolism. PLoS One. 2013;8(9):e75961.

8. Abrams ST, Zhang N, Manson J, Liu T, Dart C, Baluwa F, Wang SS, Brohi K, Kipar A, Yu W, et al. Circulating histones are mediators of trauma-associated lung injury. Am J Respir Crit Care Med. 2013;187(2):160–9.
9. Ito T, Nakahara M, Masuda Y, Ono S, Yamada S, Ishikura H, Imaizumi H, Kamikokuryo C, Kakihana Y, Maruyama I. Circulating histone H3 levels are increased in septic mice in a neutrophil-dependent manner: preclinical evaluation of a novel sandwich ELISA for histone H3. J Intensive Care. 2018; 6:79. https://doi.org/10.1186/s40560-018-0348-y.
10. Wildhagen KC, Wiewel MA, Schultz MJ, Horn J, Schrijver R, Reutelingsperger CP, van der Poll T, Nicolaes GA. Extracellular histone H3 levels are inversely correlated with antithrombin levels and platelet counts and are associated with mortality in sepsis patients. Thromb Res. 2015;136(3):542–7.
11. Moreno R, Vincent JL, Matos R, Mendonca A, Cantraine F, Thijs L, Takala J, Sprung C, Antonelli M, Bruining H, et al. The use of maximum SOFA score to quantify organ dysfunction/failure in intensive care. Results of a prospective, multicentre study. Working group on Sepsis related problems of the ESICM. Intensive Care Med. 1999;25(7):686–96.
12. Gando S, Iba T, Eguchi Y, Ohtomo Y, Okamoto K, Koseki K, Mayumi T, Murata A, Ikeda T, Ishikura H, et al. A multicenter, prospective validation of disseminated intravascular coagulation diagnostic criteria for critically ill patients: comparing current criteria. Crit Care Med. 2006;34(3):625–31.
13. Fuchs TA, Kremer Hovinga JA, Schatzberg D, Wagner DD, Lammle B. Circulating DNA and myeloperoxidase indicate disease activity in patients with thrombotic microangiopathies. Blood. 2012;120(6):1157–64.
14. Dwivedi DJ, Toltl LJ, Swystun LL, Pogue J, Liaw KL, Weitz JI, Cook DJ, Fox-Robichaud AE, Liaw PC. Prognostic utility and characterization of cell-free DNA in patients with severe sepsis. Crit Care. 2012;16(4):R151.
15. Margraf S, Logters T, Reipen J, Altrichter J, Scholz M, Windolf J. Neutrophil-derived circulating free DNA (cf-DNA/NETs): a potential prognostic marker for posttraumatic development of inflammatory second hit and sepsis. Shock. 2008;30(4):352–8.
16. Lui YY, Chik KW, Chiu RW, Ho CY, Lam CW, Lo YM. Predominant hematopoietic origin of cell-free DNA in plasma and serum after sex-mismatched bone marrow transplantation. Clin Chem. 2002;48(3):421–7.

14

Guideline compliance for bridging anticoagulation use in vitamin-K antagonist patients: Practice variation and factors associated with non-compliance

M. J. Moesker[1*], J. F. de Groot[2], N. L. Damen[3], N. R. Bijsterveld[4], J. W. R. Twisk[5], M. V. Huisman[6], M. C. de Bruijne[1] and C. Wagner[1,2]

Abstract

Background: Bridging anticoagulation is used in vitamin-K antagonist (VKA) patients undergoing invasive procedures and involves complex risk assessment in order to prevent thromboembolic and bleeding outcomes.

Objectives: Our aim was to assess guideline compliance and identify factors associated with bridging and especially, non-compliant bridging.

Methods: A retrospective review of 256 patient records in 13 Dutch hospitals was performed. Demographic, clinical, surgical and care delivery characteristics were collected. Compliance to the American College of Chest Physicians ninth edition guideline (AT9) was assessed. Multilevel regression models were built to explain bridging use and predict non-compliance.

Results: Bridging use varied from 15.0 to 83.3% (mean = 41.8%) of patients per hospital, whereas guideline compliance varied from 20.0 to 88.2% (mean = 68.5%) per hospital. Both established thromboembolic risk factors and characteristics outside thromboembolic risk assessment were associated with bridging use. Predictors for overuse were gastrointestinal surgery (OR 14.85, 95% CI 2.69–81.99), vascular surgery (OR 13.01, 95% CI 1.83–92.30), non-elective surgery (OR 8.67, 95% CI 1.67–45.14), lowest 25th percentile socioeconomic status (OR 0.33, 95% CI 0.11–1.02) and use of VKA reversal agents (OR 0.22, 95% CI 0.04–1.16).

Conclusion: Bridging anticoagulation practice was not compliant with the AT9 in 31.5% of patients. The aggregated AT9 thromboembolic risk was inferior to individual thromboembolic risk factors and other characteristics in explaining bridging use. Therefor the AT9 risk seems less important for the decision making in everyday practice. Additionally, a heterogeneous implementation of the guideline between hospitals was found. Further research and interventions are needed to improve bridging anticoagulation practice in VKA patients.

Keywords: Anticoagulants, Perioperative care, Practice guideline, Quality of healthcare, Coumarins

* Correspondence: m.moesker@amsterdamumc.nl
[1]Department of Public and Occupational Health, Amsterdam Public Health Research Institute, Amsterdam UMC, Vrije Universiteit Amsterdam, De Boelelaan 1117, 1081BT Amsterdam, The Netherlands
Full list of author information is available at the end of the article

Background

Long-term use of oral anticoagulants such as vitamin-K antagonists (VKA) reduces the risk of thromboembolic events in patients with atrial fibrillation, venous thromboembolism or mechanical heart valves [1–3]. When these patients undergo invasive procedures, such as surgery, the anticoagulant therapy often needs interruption to reduce bleeding. This interruption can increase the risk of thromboembolic complications [4]. In an effort to reduce this risk, short-acting low molecular weight heparin (LMWH) or unfractionated heparin (UFH) are temporarily administered. This is known as 'bridging anticoagulation' [5–7].

In general, anticoagulants are consistently identified in adverse event studies as factors involved in preventable adverse events [8, 9], partially occurring in the context of bridging [10].

Due to the risks involved, bridging anticoagulation urges a careful trade-off between thromboembolic and bleeding risk [11, 12]. Consequently, clinicians are required to perform a thorough risk assessment as part of the decision-making in perioperative VKA management.

The American College of Chest Physicians' *Antithrombotic Therapy and Prevention of Thrombosis, Ninth Edition* guideline (AT9) published in 2012 includes recommendations for this risk assessment by classifying patients in low, moderate or high thromboembolic risk [4]. Bridging is only explicitly recommended for high-risk patients, but might be considered for moderate risk patients too based on individual patient and surgical factors.

Compliance to the AT9 risk stratification and similar guidelines related to bridging is suboptimal [13–15]. Non-compliant bridging can be differentiated in underuse or overuse of bridging anticoagulation. Underuse refers to withholding bridging anticoagulation in high thromboembolic risk patients and overuse refers to unnecessarily administering bridging anticoagulation in low thromboembolic risk patients (Fig. 1). Underuse exposes patients to a higher risk of thromboembolic complications whereas overuse exposes patients to a higher risk for bleeding complications [13–16].

Both bleeding and thromboembolic complications can have serious consequences for patients' mortality and morbidity [3, 17]. Keeping non-compliant bridging strategies at a minimum should therefore be pursued. Which patients are at risk for non-compliant bridging strategies is relatively unknown. Together with the risks involved around non-compliant bridging, and accumulating evidence reporting up to a 5-fold increased bleeding incidence when bridging is used, identifying patients at risk for a non-compliant bridging strategy is important in reducing preventable mortality and morbidity [11, 12, 18].

Therefore, this study aims to determine guideline compliance of bridging anticoagulation in everyday practice and identify factors associated with bridging use, especially predictors for non-compliant under- and overuse of bridging anticoagulation in Dutch hospitals.

Methods

Study design and population

Our current study is part of a larger study evaluating the quality of anticoagulant management in Dutch hospitals by retrospectively reviewing patient records [19]. The hospital sample was stratified by type: university, tertiary teaching, and general hospitals. Within these strata a random selection of hospitals was made while accounting for a proper representation of urban and rural based hospitals. In total, 25 hospitals were invited in two waves of which 13 hospitals participated including two university, four tertiary teaching and seven general hospitals (Fig. 2). Twenty records of patients on long-term VKA, admitted in three consecutive months between June to December 2015 were randomly selected for reviewing the bridging anticoagulation policy. Randomisation of eligible patient records was executed by hospital or research personnel using a random number generator available in common spreadsheet applications.

In case of the absence of a required (section of a) health record, a replacement was randomly selected instead. Inclusion criteria were: age ≥ 18 years, length of stay ≥24 h, undergoing acute or elective surgical procedure using general and/or spinal/epidural anaesthesia. Exclusion criteria were: psychiatric or gynaecologic/obstetric ward admission, admission from or discharge to other hospitals, trauma other than hip fractures on admission, pregnancy or six weeks postpartum and palliative care admission. We excluded patients from

	Compliance type	Description[a]	Potential avoidable consequences
Compliance	Compliant use	Not bridging low-risk patients Bridging high-risk patients	n/a
Non-compliance	Overuse	Bridging low-risk patients	Bleeding complications
	Underuse	Not bridging high-risk patients	Thromboembolic complications

[a] Recommendations for moderate-risk patients involve both bridging or not bridging, so both strategies are considered compliant

Fig. 1 A typology of guideline compliance in perioperative VKA management based on the American College of Chest Physicians' *Antithrombotic Therapy and Prevention of Thrombosis, Ninth Edition guideline*

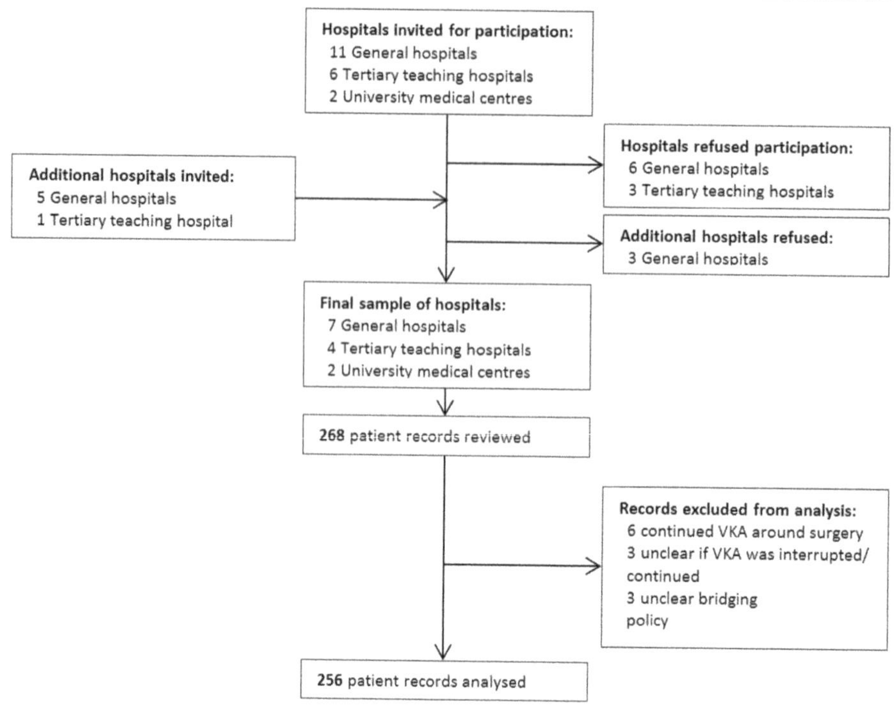

Fig. 2 Hospital sample and patient record flowchart

analysis if the bridging policy was not recorded, preventing the bridging classification or in case of continued VKA during surgery, making bridging unnecessary (Fig. 2).

Guideline selection

At the time of data collection in 2015 the Dutch guideline that encompassed bridging anticoagulation in VKA patients was the *Guideline for Diagnostics, Prevention and Treatment of Venous Thromboembolism and Secondary Prevention of Arterial Thrombosis* released by the former Dutch Quality Institute for Healthcare (CBO) in 2008 [20]. This guideline however was an adoption of the ACCP guideline for warfarin patients released back in 2004 [21]. During study preparations it became apparent that in 2015, current practice had moved on and the CBO guideline, at least partially, reflected outdated evidence regarding bridging anticoagulation. Especially since the ACCP updated their guidelines in 2008 and 2012. Several hospitals that were included in our study already pointed out that the AT9 recommendations regarding bridging were incorporated in local protocols. Taken altogether, using the AT9 as a frame of reference for the current study was regarded as the most appropriate.

Patient record review and compliance assessment

The patient record review consisted of two phases. Phase one involved the extraction of all data from patient records. Phase two involved the actual bridging anticoagulation evaluation. A panel of five experts in the thrombosis and haemostasis field, all of whom participated in guideline development on antithrombotic care, were consulted throughout the two phases. The panel contributed in developing standardized case report forms for phase one and classification models for determining guideline compliance in phase two.

Phase one: data extraction

In phase one, LMWH and UFH administration data was extracted from the patient records. Other data extracted were: demographic, clinical, surgery, and care delivery characteristics (Additional file 1: Tables S2-S3). For demographic characteristics variables such as age, sex and socioeconomic status (SES) were collected. SES was extracted from open source data available from the Netherlands Institute for Social Research and matched with our data using the patients four-digit zip code [22]. Clinical characteristics primarily included risk factors used for determining the AT9 thromboembolic risk classification [4]. These were supplemented with characteristics used in thrombo-prophylaxis risk assessment [23, 24] and patient related risk factors for bleeding as well as surgical bleeding [25, 26]. A previous bleeding event was defined as any bleeding coming to the attention of the treating physician.

In absence of an alternative validated instrument, determination of surgical bleeding risk was based on a Dutch expert consensus classification of procedures in low-, medium- or high-risk strata [27]. Other surgical characteristics extracted were: type, duration, whether a second surgery was performed and type of anaesthesia. Lastly, care delivery characteristics based on adverse event studies, such as weekend admission or surgery, were extracted [28–30].

Data extraction took place from January to August 2016. Trained research assistants and one researcher (MM) extracted all patient record data. The study protocol was approved by the medical ethics committee of the VU University Medical Centre, Amsterdam, The Netherlands and the informed consent was waived because of the use of patient record data only (protocol number: 2015/430).

Phase 2: classification of guideline compliance

In phase two, patients were classified on thromboembolic risk according to the AT9 (Additional file 1: Table S1) and bridging anticoagulation use. In case of multiple indications for VKA use (e.g. atrial fibrillation and mechanical heart valve), the indication associated with the highest thromboembolic risk was used for determining guideline compliance.

The bridging anticoagulation classification was based on postoperative administration of LMWH or use of continuous intravenous UFH infusion. Prophylactic LMWH regimens were not classified as bridging. See Additional file 1: Table S2 for details on LMWH dosages classified as bridging anticoagulation. Compliance with the guideline was defined as withholding bridging anticoagulation in low thromboembolic risk patients and administering bridging anticoagulation in high thromboembolic risk patients. *Underuse* was defined as not bridging high thromboembolic risk patients. *Overuse* was defined as bridging low thromboembolic risk patients (Fig. 1). For moderate-risk patients, both bridging and non-bridging were defined as compliant, since the AT9 does not recommend a specific approach for this patient group.

Statistics and model development

To describe the study population regarding demographic, clinical, surgical and care delivery characteristics we used descriptive statistics. Characteristics associated with bridging use were analysed with univariable and multivariable logistic regression. The dependent variable in our first model was: *bridging* versus *no bridging*. Independent variables considered for entry in the model were the aforementioned demographic, clinical, surgical and care delivery characteristics.

To predict a guideline discordant bridging decision in relation to the AT9 guideline, we created two separate models. One to identify predictors for *overuse* in the low thromboembolic risk population and one to identify predictors for *underuse* in the high thromboembolic risk population. Independent variables in this second and third model were slightly different compared to the first model. We excluded variables for which the exposure to the independent variable did not precede the measurement of the dependent variable (admission on critical or cardiac care unit, length of stay, presence of central venous or spinal and epidural catheters, second surgery performed, laboratory tests during admission). Furthermore, the AT9 thromboembolic risk was not considered as an independent variable since thromboembolic risk served as the foundation for the classification of, under- and overuse of bridging, therefore not being informative.

Univariable logistic regression results are presented as odds ratios (OR) and 95% confidence intervals (95% CI). Results were considered significant if the 95% CI did not intersect unity. Following the univariable analyses, a p-value entry level set at < 0.10 was used for a multivariable forward selection procedure. The maximum number of independent variables allowed in the models was based on the 10:1 rule to prevent overfitting [31]. Cases with missing values for independent variables were excluded from the regression analyses. Furthermore, variables with more than 10% missing values were not considered for multivariable modelling.

To enable our models to estimate predictor coefficients independent of possible practice variation between hospitals, we applied a multilevel approach in all regression analyses. Because the patient data were clustered within hospitals a random intercept on hospital level was allowed. C-statistics were calculated to evaluate the discriminative power of the models. A c-statistic of 0.5 to 0.7 is interpreted as a low discriminative power, 0.7–0.9 as moderate and > 0.9 as high. Statistical analyses were performed with SPSS version 22 (IBM, Chicago, IL).

Results

Study population

In total, 268 records were reviewed of which 256 records were eligible for bridging anticoagulation analyses (Fig. 2). The mean age of patients was 74.8 (SD = 10.6) years, 55.9% were male. Other characteristics are displayed in Table 1. Atrial fibrillation was the most common indication for VKA use with (74.2%). Thromboembolic risk was low, moderate or high in 52.7, 14.8 and 15.6% of patients respectively. 33 (12.9%) patients used VKA for other indications than AT9 provides recommendations and could thus not be classified according to AT9 thromboembolic risk. In 10 (3.9%) patients the records provided insufficient information for thromboembolic risk classification.

Table 2 displays the AT9 thromboembolic risk of the patients for each of the indications for VKA use. The most prevalent thromboembolic risk category was low for atrial fibrillation patients (69%), moderate for venous

Table 1 Demographic, clinical and surgical characteristics for the overall population

	Patients $N = 256$
Demographic characteristics	
Male sex	143 (55.9)
Age (years), mean (SD)	74.76 (10.59)
Clinical characteristics	
AT9 Thromboembolic risk	
Low	135 (52.7)
Moderate	38 (14.8)
High	40 (15.6)
Other VKA indication [a]	33 (12.9)
Risk factors unknown	10 (3.9)
Atrial fibrillation	190 (74.2)
Mechanical heart valve	20 (7.8)
Venous thromboembolism	34 (13.3)
Previous thromboembolic event during VKA interruption	3 (1.2)
iCVA/TIA	37 (14.5)
Thrombophilia	7 (2.7)
Coronary heart disease	74 (28.9)
Heart failure	20 (7.8)
Hypertension	129 (50.4)
Diabetes mellitus	62 (24.2)
Active cancer/malignancy	54 (21.5)
Previous bleeding[b]	13 (5.1)
VKA regimen	
Acenocoumarol	203 (79.3)
Phenprocoumon	53 (20.7)
Length of stay (days): median (IQR)	6 (3–10)
Surgery characteristics	
Elective	181 (70.7)
Type of 1st surgery	
Urologic	40 (15.6)
Orthopaedic	89 (34.8)
Gastrointestinal	52 (20.3)
Vascular	36 (14.1)
Other	39 (15.2)
Surgical bleeding risk	
High	209 (81.6)
Moderate	44 (17.2)
Low	3 (1.2)

Results are expressed as n (%) unless stated otherwise
AT9 Antithrombotic Therapy and Prevention of Thrombosis, Ninth Edition guideline, *iCVA* ischaemic cerebrovascular accident, *IQR* inter quartile range, *SD* standard deviation, *TIA* transient ischaemic attack, *VKA* vitamin-K antagonist
[a] No AT9 risk classification is available for VKA indications other than atrial fibrillation, mechanical heart valves and venous thromboembolism
[b] Any previous bleeding event annotated in the medical record

Table 2 AT9 thromboembolic risk for each of the VKA indication groups

	Indication group: n (column %) [a]		
AT9 Thromboembolic risk	Atrial fibrillation	Mechanical heart valve	Venous thromboembolism
Low	131 (69)	1 (5)	6 (18)
Moderate	19 (10)	4 (20)	22 (65)
High	35 (18)	9 (45)	4 (12)
Unknown [b]	5 (3)	6 (30)	2 (6)

AT9: Antithrombotic Therapy and Prevention of Thrombosis, Ninth Edition guideline; VKA: vitamin-K antagonist
[a] Multiple indications are possible
[b] Insufficient documentation of risk factors in the records, so the AT9 risk could not be determined

thromboembolism patients (65%) and high for mechanical heart valve patients (45%).

Bridging use and guideline compliance

In 107 (41.8%) patients, bridging anticoagulation was used. Bridging rates between hospitals ranged from 15 to 83% of all patients per hospital (Fig. 3a). Based on the AT9 thromboembolic risk recommendations, the decision to apply or withhold bridging anticoagulation was compliant with the guideline in 68.5% of all patients for which the thromboembolic risk could be determined ($N = 213$). Compliance rates for each AT9 risk and VKA indication strata are given in Table 3. Compliance was lowest for high risk atrial fibrillation patients (46%, $N = 35$), low risk venous thromboembolism (50%, $N = 6$) and low risk mechanical heart valve patients (0%, $N = 1$), however the latter two were very small strata. Low risk atrial fibrillation patients on the other hand, comprised the largest stratum in our study (51.2% of the total population), with a compliance of 67%.

Comparing hospitals, the compliance rate ranged from 20 to 88% of all patients per hospital (Fig. 3b).

Factors associated with use of bridging anticoagulation

Univariable logistic regression results for the application of bridging are presented in Additional file 1: Table S3. Compared to low-risk patients, moderate thromboembolic risk patients had a significant increased odds (OR 3.36, 95% CI 1.52–7.41) and high-risk patients a borderline insignificant increased odds (OR 2.05, 95% CI 0.95–4.21) for receiving bridging anticoagulation. Furthermore, all three main indications for VKA use were significantly associated with bridging: mechanical heart valve (OR 3.69, 95% CI 1.34–10.20) and venous thromboembolism (OR 2.35, 95% CI 1.09–5.07) patients were more likely to receive bridging anticoagulation while atrial fibrillation patients (OR 0.50, 95% CI 0.27–0.92) were less likely to be bridged.

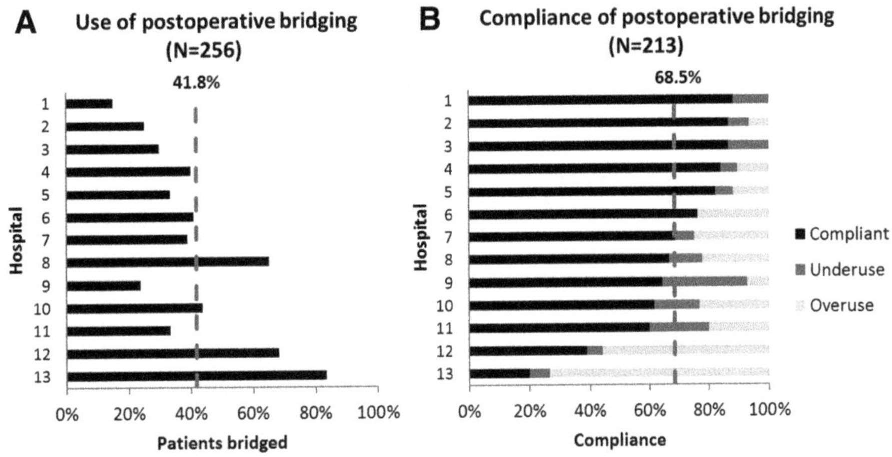

Fig. 3 Barcharts displaying the use (**a**) and compliance (**b**) of postoperative bridging anticoagulation per hospital and on average. The dashed vertical lines represent the average

Characteristics outside the AT9 thromboembolic risk assessment associated with bridging were length of hospital stay (OR 1.07 per day, 95% CI 1.03–1.11), critical or cardiac care unit admission (OR 3.80, 95% CI 1.80–8.05), second surgery (OR 6.45, 95% CI 1.96–21.21), and admission to a university hospital (OR 3.94, 95% CI 1.16–13.35). Lastly, gastrointestinal (OR 3.83, 95% CI 1.50–9.74), vascular (OR 3.74, 95% CI 1.36–10.29) and other (OR 3.06, 95% CI 1.12–8.39) surgery types were positively associated with bridging.

Our multivariable logistic regression analysis included 249 patients and resulted in a model with critical or cardiac care unit admission, second surgery, mechanical heart valve, surgery type, venous thromboembolism, ischeamic CVA or TIA and previous bleeding, as explanatory variables for bridging use. Regression parameters are displayed in Table 4. The model's power to discriminate between bridged and non-bridged patients was moderate (c-statistic 0.85, 95% CI 0.80–0.90).

Table 3 Compliance of postoperative bridging per indication and AT9 thromboembolic risk group

	Compliance of postoperative bridging per indication group: n(%) [a]		
AT9 Thromboembolic risk	Atrial fibrillation	Mechanical heart valve	Venous thromboembolism
Low	88 (67)	0 (0)	3 (50)
Moderate	19 (100)	4 (100)	22 (100)
High	16 (46)	5 (56)	3 (75)

AT9: Antithrombotic Therapy and Prevention of Thrombosis, Ninth Edition guideline; VKA: vitamin-K antagonist

[a] Multiple indications are possible

Predictors of over- and underuse of bridging anticoagulation

Overuse of bridging anticoagulation occurred in 34.1% of low thromboembolic risk patients and underuse occurred in 52.5% of high thromboembolic risk patients. Univariable logistic regression results for both over- and under use are presented in Additional file 1: Table S4. Within low risk patients, positive associations for overuse of bridging were found for non-elective surgery (OR 2.72, 95% CI 1.03–7.19), gastrointestinal (OR 15.87, 95% CI 3.02–83.42), vascular (OR 9.58, 95% CI 1.49–61.42) and other (OR 27.43, 95% CI 3.49–215.38) surgery types, and admission to a university medical centre (OR 9.01, 95% CI 1.05–77.57).

The high risk patient strata was of limited size (40 patients). Hence, the power to capture a significant association for underuse within this population was limited. Only a borderline insignificant effect for surgery duration was observed (OR 0.98 per minute, 95% CI 0.96–1.00).

The multivariable logistic regression parameters for predicting overuse of bridging are presented in Table 4. Surgery type and non-elective surgery were positive predictors whereas membership of the lowest 25th percentile SES and VKA reversal agent use were negative predictors for overuse of bridging. The discriminative power for predicting overuse of bridging was high (c-statistic 0.91, 95% CI 0.86–0.97).

Discussion

Bridging use and guideline compliance

In 31.5% of the patients in our sample the bridging anticoagulation policy was not compliant with the American College of Chest Physicians' *Antithrombotic Therapy and Prevention of Thrombosis, Ninth Edition* recommendations. Bridging was used during 41.8% of

Table 4 Multivariable logistic regression models for bridging use and overuse of bridging, adjusted for clustering at hospital level

	OR (95% CI)[a]
Model 1, All patients	
Bridging used (reference: no bridging used)	
ICU/CCU stay during admission	4.45 (1.72–11.51)
Second surgery performed	3.21 (0.83–12.49)
Mechanical heart valve	8.10 (2.38–27.50)
Type of 1st surgery (reference category: urologic)	
Orthopaedic	1.10 (0.42–2.91)
Gastrointestinal	3.45 (1.21–9.87)
Vascular	3.21 (1.01–10.21)
Other	3.57 (1.14–11.21)
Venous thromboembolism	3.91 (1.57–9.74)
iCVA/TIA	2.49 (1.02–6.11)
Previous bleeding[b]	3.59 (0.80–16.17)
Model 2, Low TE risk patients:	
Overuse of bridging (reference: compliant use)	
Type of 1st surgery (reference category: urologic)	
Orthopaedic	3.18 (0.60–16.71)
Gastrointestinal	14.85 (2.69–81.99)
Vascular	13.01 (1.83–92.30)
Other	57.30 (5.27–623.62)
Non-elective surgery	8.67 (1.67–45.14)
Lowest 25th percentile SES	0.33 (0.11–1.02)
VKA reversal agent used	0.22 (0.04–1.16)

CCU cardiac care unit, *ICU* intensive care unit, *iCVA* ischaemic cerebrovascular accident, *TIA* transient ischaemic attack, *SES* Socioeconomic status, *VKA* Vitamin-K antagonist

[a] Adjusted for clustering at hospital level

[b] Any previous bleeding event annotated in the medical record

VKA interruptions, lower than reported in existing literature [13, 14, 16]. As a result, the 52.5% underuse of bridging was higher in our study compared with 36.8 and 13.0% reported in other studies [13, 14]. Conversely, the 34.1% overuse of bridging in low risk patients is on the lower side of the spectrum of overuse rates reported by others that ranged between 28.7 and 84.3% [13–15].

However, these low-risk patients represent over 50% of the VKA patient population in our study and are mostly patients with atrial fibrillation. Although the exact number of VKA patients undergoing surgery in the Netherlands is unavailable, there are over 460.000 VKA patients present [32]. Based on our findings, overuse of bridging is likely to occur in a substantial amount.

In light of accumulating evidence towards increased bleeding risk among bridged patients this overuse warrants attention. In a meta-analysis of predominantly observational studies, Siegal et al. 2012 found that bridged patients had a 5-fold increased risk for overall, and a 3-fold increased risk for major bleeding [11, 12]. This was confirmed by Douketis et al. in 2015 in the BRIDGE-trial, where the risk of major bleeding was 0.41; 95% CI 0.20–0.78 for non-bridged patients relative to bridged patients [12]. Bleeding complications occurring in bridged patients have been found to increase the risk for reoperation and prolonged hospitalisation [33, 34]. Moreover, the BRIDGE-trial also found that non-bridging was not associated with an increased incidence of thromboembolic complications, which contradicts the rationale behind bridging anticoagulation.

Given this evidence and our study results, low risk atrial fibrillation patients undergoing surgery comprise a large group of patients who might benefit the most from improvement efforts to reduce bridging overuse and reduce adverse bleeding outcomes.

Factors associated with bridging anticoagulation, predicting over- and underuse

To understand why current bridging practice is not always in line with guideline recommendations we aimed to identify characteristics associated with bridging use. The associations found for atrial fibrillation, mechanical heart valve and venous thromboembolism patients correspond with the findings of others where most atrial fibrillation patients did not receive bridging and most mechanical heart valve and venous thromboembolism patients were at least at moderate thromboembolic risk justifying bridging anticoagulation use [14, 35, 36]. Regarding, the aggregated AT9 thromboembolic risk strata, the moderate and high risk strata were more likely to receive bridging, which is to be expected. However, the introduction of individual thromboembolic risk factors and other characteristics in our multivariable analysis rendered the association insignificant. Translating this to practice, it can be argued that awareness to the aggregated AT9 thromboembolic risk might be limited to individual risk factors that make up the AT9 risk strata. Also, patient characteristics outside the AT9 thromboembolic risk assessment may be involved in the decision to apply bridging. Our study points to several of these.

First, a history of bleeding showed a positive association with bridging. This seems contradictory, and is difficult to explain. One would expect a more conservative approach to using bridging anticoagulation in patients with signs of a previous bleeding. However, only recently the risks of bridging versus uninterrupted anticoagulation were supported with high quality data. Before this, bridging with fast onset and offset heparins seemed the safest option.

Second, bridging use and overuse occurred more frequently in gastrointestinal, vascular and other surgery types compared with urologic and orthopaedic surgery.

Perceived thromboembolic risks relative to the surgical procedures can play a role. The AT9 thromboembolic risk classification does not formally include this but designates certain high thromboembolic risk procedures [4]. Furthermore, heterogeneous practice and preferences between medical specialties related to the studied surgery types might be responsible for our findings. A recent survey study underscores this. Flaker et al. (2016), found different perioperative management strategies between medical specialties [37].

Third, Intensive or cardiac care unit admission and a second surgery were associated with higher bridging rates. We think this is possibly explained by factors related with the severity of the patient's disease and clinical course that we were unable to correct for, such as the inability to take oral medication. In these cases parenteral heparins are a feasible alternative to oral VKAs.

Regarding bridging overuse specifically, primarily surgical characteristics such as type and urgency were predictive for non-compliant use of bridging. Based on these characteristics, the population at which further investigation and improvement efforts should aim for can be narrowed down. Additionally we found that membership of the lowest 25th percentile socioeconomic status was a significant negative predictor for overuse. Besides socioeconomic status being a well-established determinant for health and access to health services [38, 39], associations with guideline compliance have also been found before [40, 41].

Altogether, our exploratory analyses indicate that current bridging anticoagulation practice is not explained by the ACCP's thromboembolic risk assessment recommendations alone. Our study therefore confirms the findings of several other studies [14, 15, 42, 43]. Why practice is not in accordance with bridging recommendations is relatively unknown. Whether the other associated clinical and surgical characteristics identified, are the result of a conscious assessment in everyday bridging practice, cannot be concluded based on our results.

Practice variation

Our results also revealed variation between hospitals. Bridging varied from 15 to 83% of patients, similar to a US study where rates ranged from 10 to 88% [36]. Furthermore, hospitals that bridged more frequently had lower compliance rates and higher overuse rates. Thus, higher bridging rates cannot solely be explained by case-mix differences regarding thromboembolic risk.

More likely, a heterogeneous implementation or embedment of guidelines into local processes and protocols results in variations in practice. For example, differences in responsible professionals in terms of specialty or experience might affect the risk assessment for bridging anticoagulation. The existence of variation like this was endorsed in a Dutch report revealing substantial differences between hospital's adaptations of an integrated anticoagulant care guideline. This guideline predominantly contains recommendations regarding care processes, responsibilities and communication for anticoagulant care. Among others, a major difference observed was the instalment of dedicated anticoagulation committee's or case managers while other hospitals were less progressive [44].

Strengths and limitations

Our study has several strengths and limitations. Our multi-centre design is a strength and informs us on bridging anticoagulation practice in a variety of hospitals while the entire sample was representative for the Dutch hospital distribution. The voluntary hospital participation can be regarded as a limitation that could have introduced selection bias on hospitals' awareness or priority regarding anticoagulant care. The retrospective approach is another strength in ensuring results not being influenced by carrying out the study but rather reflect everyday care. On the other hand, the dependency of routinely recorded medical data might be a limitation. Although efforts were made to retrieve all required information, some records were found to be too incomplete to include and others were prone to missing information, especially details required for thromboembolic risk classification of mechanical heart valve patients. While this might have introduced some bias to our results, it also stresses the importance of adequate record quality.

Additionally, we wish to nuance non-compliance. First, our study was carried out in a transition period between an outdated guideline and the adoption of the AT9. Second, the reasoning behind informed guideline deviations were not collected from the medical records. Hence, we wish to point out that non-compliance with the guidelines does not necessarily reflect poor care.

Lastly, the limited amount of high risk patients in our sample prevented a multivariable analysis for bridging underuse.

Conclusions and implications

In 31.5% of the patients the bridging anticoagulation policy was not in line with the AT9 recommendations.

Improvement efforts targeted at low-risk atrial fibrillation patients are expected have the biggest effect on overall compliance and potentially adverse outcomes since these patients represented over 50% of our study population. Bridging was predominantly related with individual clinical and surgical characteristics rather than the aggregated AT9 thromboembolic risk. Overuse of bridging, was the most prevalent form of non-compliance. Gastrointestinal, vascular and non-elective surgery were risk factors for overuse. Underuse of bridging in high-risk patients was less prevalent and no significant risk factors were identified. Our results raise the question whether AT9 risk assessment sufficiently reflects the risks that are perceived in everyday practice or if they are deviant for other reasons. Also a large variation in bridging practice between hospitals was observed, where hospitals with high bridging rates had lower compliance rates and vice versa.

Based on our study, several implications can be thought of to improve bridging anticoagulation practice. 1) Qualitative research can inform us on the reasons and mechanisms leading to differences between everyday practice and what is advocated in the guidelines. 2) The characteristics associated with non-compliant bridging, should be taken in to account in interventions aimed at improving decision making in bridging anticoagulation, e.g. electronic decision support systems. 3) Variation between hospitals regarding the implementation and embedment of guidelines in local practice should be studied to identify factors related with practice variation.

Abbreviations

AT9: The American College of Chest Physicians' *Antithrombotic Therapy and Prevention of Thrombosis, Ninth Edition*; LMWH: Low molecular weight heparin; SES: Socioeconomic status; UFH: Unfractionated heparin; VKA: Vitamin-K antagonist

Acknowledgements

This work was funded by the Dutch Ministry of Healthcare Welfare and Sport.

We wish to acknowledge F.W.G. Leebeek from the Erasmus University Medical Center, T.P.W. Kamphuisen from the University Medical Center Groningen, and S. Middeldorp from the Academic Medical Center Amsterdam for their contribution to this study as members of the expert panel.

Additionally, we acknowledge W. van de Schors, M. Tanaka and L.M. Bouma for assisting with the data collection.

Authors' contributions

MM, ND, MdB and CW designed the study and developed the study protocol. MH chaired the expert panel and contributed to the study design. NB was a member of the expert panel and contributed to study design. MM and ND organised the data collection. MM collected the data. MM and JT designed the statistical analyses. MM performed the statistical analyses. MM, JdG, MH and NB interpreted the results. MM wrote the manuscript. All authors made critical revisions and approved the final version of the manuscript.

Author details

[1]Department of Public and Occupational Health, Amsterdam Public Health Research Institute, Amsterdam UMC, Vrije Universiteit Amsterdam, De Boelelaan 1117, 1081BT Amsterdam, The Netherlands. [2]Netherlands Institute for Health Services Research (NIVEL), Utrecht, The Netherlands. [3]Department of Quality and Safety, Elizabeth Tweesteden Ziekenhuis, Tilburg, The Netherlands. [4]Department of Cardiology, Flevoziekenhuis, Almere, The Netherlands. [5]Department of Epidemiology and Biostatistics, Amsterdam UMC, Vrije Universiteit Amsterdam, Amsterdam, The Netherlands. [6]Department of Thrombosis and Hemostasis, Leiden University Medical Centre, Leiden, The Netherlands.

References

1. Kearon C, Akl EA, Comerota AJ, Prandoni P, Bounameaux H, Goldhaber SZ, et al. Antithrombotic therapy for VTE disease: antithrombotic therapy and prevention of thrombosis, 9th ed: American College of Chest Physicians Evidence-Based Clinical Practice Guidelines. Chest. 2012;141(2 Suppl):e419S–94S.
2. Baumgartner H, Falk V, Bax JJ, De Bonis M, Hamm C, Holm PJ, et al. 2017 ESC/EACTS guidelines for the management of valvular heart disease. Eur Heart J. 2017;38(36):2739–91.
3. Kirchhof P, Benussi S, Kotecha D, Ahlsson A, Atar D, Casadei B, et al. 2016 ESC guidelines for the management of atrial fibrillation developed in collaboration with EACTS. Eur Heart J. 2016;37(38):2893–962.
4. Douketis JD, Spyropoulos AC, Spencer FA, Mayr M, Jaffer AK, Eckman MH, et al. Perioperative management of antithrombotic therapy: antithrombotic therapy and prevention of thrombosis, 9th ed: American College of Chest Physicians Evidence-Based Clinical Practice Guidelines. Chest. 2012;141(2 Suppl):e326S–e50S.
5. Douketis JD, Johnson JA, Turpie AG. Low-molecular-weight heparin as bridging anticoagulation during interruption of warfarin: assessment of a standardized periprocedural anticoagulation regimen. Arch Intern Med. 2004;164(12):1319–26.
6. Wysokinski WE, McBane RD, Daniels PR, Litin SC, Hodge DO, Dowling NF, et al. Periprocedural anticoagulation management of patients with nonvalvular atrial fibrillation. Mayo Clin Proc. 2008;83(6):639–45.
7. Spyropoulos AC, Turpie AG, Dunn AS, Spandorfer J, Douketis J, Jacobson A, et al. Clinical outcomes with unfractionated heparin or low-molecular-weight heparin as bridging therapy in patients on long-term oral anticoagulants: the REGIMEN registry. J Thromb Haemost. 2006;4(6):1246–52.
8. Hoonhout LH, de Bruijne MC, Wagner C, Asscheman H, van der Wal G, van Tulder MW. Nature, occurrence and consequences of medication-related adverse events during hospitalization: a retrospective chart review in the Netherlands. Drug Saf. 2010;33(10):853–64.
9. Damen NL, Baines R, Wagner C, Langelaan M. Medication-related adverse events during hospitalization: a retrospective patient record review study in the Netherlands. Pharmacoepidemiol Drug Saf. 2017;26(1):32–9.
10. Henriksen JN, Nielsen LP, Hellebek A, Poulsen BK. Medication errors involving anticoagulants: data from the Danish patient safety database. Pharmacol Res Perspect. 2017;5(3):e00307.
11. Siegal D, Yudin J, Kaatz S, Douketis JD, Lim W, Spyropoulos AC. Periprocedural heparin bridging in patients receiving vitamin K antagonists: systematic review and meta-analysis of bleeding and thromboembolic rates. Circulation. 2012;126(13):1630–9.
12. Douketis JD, Spyropoulos AC, Kaatz S, Becker RC, Caprini JA, Dunn AS, et al. Perioperative bridging anticoagulation in patients with atrial fibrillation. N Engl J Med. 2015;373(9):823–33.
13. Clark NP, Witt DM, Davies LE, Saito EM, McCool KH, Douketis JD, et al. Bleeding, recurrent venous thromboembolism, and mortality risks during warfarin interruption for invasive procedures. JAMA Intern Med. 2015;175(7):1163–8.
14. Steib A, Mertes PM, Marret E, Albaladejo P, Fusciardi J. Compliance with guidelines for the perioperative management of vitamin K antagonists. Thromb Res. 2014;133(6):1056–60.
15. Eijgenraam P, ten Cate H, ten Cate-Hoek AJ. Practice of bridging anticoagulation: guideline adherence and risk factors for bleeding. Neth J Med. 2014;72(3):157–64.
16. Perrin MJ, Vezi BZ, Ha AC, Keren A, Nery PB, Birnie DH. Anticoagulation bridging around device surgery: compliance with guidelines. Pacing Clin Electrophysiol. 2012;35(12):1480–6.
17. Wysowski DK, Nourjah P, Swartz L. Bleeding complications with warfarin use: a prevalent adverse effect resulting in regulatory action. Arch Intern Med. 2007;167(13):1414–9.
18. Douketis JD, Healey JS, Brueckmann M, Eikelboom JW, Ezekowitz MD, Fraessdorf M, et al. Perioperative bridging anticoagulation during dabigatran or warfarin interruption among patients who had an elective surgery or procedure. Substudy of the RE-LY trial. Thromb Haemost. 2015;113(3):625–32.
19. Moesker MJ, Damen NL, de Groot JF, de Bruijne MC, Wagner C. Antistollingszorg in Nederlandse ziekenhuizen. Evaluatie van tromboseprofylaxe en perioperatief antistollingsbeleid in vergelijking met geldende richtlijnen. July 26 2017 ed. Utrecht/Amsterdam: NIVEL (Netherlands Institute for Health Services Research) / EMGO+ (EMGO Institute for Health and Care Research). p. 2017.

20. Diagnostiek R. Preventie en Behandeling van Veneuze Trombo-embolie en Secundaire Preventie Arteriële Trombose. Utrecht: CBO (Kwaliteitsinstituut voor de Gezondheidszorg); 2008.
21. Ansell J, Hirsh J, Poller L, Bussey H, Jacobson A, Hylek E. The pharmacology and management of the vitamin K antagonists: the seventh ACCP conference on antithrombotic and thrombolytic therapy. Chest. 2004;126(3 Suppl):204s–33s.
22. The Netherlands Institute for Social Research: Statusscores 2014: The Netherlands Institute for Social Research; 2014 [Available from: https://www.scp.nl/Onderzoek/Lopend_onderzoek/A_Z_alle_lopende_onderzoeken/Statusscores.
23. Barbar S, Noventa F, Rossetto V, Ferrari A, Brandolin B, Perlati M, et al. A risk assessment model for the identification of hospitalized medical patients at risk for venous thromboembolism: the Padua prediction score. Journal of thrombosis and haemostasis : JTH. 2010;8(11):2450–7.
24. Spyropoulos AC, Anderson FA Jr, FitzGerald G, Decousus H, Pini M, Chong BH, et al. Predictive and associative models to identify hospitalized medical patients at risk for VTE. Chest. 2011;140(3):706–14.
25. Decousus H, Tapson VF, Bergmann JF, Chong BH, Froehlich JB, Kakkar AK, et al. Factors at admission associated with bleeding risk in medical patients: findings from the IMPROVE investigators. Chest. 2011;139(1):69–79.
26. Klok FA, Barco S, Konstantinides SV. External validation of the VTE-BLEED score for predicting major bleeding in stable anticoagulated patients with venous thromboembolism. Thromb Haemost. 2017;117(6):1164–70.
27. Hematology folder Leiden 2006. Available from: https://www.lumc.nl/org/hematologie/hematologieklapper/. Accessed 1 Apr 2017.
28. Van De Steeg L, Langelaan M, Wagner C. Can preventable adverse events be predicted among hospitalized older patients? The development and validation of a predictive model. Int J Qual Health Care. 2014;26(5):547–52.
29. Baines RJ, Langelaan M, de Bruijne MC, Asscheman H, Spreeuwenberg P, van de Steeg L, et al. Changes in adverse event rates in hospitals over time: a longitudinal retrospective patient record review study. BMJ quality & safety. 2013;22(4):290–8.
30. Shahian DM, Nordberg P, Meyer GS, Mort E, Atamian S, Liu X, et al. Predictors of nonadherence to national hospital quality measures for heart failure and pneumonia. Am J Med. 2011;124(7):636–46.
31. Harrell FE Jr, Lee KL, Mark DB. Multivariable prognostic models: issues in developing models, evaluating assumptions and adequacy, and measuring and reducing errors. Stat Med. 1996;15(4):361–87.
32. Samenvatting Medische Jaarverslagen 2015. Federatie van Nederlandse Trombosediensten FNT; 2016.
33. Birnie DH, Healey JS, Wells GA, Verma A, Tang AS, Krahn AD, et al. Pacemaker or defibrillator surgery without interruption of anticoagulation. N Engl J Med. 2013;368(22):2084–93.
34. Rechenmacher SJ, Fang JC. Bridging anticoagulation: Primum non Nocere. J Am Coll Cardiol. 2015;66(12):1392–403.
35. Delate T, Meisinger SM, Witt DM, Jenkins D, Douketis JD, Clark NP. Bridge therapy outcomes in patients with mechanical heart valves. Clin Appl Thromb Hemost. 2017;23(8):1036–41.
36. Jaffer AK, Brotman DJ, Bash LD, Mahmood SK, Lott B, White RH. Variations in perioperative warfarin management: outcomes and practice patterns at nine hospitals. Am J Med. 2010;123(2):141–50.
37. Flaker GC, Theriot P, Binder LG, Dobesh PP, Cuker A, Doherty JU. Management of Periprocedural Anticoagulation: a survey of contemporary practice. J Am Coll Cardiol. 2016;68(2):217–26.
38. Erreygers G. A dual Atkinson measure of socioeconomic inequality of health. Health Econ. 2013;22(4):466–79.
39. Bashinskaya B, Nahed BV, Walcott BP, Coumans JV, Onuma OK. Socioeconomic status correlates with the prevalence of advanced coronary artery disease in the United States. PLoS One. 2012;7(9):e46314.
40. Hodeib M, Chang J, Liu F, Ziogas A, Dilley S, Randall LM, et al. Socioeconomic status as a predictor of adherence to treatment guidelines for early-stage ovarian cancer. Gynecol Oncol. 2015;138(1):121–7.
41. Kuijer A, Verloop J, Visser O, Sonke G, Jager A, van Gils CH, et al. The influence of socioeconomic status and ethnicity on adjuvant systemic treatment guideline adherence for early-stage breast cancer in the Netherlands. Ann Oncol. 2017;28(8):1970–8.
42. Krahn AD, Healey JS, Simpson CS, Essebag V, Sivakumaran S, Birnie DH. Anticoagulation of patients on chronic warfarin undergoing arrhythmia device surgery: wide variability of perioperative bridging in Canada. Heart Rhythm. 2009;6(9):1276–9.
43. Skolarus LE, Morgenstern LB, Froehlich JB, Lisabeth LD, Brown DL. Guideline-discordant periprocedural interruptions in warfarin therapy. Circ Cardiovasc Qual Outcomes. 2011;4(2):206–10.
44. Damen N, Damen P, Dijk van L, Bakker de D, Wagner C. Implementatie Landelijke Standaard Ketenzorg Antistolling 2.0: stand van zaken. Utrecht: NIVEL; 2015.

15

Sepsis and septic shock: Endothelial molecular pathogenesis associated with vascular microthrombotic disease

Jae C. Chang

Abstract

In addition to protective "immune response", sepsis is characterized by destructive "endothelial response" of the host, leading to endotheliopathy and its molecular dysfunction. Complement activation generates membrane attack complex (MAC). MAC causes channel formation to the cell membrane of pathogen, leading to death of microorganisms. In the host, MAC also may induce channel formation to innocent bystander endothelial cells (ECs) and ECs cannot be protected. This provokes endotheliopathy, which activates two independent molecular pathways: inflammatory and microthrombotic. Activated inflammatory pathway promotes the release of inflammatory cytokines and triggers inflammation. Activated microthrombotic pathway mediates platelet activation and exocytosis of unusually large von Willebrand factor multimers (ULVWF) from ECs and initiates microthrombogenesis. Excessively released ULVWF become anchored to ECs as long elongated strings and recruit activated platelets to assemble platelet-ULVWF complexes and form "microthrombi". These microthrombi strings trigger disseminated intravascular microthrombosis (DIT), which is the underlying pathology of endotheliopathy-associated vascular microthrombotic disease (EA-VMTD). Sepsis-induced endotheliopathy promotes inflammation and DIT. Inflammation produces inflammatory response and DIT orchestrates consumptive thrombocytopenia, microangiopathic hemolytic anemia, and multiorgan dysfunction syndrome (MODS). Systemic inflammatory response syndrome (SIRS) is a combined phenotype of inflammation and endotheliopathy-associated (EA)-VMTD. Successful therapeutic design for sepsis can be achieved by counteracting the pathologic microthrombogenesis.

Keywords: Anti-microthrombotic therapy, C5b-9 (membrane attack complex [MAC]), Disseminated intravascular coagulation ("DIC", ill-founded DIC), Disseminated intravascular microthrombosis (DIT), Endotheliopathy, Microthrombogenesis, Multiorgan dysfunction syndrome (MODS), Therapeutic plasma exchange (TPE), TTP-like syndrome, Unusually large von Willebrand factor multimers (ULVWF), Vascular microthrombotic disease (VMTD)

Background

Sepsis is a serious life-threatening systemic physical illness associated with toxicity due to invasion of the bloodstream by pathogen. Invading pathogen can be bacteria, viruses, rickettsia, fungi or parasites. Early clinical manifestations begin with inflammation and progress to circulatory organ dysfunction associated with significant hematopathologic changes. Well-defined clinical phenotypic features include consumptive thrombocytopenia [1, 2], hemolytic anemia [1], vascular microthrombosis [1–4], multiorgan dysfunction syndrome (MODS) [3], coagulopathy [5] and septic shock. While natural defensive mechanisms of innate and adaptive immune systems of the host (Fig. 1) and effective antimicrobial therapy can favorably influence the course of sepsis, it is still accountable for roughly 15% of in-hospital deaths and 6.2% of discharges to hospice [6].

More recently, pathologic destructive mechanism through endothelial system (Fig. 1) has been suspected to be detrimental to the recovery of septic patient and contribute to the high morbidity and mortality [7–10]. Although the role of the endothelium in the pathogenesis of sepsis is not clearly established yet, coagulopathy has been proposed to play a key role through crosstalk mechanism between inflammation and coagulation as a result of systemic endothelial injury [11–13]. This theory

Correspondence: jaec@uci.edu
Department of Medicine, University of California Irvine School of Medicine, Irvine, CA, USA

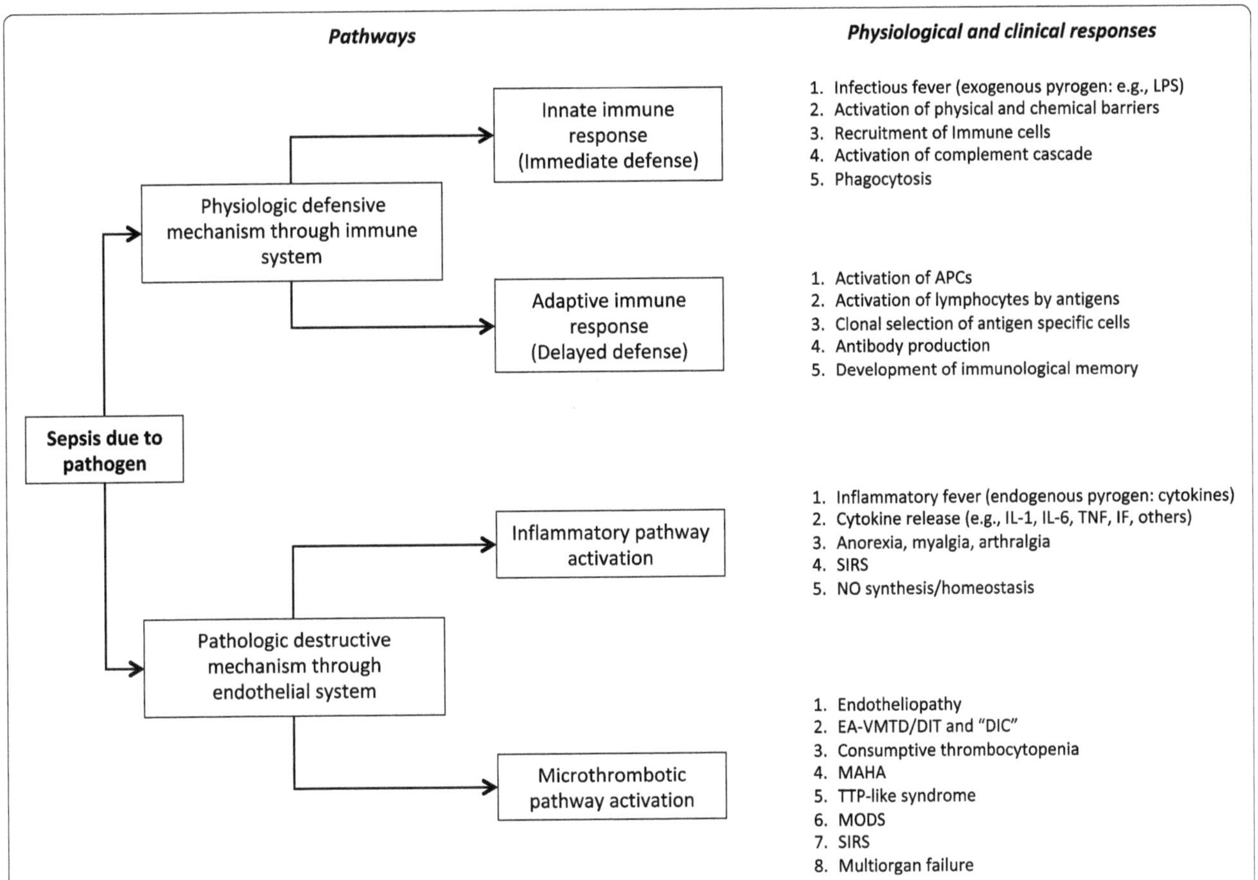

Fig. 1 Physiological and pathological response mechanisms in sepsis. In sepsis, host response is characterized by two mechanisms. One is physiologic defensive mechanism through immune system, and the other is pathologic destructive mechanism through endothelial system. The physiologic response and pathologic clinical syndromes are notated in the Figure. Now, we know that complement system, protecting the host through innate immune system, could trigger harmful endothelial pathogenesis. This dual role of the complement must be nature's rule just like normal hemostasis, which protects human lives in external bodily injury, but also may harm human lives in intravascular injury through thrombogenesis. Abbreviations: *APC* antigen presenting cell, *"DIC"* disseminated intravascular coagulation, *DIT* disseminated intravascular microthrombosis, *EA-VMTD* endotheliopathy-associated vascular microthrombotic disease, *MAHA* microangiopathic hemolytic anemia, *MODS* multiorgan dysfunction syndrome, *MOF* multiorgan failure, *NO* nitric oxide, *IF* interferon, *IL* interleukin, *LPS* lipopolysaccharide, *SIRS* systemic inflammatory response syndrome, *TNF* tumor necrosis factor, *TTP* thrombotic thrombocytopenic purpura

has been particularly attractive because disseminated intravascular coagulation ("DIC") commonly occurs in sepsis [14–16]. However, a quotation mark has been placed on "DIC" to note that it has been founded on the basis of poorly established concept of hemostasis [1–4]. Until recent controversy, "DIC" had been claimed to occur due to uncontrolled systemic activation of coagulation through tissue factor (TF)/FVIIa-initiated hemostasis, manifesting with thrombocytopenia, consumption coagulopathy, microvascular thrombosis and MODS [17, 18].

Now, a recurring question has come to light about the true character of "DIC". Although it is a hemostatic disease, "DIC" is defined as microthrombotic disease different from macrothrombotic disease seen in arterial thrombosis and deep vein thrombosis [1–4]. This author has made a reinterpretation of "DIC" utilizing clinical, pathological, pathophysiological characteristics and molecular pathogenesis, and has determined it to be identical to disseminated intravascular microthrombosis (DIT) seen in thrombotic thrombocytopenic purpura (TTP)-like syndrome in recent publications [1–4, 19]. For the time being, both terms "DIC" and ill-founded DIC will be used interchangeably throughout this article.

DIC and DIT in hemostasis and sepsis

To embrace the genuine characters between "DIC" and DIT, the pathophysiological mechanism of true DIC, "DIC" and DIT occurring as hemostasis should be clearly established. Indeed, through the better understanding of sepsis-associated coagulopathy, this has become possible to construct "two-path unifying theory" of hemostasis derived from the insights of endothelial pathogenesis that leads to microthrombogenesis and

TTP-like syndrome. This novel theory, which is simple and easily understandable, was proposed and published [20] and updated in Fig. 2. This effort has allowed us to use the term disseminated intravascular microthrombosis (DIT) instead of "DIC" and endotheliopathy-associated vascular microthrombotic disease (EA-VMTD/DIT) as a distinct disease entity that is associated with TTP-like syndrome [1–4, 19, 20]. This new identity of "DIC" could provide a paradigm shift in the management of sepsis by utilizing targeted therapies to improve the outcome. In addition, this thesis along with "two-activation theory of the endothelium" (Fig. 3) identifies the true character of "DIC" [1–4, 20] as well as the mechanisms of thrombogenesis [4] and different phenotypes of the thrombotic disorder [3, 4, 20].

The hemostasis in sepsis-induced endotheliopathy has been poorly understood even though research scientists have known that endothelial injury plays a prominent role in sepsis and many other critical illnesses such as trauma, pregnancy, autoimmune disease, cancer and

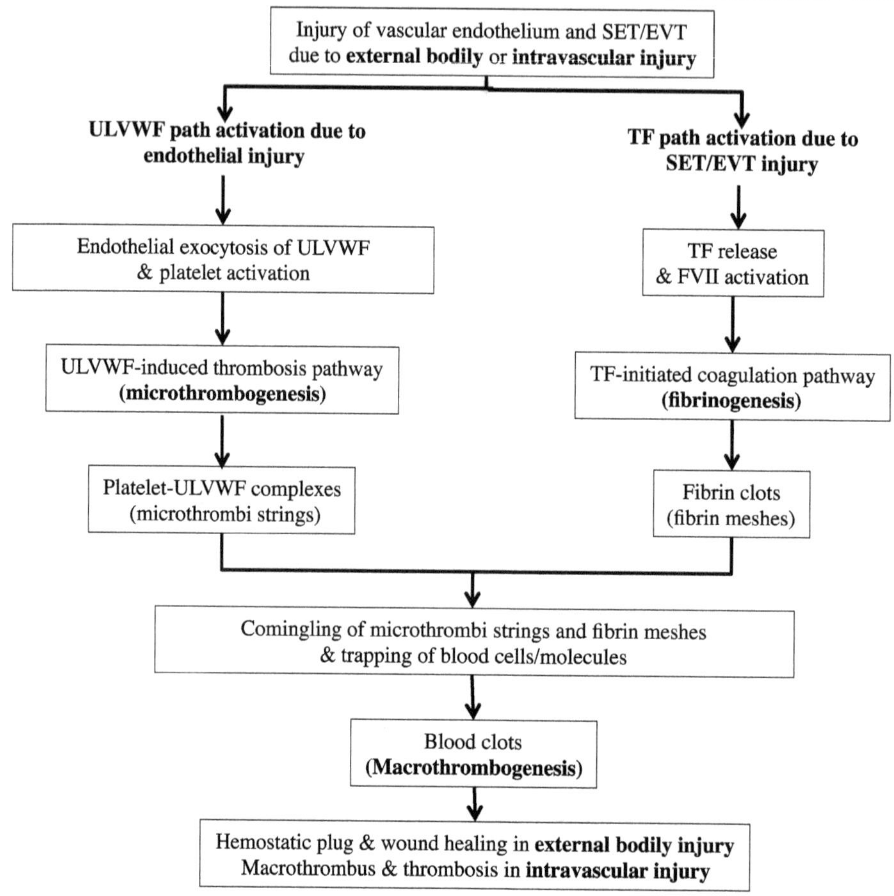

Fig. 2 Three different paths of thrombogenesis that can occur within normal hemostasis. (Reproduced and updated with permission from Chang JC. Blood Coagul Fibrinoplysis. 2018; 29:573–84). Two different thrombotic paths, microthrombotic (ULVWF) and fibrinogenic (TF), are initiated in normal hemostasis, but later the two paths must unify to conclude normal hemostasis with passive role of NETs; it stops the bleeding in external bodily injury and produce the thrombosis in intravascular injury. However, in the different level (depth) of intravascular injury, thrombogenesis takes two different paths. If the level of intravascular injury is confined to the endothelium, lone ULVWF path become activated and causes microthrombosis (i.e., VMTD) because TF path is not activated. On the other hand, if the level of intravascular injury extends from the endothelium to SET/EVT, TF path becomes also activated and causes macrothrombosis (e.g., DVT). In one theoretical situation, if only SET/EVT is injured, available TF is supposed to activate TF path, but in reality this injury does not cause thrombosis without breached endothelium. However, in pathologic hemostasis, aberrant TF activation occurs and produces fibrin clots (i.e., true DIC) in APL due to TF expression in intravascular space from leukemic promyelocytes. APL is a consumption coagulopathy due to lone activation of TF path. This logic is based on "two-path unifying theory". Please see Figure 2, showing 3 different thrombosis disorders via microthrombogenesis, fibrinogenesis, macrothrombogenesis, which are annotated in bold face. Each thrombognesis occurs when ULVWF path, TF path or combined paths are activated depending upon the levels of damage in intravascular injury (endothelium and SET/EVT). The characters of microthrombi, fibrin clots and macrothrombus from different paths are very different and produce distinctly different clinical thrombotic disorders [20]. Abbreviations: *APL* acute promyelocytic leukemia, *DIC* disseminated intravascular coagulation, *DVT* deep vein thrombosis, *EVT* extravascular tissue, *NET* neutrophil extracellular traps, *SET* subendothelial tissue, *TF* tissue factor, *ULVWF* unusually large von Willebrand factor multimers, *VMTD* vascular microthrombotic disease

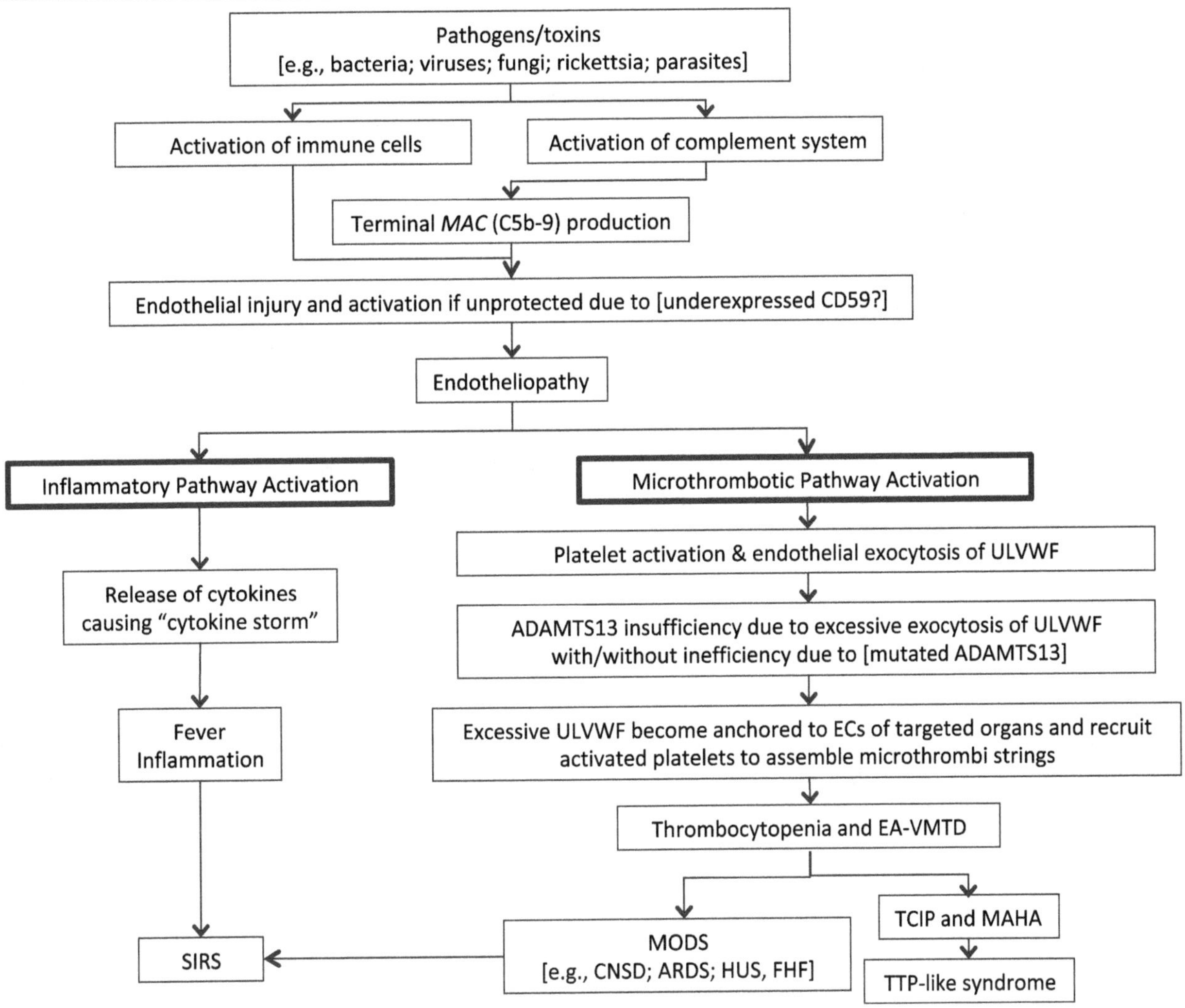

Fig. 3 Endothelial molecular pathogenesis and microthrombogenesis in sepsis. (Based on "two-activation theory of the endothelium"). (Reproduced and modified from Thrombosis Journal 2018;16:20). Endothelial molecular pathogenesis is succinctly illustrated. Endotheliopathy activates two main pathways. Activation of inflammatory pathway produces cytokines, which main function is the modulation of inflammation, including fever and myalgia. Activation of microthrombotic pathway causes much more deadly septic syndromes via generalized EA-VMTD/DIT. Abbreviations: *CNSD* central nervous system dysfunction, *DIC* disseminated intravascular coagulation, *DIT* disseminated intravascular microthrombosis, *EA-VMTD* endotheliopathy-associated VMTD, *FHF* fulminant hepatic failure, *MAHA* microangiopathic hemolytic anemia, *SIRS* systemic inflammatory response syndrome, *TCIP* thrombocytopenia in critically ill patients, *TTP* thrombotic thrombocytopenic purpura, *ULVWF* unusually large von Willebrand factor multimers, * cell-mediated immune cells are T lymphocyte, macrophage, monocyte, neutrophil, and dendritic cell

drug/toxin, leading to poorly defined but serious coagulopathy. The frequently noticed early hematologic manifestation is occurrence of unexplained "thrombocytopenia in critically ill patients" (TCIP), which until recently has been a clinical mystery [1, 2]. Now, it has become evident that thrombocytopenia develops due to contribution of platelets in microthrombogenesis and their consumption. Microthrombogenesis is the process of forming microthrombi composed of platelet-unusually large von Willebrand factor (ULVWF) complexes following platelet activation and endothelial exocytosis of ULVWF as a result of endotheliopathy [1–3].

EA-VMTD/DIT in sepsis is a unique hemostatic disorder developing via lone activation of ULVWF path without simultaneous activation of TF path according to "two-path unifying theory of hemostasis" [4, 20]. In this article, the pathophysiological mechanism promoting EA-VMTD/DIT in sepsis will be reviewed; the molecular pathogenesis of endotheliopathy and hemostasis will be discussed and the clinical phenotypes of various septic syndromes be analyzed. Finally, we should be able to recognize that EA-VMTD/DIT is the underlying clinical disease associated with sepsis-associated coagulopathy and oftentimes presents with hematologic phenotype of TTP-like syndrome.

Endothelial pathogenetic mechanisms involved in sepsis

Activation of complement system

The activation of complement system is normally one of the key events in defensive mechanism against pathogen in sepsis. Its protective function for the host rapidly identifies and eliminates invading pathogen whenever possible. Opsonization of foreign surfaces by covalently attached C3b fulfills three major functions: pathogen clearance by phagocytosis; amplification of complement activation by the formation of surface-bound C3 convertase; and assembly of C5 convertase. Cleavage of C5 induces the formation of the multi-protein pore complex C5b-9 (i.e., membrane attack complex [MAC]), which leads to lysis of pathogen [21].

Although the major role of complement is protective function for the host through innate immune defense, activated complement could also cause the destructive action to the host endothelium [22], which impacts the course of sepsis as illustrated in Fig. 1. MAC exerts harmful effects to host's endothelial cells (ECs) [22] unless complement regulator CD59 is adequately expressed in ECs and protects them by inhibiting C9 polymerization into MAC [23, 24].

If CD59 is downregulated in the ECs due to either gene mutation or acquired diseases [25], activated terminal complement MAC could exert destructive effects to ECs in sepsis, trauma and other critical illnesses [26–28]. When MAC attacks innocent bystander ECs, channel (transmembrane pores) formation could occur on the endothelial membrane [22] and trigger endothelial dysfunction [3, 22, 29].

The central role of the endothelium

The endothelium, located at the interface between blood and subendothelial tissue (SET)/ extravascular tissue (EVT), is a monolayer of endothelial cells distributed in the vasculature and organ system of the entire human body. Its vital function is maintaining not only homeostasis of circulatory networks but also providing hemostasis in vascular injury. It is essential to preserve both anatomical and functional integrity of the endothelium at all cost to prevent unneeded intravascular thrombosis from the exposure to TF present in SET/ EVT [30], which could cause macrothrombosis and get involved in vascular and organ damage [20, 31].

Hemostasis plays a critical role in the pathogenesis of sepsis. Sepsis occurring due to a variety of pathogen causes generalized endothelial injury as a result of disseminated nature of the circulatory system, and leads to systemic endotheliopathy. However, sepsis-induced endotheliopathy triggers functional changes contributing to molecular dysfunction and subsequent partial hemostasis, mediating microthrombogenesis. Unlike localized traumatic intravascular injury, which initiates bleeding and hemostasis due to combined endothelial and SET/EVT damage [4, 20], anatomic disruption to the vascular endothelium is minimal in sepsis and bleeding is not consequential because endotheliopathy is confined to the endothelium and TF is not exposed [32]. Thus, complement activation due to pathogen or endotoxin provokes endothelial dysfunction promoting exocytosis of ULVWF and platelet activation, which initiates ULVWF path of hemostasis and leads to microthrombogenesis [2, 3] as shown in Figs. 2 and 3. This is a very important conception to understand sepsis-associated coagulopathy because sepsis promotes lone activation of ULVWF path without activation of TF path since SET/ EVT damage does not occur. This thesis based on microthrombogenesis refutes the current theory that sepsis-associated "DIC" occurs via activation of TF path that leads to fibrinogenesis. Now, "DIC" is reinterpreted to be DIT that occurs via activated ULVWF-initiated thrombogenesis associated with endotheliopathy as illustrated in Fig. 2 [2–4, 20].

"DIC" is still the result of hemostasis even though TF is not involved [4, 20]. The exocytosis of ULVWF from Weibel-Palade bodies activates ULVWF path and platelets are easily recruited because ULVWF show extremely high affinity to the platelet. This lone activation of ULVWF path in endotheliopathy promotes the formation of microthrombi and lead to EA-VMTD/DIT [2, 3, 20]. Thus, microthombi of "DIC" are the same as those of TTP and TTP-like syndrome. "DIC" has been inappropriately conceptualized as a fibrin clot disease produced via activated TF/FVIIa-initiated cascade/cell-based coagulation. This ill-founded DIC must be properly renamed as EA-VMTD/DIT, which hematologic phenotype is TTP-like syndrome. On the other hand, consumption coagulopathy in acute promyelocytic leukemia (APL) that occurs due to pathologic activation of aberrant TF path caused by TF released from leukemic promyelocytes should be called true DIC [20]. True DIC in APL is made of disseminated fibrin clots that occur without vascular injury. Its hematologic phenotype is always characterized by hemorrhagic syndrome [2–4, 20].

Endotheliopathy-associated microthrombosis as the crux of the clinical phenotypes

A hallmark of advancing sepsis is microvascular dysfunction [1, 8, 9] due to disseminated microthrombi composed of platelet-ULVWF complexes in multiorgans, which is promoted via microthrombogenesis [1, 2]. The pathophysiological mechanism inducing circulatory dysfunction that leads to organ ischemia has been well defined in clinical medicine as vascular microthrombosis [33, 34] or VMTD [1–4].

In addition to hypoxic microvascular dysfunction, endotheliopathy causes the release of various inflammatory cytokines [7, 35, 36] and bioactive biomarkers from ECs when infected by different types of pathogen [37]. Unlike adaptive immune mechanism that produces antigen-specific protective antibody response from each pathogen to each host, sepsis-induced endotheliopathy triggers similar, if not the same, endothelial molecular response without specificity among different types of pathogen (Fig. 3 and Table 1). This endothelial dysfunction causes independent inflammation different from vascular microthrombosis. The manifestations of inflammation and microthrombosis are universally similar among each class and each type of pathogens and display no specificity.

However, an intriguing question is why then the clinical organ phenotype expression of sepsis is so different among the different hosts and also among the different pathogens. For example, the organ phenotypic features of sepsis could be encephalopathy, myocarditis, pancreatitis, acute respiratory distress syndrome, adrenal insufficiency, fulminant hepatic failure, hemolytic-uremic syndrome (HUS) and others, or any combination of them. This is not due to different kinds of VMTD, but is due to at least two underlying genetic variables governing individual organ susceptibility of the host and affinity of the pathogen to specific organ(s).

Endothelial heterogeneity and organotropism

In endothelial pathogenesis of sepsis, the organ phenotype expression is variable among different hosts by the same pathogen as well as different pathogens. Organ phenotypes are determined by two main endowed biological mechanisms: endothelial heterogeneity of host [38, 39], and organotropism of pathogen [40–43]. Variable clinical organ phenotypic syndromes occur as seen in the same type of pathogen. Examples are hanta virus, causing cardio-pulmonary syndrome in the heart and lungs, Shiga toxin-producing *E. coli*, presenting with HUS (STEC-HUS) in the brain, bowels and kidneys, and Neisseria meningitides, inciting Waterhouse-Friderichsen syndrome and meningitis in the adrenals and meninges. Also, the same organ phenotype can occur in different types of pathogen.

Clinicians have used the mysterious combined clinical terms such as hepato-renal syndrome [44], hepatic encephalopathy [45], and cardio-pulmonary syndrome [46] often without identifying the involved pathogen even though sepsis might have existed. Perhaps, this oversimplified designation of organ phenotypes has impeded detecting the underlying etiology of organ syndromes. In modern medical literature, for example, many authors claimed one organ phenotypic syndrome causes several other organ dysfunctions such as acute renal failure,

Table 1 Examples of thrombocytopenia (TCIP) in sepsis-associated coagulopathy (VMTD)

	Specific pathogens	Commonly involved organs	Associated clinical syndromes
Bacteria	Neisseria meningitides	adrenals; meninges	Waterhouse-Friderichsen syndrome
	E. coli O157:H7	bowels; kidneys; brain	GE; HUS; MODS
	MRSA	lung; multiorgans	ARDS
	Klebsiella pneumonia	lungs; multiorgans	ARDS; TAMOF
	Various bacterial sepsis	multiorgans	
Viruses	Ebola	lungs; liver; multiorgans	ARDS; hepatic necrosis
	H1N1 influenza	brain; lungs; multiorgans	Encephalopathy; ARDS
	MERS-CoV	lungs	ARDS
	SARS-CoV	lungs	ARDS
	Hantavirus	heart; lungs; kidneys	HCPS; HPS; HFRS
	Dengue	adrenals; multiorgans	DSS
	SFTS virus	multiorgans	SFTS
	Hepatitis A virus	liver	Fulminant hepatic failure
Fungi	*Candida albicans*	multiorgans	ARDS
Rickettsia	Rickettsia rickettsii	skin; multiorgans	RMSF
Parasites	Plasmodium falciparum	brain; multiorgans	Cerebral malaria; ARDS
	Plasmodium vivax	lungs; multiorgans	ARDS

Abbreviations: *ARDS* acute respiratory distress syndrome, *DSS* dengue shock syndrome, *GE* gastroenteritis, *HUS* hemolytic uremic syndrome, *MERS-CoV* Middle East respiratory syndrome-corona virus, *MODS* multiorgan dysfunction syndrome, *MRSA* methicillin resistant *Staphylococcus aureus*, *SARS-CoV* severe acute respiratory syndrome-corona virus, *SFTS* severe fever with thrombocytopenia syndrome, *TAMOF* thrombocytopenia and multiple organ failure, *HCPS* hantavirus cardiopulmonary syndrome, *HPS* hantavirus pulmonary syndrome, *HFRS* hemorrhagic fever with renal syndrome, *DSS* dengue shock syndrome, *RMSF* Rocky mountain spotted fever encephalopathy, and hepatic failure [47–49]. In sepsis, the concept of endothelial heterogeneity and organotropism for organ localization and the phenotype of VMTD as expression of underlying pathology support that biorgan or multiorgan manifestations are just different phenotypes of the same disease EA-VMTD [29]. Likewise, additional organ phenotypes that are often designated as extra-organ manifestations (e.g., extra-renal phenotypes in HUS, extra-pancreatic phenotypes in pancreatitis and extra-pulmonary phenotypes in acute respiratory distress syndrome [ARDS]) are expression of the same hemostatic disease EA-VMTD. Indeed, extra-organ phenotypes are misrepresentation.

The organ phenotype expression associated with EA-VMTD is determined by combined mechanism of underlying endothelial heterogeneity of each host [29, 38, 39] and organ/tissue tropism of each pathogen/toxin [40]. For examples, the phenotypes of STEC-HUS [29, 41] are the combined results of genotypic endothelial heterogeneity of the host and genotypic tropism of STEC on the gastrointestinal tracts, kidneys, lungs, brain and others. Likewise, the phenotypes of Waterhouse-Friderichsen syndrome [50] associated with Neisseria meningitides are expressed as a result of combined endothelial heterogeneity of the host and tissue tropism of the bacteria in the adrenals and meninges. These concepts are very important in the understanding of genesis of MODS.

Molecular pathogenesis based on "two-activation theory of the endothelium"

Although endotheliopathy is proposed to be the main pathology promoting crosstalk mechanism between inflammation and coagulation in sepsis, both the character of blood clots and thrombogenetic mechanism producing coagulopathy could not be comprehensible by this crosstalk theory [4]. However, because complement activation in critical illnesses affects several molecular functions of ECs, it fits well with endothelial pathogenesis of sepsis.

The "two-activation theory of the endothelium" (Fig. 3) [1–4, 19, 20] has been proposed based on the role of endotheliopathy not only in sepsis, but also in other critical illnesses. Once endotheliopathy occurs in sepsis, endothelial dysfunction promotes the activation of two independent endothelial pathways; one is inflammatory and the other is microthrombotic. In short, two main molecular events are: 1) release of inflammatory cytokines such as interleukin (IL)-1, IL-6, tumor necrosis factor-α, and others [1–3, 7, 35, 36], and 2) activation of the platelet [51, 52] and exocytosis of ULVWF [53, 54]. The former triggers inflammation through "activated inflammatory pathway", and the latter mediates microthrombogenesis via "activated microthrombotic pathway". Inflammation promotes inflammatory symptoms, but microthrombogenesis mediates thrombosis to produce microthrombi strings composed of platelet-ULVWF complexes [54–56]. Micothrombogenesis is activated when the protease ADAMTS13 is insufficient to cleave the excess of exocytosed ULVWF with or without mild to moderate underlying enzyme deficiency that occurs due to polymorphism or heterozygous mutation of ADAMTS13 [57–60]. These microthrombi strings produce DIT in the smaller and larger vasculatures of various organs [1–4].

Mechanism of sepsis-associated coagulopathy

According to novel "two-path unifying theory" (Fig. 2), in intravascular injury normal hemostasis begins with simultaneous but independent activation of ULVWF path and TF path [4, 20]. However, in endotheliopathy associated with sepsis, only ULVWF path of normal hemostasis becomes activated because TF is not available in the endothelium [32]. Endotheliopathy in sepsis triggers exocytosis of ULVWF and activates platelets, and produces microthrombi strings. The microthrombi strings composed of platelet-ULVWF complexes must be the intrinsic character of the blood clots in "DIC" of sepsis that occurs as the result of lone activation of ULVWF path illustrated in Fig. 2. Therefore, the concept of "DIC" in sepsis-associated coagulopathy is due to activated TF path producing "fibrin clots" is an incorrect interpretation. This interpretative mistake has been very costly. Thus, according to "two-path unifying theory", "DIC" is the same to EA-VMTD/DIT and acute "DIC" is EA-VMTD/DIT with hepatic coagulopathy/hemorrhagic syndrome. EA-VMTD/DIT without hemorrhagic syndrome had been called chronic "DIC" as previously detailed [2–4, 20].

The endothelium positioned at the interface between the blood and SET/EVT functions as the initiating site of the foundry producing thrombus following vascular injury [4]. Because sepsis-induced vascular injury is typically limited to the endothelium, only ULVWF path becomes activated and causes vascular microthrombosis in the smaller and larger vasculatures [3]. Sepsis-associated coagulopathy has shown to express the changes in antithrombotic and prothrombotic markers, which include TF path-related markers such as thrombin, antithrombin, thrombin-antithrombin complexes, activated protein C (APC), tissue factor pathway inhibitor (TFPI), and thrombomodulin (TM). These expressed TF path markers have indirectly supported the role of activated TF path in the pathogenesis of "DIC". However, these TF path-related markers can be explained by alternative mechanisms, including acute DIC with hepatic coagulopathy, combined micro-macrothrombotic syndrome, concomitant vascular injury as a result of surgery and vascular access, and perhaps more commonly MODS resulting in tissue necrosis, especially of the liver, kidneys and lungs, in advancing severe sepsis. In retrospect, expressed TF path markers in sepsis can be ascribed to secondary events that are unrelated to primary pathogenesis of activated ULVWF path as illustrated in Fig. 2.

In patients with severe sepsis, thrombin generation, APC, TFPI and TM are not the true markers of "DIC", but are interpreted as secondary markers of TF path. The true markers of "DIC" are activated platelet, increased FVIII activity, and increased VWF from, excessive release of ULVWF, and upregulated collagen binding that result from activated ULVWF path [4, 20]. It is no wonder why clinical trials using antithrombin agents, APC, recombinant TFPI and recombinant TM was unsuccssful in sepsis-associated coagulopathy. It was likely due to the fact that sepsis-associated "DIC" is not true DIC, but is EA-VMTD (i.e., TTP-like syndrome) caused by activated ULVWF path. Further, endothelial injury alone cannot induce fibrinogenesis, but only provokes partial hemostasis, leading to microthrombogenesis, and produces EA-VMTD. On the other hand, APL-associated DIC that occurs due to activation of aberrant TF path is true DIC [20].

In sepsis-associated coagulopathy, this author believes neutrophil extracellular traps (NETs) participate in thrombosis in the unifying stage of macrothrombus, but the problem of NETosis is with the logic of hemostatic principle and inconsistent characters of thrombosis in vivo and animal models in the reported literature. The essential logic in the formation of thrombosis is normal hemostasis must be accompanied by intravascular injury [4, 20]. Without it, no thrombosis can be formed because activation of ULVWF path and/or TF path is sine qua non in thrombogenesis. Further, only three exceptions among thrombotic disorders occur without vascular injury. They are (1) TTP, which is aberrant thrombosis because vessel is not injured, but still utilizes "ULVWF" path in ADAMTS13 deficiency, (2) APL, which is aberrant thrombosis because vessel is not injured, but still utilizes "TF" path due to overexpressed TF from leukemic promyelocytes, and (3) heparin-induced thrombocytopenia with thrombosis syndrome, which is pathologic thrombosis because vessel is not injured, and also utilizes neither ULVWF path nor TF path [4, 20]. However, we know cellular and molecular traps, and perhaps adhesion molecules participate in the hemostatic plug and thrombus formation, which is illustrated in Fig. 2, where blood cells/molecules are trapped in comingling process during blood clots formation (macrothrombosis) in the unifying stage of microthrombi strings and fibrin clots. This author believes that NETosis is not active hemostatic processes, but is a passive one associated with secondary event trapping blood cells and molecules such as DNAs and histones in the process of thrombogenesis.

Now, supported by "two-activation theory of the endothelium" and "two-path unifying theory of hemostasis" in the understanding of sepsis and septic shock, we can easily define all clinical phenotypes of the sepsis-associated syndromes via two independent endothelial pathways promoting inflammation and microthrombogenesis.

Sepsis-associated clinical syndromes

Inflammation

Inflammation is a clinical phenotype of the endothelial pathogenesis in sepsis, which presents with inflammatory symptoms such as fever, chills, headache, myalgia and arthralgia. Inflammatory fever occurs due to endogenous pyrogens (e.g., cytokines IL-1, IL-6, TNF, and others) and is different from infectious fever, which occurs due to exogenous pyrogens (e.g., lipopolysaccharide) [61–63] as physiologic defensive mechanism through immune system as shown in Fig. 1. Clearly there are two kinds of fever combined in sepsis. One is infectious fever due to exogenous pyrogen and the other is inflammatory fever caused by endotheliopathy [61, 62]. In sepsis, fever develops due to combined effects of exogenous pyrogens from pathogen and endogenous pyrogens from endotheliopathy. On the other hand, other non-septic (and non-infectious) critical illnesses such as trauma and cancer present with inflammatory fever [62, 63] only via endothelial pathogenesis.

It should be emphasized that in sepsis inflammation is not the cause of endotheliopathy, but is the result of endotheliopathy, which is promoted by activated complement system (Fig. 3). Even though inflammation and VMTD occurs simultaneously in endotheliopathy, inflammation does not sufficiently alter activation of coagulation system or cause microthrombogenesis because they are the result of independent molecular processes, and their interaction is not established at molecular level. Although simultaneous occurrence of inflammation and coagulation in endotheliopathy-associated sepsis seems to be consistent with a high profile crosstalk theory, no persuasive evidence based on hemostatic mechanism has been offered for inflammation promoting microthrombosis to cause "DIC" [11–13], Additionally, it should be noted that inflammation does not occur in AA-VMTD even though platelet-ULVWF complexes are the same microthrombi produced by microthrombogenesis in EA-VMTD. This observation also supports inflammation only occurs in endotheliopathy, and inflammation is not the essential component leading to coagulation/microthrombogenesis.

Different from non-septic critical illnesses, sepsis presents with much severer inflammation, which is perhaps due to increased cytokine expression through additional loop of activated circulating immune cell pathway (e.g., T lymphocyte, macrophage, monocyte, neutrophil, and dendritic cell) suggested by another arrow line in Fig. 3 [64]. This pathway could explain why severe inflammation and SIRS are more common in sepsis than in non-septic critical illnesses [65, 66]. Inflammation provokes inflammatory symptoms in sepsis, but the major endothelial pathogenesis determining the severity and outcome of sepsis is elicited by microthrombosis of EA-VMTD/DIT.

Consumptive thrombocytopenia in critically ill patients (TCIP)

In sepsis, unexplained thrombocytopenia is a very important early telltale sign that endotheliopathy has occurred and on-going microthrombogenesis is generating consumptive thrombocytopenia (Table 1). Thrombocytopenia has been blamed too often to heparin in the critical care setting [67]. The term "thrombocytopenia in critically ill patients" (TCIP) has been used for etiology-undetermined thrombocytopenia after exclusion of known causes of thrombocytopenia (e.g., heparin-induced, drug and transfusion-associated, "DIC"-associated, bone marrow suppression-caused, hypersplenism-related, idiopathic autoimmune-induced, and others.) [68–70]. Often, it was blamed to multifactorial causes, including consumption associated with thrombin-mediated platelet activation [68]. The significance of mild to moderate TCIP in early stage of sepsis has been commonly downplayed in the care of very sick septic patients because its pathogenesis is not clearly understood and hemorrhagic syndrome is uncommon [1]. Now, it is established TCIP is common and predictable hematologic feature in sepsis-associated coagulopathy as noted in Table 1, occurring as a result of consumptive thrombocytopenia via activated ULVWF path [1–4, 20].

Increasing degree of TCIP is typically associated with advancing VMTD in the septic patient. Since "DIC" is associated with thrombocytopenia, hypofibrinogenemia, prolonged prothrombin time, and activated partial thromboplastin time and positive fibrin degradation products, TCIP has been attributed to consumptive thrombocytopenia as well as consumption of coagulation factors contributing to fibrin clots. However, activated TF path alone does not consume platelets in the formation of fibrin meshes/fibrin clots via the extrinsic coagulation cascade. The above abnormal coagulation profile is also consistent with hepatic coagulopathy. The main differentiating laboratory features are increased FVIII and markedly decreased FVII in EA-VMTD/DIT (i.e., "DIC")–associated hepatic coagulopathy but markedly decreased FVIII and normal FVII in true DIC (e.g., APL). These coagulation characteristics can serve as very useful laboratory diagnostic features between true DIC associated with APL and "DIC" associated with critical illnesses.

In the critical care setting, significant correlations have been observed between the onset, duration and degree of thrombocytopenia, and severity and outcome of critical illnesses [71, 72]. This observation supports that TCIP is a key hematologic phenotype and plays its important role in the endothelial pathogenesis of sepsis.

TTP-like syndrome

TTP-like syndrome is the hematologic phenotype of EA-VMTD/DIT characterized by TCIP and MAHA associated with endotheliopathy after exclusion of gene mutation associated TTP (GA-VMTD) and antibody-associated TTP (AA-VMTD). It is an acquired VMTD of non-immune type and develops in sepsis and other critical illnesses [3, 73–77]. TTP-like syndrome is very common but often masked because its pathogenesis is no well appreciated by clinicians [3] and it is often classified and reported in the medical literature as TTP, thrombotic microangiopathy, "DIC", HUS and unclassifiable microthrombosis. Most commonly, TTP-like syndrome in sepsis has been misdiagnosed as DIC, and the hematologic feature of MAHA is often not uncovered because mild to modest anemia brings very little attention in critically ill patients.

TTP is very similar to hematologic phenotype of EA-VMTD (i.e., TTP-like syndrome) although the pathophysiological mechanism is diametrically different from TTP-like syndrome. TTP occurs due to ADAMTS13 deficiency and typically provokes "microvascular thrombosis" in capillaries and arterioles of the kidneys and brain because microthrombi are formed in microcirculation [78, 79] and become trapped in the microvasculature in situ, promoting activation of "aberrant" ULVWF path of hemostasis without intravascular injury. However, TTP-like syndrome (e.g., sepsis-associated coagulopathy) occurs due to microthrombi that are anchored to ECs of smaller and larger vasculatures as well as capillaries and arterioles and provokes "vascular microthrombosis", promoting activation of "normal" ULVWF path of hemostasis due to endothelial injury.

Unlike TTP that creates higher stress force when red cells pass through severely occluded microvascular space of the bloodstream [80], perhaps microthrombi strings of EA-VMTD anchored to ECs that create milder shear stress in smaller and larger vasculatures and produce fewer schistocytes that lead to mild MAHA, resulting in less apparent TTP-like syndrome. Milder shear stress is the likely reason why schistocytes are rare and hemolytic anemia is more covert in sepsis. Thus, MAHA is less commonly unmasked, but hemolytic anemia can be an important clue establishing the sepsis-associated VMTD if an experienced hematologist carefully evaluates the thrombocytopenic patient in sepsis with a high index of suspicion. This will save many lives if the diagnosis of

TTP-like syndrome can be uncovered in the earliest stage of sepsis. In regard to shear stress, exception may be hemolytic-uremic syndrome associated with EA-VMTD, in which schistocytes are encountered in a large number and produces severe MAHA. It could due to rich microvasculatures present within the kidneys.

Even though both TTP-like syndrome and TTP are pathologically characterized by DIT and clinically by VMTD, the genesis and clinical features between two different types of VMTD [1–4] are very different as presented in Table 2. Variable phenotypic features are the reason why TTP-like syndrome can be easily overlooked when it occurs as a result of EA-VMTD in sepsis and critical illnesses [1–3, 19, 20, 76, 77]. This is particularly true in sepsis, trauma, postoperative TTP [81], and cancer [82, 83]. Because microthrombosis may occur in both smaller and larger vasculatures of various organs in EA-VMTD [84, 85], exotic and fanciful phenotypes are not uncommon as shown in Fig. 4 [3].

Unexplained thrombocytopenia and hemolytic anemia could be the first clue of serious sepsis, which should be evaluated immediately to determine any evidence of mild MAHA and TTP-like syndrome [86–89]. If a patient presents with minimal number of schistocytes [80–83, 88, 89] and TCIP in sepsis, MODS might be developing in unusual organs such as the lungs, heart, pancreas, adrenals, digits, liver, muscles and others as well as the brain and kidneys [50, 90–93]. In those cases, the evidence of hemolysis with reticulocytosis, hypohaptoglobinemia, indirect hyperbilirubinemia and elevated lactic acid dehydrogenase [1–4] should confirm the diagnosis of TTP-like syndrome as well as EA-VMTD/DIT.

Multiorgan dysfunction syndrome (MODS)

Mortality of sepsis is highly correlated to the development of MODS. The pathogenetic mechanism of MODS is circulatory dysfunction caused by vascular microthrombosis, obstructing blood flow and decreasing oxygenation in various organs [1–3, 33]. Further progression to hypoxia would lead to organ failure and hasten the demise of the patient.

The mechanism developing different organ phenotypes of MODS in EA-VMTD is the result of complex interaction involving endothelial heterogeneity of the host and tropism of the pathogen. Endothelial heterogeneity is governed by endowed molecules in the endothelium and tropism in different types of pathogen determines the invasiveness of pathogen in each patient. Unique expressivity in the host endothelium and specific affinity of the pathogen to different host organs appears to create variable organ syndromes. Figure 4 summarizes the examples of MODS in various organs that had been known to occur in EA-VMTD. The organ examples showing affinity to pathogens could be the brain with encephalopathy [94] due to various encephalitis viruses, lungs with acute respiratory distress syndrome (ARDS) [70] due to severe acute respiratory syndrome (SARS)-CoV or middle east respiratory syndrome (MERS)-CoV, kidneys with hemolytic-uremic syndrome (HUS) [29] due to STEC, and others.

Table 2 Pathogenetic difference of VMTD between TTP and sepsis-associated TTP-like syndrome

	TTP (GA-VMTD and AA-VMTD)	TTP-like syndrome in sepsis (EA-VMTD)
Primary event	Hereditary TTP (GA-VMTD)	Sepsis due to pathogens
	(due to mutation of ADAMTSD13 gene)	(e.g., bacteria; viruses; fungi; rickettsia; parasites)
	Acquired TTP (AA-VMTD)	↓
	(due to anti-ADAMTS13 antibody)	Complement activation → MAC (C5b-9) formation (?)
	↓	↓
Secondary event	Excess of circulating ULVWF & platelets	C5b-9-induced endotheliopathy
	↓	↓
	Microthrombogenesis, leading to ULVWF-platelet complex formation in microcirculation	ULVWF released from ECs & anchored to endothelial membrane to recruit platelets → microthrombogenesis
	↓	↓
Tertiary event	Microthrombi in microvasculatures	Microthrombi strings in smaller and larger vasculatures
	↓	↓
Final event	Microvascular thrombosis	Vascular microthrombosis
	↓	↓
	GA-VMTD; AA-VMTD	EA-VMTD
	↓	↓
	TTP	TTP-like syndrome

Abbreviations: *ECs* endothelial cells, *MAC* membrane attack complex, *TTP* thrombotic thrombocytopenic purpura, *VMTD* vascular microhrombotic disease, *AA-VMTD* antibody-associated VMTD, *EA-VMTD* endotheliopathy-associated VMTD, *ULVWF* unusually large von Willebrand factor multimers

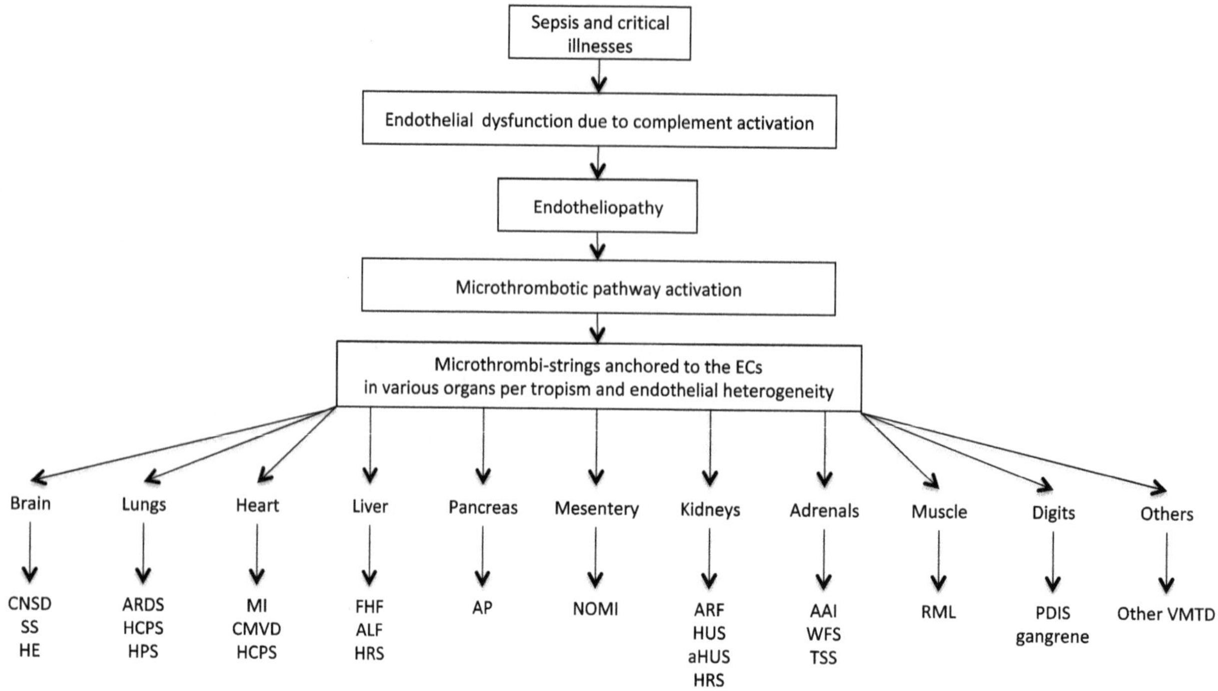

Fig. 4 Pathogenesis of multiorgan dysfunction syndrome in sepsis-associated vascular microthrombotic disease. The pathogenesis of MODS in sepsis is summarized. Any organ can be involved by VMTD. However, MODS is much more common in vital organs such as the lungs with ARDS, the brain with CNSD, and the kidneys with acute renal failure. The illustration is self-explanatory. Abbreviations: *AAI* acute adrenal insufficiency, *ALF* acute liver failure, *aHUS* atypical hemolytic uremic syndrome, *AP* acute pancreatitis, *ARDS* acute respiratory distress syndrome, *ARF*, acute renal failure, *CNSD* central nervous system dysfunction, *CMVD* coronary microvascular disease, *FHF* fulminant hepatic failure, *HCPS* hantavirus cardio-pulmonary syndrome, *HE* hepatic encephalopathy, *HPS* hantavirus pulmonary syndrome, *HRS* hepato-renal syndrome, *HUS* hemolytic-uremic syndrome, *NOMI* non-occlusive mesenteric ischemia syndrome, *PDIS* peripheral digit ischemic syndrome, *RML* rhabdomyolysis, *SS*, stroke syndrome, *TSS* toxic and septic shock syndrome, *WFS* Waterhouse-Friderichsen syndrome

Currently the pathophysiological mechanism of organ dysfunction is thought to be the results of the direct invasion of pathogen into the tissue of the involved organs. This mechanism may lead to combined compromise of the endothelium and SET/EVT, triggering cellulitis or abscess and macrothrombsis. However, this endothelial breach occurs uncommonly because although sepsis-induced endotheliopathy is universal, typically SET/EVT is not compromised in sepsis. MODS occur as a result of microthrombosis that causes reversible hypoxia in organs unless it advances to terminal multiorgan failure. Organ dysfunction syndrome in sepsis, including fulminant hepatic failure (FHF)/acute liver failure [92], acute necrotizing pancreatitis [93], myocardial infarction [95], acute adrenal insufficiency [50, 96, 97], and rhabdomyolysis [98] may be reversible with improvement of sepsis unless necrotic and gangrenous change develops as seen in tissue gangrene [90] and peripheral digit ischemic syndrome [91]. More serious combined phenotypic syndromes, which underlying pathogenesis is still being debated, include hepatic encephalopathy [92], hepato-renal syndrome [99, 100], hepatic coagulopathy [1–3, 101], and cardio-pulmonary syndrome [102]. Clinicians should take the advantage of hemostatic nature of VMTD when therapeutic approach is considered.

Since the phenotypes of MODS are clinical syndromes occurring due to vascular microthrombosis, it should be stressed that ARDS, HUS, FHF, rhabdomyolysis or acute pancreatitis does not cause sepsis-associated coagulopathy or endotheliopathy. Instead, these organ phenotypes are the manifestations of sepsis-associated coagulopathy that develop due to complement-induced endotheliopathy and subsequent microthrombogenesis. For example, case reports in medical literature have often cited that pancreatitis was responsible for TTP-like syndrome [93] and caused acute renal failure [47], encephalopathy [48] and FHF [49]. These articles should have stated that TTP-like syndrome (i.e., EA-VMTD) caused pancreatitis as well as other organ dysfunction.

Septic shock due to adrenal insufficiency

Septic shock is a serious and well-known sepsis-associated syndrome, presenting with altered mental state and hypotension associated with hyperthermia/hypothermia, tachycardia/tachypnea, and oliguria. Precipitating circulatory failure with hypotension suggests it is an advancing state of multiple vital organ failure. Therefore, septic shock had been ascribed to underlying circulatory and cellular metabolic abnormalities profound enough to substantially increase mortality [103]. Although multiple organ dysfunctions take place due to hypoxia induced by DIT, the pathophysiological mechanism of septic shock is still an unsolved riddle. There are no answers to the questions on how hypotension is incited by DIT and what kinds of cellular metabolic abnormalities develop in septic shock.

According to "two-activation theory of the endothelium", microthrombi strings anchored to ECs within the vasculature trigger circulatory dysfunction, but it is difficult to incriminate the onset of hypotension and "shock" to hypoxia, inflammation or yet unidentified metabolic molecules. However, enough evidences have been accumulated that septic shock represents adrenal crisis/acute adrenal insufficiency (AAI) associated with adrenal hemorrhage and/or necrosis due to DIT. AAI is a more serious form of MODS [104–106]. For example, the tropism of pathogen Neisseria meningitides to the adrenal glands causes Waterhouse-Friderichsen syndrome [50, 96, 97, 106]. Indeed, "DIC", which presently is being renamed as TTP-like syndrome or EA-VMTD/DIT [1–4, 20], has been associated with Waterhouse-Friderichsen syndrome [96, 97, 106, 107]. The role of AAI in septic shock has been established [107–110]. Its proposed pathogenesis is illustrated in detail in Fig. 5. This proposition of adrenal crisis/AAI resulting from microthrombosis would have a significant ramification in the management of septic shock [111].

"DIC" and DIT

When we look back to the enigmatic nature of true DIC (i.e., APL-associated coagulopathy), "DIC" (e.g., "DIC", acute "DIC" and chronic "DIC" in critical illnesses), DIT (e.g., VMTD: TTP of GA-VMTD and AA-VMTD, TTP-like syndrome of EA-VMTD), HUS, thrombotic microangiopathy, their concepts are ambiguous and contradictive one another. Additionally, many of these disorders can occur as a serious life-threatening complication of sepsis or other critical illnesses and are hemostatic diseases presenting as microthrombosis or microthrombo-coagulopathy [20] without clear distinction. This complexity has contributed to the confusion on the exact nature of coagulopathy and thrombopathy in sepsis. In the medical literature, poorly defined sepsis-associated coagulopathy has often been called to be DIC. Sometimes, even proficient coagulation specialists have expressed uneasiness using the term DIC because the diagnosis and pathogenesis could not be unequivocally established because fatality be high and no known effective treatment available. This ambiguity in the concept of DIC must have contributed to the poor

Sepsis due to pathogen Neisseria meningitides or other bacteria
↓
Complement activation producing MAC due to disendowed CD59 in the host endothelium
↓
Neisseria miningitides tropism to adrenals
↓
Host ECs injury of adrenals due to MAC
↓
Endotheliopathy and endothelial molecular activation
↓
Microthrombogenesis to form platelet-ULVWF complexes (microthrombi strings)
↓
Microthrombi strings anchored to ECs of adrenals
↓
DIT ("DIC"), leading to thrombocytopenia and atypical MAHA (TTP-like syndrome)
↓
EA-VMTD/DIT
↓
Adrenal hemorrhage and necrosis
↓
AAI (Waterhouse-Friderichsen syndrome)
↓
Septic shock

Fig. 5 Pathogenesis of septic shock: acute adrenal insufficiency model. Abbreviations: *AAI* acute adrenal insufficiency, *DIT* disseminated intravascular microthrombosis, *ECs* endothelial cells, *MAC* membrane attack complex, *MAHA* microangiopathic hemolytic anemia, *MODS* multiorgan dysfunction syndrome, *TTP* thrombotic thrombocytopenic purpura, *ULVWF* unusually large von Willebrand factor multimers outcome of the treatment in sepsis, including unsatisfactory clinical trials and inappropriate anticoagulation, antithrombotic therapy and plasma therapy.

Recently, the concept of "DIC" has been reappraised as DIT, which clinical disease is EA-VMTD/DIT [1–4, 20]. If we move "DIC" from the column of DIC to that of EA-VMTD/DIT, the leftover is true DIC occurring in APL that is coagulopathy associated with hemorrhagic syndrome [2, 3, 20], in which fibrin clots are formed by fibrinogenesis via extrinsic coagulation cascade from activated aberrant TF path [20]. In true DIC associated with APL, the intrinsic character of blood clots is fibrin clots made of fibrin meshes via activated TF path, but in "DIC" of sepsis-associated coagulopathy, the intrinsic character of blood clots is microthrombi composed of platelet-ULVWF strings via activated ULVWF path. Therefore, sepsis-associated coagulopathy is "DIC", and chronic "DIC" is EA-VMTD/DIT (i.e., TTP-like syndrome). Thus, acute "DIC" is EA-VMTD/DIT with hepatic coagulopathy.

Now, we have hope that effective treatments such as anti-microthrombotic therapy could become available to save lives in sepsis-associated EA-VMTD/DIT. Since "DIC" coined as DIC by Donald McKay [112] in 1950 and with the works of clinicians and coagulation specialists [113–116], the pathophysiological mechanism of "DIC" has been eventually identified to be endotheliopathy-associated microthrombosis. Finally, the analysis of the molecular event of hemostasis in the pathogenesis of EA-VMTD can explain the mechanisms of microvascular thrombosis/vascular microthrombosis, thrombo-hemorrhagic syndrome, abnormal coagulation profile in "DIC", coexistence of "DIC" and liver disease in acute "DIC", endothelial contribution to hemostasis, and role of ULVWF path in thrombogenesis. It also explains the meaning of confusing "DIC" markers such as thrombin generation, thrombin-antithrombin complex, tissue factor pathway inhibitor, activated protein C, and thrombomodulin. These markers are not from "DIC" but from secondary events of VMTD such as tissue damage from MODS, vascular accesses during hospitalization, hepatic coagulopathy or combined micro-macrothrombotic syndrome.

The differential features of sepsis-associated "DIC" from other thrombo-coagulopathies are summarized in Table 3. Since "DIC" in sepsis is not coagulation disorder resulting in fibrin clots, but is microthrombotic disorder (DIT) resulting in microthrombi, this paradigm shift from "DIC" to EA-VMTD/DIT has been able to identify the mechanism of microthrombogenesis and true character of microthrombi. While continuing on "DIC" debates to date, nature finally has endowed us the insights that have helped to construct the pathophysiological mechanism of normal hemostasis and thrombogenesis (Fig. 2) [4, 20].

Systemic inflammatory response syndrome (SIRS)

SIRS has been defined as severe febrile and toxic clinical syndrome associated with variable organ dysfunction to the nonspecific but dreadful insult of septic and/or non-septic origin. The clinical description of this syndrome has included severe inflammation and MODS [117–119] associated with various organ phenotypic manifestations due to disseminated microthrombosis and vascular hypoxia such as ARDS, HUS/acute renal failure, and FHF/acute liver failure.

According to the "two-activation theory of the endothelium", SIRS can be easily recognized in the context of ultimate manifestation of severe inflammation as presented in Fig. 3. SIRS represents intense clinical syndrome due to combined activation of inflammatory pathway, promoting cytokine release and storm, and microthrombotic pathway, triggering platelet activation and endothelial exocytosis, which results in EA-VMTD/DIT. SIRS is more common in severe sepsis compared to non-septic critical illnesses. TCIP occurs often due to platelet consumption and is an early marker of SIRS as in sepsis. Severer thrombocytopenia is more likely associated with combined syndrome of SIRS and MODS [119, 120].

Combined micro-macrothrombotic syndrome

It is uncommon, but a unique and mysterious thrombosis syndrome often occurring in sepsis with well-demarcated symmetrical peripheral gangrene shown in Fig. 6 [90, 91, 121, 122]. This gangrene developing in "DIC" has been well documented in the literature [91, 123–125], which has been an unexplainable complication of TTP-like syndrome because microthrombosis alone does not cause gangrene. The pathogenesis is unknown, but this author attributes this to combined micro-macrothrombotic syndrome, which becomes obvious if we understand "two-path unifying theory" of hemostasis.

Since EA-VMTD/DIT is composed of platelet-ULVWF complexes via lone activation of ULVWF path, TF path-related thrombin and fibrin markers should not be expressed in EA-VMTD/DIT. However, in sepsis-associated coagulopathy, additional TF path activation may occur due to incidental vascular damage extending from the endothelium to SET/EVT. These may be seen in advanced stage of MODS and hepatic coagulopathy. A more serious condition is multiple macrothrombi associated with gangrene developing in sepsis-associated coagulopathy (EA-VMTD). This disorder typically presents with well-demarcated symmetrical peripheral gangrene involving digits and extremities. This disorder has been termed symmetrical peripheral gangrene [121–125], or peripheral digit ischemic syndrome in the post-operative patient [91]. Also, extensive purpura and

Table 3 Differential features of sepsis-associated "DIC" and other thrombo-coagulopathies

	"DIC" without hepatic coagulopathy	"DIC" with hepatic coagulopathy	EA-VMTD/DIT without hepatic coagulopathy	True DIC
Associated disease	Sepsis Other critical illnesses	Sepsis Other critical illnesses FHF	Sepsis Other critical illnesses	APL
Mechanism				
Complement activation	+	+	+	No evidence
Endotheliopathy	+	+	+	No evidence
Inflammatory path	Activated	Activated	Activated	No evidence
Activated hemostatic path	ULVWF	ULVWF	ULVWF	Aberrant TF path
Thrombogenesis	Via micothrombogenesis	Via micothrombogenesis	Via micothrombogenesis	Via fibrinogenesis
Liver involvement	None	FHF	None	Unlikely to occur
Pathology				
Coagulation disorder	VMTD	VMTD with FHF	VMTD	Hemorrhagic disorder
Character of blood clots	Microthrombi	Microthrombi	Microthrmbi	Fibrin clots
Nature of blood clot	Platelet + ULVWF	Platelet +ULVWF	Platelet +ULVWF	Fibrin meshes
Hematology				
Platelet	Decreased	Decreased	Decreased	Decreased due to APL
MAHA	+	+	+	–
FVIII	Normal/increased	Markedly increased	Normal/increased	Markedly decreased
PT/aPTT	Normal	Prolonged	Normal	Prolonged
Fibrinogen	Normal	Decreased	Normal	Decreased
Clinical Phenotype	MODS	MODS	MODS	Hemorrhagic syndrome
	TTP-like syndrome	TTP-like syndrome	TTP-like syndrome	
	Chronic "DIC"	Acute "DIC"	EA-VMTD/DIT	
		FHF		
Correct diagnosis	EA-VMTD/DIT	EA-VMTD/DIT with hepatic coagulopathy	EA-VMTD/DIT	True DIC
Disease designation	EA-VMTD	EA-VMTD with hepatic coagulopathy	EA-VMTD	Consumption coagulopathy

Abbreviations: *APL* acute promyelocytic leukemia, *DIC* disseminated intravascular coagulation, *"DIC"* ill-founded DIC, *DIT* disseminated intravascular microthrombosis, *FHF* fulminant hepatic failure, *MAHA* microangiopathic hemolytic anemia, *MODS* multiorgan dysfunction syndrome, *PT* prothrombin time, *aPTT* activated partial thromboplastin time, *TTP* thrombotic thrombocytopenic purpura, *ULVWF* unusually large von Willebrand factor multimers, *EA-VMTD* endotheliopathy-associated vascular microthrombotic disease

[a]Please note that "DIC" without hepatic coagulopathy is exactly the same to EA-VMTD/DIT without hepatic coagulopathy, which is also consistent with the thesis that "DIC" is DIT

ischemic gangrene seen in septic patient typically occurring in thrombophilic neonates (purpura fulminans) is consistent with combined micro-macrothrombotic syndrome. These severe sepsis-associated complex thrombopathic disorders suggest "additional vascular injury" and/or underlying "thrombophilic state" such as protein C deficiency, protein S deficiency, or FV-Leiden play the role in provoking combined micro-macrothrombotic syndrome.

Currently, symmetrical peripheral gangrene in sepsis is suspected to occur due to vasopressor agents used in septic shock. However, a better explanation rests on combined hemostasis of microthrombogenesis and macrothrombogenesis although the latter is secondary event occurring as a complication of EA-VMTD/DIT. The pathophysiological mechanism of combined micro-macrothrombotic syndrome can be easily supported by "two-path unifying theory". Typically, first set-up is the hospital admission of a patient with sepsis that provokes EA-VMTD/DIT. The patient would need multiple venous and arterial accesses and/or perhaps surgery during hospitalization [91]. This vascular intervention causes intravascular injury, leading to damage of SET/EVT and release of TF. Activated TF path leads to fibrinogenesis

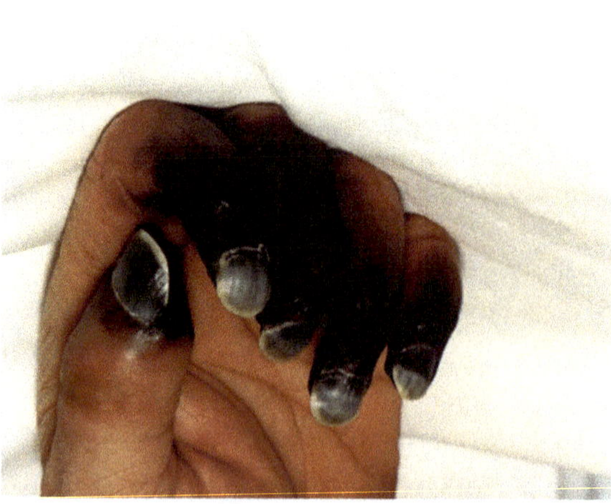

Fig. 6 Combined micro-macrothrombotic syndrome in EA-VMTD showing well demarcated symmetrical peripheral gangrene. This photo demonstrates peripheral digit ischemic syndrome in a young man with severe meningococcemia who developed classical "DIC" (TTP-like syndrome). Following a surgical procedure. Severe symmetrical well-demarcated dry gangrenes developed in the fingers of both hands. He survived, but lost gangrenous parts of fingers. This syndrome can be explained by combined micro-macrothrombotic syndrome per "two-path unifying theory" as noted in the text

and forms fibrin clots in circulation. Since on-going ULVWF path is already active from endotheliopathy and microthrombi strings are being formed, fibrin clots and some of microthrombi would be unified and produce intravascular macrothrombi via macrothrombogenesis [20]. Macrothrombi could travel to the peripheral arterial vasculatures in small and large arteries of the digits and extremities. This combined micro-macrothrombotic syndrome could lead to multiple well-demarcated peripheral symmetrical gangrene at distal vascular trees.

Therapeutic approaches for sepsis-induced endothelial pathogenesis

Normal protective physiologic immune systems eradicate pathogen in early stage of sepsis with help of prompt use of effective antimicrobial therapy. However, if immune system is or becomes ineffective, sepsis advances to endotheliopathy, orchestrating multiple septic syndromes. If this serious pathogenesis is activated, the phase of sepsis may be shifting from protective role of the immune system to destructive role of the endothelial system.

Although inflammation is a serious clinical syndrome, it is not the main contributor leading to fatality of the host, but microthrombogenesis is the culprit. Since EA-VMTD/DIT and "DIC" are conceptually identical and can be understood best as TTP-like syndrome caused by oversupply of ULVWF from endothelial exocytosis, the term TTP-like syndrome will be used in the discussion of therapeutic modalities in sepsis for the rest of this article.

Therapeutic plasma exchange

Therapeutic plasma exchange (TPE) has been effective treatment for TTP-like syndrome if the therapy had been promptly applied [76, 78, 81, 89, 91, 126, 127]. Perhaps untimely treatment with TPE due to the delay in establishing the diagnosis has been the main factor for poor outcome of EA-VMTD [78, 81, 89, 91]. The rationale utilizing TPE in TTP-like syndrome is to cleave and/or remove the excess of ULVWF in the patient plasma with replacement of normal donor plasma that contains adequate amounts of ADAMTS13 [19]. Indeed, TPE has been extensively utilized by clinicians when a patient was thought to have atypical TTP, TTP-like syndrome or "thrombotic microangiopathy". In retrospective review of many reports and case studies, most of them were found to have the clinical features of EA-VMTD/DIT. We know that TPE is effective in TTP-like syndrome, including "DIC" [128], and is being recommended, in patients with thrombocytopenia and MAHA with/without MODS. TPE is a life-saving measure in TTP-like syndrome at this time.

As far as sepsis is concerned, TPE was utilized in the treatment of sepsis in few respectable clinical trials [129, 130] and pilot studies as well as case studies [131–134]. These studies were done even without the understanding of the concept of microthrombogenesis and mechanism of endothelial pathogenesis. Interestingly, TPE seemed to be more effective than plasma filtration therapy in sepsis [135, 136]. Additionally, TPE was also effective for "DIC" [128, 134] and MODS [131, 134]. The response of sepsis to TPE also supports septic syndromes are the manifestations of DIT. TPE is expected to be a very effective therapy if utilized in the earliest possible time in the septic syndromes.

Anti-complement therapy

Atypical HUS (aHUS) is a well-recognized syndrome associated with "dysfunctional" or more likely with "unprotected" complement activation [22, 29]. The effectiveness of anti-complement therapy is established for aHUS [137, 138]. Since complement activation in sepsis to kill invading pathogen can be inhibited by anti-complement therapy in early immune defense system, theoretically anticomplement therapy such as eculizumab should be contraindicated in sepsis at this time until we know more about the precise role of complement in advancing stage of endotheliopathy. However, potential benefit in non-pathogen related EA-VMTD such as trauma, toxin and envenomation is an interesting speculation.

Anti-microthrombotic therapy

TPE as a surrogate form of antimicrothrombotic therapy is highly time-sensitive, but is an effective treatment. However, if the treatment is delayed due to technical limitation and/or becomes complicated by vascular volume overload or electrolyte imbalance, the outcome from TPE is likely to be unfavorable in the septic syndromes and other critical illnesses.

Theoretically, the ideal treatment is targeted therapy with one of antimicrothrombotic agents such as recombinant ADAMTS13 (rADAMTS13) at the earliest possible time. The studies have shown ADAMTS13 is effective in cleaving platelet-ULVWF complexes in vitro. Antimicrothrombotic therapy has been advocated in vivo use for potential benefit in acquired TTP (TTP-like syndrome) [139–142]. Recombinant ADAMTS13 is being used only in Phase I study for GA-VMTD (i.e., hereditary TTP) at this time. Since its proteolytic activity on ULVWF is well established, this investigational agent should be available for clinical trials in EA-VMTD, including septic syndromes and "DIC" as well as MODS caused by EA-VMTD. In a latest study on acquired TTP, anti-microthrombotic agent caplacizumab, which is anti-von Willebrand factor (VWF) nanobody [143], has shown significant clinical benefit. This positive result also supports the role of ULVWF path in TTP-like syndrome as well as EA-VMTD/DIT.

Anti-microthrombotic therapy can be utilized with a close monitoring of the platelet count, intravascular hemolysis, and evaluation of organ dysfunctions. Streamlined therapeutic trials with recombinant ADAMTS13 should be proposed for the patients with septic syndromes associated with EA-VMTD/DIT as soon as possible.

Conclusion

The endothelial pathogenesis of sepsis and septic shock is due to lone activation of ULVWF path without activation of TF path of hemostasis. The molecular pathogenesis of sepsis is clearly established from the dual concepts based on "two-path unifying theory of hemostasis" and "two-activation theory of the endothelium". In sepsis, complement activation that is destructive to ECs of the host can provoke endotheliopathy, which promotes two independent molecular events, leading to activation of inflammatory pathway and microthrombotic pathway. The result is endotheliopathy associated inflammation and EA-VMTD presenting with TTP-like syndrome. In a big picture, physiologic protective immune system guards the host and eliminates invading pathogen, but pathologic destructive endothelial system is detrimental to the host when protective immune systems are compromised and antibiotics are not optimally effective. With better understanding of sepsis-associated coagulopathy (microthrombopathy), effective treatments could be designed by targeting the microthrombotic pathway of endothelial pathogenesis.

Abbreviations

"DIC": Ill-founded disseminated intravascular coagulation; AAI: Acute adrenal insufficiency; AA-VMTD: Antibody-associated VMTD; ADAMTS13: A disintegrin and metalloproteinase with a thrombospondin type 1 motif, member 13; aHUS: Atypical HUS; ALF: Acute liver failure; aMAHA: Atypical MAHA; AP: Acute pancreatitis; APC: Activated protein C; APL: Acute promyelocytic leukemia; ARDS: Acute respiratory distress syndrome; ARF: Acute renal failure; CMVD: Coronary microvascular dysfunction; CNSD: Central nervous system dysfunction; DIC: Disseminated intravascular coagulation; DIT: Disseminated intravascular microthrombosis; DSS: Dengue shock syndrome; EA-VMTD: Endotheliopathy-associated VMTD; ECs: Endothelial cells; EVT: Extravascular tissue; FHF: Fulminant hepatic failure; FVIIa: Activated factor VII; GA-VMTD: Gene mutation-associated VMTD; GE: Gastroenteritis; HCPS: Hantavirus cardiopulmonary syndrome; HE: Hepatic encephalopathy; HFRS: Hemorrhagic fever with renal syndrome; HPS: Hantavirus pulmonary syndrome; HRS: Hepato-renal syndrome; HUS: Hemolytic-uremic syndrome; IF: Interferon; IL: Interleukin; LPS: Lipopolysaccharide; MAC: Membrane attack complex; MAHA: Microangiopathic hemolytic anemia; MI: Myocardial infarction; MODS: Multi-organ dysfunction syndrome; NETs: Neutrophil extracellular traps; NO: Nitric oxide; NOMI: Non-occlusive mestenteric ischemia; PDIS: Peripheral digit ischemic syndrome; rADAMTS13: Recombinant ADAMTS13; RML: Rhabdomyolysis; RMSF: Rocky mountain spotted fever; SET: Subendothelial tissue; SFTS: Severe fever with thrombocytopenia syndrome; SIRS: Systemic inflammatory response syndrome; SS: Stroke syndrome; STEC: Shiga toxin-producing *Escherichia coli*; TCIP: Thrombocytopenia in critically ill patients; TF: Tissue factor; TFPI: Tissue factor pathway inhibitor; TM: Thrombomodulin; TMOF: Thrombocytopenia and multiple organ failure; TNF: Tumor necrosis factor; TPE: Therapeutic plasma exchange; TSS: Toxic shock syndrome; TTP: Thrombotic thrombocytopenic purpura; ULVWF: Unusually large von Willebrand factor; VMTD: Vascular microthrombotic disease; WFS: Waterhouse Friderichsen syndrome

Acknowledgements
Not applicable

References

1. Chang JC. Thrombocytopenia in critically ill patients due to vascular microthrombotic disease: pathogenesis based on "two activation theory of the endothelium". Vascul Dis Ther. 2017;2:1–7.
2. Chang JC. Disseminated intravascular coagulation (DIC): is it fact or fancy? Blood Coagul Fibrinolysis. 2018;29:330–7.
3. Chang JC. TTP-like syndrome: novel concept and molecular pathogenesis of endotheliopathy-associated vascular microthrombotic disease. Thromb J. 2018;16:20.
4. Chang JC. Thrombogenesis and thrombotic disorders based on 'two-path unifying theory of hemostasis': philosophical, physiological and phenotypical interpretation. Blood Coagul Fibrinolysis. 2018;29:585–95.
5. Dempfle CE. Coagulopathy of sepsis. Thromb Haemost. 2004;91:213–24.
6. Rhee C, Dantes R, Epstein L, et al. Incidence and trends of sepsis in US hospitals using clinical vs claims data, 2009-14. JAMA. 2017;318:1241–9.
7. Aird WC. The role of the endothelium in severe sepsis and multiple organ dysfunction syndrome. Blood. 2003;101:3765–77.
8. Peters K, Unger RE, Brunner J, Kirkpatrick CJ. Molecular basis of endothelial dysfunction in sepsis. Cardiovasc Res. 2003;60:49–57.
9. Lush CW, Kvietys PR. Microvascular dysfunction in sepsis. Microcirculation. 2000;7:83–101.
10. Ince C, Mayeux PR, Nguyen T, et al. The endothelium in sepsis. Shock. 2016; 45:259–70.
11. Esmon CT. Crosstalk between inflammation and thrombosis. Maturitas. 2004; 15(47):305–14.
12. van der Poll T, Levi M. Crosstalk between inflammation and coagulation: the lessons of sepsis. Curr Vasc Pharmacol. 2012;10:632–8.

13. O'Brien M. The reciprocal relationship between inflammation and coagulation. Top Companion Anim Med. 2012;27:46–52.
14. Semeraro N, Ammollo CT, Semeraro F, Colucci M. Coagulopathy of Acute Sepsis. Semin Thromb Hemost. 2015;41:650–8.
15. Okamoto K, Tamura T, Sawatsubashi Y. Sepsis and disseminated intravascular coagulation. J Intensive Care. 2016;4:23.
16. Umemura Y, Yamakawa K, Hayakawa M, Hamasaki T, Fujimi S. Screening itself for disseminated intravascular coagulation may reduce mortality in sepsis: A nationwide multicenter registry in Japan. Thromb Res. 2018;161:60–6.
17. Levi M. Disseminated intravascular coagulation. Crit Care Med. 2007;35: 2191–5.
18. Østerud B, Bjørklid E. The tissue factor pathway in disseminated intravascular coagulation. Semin Thromb Hemost. 2001;27:605–17.
19. Chang JC. A thought on possible pathogenesis of ebola viral hemorrhagic disease and potential treatments: could it be thrombotic thrombocytopenic purpura-like syndrome? Ther Apher Dial. 2015;20:93–8.
20. Chang JC. Hemostasis based on "two-path unifying theory" and classification of hemostatic disorders. Blood Coagul Fibrinolysis. 2018;29:573–84.
21. Lambris JD, Ricklin D, Geisbrecht BV. Complement evasion by human pathogens. Nat Rev Microbiol. 2008;6:132–42.
22. Kerr H, Richards A. Complement-mediated injury and protection of endothelium: lessons from atypical haemolytic uraemic syndrome. Immunobiology. 2012;217:195–203.
23. Lehto T, Meri S. Interactions of soluble CD59 with the terminal complement complexes. CD59 and C9 compete for a nascent epitope on C8. J Immunol. 1993;151:4941–9.
24. Brooimans RA, Van der Ark AA, Tomita M, Van Es LA, Daha MR. CD59 expressed by human endothelial cells functions as a protective molecule against complement-mediated lysis. Eur J Immunol. 1992;22:791–7.
25. Sugita Y, Masuho Y. CD59: its role in complement regulation and potential for therapeutic use. Immunotechnology. 1995;1:157–68.
26. Helling H, Stephan B, Pindur G. Coagulation and complement system in critically ill patients. Clin Hemorheol Microcirc. 2015;61:185–93.
27. McPhaden AR, Whaley K. The complement system in sepsis and trauma. Br Med Bull. 1985;41:281–6.
28. Cross AR, Glotz D, Mooney N. The role of the endothelium during antibody-mediated rejection: from victim to accomplice. Front Immunol. 2018;9:106.
29. Chang JC. Molecular pathogenesis of STEC-HUS caused by endothelial heterogeneity and unprotected complement activation, leading to endotheliopathy and impaired ADAMTS13 activity: based on two-activation theory of the endothelium and vascular microthrombotic disease. Nephrol Renal Dis. 2017;2:1–8.
30. Osterud B, Bjorklid E. Tissue factor in blood cells and endothelial cells. Front Biosci (Elite Ed). 2012;4:289–99.
31. Pries AR, Kuebler WM. Normal endothelium. Handb Exp Pharmacol. 2006; 176(176 Pt 1):1–40.
32. Mackman N. The role of tissue factor and factor VIIa in hemostasis. Anesth Analg. 2009;108:1447–52.
33. Vincent JL, De Backer D. Microvascular dysfunction as a cause of organ dysfunction in severe sepsis. Crit Care. 2005;9(Suppl 4):S9–12.
34. Levi M, Schultz M, van der Poll T. Sepsis and thrombosis. Semin Thromb Hemost. 2013;39:559–66.
35. Chousterman BG, Swirski FK, Weber GF. Cytokine storm and sepsis disease pathogenesis. Semin Immunopathol. 2017;39:517–28.
36. Ulloa L, Tracey KJ. The "cytokine profile": a code for sepsis. Trends Mol Med. 2005;11:56–63.
37. Xing K, Murthy S, Liles WC, Singh JM. Clinical utility of biomarkers of endothelial activation in sepsis--a systematic review. Crit Care. 2012;16:R7.
38. Aird WC. Endothelial cell heterogeneity. Cold Spring Harb Perspect Med. 2012;2:a006429.
39. Nolan DJ, Ginsberg M, Israely E, et al. Molecular signatures of tissue-specific microvascular endothelial cell heterogeneity in organ maintenance and regeneration. Dev Cell. 2013;26:204–19.
40. McCall LI, Siqueira-Neto JL, McKerrow JH. Location, location, location: five facts about tissue tropism and pathogenesis. PLoS Pathog. 2016;12(5): e1005519.
41. Obrig TG, Louise CB, Lingwood CA, Boyd B, Barley-Maloney L, Daniel TO. Endothelial heterogeneity in Shiga toxin receptors and responses. J Biol Chem. 1993;268:15484 8.
42. Zeng H, Goldsmith CS, Maines TR, et al. Tropism and infectivity of influenza virus, including highly pathogenic avian H5N1 virus, in ferret tracheal differentiated primary epithelial cell cultures. J Virol. 2013;87:2597–607.
43. Caugant DA, Maiden MC. Meningococcal carriage and disease--population biology and evolution. Vaccine. 2009;27(Suppl 2):B64–70.
44. Al-Khafaji A, Nadim MK, Kellum JA. Hepatorenal Disorders. Chest. 2015;148: 550–8.
45. Hadjihambi A, Arias N, Sheikh M. Jalan Hepatic encephalopathy: a critical current review. Hepatol Int. 2018;12(Suppl 1):135–47.
46. Kalaiselvan S, Sankar S, Ramamurthy M, Ghosh AR, Nandagopal B, Sridharan G. Prediction of pan-specific b-cell epitopes from nucleocapsid protein of hantaviruses causing hantavirus cardiopulmonary syndrome. J Cell Biochem. 2017;118:2320–4.
47. Petejova N, Martinek A. Acute kidney injury following acute pancreatitis: A review. Biomed Pap Med Fac Univ Palacky Olomouc Czech Repub. 2013; 157:105–13.
48. Zhang XP, Tian H. Pathogenesis of pancreatic encephalopathy in severe acute pancreatitis. Hepatobiliary Pancreat Dis Int. 2007;6:134–40.
49. Bernal Monterde V, Campillo Arregui A, Sostres Homedes C, García Gil A, Simón Marco MA, Serrano Aulló MT. Acute necrotizing pancreatitis and severe hepatic failure: description of three cases. Transplant Proc. 2008;40:3117–20.
50. Sonavane A, Baradkar V, Salunkhe P, Kumar S. Waterhouse-Friderichsen syndrome in an adult patient with meningococcal meningitis. Indian J Dermatol. 2011;56:326–8.
51. Schäfer A, Bauersachs J. Endothelial dysfunction, impaired endogenous platelet inhibition and platelet activation in diabetes and atherosclerosis. Curr Vasc Pharmacol. 2008;6:52–60.
52. Coenen DM, Mastenbroek TG, Cosemans JMEM. Platelet interaction with activated endothelium: mechanistic insights from microfluidics. Blood. 2017; 130:2819–28.
53. Turner NA, Moake J. Assembly and activation of alternative complement components on endothelial cell-anchored ultra-large von Willebrand factor links complement and hemostasis-thrombosis. PLoS One. 2013;8:e59372.
54. Valentijn KM, van Driel LF, Mourik MJ, et al. Multigranular exocytosis of Weibel-Palade bodies in vascular endothelial cells. Blood. 2010;116:1807–16.
55. Bernardo A, Ball C, Nolasco L, Choi H, Moake JL, Dong JF. Platelets adhered to endothelial cell-bound ultra-large von Willebrand factor strings support leukocyte tethering and rolling under high shear stress. J Thromb Haemost. 2005;3:562–70.
56. Chauhan AK, Goerge T, Schneider SW, Wagner DD. Formation of platelet strings and microthrombi in the presence of ADAMTS-13 inhibitor does not require P-selectin or beta3 integrin. J Thromb Haemost. 2007;5:583–9.
57. Feng S, Eyler SJ, Zhang Y, et al. Partial ADAMTS13 deficiency in atypical hemolytic uremic syndrome. Blood. 2013;122:1487–93.
58. Pourrat O, Coudroy R, Pierre F. ADAMTS13 deficiency in severe postpartum HELLP syndrome. Br J Haematol. 2013;163:409–10.
59. Gandhi C, Motto DG, Jensen M, Lentz SR, Chauhan AK. ADAMTS13 deficiency exacerbates VWF-dependent acute myocardial ischemia/ reperfusion injury in mice. Blood. 2012;120:5224–30.
60. Fujioka M, Hayakawa K, Mishima K, et al. ADAMTS13 gene deletion aggravates ischemic brain damage: a possible neuroprotective role of ADAMTS13 by ameliorating postischemic hypoperfusion. Blood. 2010;115:1650–3.
61. Dinarello CA. Infection, fever, and exogenous and endogenous pyrogens: some concepts have changed. J Endotoxin Res. 2004;10:201–22.
62. Chang JC, Gross HM. Utility of naproxen in the differential diagnosis of fever of undetermined origin in patients with cancer. Am J Med. 1984;76:597–603.
63. Zell JA, Chang JC. Neoplastic fever: a neglected paraneoplastic syndrome. Support Care Cancer. 2005;13:870–7.
64. Joyce DE, Nelson DR, Grinnell BW. Leukocyte and endothelial cell interactions in sepsis: relevance of the protein C pathway. Crit Care Med. 2004;32:S280–6.
65. Kaukonen KM, Bailey M, Pilcher D, Cooper DJ, Bellomo R. Systemic inflammatory response syndrome criteria in defining severe sepsis. N Engl J Med. 2015;372:1629–38.
66. Brun-Buisson C. The epidemiology of the systemic inflammatory response. Intensive Care Med. 2000;26(Suppl 1):S64–74.

67. Warkentin TE. Heparin-induced thrombocytopenia in critically ill patients. Semin Thromb Hemost. 2015;41:49–60.
68. Thachil J, Warkentin TE. How do we approach thrombocytopenia in critically ill patients? Br J Haematol. 2017;177:27–38.
69. Venkata C, Kashyap R, Farmer JC, Afessa B. Thrombocytopenia in adult patients with sepsis: incidence, risk factors, and its association with clinical outcome. J Intensive Care. 2013;1:9.
70. Greco E, Lupia E, Bosco O, Vizio B, Montrucchio G. Platelets and multi-organ failure in sepsis. Int J Mol Sci. 2017;18(10):2200.
71. Nguyen TC, Carcillo JA. Bench-to-bedside review: thrombocytopenia-associated multiple organ failure--a newly appreciated syndrome in the critically ill. Crit Care. 2006;10:235.
72. Levi M, Löwenberg EC. Thrombocytopenia in critically ill patients. Semin Thromb Hemost. 2008;34:417–24.
73. Schriber JR, Freedman JJ, Brandwein JM. Thrombotic thrombocytopenic purpura associated with pneumococcal sepsis. Can J Infect Dis. 1993;4:145–7.
74. Centers for Disease Control and Prevention (CDC). Thrombotic thrombocytopenic purpura (TTP)-like illness associated with intravenous Opana ER abuse--Tennessee, 2012. MMWR Morb Mortal Wkly Rep. 2013; 62:1–4.
75. Wang Z, Yu Z, Su J, Cao L, Zhao X, Ruan C. Sepsis-induced disseminated intravascular coagulation with features of thrombotic thrombocytopenic purpura: a fatal fulminant syndrome. Clin Appl Thromb Hemost. 2011;17: 251–3.
76. Booth KK, Terrell DR, Vesely SK, George JN. Systemic infections mimicking thrombotic thrombocytopenic purpura. Am J Hematol. 2011;86:743–51.
77. Lopes da Silva R. Viral-associated thrombotic microangiopathies. Hematol Oncol Stem Cell Ther. 2011;4(2):51–9.
78. Tsai HM. Pathophysiology of thrombotic thrombocytopenic purpura. Int J Hematol. 2010;91:1–19.
79. Sadler JE, Moake JL, Miyata T, George JN. Recent advances in thrombotic thrombocytopenic purpura. Hematology Am Soc Hematol Educ Program. 2004;2004:407–23.
80. Chang JC, Shipstone A, Llenado-Lee MA. Postoperative thrombotic thrombocytopenic purpura following cardiovascular surgeries. Am J Hematol. 1996;53:11–7.
81. Naqvi TA, Baumann MA, Chang JC. Post-operative thrombotic thrombocytopenic purpura: a review. Int J Clin Pract. 2004;58:169–72.
82. Chang JC, Naqvi T. Thrombotic thrombocytopenic purpura associated with bone marrow metastasis and secondary myelofibrosis in cancer. Oncologist. 2003;8:375–80.
83. Scully M. Cancer and thrombotic microangiopathy. J Oncol Pract. 2016;12: 531–2.
84. Yu WL, Leung T, Soo Y, Lee J, Wong KS. Thrombotic thrombocytopenic purpura with concomitant small- and large-vessel thrombosis, atypical posterior reversible encephalopathy syndrome and cerebral microbleeds. Oxf Med Case Reports. 2015;91:179–82.
85. Ibernon M, Moreso F, Carreras L, et al. Thrombotic thrombocytopenic purpura with severe large artery branch involvement. Nephrol Dial Transplant. 2005;20:467–8.
86. Scully M, Hunt BJ, Benjamin S, et al. Guidelines on the diagnosis and management of thrombotic thrombocytopenic purpura and other thrombotic microangiopathies. Br J Haematol. 2012;158:323–35.
87. Sarode R. Atypical presentations of thrombotic thrombocytopenic purpura: a review. J Clin Apher. 2009;24:47–52.
88. Nguyen TC, Liu A, Liu L, et al. Acquired ADAMTS-13 deficiency in pediatric patients with severe sepsis. Haematologica. 2007;92:121–4.
89. Chang JC, Kathula SK. Various clinical manifestations in patients with thrombotic microangiopathy. J Investig Med. 2002;50:201–6.
90. Warkentin TE, Pai M. Shock, acute disseminated intravascular coagulation, and microvascular thrombosis: is 'shock liver' the unrecognized provocateur of ischemic limb necrosis? J Thromb Haemost. 2016;14:231–5.
91. Chang JC, Ikhlaque N. Peripheral digit ischemic syndrome can be a manifestation of postoperative thrombotic thrombocytopenic purpura. Ther Apher Dial. 2004;8:413–8.
92. Panackel C, Thomas R, Sebastian B, Mathai SK. Recent advances in management of acute liver failure. Indian J Crit Care Med. 2015;19:27–33.
93. McDonald V, Laffan M, Benjamin S, Bevan D, Machin S, Scully MA. Thrombotic thrombocytopenic purpura precipitated by acute pancreatitis: a report of seven cases from a regional UK TTP registry. Br J Haematol. 2009; 144:430–3.
94. Dal-Pizzol F, Tomasi CD, Ritter C. Septic encephalopathy: does inflammation drive the brain crazy? Rev Bras Psiquiatr. 2014;36:251–8.
95. Balasubramaniyam N, Kolte D, Palaniswamy C, et al. Predictors of in-hospital mortality and acute myocardial infarction in thrombotic thrombocytopenic purpura. Am J Med. 2013;126:1016.e1–7.
96. Mele JA 3rd, Linder S, Capozzi A. Treatment of thromboembolic complications of fulminant meningococcal septic shock. Ann Plast Surg. 1997;38:283–90.
97. Hale AJ, LaSalvia M, Kirby JE, Kimball A, Baden R. Fatal purpura fulminans and Waterhouse-Friderichsen syndrome from fulminant Streptococcus pneumoniae sepsis in an asplenic young adult. IDCases. 2016;6:1–4.
98. Qahtani SA. Acute renal failure and severe rhabdomyolysis in a patient with resistant thrombotic thrombocytopenic purpura. Int J Gen Med. 2011;4:687–9.
99. Bejan C, Constantinescu S, Dorobăţ G, Dorobăţ C. Hepatorenal dysfunction in sepsis: epidemiological, clinical and laboratory aspects. Rev Med Chir Soc Med Nat Iasi. 2014;118:759–63.
100. van Dongen PW, Eskes TK, Gimbrère JS, Snel P. Maternal mortality due to the hepatorenal syndrome of pre-eclampsia. A case report. Eur J Obstet Gynecol Reprod Biol. 1979;9:299–306.
101. Levi M, Scully M. How I treat disseminated intravascular coagulation. Blood. 2018;131:845–54.
102. Denecke B, Bigalke B, Haap M, Overkamp D, Lehnert H, Haas CS. Hantavirus infection: a neglected diagnosis in thrombocytopenia and fever? Mayo Clin Proc. 2010;85:1016–20.
103. Singer M, Deutschman CS, Seymour CW, et al. The third international consensus definitions for sepsis and septic shock (Sepsis-3). JAMA. 2016;315: 801–10.
104. Annane D. Adrenal insufficiency in sepsis. Curr Pharm Des. 2008;14:1882–6.
105. den Brinker M, Joosten KF, Liem O, et al. Adrenal insufficiency in meningococcal sepsis: bioavailable cortisol levels and impact of interleukin-6 levels and intubation with etomidate on adrenal function and mortality. J Clin Endocrinol Metab. 2005;90:5110–7.
106. Hamilton D, Harris MD, Foweraker J, Gresham GA. Waterhouse–Friderichsen syndrome as a result of non-meningococcal infection. J Clin Pathol. 2004;57: 208–9.
107. Harms D, Pape GR, Bohle A. Pathological anatomy of Waterhouse-Friderichsen syndrome with special reference to disseminated intravascular coagulation. Dtsch Med. 1973;98:542–7.
108. CF WP, Troster EJ, Damiani D, Carcillo JA. Absolute and relative adrenal insufficiency in children with septic shock. Crit Care Med. 2005;33:855–9.
109. Annane D, Maxime V, Ibrahim F, Alvarez JC, Abe E, Boudou P. Diagnosis of adrenal insufficiency in severe sepsis and septic shock. Am J Respir Crit Care Med. 2006;174:1319–26.
110. Maxime V, Lesur O, Annane D. Adrenal insufficiency in septic shock. Clin Chest Med. 2009;30:17–27.
111. Marik PE, Zaloga GP. Adrenal insufficiency during septic shock. Crit Care Med. 2003;31:141–5.
112. McKay DG. Progress in disseminated intravascular coagulation. Calif Med. 1969;111:186–98.
113. Levi M, van der Poll T. A short contemporary history of disseminated intravascular coagulation. Semin Thromb Hemost. 2014;40:874–80.
114. Ten Cate H, Hackeng TM, García de Frutos P. Coagulation factor and protease pathways in thrombosis and cardiovascular disease. Thromb Haemost. 2017;117:1265–71.
115. Levi M, van der Poll T. Inflammation and coagulation. Crit Care Med. 2010; 38(2 Suppl):S26–34.
116. Gando S, Levi M, Toh CH. Disseminated intravascular coagulation. Nat Rev Dis Primers. 2016;2:16037.
117. Bone RC, Balk RA, Cerra FB, Dellinger RP, Fein AM, Knaus WA, Schein RM, Sibbald WJ, The ACCP/SCCM Consensus Conference Committee. American College of Chest Physicians/Society of Critical Care Medicine Definitions for sepsis and organ failure and guidelines for the use of innovative therapies in sepsis. Chest. 1992;101:1644–55.
118. Balk RA. Systemic inflammatory response syndrome (SIRS): where did it come from and is it still relevant today? Virulence. 2014;5:20–6.
119. Ogura H, Gando S, Iba T, et al. SIRS-associated coagulopathy and organ dysfunction in critically ill patients with thrombocytopenia. Shock. 2007;28:411–7.
120. Stravitz RT, Ellerbe C, Durkalski V, Reuben A, Lisman T, Lee WM. Thrombocytopenia is associated with multi-organ system failure in patients with acute liver failure. Clin Gastroenterol Hepatol. 2016;14:613–20.

121. Camous L, Veyradier A, Darmon M, et al. Macrovascular thrombosis in critically ill patients with thrombotic micro-angiopathies. Intern Emerg Med. 2014;9:267–72.
122. Reid VL, McLay J. Symmetrical peripheral gangrene and disseminated intravascular coagulation associated with pelvi-ureteric junction obstruction. BMJ Case Rep. 2013;2013:bcr2012006519.
123. Ruffin N, Vasa CV, Breakstone S, Axman W. Symmetrical peripheral gangrene of bilateral feet and unilateral hand after administration of vasopressors during septic shock. BMJ Case Rep. 2018;2018:bcr-2017-223602.
124. Molos MA, Hall JC. Symmetrical peripheral gangrene and disseminated intravascular coagulation. Arch Dermatol. 1985;121:1057–61.
125. Liechti ME, Zumsteg V, Hatz CF, Herren T. Plasmodium falciparum cerebral malaria complicated by disseminated intravascular coagulation and symmetrical peripheral gangrene: case report and review. Eur J Clin Microbiol Infect Dis. 2003;22:551–4.
126. Adamski J. Thrombotic microangiopathy and indications for therapeutic plasma exchange. Hematology Am Soc Hematol Educ Program. 2014;2014: 444–9.
127. Nguyen TC, Han YY. Plasma exchange therapy for thrombotic microangiopathies. Organogenesis. 2011;7:28–31.
128. Oral M, Yilmaz A, Yilmaz D, Ünal N, Tulunay M. The place of plasmapheresis in septic patients complicated with disseminated intravascular coagulation. Crit Care. 2004;8(Suppl 1):P150.
129. Busund R, Koukline V, Utrobin U, Nedashkovsky E. Plasmapheresis in severe sepsis and septic shock: a prospective, randomised, controlled trial. Intensive Care Med. 2002;28:1434–9.
130. Nguyen TC, Han YY, Kiss JE, Hall MW, Hassett AC, Jaffe R, Orr RA, et al. Intensive plasma exchange increases a disintegrin and metalloprotease with thrombospondin motifs-13 activity and reverses organ dysfunction in children with thrombocytopenia-associated multiple organ failure. Crit Care Med. 2008;36:2878–87.
131. Qu L, Kiss JE, Dargo G, Carcillo JA. Outcomes of previously healthy pediatric patients with fulminant sepsis-induced multisystem organ failure receiving therapeutic plasma exchange. J Clin Apher. 2011;26(4):208–13.
132. Hadem J, Hafer C, Schneider AS, Wiesner O, Beutel G, Fuehner T, et al. Therapeutic plasma exchange as rescue therapy in severe sepsis and septic shock: retrospective observational single-centre study of 23 patients. BMC Anesthesiol. 2014;14:24.
133. De Simone N, Racsa L, Bevan S, Matevosyan K, Valley T, Girod C, Sarode R. Therapeutic plasma exchange in the management of sepsis and multiple organ dysfunction syndrome: a report of three cases. J Clin Apher. 2014;29: 127–31.
134. Stegmayr B. Apheresis in patients with severe sepsis and multiorgan dysfunction syndrome. Transfus Apher Sci. 2008;38:203–8.
135. Rimmer E, Houston BL, Kumar A, Abou-Setta AM, Friesen C, Marshall JC, et al. The efficacy and safety of plasma exchange in patients with sepsis and septic shock: a systematic review and meta-analysis. Crit Care. 2014;18:699.
136. Reeves JH, Butt WW, Shann F, Layton JE, Stewart A, Waring PM, Presneill JJ. Continuous plasmafiltration in sepsis syndrome. Plasmafiltration in Sepsis Study Group. Crit Care Med. 1999;27:2096–104.
137. Palma LM, Langman CB. Critical appraisal of eculizumab for atypical hemolytic uremic syndrome. J Blood Med. 2016;7:39–72.
138. Kaplan BS, Ruebner RL, Spinale JM, Copelovitch L. Current treatment of atypical hemolytic uremic syndrome. Intractable Rare Dis Res. 2014;3:34–45.
139. Schwameis M, Schörgenhofer C, Assinger A, Steiner MM, Jilma B. VWF excess and ADAMTS13 deficiency: a unifying pathomechanism linking inflammation to thrombosis in DIC, malaria, and TTP. Thromb Haemost. 2015;113:708–18.
140. Zheng XL. ADAMTS13 and von Willebrand factor in thrombotic thrombocytopenic purpura. Annu Rev Med. 2015;66:211–25.
141. Tersteeg C, Schiviz A, De Meyer SF, Plaimauer B, Scheiflinger F, Rottensteiner H, Vanhoorelbeke K. Potential for Recombinant ADAMTS13 as an Effective Therapy for Acquired Thrombotic Thrombocytopenic Purpura. Arterioscler Thromb Vasc Biol. 2015;35:2336–42.
142. Plaimauer B, Kremer Hovinga JA, et al. Recombinant ADAMTS13 normalizes von Willebrand factor-cleaving activity in plasma of acquired TTP patients by overriding inhibitory antibodies. J Thromb Haemost. 2011;9:936–44.
143. Scully M, Cataland SR, Peyvandi F, et al. Caplacizumab treatment for acquired thrombotic thrombocytopenic purpura. N Engl J Med. 2019;380: 335–46.

SARS-CoV-2 and pulmonary embolism: Who stole the platelets?

Michael Tran[1*], Chirag Sheth[1], Rohan Bhandari[1], Scott J. Cameron[2] and Deborah Hornacek[1]

Abstract

Background: Patients infected with SARS-CoV-2 often develop venous and arterial thrombosis. The high patient mortality is partly attributed to thrombotic events. An emerging trend is the presence of immunological phenomena including antiphospholipid antibodies which may promote thrombosis. The mechanism for these observations is not clear though many patients with SARS-CoV-2 develop thrombocytopenia.

Case presentation: We describe a patient with SARS-CoV-2 pneumonitis who presented with intermediate risk pulmonary embolism (PE). Careful attention to his daily platelet count suggested the possibility of immune mediated heparin-induced thrombocytopenia (HIT) which was confirmed by laboratory testing and resolved when anticoagulation was switched to a direct thrombin inhibitor.

Conclusions: Since excessive platelet activation and in situ thrombosis occur in HIT, this case underscores the need to consider that thrombocytopenia in patients with SARS-CoV-2—most of whom receive heparinoids—may be unrecognized HIT. A central role for the platelet in the etiology of thrombosis during the COVID-19 pandemic should be explored.

Keywords: Pulmonary embolism, HIT, SARS-CoV-2, COVID-19, Thrombosis, Heparin

Background

Critically ill patients with COVID-19 infection often have multiple abnormalities in hemostasis and thrombosis. Recent literature documents hematologic derangements including mild thrombocytopenia [1], elevated d-dimer [2], prolonged activated partial-thromboplastin time (aPTT), and disseminated intravascular coagulation (DIC) [3]. It is unclear whether these changes reflect SARS-CoV-2 infection, or an inflammatory state of acute illness. The recent medical literature reports multiple anticoagulation strategies to prevent thrombotic events in patients infected with SARS-CoV-2. Emerging reports suggest the possibility of HIT developing in SARS-CoV-2 patients receiving heparin anticoagulation [4, 5]. This case was a diagnostic dilemma since both thrombocytopenia and in situ pulmonary thrombosis are common features of SARS-CoV-2 infection [6], making less common diagnoses, such as HIT, which shares similar features, more challenging to diagnose.

* Correspondence: TRANM2@ccf.org
[1]Heart Vascular and Thoracic Institute, Department of Cardiovascular Medicine, Section of Vascular Medicine, Cleveland Clinic Foundation, Desk J-35, Cleveland Clinic Foundation, Cleveland, OH 44195, USA
Full list of author information is available at the end of the article

Case presentation

A 62-year-old man with type 2 diabetes mellitus presented to the emergency department (ED) with 4-day history fever, cough, and dyspnea. The patient's vitals in the ED were as follows: Temperature 39.2 °C (102.6 °F), blood pressure 167/67 mmHg, heart rate 135 beats per min, respiratory rate 22 breaths per minute, oxygen saturation 74% on room air. The SaO_2 improved to 96% with oxygen therapy at 10 L/minute by non-rebreather mask. Relevant laboratory data was as follows: white blood count 13.9 K/uL, platelet 412 K/uL, sodium 126 mmol/L, creatinine 0.7 mg/dL. Chest radiography showed bilateral diffuse patchy airspace opacities. There was concern for COVID-19 which was confirmed by polymerase chain reaction (PCR) for the SARS-CoV-2 amplicon. The patient developed hypoxemic respiratory failure the following day and was placed on mechanical ventilation. Over a 10-day time period, the patient was treated with investigational therapies for COVID-19 including azithromycin, hydroxychloroquine, and convalescent plasma. The anticoagulation regimen consisted of subcutaneous enoxaparin (60 mg once daily) for venous thromboembolism (VTE) prophylaxis. Intravenous unfractionated heparin (UFH) flushes were used to maintain patency of vascular access. The patient was subsequently transferred to our tertiary medical center on hospital day 12 for further management.

Bedside echocardiography suggested right ventricular (RV) dilation, raising the possibility of pulmonary embolism (PE). Venous Duplex of the lower extremities was unremarkable. Chest computed tomography angiography (CTA) demonstrated right upper lobe lobar and segmental PE (Fig. 1). Relevant laboratory data included hemoglobin 11.1 g/dL, platelets 487 k/uL, creatinine 0.93 mg/dL, pro-BNP 7600 pg/mL, troponin T < 0.01 ng/mL, d-dimer 11,040 ng/mL, activated partial thromboplastin time (aPTT) 24.2 s, prothrombin time (PT) 14 s, fibrinogen 661 mg/dL. The patient had an equivocal cardiolipin IgM antibody of 15 IgG phospholipid units (MPL) since our reference range is < 12 MPL, and beta-2 glycoprotein antibodies were negative. Using the patient's PT, d-dimer, and measured fibrinogen concentration of 661 mg/dL his DIC score is 6 which was compatible with overt DIC.

A multidisciplinary pulmonary embolism response team (PERT) reviewed the patient's case and recommended therapeutic anticoagulation with intravenous UFH dosed at 18 units/kg/hr and monitored by the anti-Xa chromogenic assay. The patient's platelet count decreased from 487 k/uL to a nadir of 91 k/uL over the following 4 days, raising the concern for heparin induced thrombocytopenia (HIT) with an intermediate pretest probability by the 4Ts score of 4 (Table 1). Heparin products were exchanged for the direct thrombin inhibitor (DTI) bivalirudin dosed at 0.19 mg/kg/hr and monitored with activated partial thromboplastin time with a goal of 46 to 65 s. IgG specific anti-platelet factor 4 (PF4)-heparin enzyme linked immunosorbent assay (ELISA) was quite positive (optical density 1.08, normal value < 0.4). The heparin induced platelet aggregation (HIPA) functional assay was also positive, confirming the diagnosis of HIT. The patient's platelet count increased to 279 k/uL three days after the discontinuation of UFH (Fig. 2). The patient responded favorably to anticoagulation with a DTI without new thrombotic complications.

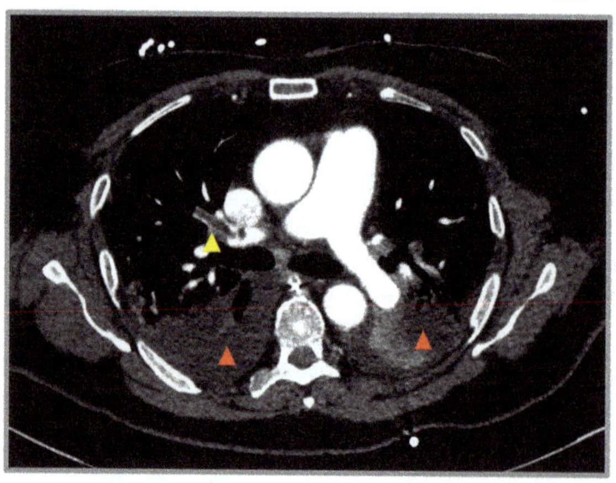

Fig. 1 Computed tomography (CT) of chest showing a filling defect in the right upper labor pulmonary artery extending into the segmental and subsegmental pulmonary branches consistent with an acute pulmonary embolism (yellow arrowhead). Patchy infiltrates are indicated by the red arrowhead

HIT is an acquired immune-mediated complication associated with UFH or LMWH administration in which platelets become haptenized [7, 8], creating a prothrombotic state that can be limb- or life-threatening due to venous or arterial thrombosis. The most common clinical manifestation of HIT is thrombocytopenia (platelet count < 150 k/uL) in 85–90% of patients between 5 and 10 days after initiating heparin products. The incidence of HIT is < 0.1–7% depending on clinical context, the heparin product used, and duration and dose of exposure [8, 9].

HIT develops when immunoglobulin G (IgG) antibodies recognize PF4-heparin complexes and activate platelet surface Fcγ receptors. This step activates platelets which degranulate and aggregate as thrombi [10]. Laboratory evaluation includes an IgG-specific anti-PF4 immunoassay which has high sensitivity but sometimes low specificity for activated platelets. Interestingly, this assay also detects non-platelet activating anti-PF4/heparin antibodies [11]. The diagnosis is supported by demonstrating IgG anti-PF4-heparin mediated platelet activation using washed platelets with functional assays such as the HIPA and

Table 1 4Ts Score for Heparin-Induced Thrombocytopenia

4Ts pretest probability	Score	4
Thrombocytopenia	Platelet count fall > 50% and platelet nadir > 20	2
Timing of platelet count fall	Onset after day 10 of heparin exposure	1
Thrombosis or other sequelae	No new thrombosis	0
Other causes for thrombocytopenia	Possible other causes	1

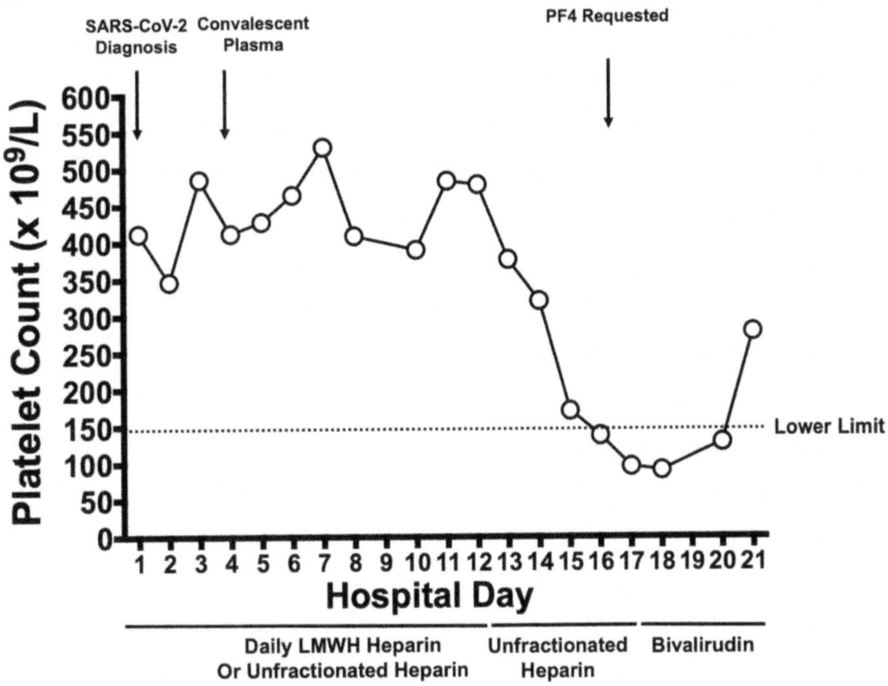

Fig. 2 Platelet count and time points for anticoagulation administration and laboratory testing

serotonin release assay (SRA) [12]. The assay performance of these two offers superior sensitivity and specificity in detection of pathogenic platelet activating antibodies. As a clinicopathologic syndrome, the diagnosis of HIT depends on taking a careful history and having temporal awareness of the exposure to heparinoids [13].

Conclusions

It is clear that patients infected with SARS-CoV-2 have unusual immunological phenomena including the presence of anti-phospholipid antibodies [14, 15]. The training of the internist is instinctually drawn to a broad differential diagnosis which is now more important than ever in the COVID-19 era. Early recognition and treatment for other immunologic phenomena such as HIT should be considered for every patient since thrombosis with thrombocytopenia often co-exist. Further study is required to determine if undiagnosed HIT contributes to the dramatically high number of thrombotic events in patients with SARS-CoV-2 infection.

Authors' contributions

All authors participated in writing of the manuscript and have reviewed and approved the submitted version.

Author details

[1]Heart Vascular and Thoracic Institute, Department of Cardiovascular Medicine, Section of Vascular Medicine, Cleveland Clinic Foundation, Desk J-35, Cleveland Clinic Foundation, Cleveland, OH 44195, USA. [2]Department of Cardiovascular and Metabolic Sciences. Cleveland Clinic Lerner College of Medicine, Cleveland, OH 44195, USA.

References

1. Lippi G, Plebani M, Henry BM. Thrombocytopenia is associated with severe coronavirus disease 2019 (COVID-19) infections: A meta-analysis. Clinica chimica acta. 2020;506:145–8.
2. Lippi G, Favaloro EJ. D-dimer is associated with severity of coronavirus disease 2019: a pooled analysis. Thromb Haemost. 2020;120(05):876–8.
3. Tang N, Li D, Wang X, Sun Z. Abnormal coagulation parameters are associated with poor prognosis in patients with novel coronavirus pneumonia. J Thromb Haemost. 2020;18(4):844–7.
4. Liu X, Zhang X, Xiao Y, Gao T, Wang G, Wang Z, et al. Heparin-induced thrombocytopenia is associated with a high risk of mortality in critical COVID-19 patients receiving heparin-involved treatment. medRxiv. 2020: 20076851.
5. Riker RR, May TL, Fraser GL, Gagnon DJ, Bandara M, Zemrak W, et al. Heparin-induced Thrombocytopenia with Thrombosis in COVID-19 Adult Respiratory Distress Syndrome. Res Pract Thromb Haemost. 2020. https://doi.org/10.1002/rth2.12390.
6. Wichmann D, Sperhake J-P, Lütgehetmann M, Steurer S, Edler C, Heinemann A, et al. Autopsy findings and venous thromboembolism in patients with COVID-19. Ann Intern Med. 2020.
7. Amiral J, Bridey F, Dreyfus M, Vissoc AM, Fressinaud E, Wolf M, et al. Platelet factor 4 complexed to heparin is the target for antibodies generated in heparin-induced thrombocytopenia. Thromb Haemost. 1992;68(1):95–6.

8. Warkentin TE, Sheppard J-AI, Horsewood P, Simpson PJ, Moore JC, Kelton JG. Impact of the patient population on the risk for heparin-induced thrombocytopenia. Blood. 2000;96(5):1703–8.
9. Martel N, Lee J, Wells PS. Risk for heparin-induced thrombocytopenia with unfractionated and low-molecular-weight heparin thromboprophylaxis: a meta-analysis. Blood. 2005;106(8):2710–5.
10. Greinacher A, Eichler P, Lubenow N, Kwasny H, Luz M. Heparin-induced thrombocytopenia with thromboembolic complications: meta-analysis of 2 prospective trials to assess the value of parenteral treatment with lepirudin and its therapeutic aPTT range. Blood. 2000;96(3):846–51.
11. WARKENTIN TE, Greinacher A, Gruel Y, Rh A, Bh C, scientific obot, et al. Laboratory testing for heparin-induced thrombocytopenia: a conceptual framework and implications for diagnosis. J Thromb Haemost. 2011;9(12): 2498–500.
12. Chong BH, Burgess J, Ismail F. The clinical usefulness of the platelet aggregation test for the diagnosis of heparin-induced thrombocytopenia. Thromb Haemost. 1993;69(04):344–50.
13. Warkentin TE. Laboratory diagnosis of heparin-induced thrombocytopenia. Int J Lab Hematol. 2019;41(S1):15–25.
14. Zhang Y, Xiao M, Zhang S, Xia P, Cao W, Jiang W, et al. Coagulopathy and Antiphospholipid antibodies in patients with Covid-19. N Engl J Med. 2020; 382(17):e38.
15. Bowles L, Platton S, Yartey N, Dave M, Lee K, Hart DP, et al. Lupus anticoagulant and abnormal coagulation tests in patients with Covid-19. N Engl J Med. 2020.

17

Protease-activated receptors (PARs): Mechanisms of action and potential therapeutic modulators in PAR-driven inflammatory diseases

Dorothea M. Heuberger[1,2] and Reto A. Schuepbach[1*]

Abstract

Inflammatory diseases have become increasingly prevalent with industrialization. To address this, numerous anti-inflammatory agents and molecular targets have been considered in clinical trials. Among molecular targets, protease-activated receptors (PARs) are abundantly recognized for their roles in the development of chronic inflammatory diseases. In particular, several inflammatory effects are directly mediated by the sensing of proteolytic activity by PARs.

PARs belong to the seven transmembrane domain G protein-coupled receptor family, but are unique in their lack of physiologically soluble ligands. In contrast with classical receptors, PARs are activated by N-terminal proteolytic cleavage. Upon removal of specific N-terminal peptides, the resulting N-termini serve as tethered activation ligands that interact with the extracellular loop 2 domain and initiate receptor signaling. In the classical pathway, activated receptors mediate signaling by recruiting G proteins. However, activation of PARs alternatively lead to the transactivation of and signaling through receptors such as co-localized PARs, ion channels, and toll-like receptors. In this review we consider PARs and their modulators as potential therapeutic agents, and summarize the current understanding of PAR functions from clinical and in vitro studies of PAR-related inflammation.

Introduction

The four mammalian members of the protease-activated receptor (PAR) family PAR1, PAR2, PAR3, and PAR4 are encoded by the genes F2R [1], F2RL1 [2], F2RL2 [3], and F2RL3 [4], respectively. Human PAR1 was discovered in 1991 as a key thrombin receptor on platelets [5, 6]. Although human and mouse PAR2 genes are homologous to PAR1 genes, PAR2 is not responsive to thrombin [2, 7, 8]. Unexpected responses of platelets to thrombin in PAR1 knockout mice lead to the discovery of the thrombin receptors PAR3 and PAR4 [4, 9, 10]. PAR regulation varies between species and tissues, with differing expression levels, protease cleaving activities, dimerization with other receptors, compartimentalization, trafficking, posttranslational modifications, and co-localization with co-receptors, as shown in Fig. 1.

Studies of PAR activation under physiological conditions are crucial for the understanding of the pathophysiological roles of PARs, such as those in inflammatory disorders.

* Correspondence: reto.schuepbach@usz.ch
[1]Institute of Intensive Care Medicine, University Hospital Zurich, University of Zurich, Zurich, Switzerland
Full list of author information is available at the end of the article

Cleavage and activation of PARs and signal transduction

PARs are specifically cleaved and irreversibly activated by various endogenous proteases, and by exogenous proteases from bacteria, plants, fungi, and insects. Proteases, soluble or cell membrane associated (bound to co-receptors or specific membrane compartments), cleave specific N-terminal peptides of PARs, resulting in exposure of new N-terminal peptides that serve as tethered activation ligands, which bind a conserved region on extracellular loop 2 (ECL2) [5, 11]. This interaction initiates conformational changes and alters affinity for

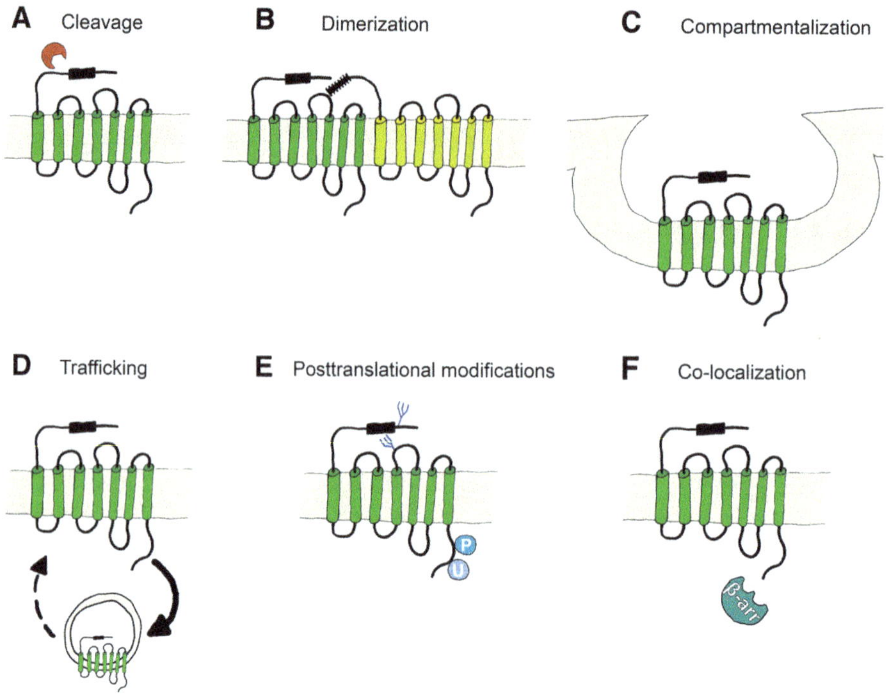

Fig. 1 Mechanisms of PAR activation. PAR activation is regulated by **a** direct proteolytic cleavage at the N-terminus, **b** homo- or heterodimerization with other PARs and transactivation through the cleaved tethered ligand, **c** compartmentalization on the cell surface, **d** degradation or recycling by endosomal trafficking, **e** posttranslational modifications such as glycosylation, phosphorylation, and ubiquitination, and **f** co-localization with other receptors and cofactors

intracellular G proteins [12]. Various N-terminal cleavage sites have been described, and these have various active conformations with specific G protein preferences. Multiple cleavage site-specific cellular responses are generally referred to as biased signaling, and the ensuing models describe how distinct proteases with distinct cleavage sites induce protease-specific responses via the same PAR [13, 14].

In contrast with PAR-activating proteases, other proteases cleave PARs at cleavage sites that are not related to signaling. Under these conditions, shedding of the PAR1 terminus, which removes the thrombin activation site, was first recognized as a mechanism for rendering platelets irresponsive to thrombin [15]. These truncated PARs can no longer be proteolyticaly activated, but remain activated by ligands from adjacent PARs [16]. Alternatively, truncated PARs bind soluble peptides with affinity for ECL2 by mimicking the tethered ligand. Both mechanisms result in receptor activation [17, 18]. Multiple ECL2-binding agonist peptides have been described and shown to induce signaling from truncated and uncleaved PARs (see agonist peptides in Tables 5, 6, 7).

PAR activation by proteolytical cleavage

PAR-cleaving proteases are a focus of many current studies. Whereas some PAR-cleaving proteases produce N-terminal components with regulatory roles, others render the receptors irresponsive to further protease exposure as shown in Fig. 2 and summarized in Tables 1, 2, 3 and 4. Important proteases are discussed below.

Mammalian proteases

Serine proteases Thrombin, the key protease of coagulation, is generated by proteolytic cleavage of zymogen prothrombin. Although thrombin production predominantly occurs on platelets and subendothelial vascular walls, extravascular thrombin has been detected in synovial fluid [19] and around tumors [20]. Thrombin has long been known to activate platelets, and the discovery of PAR1 initiated research into the underlying molecular mechanisms. PAR1 contains a hirudin-like domain, which has a high affinity thrombin binding site and recruits thrombin via exosite I. This interaction enables thrombin to specifically and efficiently activate PAR1 [6]. Similarly, PAR3 contains a hirudin-like thrombin recruitment site, which results in cleavage [9, 21]. In other studies, mouse PAR3 maintained thrombin recruitment activity but lost its receptor function, as discussed above [22–24]. Thrombin also cleaves and activates PAR4, which, in contrast with PAR1, lacks a hirudin-like domain. Thus, higher concentrations of thrombin activate PAR4 and initiate intracellular signaling [10]. PAR2 is considered the only PAR that resists cleavage or

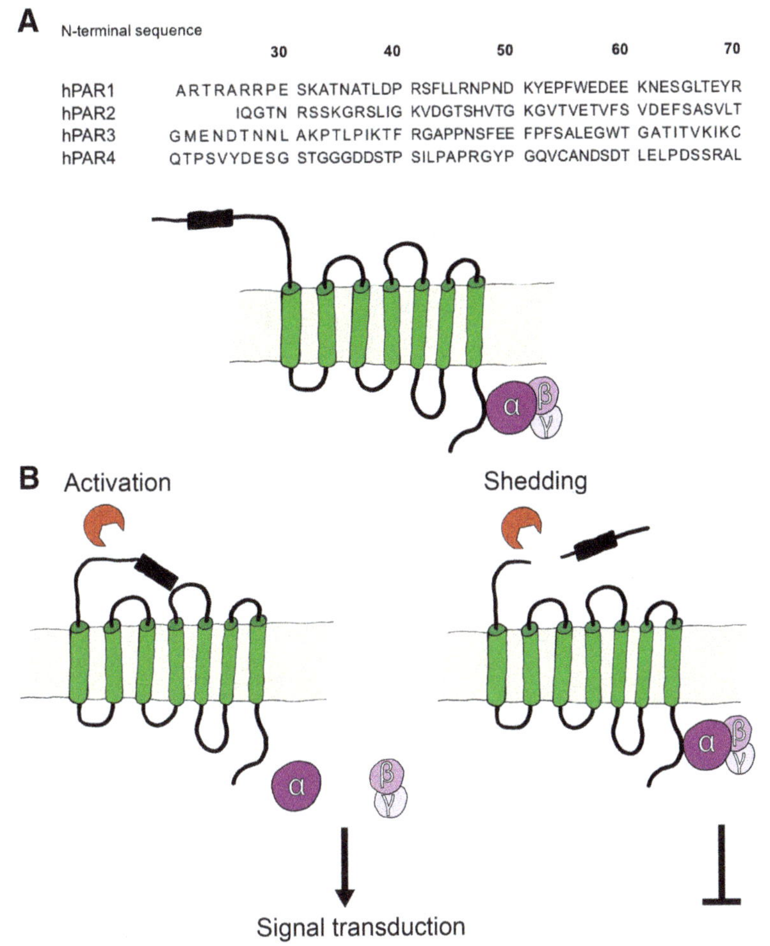

Fig. 2 Proteolytic PAR cleavage. **a** N-terminal sequences of human PARs (PAR1–4) containing potential cleavage sites. **b** Proteolytic cleavage of PARs by soluble exogenous proteases exposes new N-terminal sequences that serve as tethered ligands for G protein dependent activation of receptors. Alternatively, proteolytic cleavage at other sites destroys the function of the receptor to prevent intracellular signal transduction

activation by thrombin [4, 25], although emerging evidence suggests that at very high concentrations (100–500 nM), thrombin may directly cleave and activate PAR2 [26, 27].

In contrast with thrombin, the anticoagulant protease activated protein C (aPC) binds to the co-receptor endothelial protein C receptor (EPCR) to promote the cleavage and activation of co-localized PAR1 [28, 29] and induce anti-apoptotic and protective effects on endothelial barrier permeability [29–33]. Compartmentalization of PAR1 and co-localization with EPCR in calveolae is crucial for efficient cleavage by aPC [13]. Moreover, aPC cleaves PAR3 in humans and mice [21, 34, 35] and acts as a PAR3 shedding protease that prevents thrombin-induced barrier disruption [21]. However, the dependency of aPC cleavage of PAR3 on EPCR remains controversial [21, 35]. Similar to aPC, coagulation factor Xa binds EPCR and mediates proteolytic activation of PAR1 and PAR3 [21, 28, 36–39]. In addition, EPCR-bound factor Xa reportedly cleaves PAR2 and initiates inflammatory signaling [40]. PAR2 was also shown to be activated by tissue factor (TF)-bound coagulation factor VIIa [40–42]. Yet recent studies suggest that the TF-VIIa complex does not directly activate PAR2, and rather activates matriptase, which cleaves and activates PAR2 [42–44]. Anti-inflammatory signaling was also previously related to PAR1 cleavage by EPCR-bound VIIa [45, 46]. Taken together, these studies indicate that TF-Xa–VIIa complexes activate PAR1 and PAR2 [47].

Trypsins are PAR-activating proteases with roles as major digestive enzymes in the duodenum [48]. Trypsin is also secreted by epithelial cells, nervous system cells [49], and tumor cells [50, 51]. Trypsins may also be involved in cell growth and coagulation, as suggested by secretion from human vascular endothelial cells [52]. Trypsin cleaves human PAR1 and PAR4 at putative protease cleavage sites, and thereby prevents thrombin signaling in endothelial cells and platelets [4, 53]. Trypsin is the major PAR2 cleaving protease that initiates inflammatory signaling [2, 7].

Table 1 PAR1 cleaving proteases

	Protease	Major cleavage site	Additional cleavage sites
Mammalian proteases	Thrombin	$R^{41}S^{42}$	
	aPC	$R^{46}N^{47}$	$R^{41}S^{42}$
	FVIIa	unknown	
	FXa	$R^{41}S^{42}$	
	Trypsin	$R^{41}S^{42}$	
	Chymase	unknown	
	MMP-1	$D^{39}P^{40}$, $L^{44}L^{45}$, $F^{87}I^{88}$	$N^{47}P^{48}$, $R^{70}L^{71}$,$K^{82}Q^{83}$
	MMP-2	$L^{38}D^{39}$	
	MMP-3,-8,-9	$R^{41}S^{42}$	
	MMP-12	unknown	
	MMP-13	$S^{42}F^{43}$	$L^{38}T^{39}$, mouse
	Cathepsin G	$R^{41}S^{42}$, $F^{55}W^{56}$, $Y^{69}R^{70}$	
	Neutrophil elastase	$A^{36}T^{37}$, $V^{72}S^{73}$, $A^{86}F^{87}$	
	Proteinase-3	$A^{36}T^{37}$, $P^{48}N^{49}$, $V^{72}S^{73}$, $A^{92}S^{93}$	
	Plasmin	$K^{32}A^{33}$, $R^{41}S^{42}$, R^{70} L^{71}, K^{76} S^{77}, K^{82} Q^{83}	
	Kallikrein-4,-5,-6	unknown	
	Kallikrein-14	$R^{46}N^{47}$	
	Granzyme A,B, K	unknown	
	Calpain-1	$K^{32}A^{33}$, $S^{76}K^{77}$	
Non-mammalian proteases	PA-BJ	$R^{41}S^{42}$, $R^{46}N^{47}$	
	Thrombocytin	$R^{41}S^{42}$, $R^{46}N^{47}$	
	DerP1	unknown	
	Gingipain R	$R^{41}S^{42}$	
	SpeB	$L^{44}L^{45}$	
	LepA	unknown	
	S.pneumoniae proteases	unknown	
	Thermolysin	$F^{43}L^{44}$, $L^{44}L^{45}$	
	penC	$R^{41}S^{42}$	

Tryptase is the main protease of mast cells, and activates PAR2 by proteolytic cleavage to induce calcium signaling and proliferation [54–57]. The source tissue of tryptase reportedly plays an important role in the cleavage and induction of tryptase-activated PAR signaling, reflecting differences in posttranslational modifications, such as glycosylation and sialic acid modifications [54, 58]. Tryptase induces calcium signaling via PAR1 when PAR2 is co-expressed, but cannot activate human platelets, suggesting that tryptase does not directly cleave PAR1 [54–57]. Chymase is a mast cell serine protease that also cleaves PAR1 in human keratinocytes and fibroblasts, and thus prevents thrombin sensitivity [59]. Moreover, the epithelial serine protease matriptase cleaves and initiates inflammatory responses in human and mouse keratinocytes and in *Xenopus* oocytes overexpressing human PAR2 [44, 60–63].

PARs have been identified as substrates of kallikreins, which are serine proteases that have been related to various inflammatory and tumorigenic processes [64]. Kallikrein-4 increases intracellular calcium levels via PAR1 and PAR2, but activates PAR1 most efficiently [65]. Kallikrein-14 induces calcium signaling via PAR1, PAR2, and PAR4, but can also shed PAR1 to prevent signaling. Rat platelets are activated by kallikrein-14 via the proteolytic cleavage of PAR4, but are not activated by kallikrein-5 and kallikrein-6 [66]. Instead, neurotoxic effects of kallikrein-6 were inhibited by blocking PAR1 and PAR2, indicating a direct proteolytic role in PAR activation [67].

Neutrophils are mobilized to sites of inflammation and infection, where they modulate inflammatory signaling, in part by secreting PAR-cleaving proteases. The neutrophil serine protease cathepsin G prevents thrombin-induced effects by cleaving PAR1 into non-functional parts [68, 69]. In contrast, cathepsin G reportedly induced chemoattractant signaling via PAR1, further supporting the role

Table 2 PAR2 cleaving proteases

	Protease	Major cleavage site	Additional cleavage sites
Mammalian proteases	Thrombin	$R^{36}S^{37}$	
	aPC	unknown	
	FXa	$R^{36}S^{37}$	
	Trypsin	$R^{36}S^{37}$	$K^{34}G^{35}$, $K^{51}G^{52}$, $K^{72}L^{73}$
	Tryptase	$R^{36}S^{37}$	
	Chymase	$G^{35}R^{36}$	$L^{38}I^{39}$, mouse
	Matriptase	$R^{36}S^{37}$	
	Cathepsin G	$F^{65}S^{66}$	$F^{59}S^{60}$, $F^{64}S^{65}$
	Cathepsin S	$G^{40}K^{41}$	$E^{56}P^{57}$, mouse
	Neutrophil elastase	$A^{66}S^{67}$, $S^{67}V^{68}$	$V^{42}D^{43}$,$V^{48}T^{49}$,$V^{53}T^{54}$,$V^{58}T^{59}$,$T^{74}T^{75}$,$V^{76}F^{77}$
	Proteinase-3	$D^{62}E^{63}$	$V^{48}T^{49}$,$V^{55}E^{56}$,$T^{57}V^{58}$ $V^{61}D^{62}$,$K^{72}L^{73}$,$T^{74}T^{75}$,$T^{75}V^{76}$,$V^{76}F^{77}$
	Plasmin	$R^{36}S^{37}$	$K^{34}G^{35}$
	Testisin	unknown	
	Kallikrein-4,	unknown	
	Kallikrein-5,-6,-14	$R^{36}S^{37}$	
	Calpain-2	unknown	
Non-mammalian proteases	Der-P1,-P2,-P3,-P9	unknown	
	Cockroach E1-E3	$R^{36}S^{37}$	
	Gingipain R	unknown	
	LepA	unknown	
	EPa	$S^{37}L^{38}$	$S^{38}L^{39}$, rat
	S.pneumoniae proteases	unknown	
	Thermolysin	unknown	
	Serralysin	unknown	
	P.acnes proteases	unknown	
	aPA	unknown	
	Bromelain	unknown	
	Ficin	unknown	
	Papain	unknown	
	penC	$R^{36}S^{37}$	

of cathepsin G in PAR1 activation [70]. Another unexpected observation of cathepsin G was that cleavage sites differ between recombinant and native human PAR2 [26, 71, 72]. These discrepancies may reflect the influence of cell types and posttranslational modifications on PAR cleavage. Studies in mice and humans show that platelet activation by cathepsin G is dependent on PAR3 and PAR4 [71, 73, 74]. Cathepsin G also cleaves and activates PAR4 on endothelial cells [75]. The neutrophil proteases elastase and proteinase-3 cleave recombinant PAR1 and PAR2 at various sites [26, 72]. Recently, rat elastase was shown to cleave and activate PAR1, although sequences of rat and human PAR1 have low homology [76]. In contradiction with neutrophil proteases that prevent PAR signaling at sites of inflammation, monocytes secret the protease cathepsin S, which initiates inflammatory

Table 3 PAR3 cleaving proteases

	Protease	Major cleavage site	Additional cleavage sites
Mammalian proteases	Thrombin	$K^{38}T^{39}$	mouse PAR3 at $K^{37}S^{38}$
	aPC	$R^{41}G^{42}$	
	FXa	$R^{41}G^{42}$	
	Trypsin	unknown	

Table 4 PAR4 cleaving proteases

	Protease	Major cleavage site	Additional cleavage sites
Mammalian proteases	Thrombin	$R^{47}G^{48}$	
	Trypsin	$R^{47}G^{48}$	
	Cathepsin G	$R^{47}G^{48}$	
	Kallikrein-14	unknown	
Non-mammalian proteases	PA-BJ	$R^{47}G^{48}$	
	Thrombocytin	$R^{47}G^{48}$	
	Der-P3	unknown	
	Gingipain R	$R^{47}G^{48}$	
	LepA	unknown	
	S.pneumoniae proteases	unknown	
	Bromelain	unknown	
	Ficin	unknown	
	Papain	unknown	

signaling by cleaving PAR2 [72, 77, 78]. Low concentrations of the fibrinolytic protease plasmin prevent platelet activation by cleaving PAR1, whereas high concentrations of plasmin lead to the cleavage and activation of PAR1 [79]. Plasmin also cleaves PAR2 and prevents subsequent activation by trypsin [26, 80].

The serine proteases granzyme A and granzyme B induce intracellular signaling pathways that lead to neuronal death via PAR1 [81, 82]. Recently, granzyme K was also shown to activate PAR1 and promote inflammatory endothelial signaling [83, 84]. Few studies show activation of PAR1 by proteases of the granzyme family, and the details of this interaction remain poorly characterized.

Cysteine proteases Calpain-1 is a calcium-dependent cysteine protease that has been associated with inflammatory disorders, and initiates calcium signaling pathways by activating PAR1 [26]. At very high concentrations, calpain-2 was also shown to cleave PAR2, and the authors suggested that this cleavage event inactivated PAR2 [26]. Recently, calpain-1 was shown to be induced by thrombin-activated PAR1, and subsequently regulated the internalization of PAR1 [85].

Metalloproteases Matrix metalloproteases (MMPs) are known to be involved in various inflammatory- and cancer-related conditions. MMP-1 cleaves human PAR1 and initiates platelet activation [86–89]. MMP-1 also regulates cancer cell activities depending on PAR1 availability [90]. Similarly, MMP-2 cleaves human PAR1 and enhances platelet activation [91], and MMP-3, MMP-8, and MMP-9 were shown to induce platelet activation via PAR1 [92]. Whether these three MMPs cleave PAR2 is not clear, although PAR2 activation by trypsin induced secretion of MMP-9 in human airways, suggesting that MMP-9 is a PAR2-activating protease [93]. In mice, PAR1 expression was regulated by MMP-12, and activated PAR1 increased MMP-12 secretion [94, 95]. A similar feedback loop involving MMP-12 and PAR2 has been reported in mice [96]. Moreover, MMP-13 was shown to activate PAR1 and induce intracellular signaling [87], and thrombin-induced activation of PAR1 and PAR3 was associated with increased levels of MMP-13 in human chondrocytes [24].

In addition to coagulation and inflammation, PAR activation may play roles in human germ cells, where the serine protease testisin activates PAR2 and induces calcium signaling and ERK1/2 activation. This interaction may play roles in the regulation of ovarian and testicular cancer, as suggested previously [97, 98].

Non-mammalian proteases

Exogenous proteases from various species that modulate PAR activation are disscues in the following section and are summarized in Fig. 3.

Bacterial proteases Endogenous mammalian proteases are not the only regulators of PAR activation. Indeed, both pathogenic and commensal bacteria secret various proteases that cleave PARs and act as inflammatory modulators [99]. In this section, we describe bacterial proteases that either activate PARs, and thus allow bacteria to penetrate host barriers, or inactivate PARs to prevent inflammatory signaling by the host.

The human pathogen *Pseudomonas aeruginosa* secrets two PAR-cleaving proteases with contrasting effects. The exoprotease LepA cleaves and activates PAR1, PAR2, and PAR4, and subsequently induces nuclear factor kappa B (NFκB) promoter activity [100], whereas

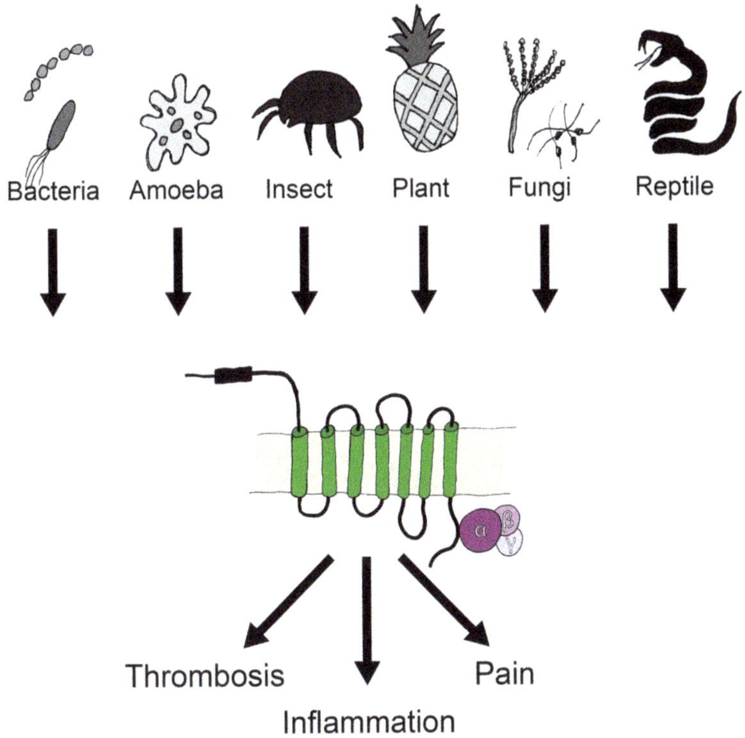

Fig. 3 Non-mammalian exogenous proteases induce PAR-driven pathological effects. Various proteases are secreted from bacteria, amoebae, insects, plants, fungi, and snakes, and can cleave PARs and modulate signal transduction, leading to inflammation, thrombosis, or pain

cleavage by elastase EPa inactivates PAR2 to prevent inflammation in lungs [101].

The streptococcal pyrogenic exotoxin B (SpeB) of Group A *Streptococcus* also inactivates PAR1 by cleaving it, and thereby renders human platelets unresponsive to thrombin [102]. In mice, proteases of *Streptococcus pneumoniae* cleaved PAR2 and facilitated the spread of the pathogen from the airways into the blood stream [103]. PAR1 has also been associated with *S. pneumonia*-mediated sepsis in mice, although direct cleavage of PAR1 was not shown [104, 105]. Pulmonary inflammation from *S. pneumoniae* infections is reduced in PAR4 knockout mice [106], further supporting this causal link.

Inflammation-associated periodontal diseases are predominantly induced by the *Porphyromonas gingivalis* cysteine protease gingipain R, which activates PAR2 [107, 108]. Subsequently, gingipain R activates PAR1 and PAR4, and thereby, human platelets [109–111]. This mechanism may also explain associations between periodontitis and cardiovascular events [112].

In addition, supernatants from *Propionibacterium acnes* cultures initiated inflammatory signaling in human keratinocytes via PAR2 [92]. The virulence of *P. acnes* was also reduced in PAR2 knockout mice [113], further suggesting that PAR2 is involved in bacterial infections.

Serralysin is a matrix metalloprotease expressed by *Serratia marcescens*, and induced inflammation in human airway cells via PAR2 in vitro [114].

Finally, *Bacillus thermoproteolyticus rokko* secretes the metalloprotease thermolysin, which cleaves and inactivates PAR1 to prevent thrombin-induced signaling in rat astrocytes [115, 116]. The in vitro effects of PAR2-cleavage by thermolysin, however, vary between cell lines [116].

Amoeba proteases In acanthamoebic keratitis, PAR2 triggers inflammation following secretion of the plasminogen activator (aPA) by *Acanthamoeba* strains, leading to induction of IL-8 in human corneal epithelial cells [117].

Reptile proteases Following snakebites, coagulation disorders in humans and mice occur due to the presence of venom proteases. In *Proatheris superciliaris* bites, venom proteases activate platelets by activating PAR1 and PAR4 [118]. *Bothrops atrox* and *B. jararaca* are snake species of the family viperidae. These snakes secrete the serine proteases PA-BJ and thrombocytin, which activate human platelets via PAR1 and PAR4 [119].

Insect proteases Several cysteine and serine proteases from insects induce inflammation-associated diseases such as asthma. For example, dust mite allergens contain the serine proteases DerP2, DerP3, and DerP9 [120] and the cysteine protease DerP1. DerP1 induces PAR2-dependent signaling, whereas thrombin-induced PAR1-signaling is prevented by these proteases in human epithelial cells [121]. DerP3 was also recently shown to activate PAR4, and this process was associated with allergies to dust mites [122].

Similar to proteases from house dust mites, three serine proteases (E1–E3) from cockroach extracts activate PAR2 and induce inflammatory signaling in mice and humans [123–125].

Fungal proteases Pen C is a serine protease from *Penicillium citrinum* that induces IL-8 in human airway cells by activating PAR1 and PAR2 [126]. Proteases from *Aspergillus fumigatus* have also been shown to prevent PAR2-dependent inflammation [127]. Moreover, serine proteases from *Alternaria alternate* induced calcium signaling in human bronchial cells and induced inflammation in mice by secreting IL-33 following PAR2 activation [128–130].

Plant proteases Bromelain is a mixture of cysteine proteases that is extracted from pineapple which is used as a PAR-independent anti-inflammatory agent [131]. Bromelain cleaves PAR2 and thereby prevents the associated inflammatory signaling [132]. In another study, however, bromelain, ficin, and papain activated PAR2 and PAR4 by proteolytic cleavage, leading to increased intracellular calcium levels [133]. Thus, further studies are required to further clarify the modes of action of pineapple proteases.

Cleavage-independent PAR activation by agonist peptides
Independent of proteolytic cleavage, PARs can be activated by synthetic soluble ligands corresponding with cleaved N-terminal sequences, or can be transactivated by cleavage-generated N-terminal regions of homo- or heterodimer partners.

Synthetic peptides that mimic the first six amino acids of tethered N-terminal ligands can act as agonist peptides that activate PARs in the absence of cleavage events [11, 18, 134]. Specific activation of PARs by a soluble agonist peptide was first shown for human PAR1 with the peptide SFLLRN [6, 18]. However, this peptide also activated PAR2 [135–137] and therefore various peptides were tested for specific PAR1 activation. Yet, PAR1 was the most specifically and efficiently activated by TFLLRN [138]. In addition to thrombin agonist peptides, other PAR1 agonist peptides have been identified. In particular, the peptide NPNDKYEPF reproduced the effects of aPC [28], and PRSFFLRN corresponds with the N-terminal peptide generated by MMP-1 [86]. SLIGKV corresponds with the trypsin cleaved N-terminal region of human PAR2. However, the corresponding rat N-terminus SLIGRL is a more specific and efficient PAR2 agonist in rodents and humans [136, 139], and only the synthetic peptide LIGRLO achieved this effect more efficiently than SLIGRL in humans [140]. The roles of ECL-2 in specific PAR activation have been shown using labeled PAR2 agonist peptides [141, 142]. Because the thrombin generated PAR3 peptide does not activate the G protein autonomously, no such agonist peptides have been identified to date [9, 143]. GYPGKF corresponds with the thrombin-cleaved human PAR4 and has weak activity as an agonist [144]. But replacement of the first amino acid glycine (G) with alanine (A) induced PAR4 by 10-fold. This peptide may be suitable as a platelet activator in humans and mice [145].

Several models of PAR–PAR interactions have been proposed and extensively studied based on PAR transactivation by agonist peptides [146]. When PAR1 is blocked on endothelial cells, however, thrombin, and not the PAR1-specific agonist peptide TFLLRN, induces signaling, reportedly by facilitating the heterodimerization of PAR1 and PAR2 [147]. Thrombin activation of the PAR1–PAR2 heterodimer leads to constitutive internalization and activation of β-arrestin by the PAR1 C-tail [146]. Accordingly, the required co-localization of PAR1 and PAR2 was shown in a human overexpression system, in mice studies of sepsis, and in PAR1–PAR2-driven cancer growth in a xenograft mouse model [148, 149]. In other studies, stable heterodimerization of human PAR1 and PAR4 was shown in platelet cells, and thrombin accelerated platelet activation under these conditions [150, 151]. Similar studies of mouse platelets showed efficient activation of platelets by thrombin in the presence of PAR3–PAR4 heterodimers [143]. Consistent with the thrombin-cleaved PAR3 peptide, which is not self-activating, PAR3 signaling was observed in the presence of PAR1 or PAR2 [22, 23, 34, 152]. Yet, heterodimerization influenced signal transduction and PAR membrane delivery due to enhanced glycosylation [153].

In addition to activation by heterodimerization, PARs interact with other receptors, such as ion channels, other G protein-coupled receptors (GPCRs), receptor tyrosine kinases (RTKs), receptor serine/threonine kinases (RSTKs), NOD-like receptors, and TLRs [154]. In particular, PAR2 initiated inflammatory signaling pathways, resulting in pain due to transactivation of the ion channels TRPV1 and TRPV4 in humans and mice [155–159]. Similar inflammatory effects follow transactivation of the RTKs EGFR and VEGFR by PAR2 and PAR4 [160–163]. Bacterial interactions with PARs suggest important roles of PARs in infectious disease. In agreement, TLRs recognize bacteria-derived molecules and contribute to innate immunity [164, 165]. Moreover, direct interactions of PAR2 with TLR3 and TLR4 were necessary for inflammatory responses to LPS in human cell lines and knockout mice and rats [166–171].

PAR signaling

Activation pathways

PARs belong to a large family of GPCRs and induce multiple signaling pathways after coupling with heterodimeric G proteins. Activation of the Gα-subunit due to the exchange of a guanine from GDP to GTP results in dissociation of the Gβγ-dimer and activation of downstream pathways [172, 173].

Following proteolytic cleavage or induction of agonist peptides, the engaged signaling pathways vary between tissues, cell lines, and the availability of co-receptors for transactivation. Depending on the ligand, specific α-subunits are activated, and these regulate subsequent cellular functions as summarized in Fig. 4. For example, thrombin-stimulated PAR1 activates the small GTPase protein RhoA via ERK1/2 kinases, but not via Rac1, whereas aPC-stimulated PAR1 induces Rac1 via Akt kinase, but not via RhoA [13, 174–176]. Moreover, in accordance with PAR1 cleavage sites, aPC prevents thrombin-induced RhoA signaling [16]. However, in contrast with thrombin-induced RhoA activation on platelets and endothelial cells, PAR1-agonist peptides and thrombin activated the inhibitory G protein G_i which leads to the inhibition of adenylyl cyclase in human fibroblasts [177, 178]. Other studies indicate that PAR2 activation is less tissue specific than PAR1 activation, and trypsin and VIIa cleaved PAR2 and activated $G_{\alpha q}$ and G_i, resulting in calcium influx, MAPK activation, and inflammatory signaling [8, 179].

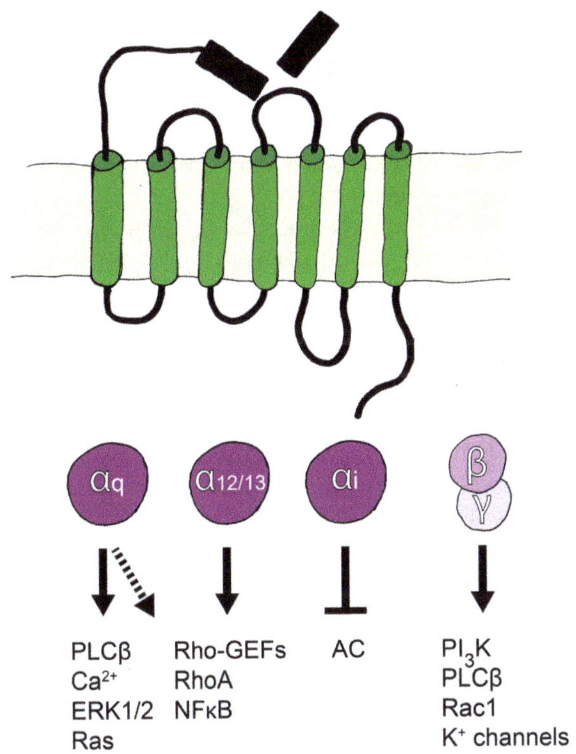

Fig. 4 G protein-coupled signaling induced by PAR activation. Depending on the tethered ligand, activated PAR couples with G protein α-subtypes. Gαq activates phospholipase Cβ (PLCβ), which mobilizes calcium. This further activates MAPKs (ERK1/2) and induces Ras signaling. Primarily, Gα12/12 and Gαq activate the Rho pathway. Gαi inhibits the activation of adenylyl cyclase, which leads to reduced production of cAMP. In contrast, the βγ-subunit functions as a negative regulator when bound to the α-subunit. After receptor activation, subunits separate, and the βγ-subunit interacts with other proteins, thereby activating or inhibiting signaling

Signaling by tethered ligands can differ from that generated by corresponding soluble agonist peptides. For example, thrombin-cleaved PAR1 activated $G_{\alpha 12/13}$ and $G_{\alpha q}$ and induced Rho and Ca^{2+} signaling, whereas the PAR1-agonist peptide activated only $G_{a12/13}$ and downstream RhoA-dependent pathways that affected endothelial barrier permeability [180]. Similar observations of human platelets suggested that platelet activation followed coupling of thrombin-activated PAR1 with multiple heterotrimeric G protein subtypes, including $G_{\alpha 12/13}$ and $G_{\alpha q}$ [181–183]. Moreover, trypsin and the PAR2-agonist peptide induced ERK1/2 signaling and inflammation by activating PAR2 [29, 180, 184–186]. β-arrestins also play major roles in PAR-induced signaling independently of G protein activation. For instance, aPC-activated PAR1 induces cytoprotective effects by recruiting β-arrestin in endothelial cells. Thus, aPC cleavage fails to protect β-arrestin deficient cells from the effects of thrombin [187, 188]. In addition, multiple studies show that activated PAR2 co-localizes with β-arrestin-1 and arrestin-2 and induces ERK1/2 signaling [77, 189–191].

Desensitization and termination

PAR activation is regulated by internalization and proteolytic desensitization, which limits the duration of signaling. For instance, PAR1 is constitutively internalized and recycled or agonist-induced internalized and degraded as described in [192, 193] and shown in the scheme of Fig. 5. As discussed above, some PAR-cleaving proteases abolish receptor responses by removing (shedding) or destroying the tethered ligands. For example, PAR1 is inactivated following cleavage by cathepsin G, and thrombin activation is hence prevented, allowing the formation of clotting under inflammatory conditions.

Depending upon proteolytic cleavage, PAR1 rapidly internalizes or accumulates on the cell surface [194, 195]. Activated PAR1 is internalized via clathrin- and dynamin-dependent mechanisms, and is sorted from early endosomes to lysosomes for degradation [196–199].

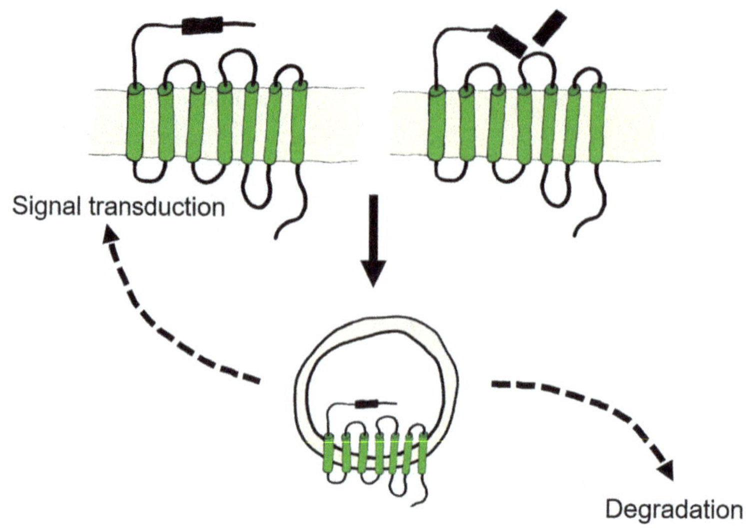

Fig. 5 PAR trafficking. Activation-independent constitutive or agonist-induced internalization regulates PAR1 signaling

Although the mechanisms that terminate PAR1 signaling are not clearly understood, this process is known to involve phosphorylation, ubiquitination, and recruitment of β-arrestin [200–204]. In contrast with PAR1, activated PAR2 is not constitutively internalized [205]. Thus, to prevent persistent signaling upon activation, PAR2 is phosphorylated and ubiquitinated and then binds β-arrestin before being internalized and degraded [206–208]. Under these conditions, the activated and internalized PAR2 is not recycled and instead induces β-arrestin-dependent endosomal ERK1/2 signaling in the cytoplasm [189, 191, 209]. Thus, large cytoplasmic stores of newly generated PAR2 are required for rapid externalization and activation on cell membranes [210]. Although less is known about how PAR4 signaling is terminated, recent observations suggest that PAR4 internalization is independent of β-arrestin and slowly occurs via clathrin- and dynamin-dependent pathways [211]. In agreement, human platelets internalized PAR4 much slower than PAR1, and exhibited prolonged PAR4 signaling activity [212]. Moreover, growing evidence indicates that PAR–PAR heterodimerization is important for internalization, and that the underlying mechanisms include PAR2-dependent glycosylation of PAR4, thus affecting membrane transport [153]. Upon internalization, endosomal PAR4 dimerizes with the purinergic receptor P2Y12 and induces Akt signaling by recruiting β-arrestin within endosomes [213].

Depending on stimuli, PAR expression patterns are regulated by complex combinations of cell surface presentation, endocytosis, vesicle born or recycled (i.e., re-exocytosed) receptors, and trafficking modes that are linked to posttranslational modifications of PAR.

Role of PARs in inflammation

With the current increases in the prevalence of inflammatory diseases, published in in vitro and in vivo studies of the roles of PARs in inflammation have become more numerous. These are reviewed below.

Systemic inflammation and inflammatory cells in the cardiovascular system

PARs are critical for the interplay between clotting proteases of platelets, endothelial cells, and vascular smooth muscle cells that regulate hemostasis, vascular barrier function, vascular tone, vascular homeostasis, cell adhesion, and inflammatory responses [150]. The roles of PARs in these processes vary significantly between species. Specifically, whereas functional PAR1 and PAR4 are expressed in human platelets [214], PAR1, PAR3, and PAR4 have been found in guinea pig platelets [215]. Whereas mouse and rat platelets lack PAR1, they are activated at low concentrations of thrombin, which is recruited by PAR3 onto the surface of platelets and then efficiently activates PAR4 [4]. Due to interspecies differences in PAR expression, mouse and rat studies of PARs are difficult to translate to humans. PARs in endothelial cells contribute positive regulatory signals for endothelial adhesion molecules such as vascular cell adhesion molecule-1 (VCAM-1), intercellular adhesion molecule-1 (ICAM-1), and E-selectin [216, 217], all of which promote vascular barrier function. As a counterpart of intravascular cells, PAR4 induces leukocyte migration [75], and PAR2 expressed on macrophages promotes inflammatory modulators such as interleukin-8 (IL-8) [218]. These modes of signaling all contribute to a complex PAR-mediated interplay of endothelial cells that is orchestrated by intravascular cells and cytokine secretion. In addition, PARs, particularly PAR1, regulate vascular barrier function, and hence, extravasation of macromolecules such as complement proteins and antibodies. In addition, thrombin-mediated activation of PAR1 increases endothelial barrier permeability by activating mitogen-activated protein kinases (MAPKs) [219]. Although this effect is reversed by activated protein C (aPC)-mediated activation of PAR1 [28, 174, 175, 220]. Thrombin further promotes prostaglandin 2 (PGE2) secretion, and consequent endothelial barrier permeability [221]. Similarly, PAR1 activation increased vascular leakage in a murine model [222]. Inflammatory mediators, such as tumor necrosis factor alpha (TNFα), were shown to regulate the expression of endothelial PAR2, and the authors suggested that these data were indicative of barrier protective effects of PAR2 [223]. Several other studies show that PAR2 activation induces endothelium-dependent relaxation in blood vessels of mice and in arteries of rats [224–228]. In contrast, dual activities of PAR2 on blood vessels were reported in a study of rats [229]. In this line, thrombin-activated PAR1 induced the expression of vascular endothelial growth factor in smooth muscle cells [230], thus revealing the relationship between coagulation and vascular growth. Although the roles of PARs in the development of arteriosclerosis are yet to be elucidated, PAR2 and PAR4 were induced in human arteries under inflammatory conditions [223], suggesting important roles of PARs in vascular inflammation.

Chronic inflammation of the gastrointestinal tract

In the gut lumen, human and bacterial proteases are both present at high concentrations. Similar to endothelial barriers, proteases regulate intestinal barrier permeability via PARs, all four of which are expressed by cells of the gastrointestinal tract [9, 224, 231, 232]. Trypsins and tryptases are prominent intestinal proteases, suggesting likely involvement of PAR2 as a major receptor of intestinal inflammation. In accordance, intestinal tight junctions are disrupted by PAR2-activating proteases, leading to inflammatory signaling in humans and rats [139, 206, 233, 234]. Although the roles of PARs in irritable bowel syndrome (IBS) and inflammatory bowel diseases remain unclear, roles of PARs in intestinal barrier function have been described. Specifically, PAR1 and PAR2 regulated permeability and chloride secretion, which are involved in diarrhea and constipation in IBS patients [234–236]. In addition, activated endosomal PAR2 caused persistent pain in a mouse model of IBS [209].

Inflammatory diseases of the respiratory system

It has long been suggested that PARs are involved in the pathophysiology of respiratory disorders, reflecting observations of elevated levels of PAR-activating proteases, such as thrombin and tryptase, in bronchoalveolar lavage fluid from patients with pulmonary inflammation [237, 238]. In a sheep asthma model and in asthmatic patients, tryptase inhibitors reduced inflammation [239, 240], further indicating important roles of PAR2 in respiratory disease. These roles of PARs are also suggested by the prominence of a variety of non-mammalian PAR-activating proteases, such as those of house dust mites and cockroaches [120, 123, 124]. Expression of PAR1, PAR2, and PAR4 on bronchial epithelial and smooth muscle cells induced inflammatory signaling in multiple studies [55, 121, 241–245]. PAR2 is also upregulated in epithelial cells of patients with asthma and chronic obstructive pulmonary syndrome (COPD) [246, 247]. Whether PAR2 activation results in bronchoconstriction or dilatation remains controversial, in part owing to interspecies differences and tissue dependencies [242, 248, 249]. In humans, however, PAR1-agonist peptides with thrombin, and a PAR2-agonist peptide with trypsin and tryptase, induced bronchoconstriction by inducing Ca^{2+} signaling in airway smooth muscle cells [241, 244]. Moreover, the long-term activation of PAR1 and PAR2 led to pulmonary fibrosis in mice models [250].

Inflammatory skin diseases

High concentrations of exogenous proteases are present on the skin of various species, and these may activate PARs to regulate epidermal permeability and barrier function [251]. Indeed, epidermal inflammation has been linked to PAR1 and PAR2 activation in keratinocytes, which comprise the epidermal barrier with sub-epidermal skin fibroblasts [179, 252, 253]. Subsequent release of IL-8, IL-6, and granulocyte macrophage colony-stimulating factor (GM-CSF) was also observed previously [254], potentially involving NFκB activation [255]. In addition, the inflammatory roles of PAR2 have been demonstrated in mice models of atopic dermatitis due to elevated tryptase and PAR2 expression levels [256, 257]. Similar to studies in mouse models, PAR2 was upregulated in patients with atopic dermatitis, and PAR2 agonists increased itch, causing irresponsiveness of sensory nerves to therapy with antihistamines [258].

Rheumatic disease

"Rheumatic disease" is a common term for autoimmune diseases that affect joints, bones, and muscles. Although rheumatic disorders are numerous, some of the common underlying symptoms include chronic joint inflammation, stiffness, and pain [259]. Currently, PAR2 is the only PAR that has been associated with the development of rheumatic diseases [260]. Direct roles of PAR2 in rheumatic diseases were first indicated in 2003 in a mouse study by Ferrell et al. [261]. In their study, a PAR2-agonist peptide induced strong inflammatory effects in wt mice, causing joint swelling and synovial hyperemia, whereas joint swelling was absent in PAR2 deficient mice [261]. Similarly, in patients with rheumatoid arthritis, PAR2 is upregulated in inflamed tissues [262]. Further increases in PAR2 expression were noted in monocytes, and the PAR2-agonist peptide upregulated IL-6. In contrast, PAR2 expression was decreased after treatments with antirheumatic drugs [263], further supporting the role of PAR2 in rheumatic disease.

PAR modulators as targets for therapy

The complexity of PAR regulation is indicated by the culmination of specific proteolytic cleavage modes (inactivating or activating), protease inhibitors, and cofactors, and with the effects of PAR glycosylation and dimerization (Fig. 1). In this section we discuss classes of agonists and antagonists that have been tested as PAR modulators for use as therapeutic agents as summarized in Fig. 6 and Tables 5, 6 and 7.

Peptide agonists and antagonists are short synthetic peptides that mimick the PAR-tethered ligand that is liberated by proteolytic cleavage, as described above. These peptides either induce signal transduction or prevent cleavage-dependent signaling following PAR rapid internalization, and some C- or N-terminal modifications of soluble ligand sequences have resulted in increased activation efficiency [18]. Peptidomimetic antagonists are small protein-like chains that mimick the tethered ligands of PARs, and were recently used as PAR modulators for the first time [264].Soon after PARs were discovered, PAR1 blocking antibodies were reported [265], and these blocked protease binding and or the cleavage site of the receptor. Non-peptide small molecules, such as the PAR1 antagonists vorapaxar [266] and atopaxar [267], also interact with PARs, mainly via ECL2.Only two classes of intracellular PAR antagonists have been developed to date. Pepducins are cell penetrating palmitoylated peptides that were derived from the intracellular loop of PAR, and these interfere with G protein binding [268]. Parmodulins, in contrast, are small molecules that bind PARs at the G protein binding pocket of the C-tail to compete with $G_{\alpha q}$ subunits, but not with other G_{α} subunits [269].

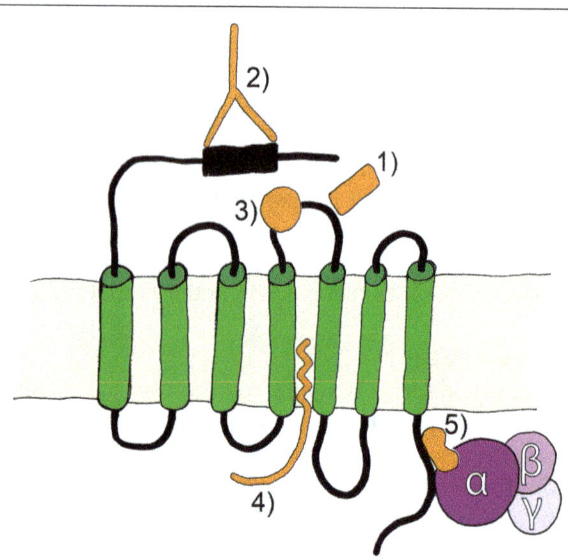

Fig. 6 PAR modulators. Pharmacological substances, such as 1) peptides and peptidomimetics, 2) blocking antibodies, 3) small molecules, 4) pepducins, and 5) parmodulins are used as therapeutic agents that affect PAR activities

Examination of agonists and antagonists in vitro and in preclinical studies (Tables 5, 6 and 7)

Clinical studies

Despite the importance of PARs in various pathophysiological conditions, few PAR modulating tools have been tested in clinical studies, and even fewer have been established for treatment. Since the identification of PAR1 as a platelet thrombin receptor, an abundance of research has been conducted to identify PAR1 antagonists that can block platelet activation and prevent thrombotic cardiovascular events. The first clinically approved PAR1 antagonist was the small-molecule antagonist vorapaxar [266]. Phase II clinical trials of this agent showed reduced risks for myocardial infarction in patients treated with vorapaxar in combination with standard antiplatelet therapy. Moreover, the risks of bleeding complications were not significantly increased [270]. Subsequently, two large-scale phase III multicenter, randomized, double-blind, placebo-controlled studies of vorapaxar (ZONTIVITY, SCH530348) were performed. In the Thrombin Receptor Antagonist in Secondary Prevention of Atherothrombotic Ischemic Events–Thrombolysis in Myocardial Infarction 50 (TRA 2°P-TIMI 50; details at www.ClinicalTrials.gov; NCT00526474) study, the rate of cardiovascular events at the second efficacy endpoint were significantly reduced by vorapaxar in combination with standard antiplatelet therapy [271]. Furthermore, in the Thrombin Receptor Antagonist for Clinical Event Reduction in Acute Coronary Syndrome (TRACER; details at www.ClinicalTrials.gov; NCT00527943) study, vorapaxar reduced the hazard of first myocardial infarction of any type in patients who were treated within 24 h of having symptoms of a cardiovascular event. However, in the TRACER study, vorapaxar failed to prevent secondary ischemic events [272]. Because vorapaxar increased bleeding complications in the clinical setting, the alternative PAR1 antagonist atopaxar (E5555) [267] was tested in a phase II clinical trial called (Lessons From Antagonizing the Cellular Effects of Thrombin-Acute

Table 5 PAR1 signaling modulators

Class	Agonist/ Antagonist	Name	Receptor/Cell/ Tissue type	Cellular response
Peptide	Agonist	SFLLRN/–NH2	Human	Induces platelet activation [6, 138, 265, 278, 279]
		TFLLRN/–NH2	Human	Induces platelet activation, enhances endothelial barrier permeability [137, 138, 265]
		NPNDKYEPF/ –NH2	Human	Induces cytoprotective signaling [28, 187]
		PRSFLLRN/–NH2	Human	Induces platelet activation [86]
			Human	Induces ERK1/2 activation [280]
	Antagonist	YFLLRN	Human	Compets with thrombin and PAR1-AP and prevents platelet activation [278, 279]
Peptidomimetic	Antagonist	RWJ-56110	Human	Blunts thrombin and PAR1-AP effects on platelets and vascular endothelial cells [264, 281, 282]
			Human	Blocks MMP-1 activaiton in SMCs [87]
		RWJ-58259	Guinea pig	Blocks thrombin and PAR1-AP platelet activation [215, 283]
			Rat	Blocks thrombin induced calcium release in AoSMC Inhibits intimal thickening [111, 215, 264, 273]
			Mouse	Prevents destruction of intestinal barrier [62, 284]
Non-peptide small molecule	Antagonist	FR17113	Human	Blocks PAR1-AP induced platelet activation [285, 286]
			Human	Inhibits thrombin and PAR1-AP induced ERK1/2 activation [287]
		ER129614–06	Human	Blocks thrombin and PAR1-AP induced platelet activation [288]
			Guinea pig	Shows antithrombotic effects [289]
		F16357, F16618	Human	Blocks PAR1-AP induced platelet activation [290]
			Rat	Shows antithrombotic effects [291]
		SCH79797	Human	Blocks thrombin and PAR1-AP induced calcium release and platelet activation [292]
			Human, Mouse	Induces NETs formation and increases bacterial killing capacity [293]
		SCH203009	Human	Blocks thrombin and PAR1-AP induced platelet activation [292]
		SCH530348 (vorapaxar)	Human, Monkey	Blocks thrombin and PAR1-AP induced platelet activation [266]
		E5555 (atopaxar)	Human	Blocks thrombin and PAR1-AP induced platelet activation and inhibits thrombus formation [267]
			Guinea pig	Bleeding time not affected [267, 294]
		Q94	Human	Blocks thrombin induced calcium release [295]
			Mouse	Blocks thrombin induced ERK1/2 activation [296]
Pepducin	Antagonist	P1pal-12	Human	Blocks thrombin induced platelet activation [268]
			Human	Blocks platelet activation [86]
			Human	Blocks MMP-1 induced endothelial damage [297]
			Mouse	Reduces lung vascular damage and sepsis lethality [297, 298]
		P1-pal7 (PZ-128)	Human	Blocks MMP-1 induced Akt signaling in cancer cells [150]
			Human	Blocks platelet activation [86]
			Mouse	Inhibits tumor growth [280]
			Guinea pig	Prevents from systemic platelet activation [86]
Parmodulin	Antagonist	ML161 (Parmodulin-2)	Human	Blocks thrombin and PAR1-AP induced platelet activation [299]
			Human	Blocks thrombin induced inflammatory signaling on endothelial cells [269]
			Mouse	Blocks thrombus formation [300]
Antibiotic	Antagonist	Doxycycline	Human	Inhibits thrombin induced cancer cell migration [301, 302]
			Human	Blocks MMP-1 cleavage [303]

Table 5 PAR1 signaling modulators *(Continued)*

Class	Agonist/ Antagonist	Name	Receptor/Cell/ Tissue type	Cellular response
Antibody	Antagonist	ATAP-2 WEDE	Human	Blocks thrombin cleavage of PAR1 and thrombin induced calcium release [147]

Table 6 PAR2 signaling modulators

Class	Agonist/ Antagonist	Name	Receptor/Cell/Tissue type	Cellular response
Peptide	Agonist	SLIGRL/–NH2	Human, Rat	Induces calcium release [2, 8, 136, 139]
		SLIGKV/–NH2	Human	Induces calcium release [136]
		2f-LIGRLO/ –NH2	Human, Rat	Induces calcium release [140]
	Antagonist	FSLLRY-NH2	Human	Blocks trypsin, not SLIGRL activation, reduces proinflammatory IL-8 and TNFα [82]
			Rat	Inhibits neuropathic pain [304]
		LSIGRL-NH2	Human	Blocks trypsin, not SLIGRL induced calcium release [305]
Peptidomimetic	Antagonist	K14585, K12940	Human	Reduces SLIGKV induced calcium release [306]
			Human	Inhibits SLIGRL induced NFkB activation [307]
		C391a	Human, Mouse	Blocks calcium release and MAPK activation [308]
Non-peptide small molecule	Agonist	GB110	Human	Induces calcium release [309]
		AC-5541, AC-264613	Human	Induces calcium release [310]
			Rat	Induces edema and hyperalgesia [310]
	Antagonist	ENMD-1068	Human	Blocks p.acnes induced calcium release and induction of IL-1a, IL-8 and TNFα [92]
			Human	Inhibited FVIIa induced cancer cell migration [311]
			Mouse	Reduces joint inflammation [260]
			Mouse	Blocks calcium release and reduces liver fibrosis [312]
		GB83	Human	Inhibits trypsin and PAR2-AP calcium release [313]
		GB88	Human	Blocks PAR2 induced calcium release [309]
			Rat	Reduces acute paw edema, inhibits PAR2-AP induced inflammation [309, 314]
		AZ8838 AZ3451	Human	Blocks PAR2-AP induced calcium release and β-arrestin recruitment [315]
Pepducin	Antagonist	P2pal-18S	Human	Blocks PAR2 induced calcium release [316]
			Mouse	Decreases risk for developing severe biliary pancreatitis [317]
		P2pal-14GQ	Human	Blocks PAR2 induced calcium release [316]
Antibiotic	Antagonist	Tetracyclines (Tetracycline, Doxycycline, Minocycline)	Human	Inhibits SLIGRL induced IL-8 release [318]
			Mouse	Topical application of tetracycline decreases PAR2 induced skin inflammation [319]
			Rat	Subantimicrobial doses of doxycycline inhibit PAR2 induced inflammation [320]
Antibody		SAM-11	Mouse	Reduces joint inflammation [260]
			Mouse	Prevents allergic inflammation [124]
		B5	Mouse	Reduces joint inflammation [260]
			Mouse	Inhibits allergic airway inflammation [124]
		MAB3949	Human	Blocks trypsin induced PAR2 activation [315]

Table 7 PAR4 signaling modulators

Class	Agonist/ Antagonist	Name	Receptor/Cell/Tissue type	Cellular response
Peptide	Agonist	GYPGQV/–NH2	Human, Rat	Induces platelet activation [144]
		GYPGKF/–NH2	Human, Rat	Induces platelet activation [144]
		AYPGKF/–NH2	Human, Mouse	Induces platelet activation [145]
Peptidomimetic	Antagonist	tc-YGPKF	Rat	Blocks thrombin and PAR4-AP induced platelets aggregation [321]
Non-peptide small molecule	Antagonist	YD-3	Human	Blocks thrombin induced platelet activation [282, 322–325]
			Mouse, Rat, Rabbit	Blocks thrombin and PAR4-AP induced platelets activation [323–325]
		ML-354	Human	Blocks PAR4-AP induced platelet activation [326–328]
		BMS-986120	Human	Blocks PAR4-AP induced calcium release and platelet activation [329]
			Human	Blocks thrombus formation at high shear stress [277]
			Monkey	Blocks platelet activation [329]
Pepducin	Antagonist	P4pal-10	Human, Mouse	Blocks thrombin and PAR4-AP induced platelet activation [268]
			Rat	Blocks thrombin and PAR4-AP induced platelets activation [330]
		P4pal-i1	Human	Blocks PAR4 induced platelets activation [150]

Coronary Syndromes (LANCELOT-ACS; details at www.ClinicalTrials.gov; NCT00548587) study [273]. Atopaxar inhibited platelet aggregation in ACS patients in a dose-dependent manner, and caused no side effects of abnormal platelet activation, such as bleeding [274, 275]. Yet, patients receiving atopaxar had dose-dependent increases in liver abnormalities [273].

To prevent the bleeding problems that arise from treatments with PAR1 antagonists, a new class of PAR1 antagonist was designed, and the member pepducin PZ-128 (P1-pal7) was tested in a phase I trial [276]. This study showed no reduction in platelet aggregation, but the platelet blocking effect of PZ-128 was reversible ex vivo in the presence of saturating concentrations of the PAR1 agonist peptide SFLLRN. Based on these promising findings, the new PAR1 blocking agent PZ-128 was considered in the coronary artery disease study Thrombin Receptor Inhibitory Pepducin-Percutaneous Coronary Intervention (TRIP-PCI). Data from this phase II trial are not yet available (details at www.ClinicalTrials.gov; NCT02561000).

As an alternative to PAR1 targeted antithrombotic drugs, the PAR4 small-peptide antagonist BMS-986120 reduced reversible thrombus formation ex vivo in a phase I trial [277]. Consequently, this promising anticoagulant PAR4 antagonist is currently being compared with a standard anticoagulant drug in a phase II study of stroke recurrence (details at www.ClinicalTrials.gov; NCT02671461).

Conclusion

Since the identification of PARs in the 1990s, studies of the complex mechanisms of PAR activation have been abundant, and these have clarified the roles of PARs in inflammatory disease. Various mammalian and non-mammalian proteases have also been recognized as PAR-mediated regulators of physiological and pathophysiological processes. Despite the development of various PAR modulators, few have been approved for therapeutic use. Obstacles to this therapeutic strategy include species differences in PAR expression and limited bioavailability of modulators in vivo and in clinical studies. Further research is needed to identify specific and efficient anti-inflammatory PAR modulators.

Abbreviations

AC: Adenylyl Cyclase; AKT: Protein kinase B; aPC: Activated Protein C; COPD: Chronic Obstructive Pulmonary Syndrome; COX: Cyclooxygenase; ECL: Extracellular Loop; EGFR: Epidermal Growth Factor Receptor; EPCR: Endothelial Protein C Receptor; ERK: Extracellular Signal-regulated Kinase; F2R: Coagulation Factor 2 Receptor; F2RL1: Coagulation Factor 2 Receptor-Like 1; F2RL2: Coagulation Factor 2 Receptor-Like 2; F2RL3: Coagulation Factor 2 Receptor-Like 3; FVIIa: Activated Coagulation Factor VII; FXa: Activated Coagulation Factor X; GM-CSF: Granulocyte Macrophage Colony-stimulating Factor; GPCR: G Protein-Coupled Receptor; IBD: Inflammatory Bowel Disease; IBS: Irritable Bowel Syndrome; ICAM: Intercellular Adhesion Molecule; IL: Interleukin; LepA: *Pseudomonas aeruginosa*-Derived Large Extracellular Protease; LPS: Lipopolysaccharides; MAPK: Mitogen-Activated Protein Kinase; MMP: Matrix Metalloprotease; NFκB: Nuclear Factor kappa B; PAR: Protease-Activated Receptor; penC: Penicillium citrinum-Derived Alkaline Serine Protease C;

PGE2: Prostaglandin E2; PI3K: Phosphatidylinositol-3-Kinase; PLCβ: Phospholipase C beta; Rac1: Ras-Related C3 Botulinum Toxin Substrate 1; RhoA: Ras homolog gene family, member A; RSTK: Receptor Serine/ Threonine Kinase; RTK: Receptor Tyrosine Kinase; SpeB: Streptococcal Pyrogenic Exotoxin B; TF: Tissue Factor; TLR: Toll-Like Receptor; TM: Thrombomodulin; TRPV: Transient Receptor Potential Channels Vanilloid Subtype; VCAM: Vascular Cell Adhesion Molecule; VEGF: Vascular Endothelial Growth Factor; VEGFR: Vascular Endothelial Growth Factor Receptor

Acknowledgements

We would like to thank Hermenegild K. Heuberger for the drawing of the figures.

We thank Enago (http://www.enago.com) for the english language review.

Author details

[1]Institute of Intensive Care Medicine, University Hospital Zurich, University of Zurich, Zurich, Switzerland. [2]Surgical Research Division, University Hospital Zurich, University of Zurich, Zurich, Switzerland.

References

1. Bahou WF, Nierman WC, Durkin AS, Potter CL, Demetrick DJ. Chromosomal assignment of the human thrombin receptor gene: localization to region q13 of chromosome 5. Blood. 1993;82:1532–7.
2. Nystedt S, Emilsson K, Wahlestedt C, Sundelin J. Molecular cloning of a potential proteinase activated receptor. Proc Natl Acad Sci U S A. 1994;91: 9208–12.
3. Schmidt VA, Nierman WC, Maglott DR, Cupit LD, Moskowitz KA, Wainer JA, Bahou WF. The human proteinase-activated receptor-3 (PAR-3) gene. Identification within a par gene cluster and characterization in vascular endothelial cells and platelets. J Biol Chem. 1998;273:15061–8.
4. Kahn ML, Zheng YW, Huang W, Bigornia V, Zeng D, Moff S, Farese RV Jr, Tam C, Coughlin SR. A dual thrombin receptor system for platelet activation. Nature. 1998;394:690–4.
5. Vu TK, Hung DT, Wheaton VI, Coughlin SR. Molecular cloning of a functional thrombin receptor reveals a novel proteolytic mechanism of receptor activation. Cell. 1991;64:1057–68.
6. Vu TK, Wheaton VI, Hung DT, Charo I, Coughlin SR. Domains specifying thrombin-receptor interaction. Nature. 1991;353:674–7.
7. Nystedt S, Larsson AK, Aberg H, Sundelin J. The mouse proteinase-activated receptor-2 cDNA and gene. Molecular cloning and functional expression. J Biol Chem. 1995;270:5950–5.
8. Nystedt S, Emilsson K, Larsson AK, Strombeck B, Sundelin J. Molecular cloning and functional expression of the gene encoding the human proteinase-activated receptor 2. Eur J Biochem. 1995;232:84–9.
9. Ishihara H, Connolly AJ, Zeng D, Kahn ML, Zheng YW, Timmons C, Tram T, Coughlin SR. Protease-activated receptor 3 is a second thrombin receptor in humans. Nature. 1997;386:502–6.
10. Xu WF, Andersen H, Whitmore TE, Presnell SR, Yee DP, Ching A, Gilbert T, Davie EW, Foster DC. Cloning and characterization of human protease-activated receptor 4. Proc Natl Acad Sci U S A. 1998;95:6642–6.
11. Gerszten RE, Chen J, Ishii M, Ishii K, Wang L, Nanevicz T, Turck CW, Vu TK, Coughlin SR. Specificity of the thrombin receptor for agonist peptide is defined by its extracellular surface. Nature. 1994;368:648–51.
12. Macfarlane SR, Seatter MJ, Kanke T, Hunter GD, Plevin R. Proteinase-activated receptors. Pharmacol Rev. 2001;53:245–82.
13. Russo A, Soh UJ, Paing MM, Arora P, Trejo J. Caveolae are required for protease-selective signaling by protease-activated receptor-1. Proc Natl Acad Sci U S A. 2009;106:6393–7.
14. Zhao P, Metcalf M, Bunnett NW. Biased signaling of protease-activated receptors. Front Endocrinol (Lausanne). 2014;5:67.
15. Renesto P, Si-Tahar M, Moniatte M, Balloy V, Van Dorsselaer A, Pidard D, Chignard M. Specific inhibition of thrombin-induced cell activation by the neutrophil proteinases elastase, cathepsin G, and proteinase 3: evidence for distinct cleavage sites within the aminoterminal domain of the thrombin receptor. Blood. 1997;89:1944–53.
16. Ludeman MJ, Kataoka H, Srinivasan Y, Esmon NL, Esmon CT, Coughlin SR. PAR1 cleavage and signaling in response to activated protein C and thrombin. J Biol Chem. 2005;280:13122–8.
17. Chen J, Ishii M, Wang L, Ishii K, Coughlin SR. Thrombin receptor activation. Confirmation of the intramolecular tethered liganding hypothesis and discovery of an alternative intermolecular liganding mode. J Biol Chem. 1994;269:16041–5.
18. Scarborough RM, Naughton MA, Teng W, Hung DT, Rose J, Vu TK, Wheaton VI, Turck CW, Coughlin SR. Tethered ligand agonist peptides. Structural requirements for thrombin receptor activation reveal mechanism of proteolytic unmasking of agonist function. J Biol Chem. 1992;267:13146–9.
19. Chang MC, Lan WH, Chan CP, Lin CP, Hsieh CC, Jeng JH. Serine protease activity is essential for thrombin-induced protein synthesis in cultured human dental pulp cells: modulation roles of prostaglandin E2. J Oral Pathol Med. 1998;27:23–9.
20. Kohli M, Williams K, Yao JL, Dennis RA, Huang J, Reeder J, Ricke WA. Thrombin expression in prostate: a novel finding. Cancer Investig. 2011;29: 62–7.
21. Burnier L, Mosnier LO. Novel mechanisms for activated protein C cytoprotective activities involving noncanonical activation of protease-activated receptor 3. Blood. 2013;122:807–16.
22. Kaufmann R, Schulze B, Krause G, Mayr LM, Settmacher U, Henklein P. Proteinase-activated receptors (PARs)--the PAR3 neo-N-terminal peptide TFRGAP interacts with PAR1. Regul Pept. 2005;125:61–6.
23. McLaughlin JN, Patterson MM, Malik AB. Protease-activated receptor-3 (PAR3) regulates PAR1 signaling by receptor dimerization. Proc Natl Acad Sci U S A. 2007;104:5662–7.
24. Huang CY, Lin HJ, Chen HS, Cheng SY, Hsu HC, Tang CH. Thrombin promotes matrix metalloproteinase-13 expression through the PKCdelta c-Src/EGFR/PI3K/Akt/AP-1 signaling pathway in human chondrocytes. Mediat Inflamm. 2013;2013:326041.
25. Coughlin SR. How the protease thrombin talks to cells. Proc Natl Acad Sci U S A. 1999;96:11023–7.
26. Loew D, Perrault C, Morales M, Moog S, Ravanat C, Schuhler S, Arcone R, Pietropaolo C, Cazenave JP, van Dorsselaer A, Lanza F. Proteolysis of the exodomain of recombinant protease-activated receptors: prediction of receptor activation or inactivation by MALDI mass spectrometry. Biochemistry. 2000;39:10812–22.
27. Mihara K, Ramachandran R, Saifeddine M, Hansen KK, Renaux B, Polley D, Gibson S, Vanderboor C, Hollenberg MD. Thrombin-mediated direct activation of proteinase-activated receptor-2 (PAR2): another target for thrombin signaling. Mol Pharmacol. 2016;89(5):606–14.
28. Schuepbach RA, Madon J, Ender M, Galli P, Riewald M. Protease activated receptor-1 cleaved at R46 mediates cytoprotective effects. J Thromb Haemost. 2012;10(8):1675–84.
29. Riewald M, Petrovan RJ, Donner A, Mueller BM, Ruf W. Activation of endothelial cell protease activated receptor 1 by the protein C pathway. Science. 2002;296:1880–2.
30. Cheng T, Liu D, Griffin JH, Fernandez JA, Castellino F, Rosen ED, Fukudome K, Zlokovic BV. Activated protein C blocks p53-mediated apoptosis in ischemic human brain endothelium and is neuroprotective. Nat Med. 2003; 9:338–42.
31. Domotor E, Benzakour O, Griffin JH, Yule D, Fukudome K, Zlokovic BV. Activated protein C alters cytosolic calcium flux in human brain endothelium via binding to endothelial protein C receptor and activation of protease activated receptor-1. Blood. 2003;101:4797–801.

32. Mosnier LO, Griffin JH. Inhibition of staurosporine-induced apoptosis of endothelial cells by activated protein C requires protease-activated receptor-1 and endothelial cell protein C receptor. Biochem J. 2003;373:65–70.
33. Ruf W, Riewald M. Tissue factor-dependent coagulation protease signaling in acute lung injury. Crit Care Med. 2003;31:S231–7.
34. Ranjan S, Goihl A, Kohli S, Gadi I, Pierau M, Shahzad K, Gupta D, Bock F, Wang HJ, Shaikh H, et al. Activated protein C protects from GvHD via PAR2/PAR3 signalling in regulatory T-cells. Nat Commun. 2017;8(1):311.
35. Madhusudhan T, Wang H, Straub BK, Grone E, Zhou Q, Shahzad K, Muller-Krebs S, Schwenger V, Gerlitz B, Grinnell BW, et al. Cytoprotective signaling by activated protein C requires protease-activated receptor-3 in podocytes. Blood. 2012;119:874–83.
36. Riewald M, Kravchenko VV, Petrovan RJ, O'Brien PJ, Brass LF, Ulevitch RJ, Ruf W. Gene induction by coagulation factor Xa is mediated by activation of protease-activated receptor 1. Blood. 2001;97:3109–16.
37. Guo H, Liu D, Gelbard H, Cheng T, Insalaco R, Fernandez JA, Griffin JH, Zlokovic BV. Activated protein C prevents neuronal apoptosis via protease activated receptors 1 and 3. Neuron. 2004;41:563–72.
38. Schuepbach RA, Riewald M. Coagulation factor Xa cleaves protease-activated receptor-1 and mediates signaling dependent on binding to the endothelial protein C receptor. J Thromb Haemost. 2010;8:379–88.
39. Stavenuiter F, Mosnier LO. Noncanonical PAR3 activation by factor Xa identifies a novel pathway for Tie2 activation and stabilization of vascular integrity. Blood. 2014;124:3480–9.
40. Camerer E, Gjernes E, Wiiger M, Pringle S, Prydz H. Binding of factor VIIa to tissue factor on keratinocytes induces gene expression. J Biol Chem. 2000; 275:6580–5.
41. Morris DR, Ding Y, Ricks TK, Gullapalli A, Wolfe BL, Trejo J. Protease-activated receptor-2 is essential for factor VIIa and Xa-induced signaling, migration, and invasion of breast cancer cells. Cancer Res. 2006;66:307–14.
42. Rothmeier AS, Liu E, Chakrabarty S, Disse J, Mueller BM, Ostergaard H, Ruf W. Identification of the integrin-binding site on coagulation factor VIIa required for proangiogenic PAR2 signaling. Blood. 2018;131:674–85.
43. Rothmeier AS, Disse J, Mueller BM, Liu EB, Ostergaard H, Ruf W. Proangiogenic TF-FVIIa-PAR2 signaling requires Matriptase-independent integrin interaction. Blood. 2016;128:3756.
44. Le Gall SM, Szabo R, Lee M, Kirchhofer D, Craik CS, Bugge TH, Camerer E. Matriptase activation connects tissue factor-dependent coagulation initiation to epithelial proteolysis and signaling. Blood. 2016;127:3260–9.
45. Sen P, Gopalakrishnan R, Kothari H, Keshava S, Clark CA, Esmon CT, Pendurthi UR, Rao LV. Factor VIIa bound to endothelial cell protein C receptor activates protease activated receptor-1 and mediates cell signaling and barrier protection. Blood. 2011;117:3199–208.
46. Kondreddy V, Wang J, Keshava S, Esmon CT, Rao LVM, Pendurthi UR. Factor VIIa induces anti-inflammatory signaling via EPCR and PAR1. Blood. 2018;131:2379–92.
47. Riewald M, Ruf W. Mechanistic coupling of protease signaling and initiation of coagulation by tissue factor. Proc Natl Acad Sci U S A. 2001;98:7742–7.
48. Rinderknecht H. Activation of pancreatic zymogens. Normal activation, premature intrapancreatic activation, protective mechanisms against inappropriate activation. Dig Dis Sci. 1986;31:314–21.
49. Koshikawa N, Hasegawa S, Nagashima Y, Mitsuhashi K, Tsubota Y, Miyata S, Miyagi Y, Yasumitsu H, Miyazaki K. Expression of trypsin by epithelial cells of various tissues, leukocytes, and neurons in human and mouse. Am J Pathol. 1998;153:937–44.
50. Koivunen E, Huhtala ML, Stenman UH. Human ovarian tumor-associated trypsin. Its purification and characterization from mucinous cyst fluid and identification as an activator of pro-urokinase. J Biol Chem. 1989;264:14095–9.
51. Koivunen E, Ristimaki A, Itkonen O, Osman S, Vuento M, Stenman UH. Tumor-associated trypsin participates in cancer cell-mediated degradation of extracellular matrix. Cancer Res. 1991;51:2107–12.
52. Koshikawa N, Nagashima Y, Miyagi Y, Mizushima H, Yanoma S, Yasumitsu H, Miyazaki K. Expression of trypsin in vascular endothelial cells. FEBS Lett. 1997;409:442–8.
53. Nakayama T, Hirano K, Shintani Y, Nishimura J, Nakatsuka A, Kuga H, Takahashi S, Kanaide H. Unproductive cleavage and the inactivation of protease-activated receptor-1 by trypsin in vascular endothelial cells. Br J Pharmacol. 2003;138:121–30.
54. Molino M, Woolkalis MJ, Reavey-Cantwell J, Pratico D, Andrade-Gordon P, Barnathan ES, Brass LF. Endothelial cell thrombin receptors and PAR-2. Two protease-activated receptors located in a single cellular environment. J Biol Chem. 1997;272:11133–41.
55. Berger P, Tunon-De-Lara JM, Savineau JP, Marthan R. Selected contribution: tryptase-induced PAR-2-mediated Ca(2+) signaling in human airway smooth muscle cells. J Appl Physiol (1985). 2001;91:995–1003.
56. Berger P, Perng DW, Thabrew H, Compton SJ, Cairns JA, McEuen AR, Marthan R, Tunon De Lara JM, Walls AF. Tryptase and agonists of PAR-2 induce the proliferation of human airway smooth muscle cells. J Appl Physiol (1985). 2001;91:1372–9.
57. Akers IA, Parsons M, Hill MR, Hollenberg MD, Sanjar S, Laurent GJ, McAnulty RJ. Mast cell tryptase stimulates human lung fibroblast proliferation via protease-activated receptor-2. Am J Physiol Lung Cell Mol Physiol. 2000;278: L193–201.
58. Compton SJ, Renaux B, Wijesuriya SJ, Hollenberg MD. Glycosylation and the activation of proteinase-activated receptor 2 (PAR(2)) by human mast cell tryptase. Br J Pharmacol. 2001;134:705–18.
59. Schechter NM, Brass LF, Lavker RM, Jensen PJ. Reaction of mast cell proteases tryptase and chymase with protease activated receptors (PARs) on keratinocytes and fibroblasts. J Cell Physiol. 1998;176:365–73.
60. Seitz I, Hess S, Schulz H, Eckl R, Busch G, Montens HP, Brandl R, Seidl S, Schomig A, Ott I. Membrane-type serine protease-1/matriptase induces interleukin-6 and -8 in endothelial cells by activation of protease-activated receptor-2: potential implications in atherosclerosis. Arterioscler Thromb Vasc Biol. 2007;27:769–75.
61. Bocheva G, Rattenholl A, Kempkes C, Goerge T, Lin CY, D'Andrea MR, Stander S, Steinhoff M. Role of matriptase and proteinase-activated receptor-2 in nonmelanoma skin cancer. J Invest Dermatol. 2009;129:1816–23.
62. Camerer E, Barker A, Duong DN, Ganesan R, Kataoka H, Cornelissen I, Darragh MR, Hussain A, Zheng YW, Srinivasan Y, et al. Local protease signaling contributes to neural tube closure in the mouse embryo. Dev Cell. 2010;18:25–38.
63. Sales KU, Friis S, Konkel JE, Godiksen S, Hatakeyama M, Hansen KK, Rogatto SR, Szabo R, Vogel LK, Chen W, et al. Non-hematopoietic PAR-2 is essential for matriptase-driven pre-malignant progression and potentiation of ras-mediated squamous cell carcinogenesis. Oncogene. 2015;34:346–56.
64. Caliendo G, Santagada V, Perissutti E, Severino B, Fiorino F, Frecentese F, Juliano L. Kallikrein protease activated receptor (PAR) axis: an attractive target for drug development. J Med Chem. 2012;55:6669–86.
65. Ramsay AJ, Dong Y, Hunt ML, Linn M, Samaratunga H, Clements JA, Hooper JD. Kallikrein-related peptidase 4 (KLK4) initiates intracellular signaling via protease-activated receptors (PARs). KLK4 and PAR-2 are co-expressed during prostate cancer progression. J Biol Chem. 2008;283:12293–304.
66. Oikonomopoulou K, Hansen KK, Saifeddine M, Vergnolle N, Tea I, Blaber M, Blaber SI, Scarisbrick I, Diamandis EP, Hollenberg MD. Kallikrein-mediated cell signalling: targeting proteinase-activated receptors (PARs). Biol Chem. 2006;387:817–24.
67. Yoon H, Radulovic M, Wu J, Blaber SI, Blaber M, Fehlings MG, Scarisbrick IA. Kallikrein 6 signals through PAR1 and PAR2 to promote neuron injury and exacerbate glutamate neurotoxicity. J Neurochem. 2013;127:283–98.
68. Molino M, Blanchard N, Belmonte E, Tarver AP, Abrams C, Hoxie JA, Cerletti C, Brass LF. Proteolysis of the human platelet and endothelial-cell thrombin receptor by neutrophil-derived Cathepsin-G. J Biol Chem. 1995;270:11168–75.
69. Molino M, Blanchard N, Belmonte E, Tarver AP, Abrams C, Hoxie JA, Cerletti C, Brass LF. Cathepsin-G cleaves the human thrombin receptor. Thromb Haemost. 1995;73:923.
70. Wilson TJ, Nannuru KC, Singh RK. Cathepsin G recruits osteoclast precursors via proteolytic activation of protease-activated Receptor-1. Cancer Res. 2009; 69:3188–95.
71. Ramachandran R, Sadofsky LR, Xiao YP, Botham A, Cowen M, Morice AH, Compton SJ. Inflammatory mediators modulate thrombin and cathepsin-G signaling in human bronchial fibroblasts by inducing expression of proteinase-activated receptor-4. Am J Phys Lung Cell Mol Phys. 2007;292: L788–98.
72. Ramachandran R, Mihara K, Chung H, Renaux B, Lau CS, Muruve DA, DeFea KA, Bouvier M, Hollenberg MD. Neutrophil elastase acts as a biased agonist for proteinase-activated receptor-2 (PAR2). J Biol Chem. 2011;286:24638–48.
73. Cumashi A, Ansuini H, Celli N, De Blasi A, O'Brien PJ, Brass LF, Molino M. Neutrophil proteases can inactivate human PAR3 and abolish the co-receptor function of PAR3 on murine platelets. Thromb Haemost. 2001;85: 533–8.
74. Sambrano GR, Huang W, Faruqi T, Mahrus S, Craik C, Coughlin SR. Cathepsin G activates protease-activated receptor-4 in human platelets. J Biol Chem. 2000;275:6819–23.

75. Vergnolle N, Derian CK, D'Andrea MR, Steinhoff M, Andrade-Gordon P. Characterization of thrombin-induced leukocyte rolling and adherence: a potential proinflammatory role for proteinase-activated receptor-4. J Immunol. 2002;169:1467–73.
76. Walsh SW, Nugent WH, Solotskaya AV, Anderson CD, Grider JR, Strauss JF 3rd. Matrix metalloprotease-1 and elastase are novel uterotonic agents acting through protease-activated receptor 1. Reprod Sci. 2017. https://doi.org/10.1177/1933719117732162.
77. Kumar VRS, Darisipudi MN, Steiger S, Devarapu SK, Tato M, Kukarni OP, Mulay SR, Thomasova D, Popper B, Demleitner J, et al. Cathepsin S cleavage of protease-activated Receptor-2 on endothelial cells promotes microvascular diabetes complications. J Am Soc Nephrol. 2016;27:1635–49.
78. Elmariah SB, Reddy VB, Lerner EA. Cathepsin S signals via PAR2 and generates a novel tethered ligand receptor agonist. PLoS One. 2014;9(6): e99702.
79. Kuliopulos A, Covic L, Seeley SK, Sheridan PJ, Helin J, Costello CE. Plasmin desensitization of the PAR1 thrombin receptor: kinetics, sites of truncation, and implications for thrombolytic therapy. Biochemistry. 1999;38:4572–85.
80. Domotor E, Bartha K, Machovich R, Adam-Vizi V. Protease-activated receptor-2 (PAR-2) in brain microvascular endothelium and its regulation by plasmin and elastase. J Neurochem. 2002;80:746–54.
81. Wang T, Lee MH, Choi E, Pardo-Villamizar CA, Lee SB, Yang IH, Calabresi PA, Nath A. Granzyme B-induced neurotoxicity is mediated via activation of PAR-1 receptor and Kv1.3 channel. PLoS One. 2012;7:e43950.
82. Lee PR, Johnson TP, Gnanapavan S, Giovannoni G, Wang T, Steiner JP, Medynets M, Vaal MJ, Gartner V, Nath A. Protease-activated receptor-1 activation by granzyme B causes neurotoxicity that is augmented by interleukin-1beta. J Neuroinflammation. 2017;14:131.
83. Cooper DM, Pechkovsky DV, Hackett TL, Knight DA, Granville DJ. Granzyme K activates protease-activated receptor-1. PLoS One. 2011;6:e21484.
84. Sharma M, Merkulova Y, Raithatha S, Parkinson LG, Shen Y, Cooper D, Granville DJ. Extracellular granzyme K mediates endothelial activation through the cleavage of protease-activated receptor-1. FEBS J. 2016;283: 1734–47.
85. Alvarez-Arce A, Lee-Rivera I, Lopez E, Hernandez-Cruz A, Lopez-Colome AM. Thrombin-induced Calpain activation promotes protease-activated receptor 1 internalization. Int J Cell Biol. 2017;2017:1908310.
86. Trivedi V, Boire A, Tchernychev B, Kaneider NC, Leger AJ, O'Callaghan K, Covic L, Kuliopulos A. Platelet matrix metalloprotease-1 mediates thrombogenesis by activating PAR1 at a cryptic ligand site. Cell. 2009;137:332–43.
87. Austin KM, Covic L, Kuliopulos A. Matrix metalloproteases and PAR1 activation. Blood. 2013;121:431–9.
88. Austin KM, Nguyen N, Javid G, Covic L, Kuliopulos A. Noncanonical matrix metalloprotease-1-protease-activated receptor-1 signaling triggers vascular smooth muscle cell dedifferentiation and arterial stenosis. J Biol Chem. 2013;288:23105–15.
89. Allen M, Ghosh S, Ahern GP, Villapol S, Maguire-Zeiss KA, Conant K. Protease induced plasticity: matrix metalloproteinase-1 promotes neurostructural changes through activation of protease activated receptor 1. Sci Rep. 2016; 6:35497.
90. Boire A, Covic L, Agarwal A, Jacques S, Sherifi S, Kuliopulos A. PAR1 is a matrix metalloprotease-1 receptor that promotes invasion and tumorigenesis of breast cancer cells. Cell. 2005;120:303–13.
91. Sebastiano M, Momi S, Falcinelli E, Bury L, Hoylaerts MF, Gresele P. A novel mechanism regulating human platelet activation by MMP-2-mediated PAR1 biased signaling. Blood. 2017;129:883–95.
92. Lee SE, Kim JM, Jeong SK, Jeon JE, Yoon HJ, Jeong MK, Lee SH. Protease-activated receptor-2 mediates the expression of inflammatory cytokines, antimicrobial peptides, and matrix metalloproteinases in keratinocytes in response to Propionibacterium acnes. Arch Dermatol Res. 2010;302:745–56.
93. Vliagoftis H, Schwingshackl A, Milne CD, Duszyk M, Hollenberg MD, Wallace JL, Befus AD, Moqbel R. Proteinase-activated receptor-2-mediated matrix metalloproteinase-9 release from airway epithelial cells. J Allergy Clin Immunol. 2000;106:537–45.
94. Hou HH, Wang HC, Cheng SL, Chen YF, Lu KZ, Yu CJ. MMP-12 activates protease activated receptor (PAR)-1, upregulates placenta growth factor and leads to pulmonary emphysema. Am J Physiol Lung Cell Mol Physiol. 2018; 315(3):L432–42.
95. Raza SL, Nehring LC, Shapiro SD, Cornelius LA. Proteinase-activated receptor-1 regulation of macrophage elastase (MMP-12) secretion by serine proteinases. J Biol Chem. 2000;275:41243–50.
96. Zang N, Zhuang JG, Deng Y, Yang ZM, Ye ZX, Xie XH, Ren L, Fu Z, Luo ZX, Xu FD, Liu EM. Pulmonary C fibers modulate MMP-12 production via PAR2 and are involved in the long-term airway inflammation and airway Hyperresponsiveness induced by respiratory syncytial virus infection. J Virol. 2016;90:2536–43.
97. Hooper JD, Nicol DL, Dickinson JL, Eyre HJ, Scarman AL, Normyle JF, Stuttgen MA, Douglas ML, Loveland KA, Sutherland GR, Antalis TM. Testisin, a new human serine proteinase expressed by premeiotic testicular germ cells and lost in testicular germ cell tumors. Cancer Res. 1999;59:3199–205.
98. Driesbaugh KH, Buzza MS, Martin EW, Conway GD, Kao JP, Antalis TM. Proteolytic activation of the protease-activated receptor (PAR)-2 by the glycosylphosphatidylinositol-anchored serine protease testisin. J Biol Chem. 2015;290:3529–41.
99. Potempa J, Pike RN. Corruption of innate immunity by bacterial proteases. J Innate Immun. 2009;1:70–87.
100. Kida Y, Higashimoto Y, Inoue H, Shimizu T, Kuwano K. A novel secreted protease from Pseudomonas aeruginosa activates NF-kappaB through protease-activated receptors. Cell Microbiol. 2008;10:1491–504.
101. Dulon S, Leduc D, Cottrell GS, D'Alayer J, Hansen KK, Bunnett NW, Hollenberg MD, Pidard D, Chignard M. Pseudomonas aeruginosa elastase disables proteinase-activated receptor 2 in respiratory epithelial cells. Am J Respir Cell Mol Biol. 2005;32:411–9.
102. Ender M, Andreoni F, Zinkernagel AS, Schuepbach RA. Streptococcal SpeB cleaved PAR-1 suppresses ERK phosphorylation and blunts thrombin-induced platelet aggregation. PLoS One. 2013;8:e81298.
103. van den Boogaard FE, Brands X, Duitman J, de Stoppelaar SF, Borensztajn KS, Roelofs J, Hollenberg MD, Spek CA, Schultz MJ, van 't Veer C, van der Poll T. Protease-activated receptor 2 facilitates bacterial dissemination in pneumococcal pneumonia. J Infect Dis. 2018;217:1462–71.
104. Schouten M, Van't Veer C, Roelofs JJ, Levi M, van der Poll T. Protease-activated receptor-1 impairs host defense in murine pneumococcal pneumonia: a controlled laboratory study. Crit Care. 2012;16:R238.
105. Asehnoune K, Moine P. Protease-activated receptor-1: key player in the sepsis coagulation-inflammation crosstalk. Crit Care. 2013;17:119.
106. de Stoppelaar SF, Van't Veer C, van den Boogaard FE, Nieuwland R, Hoogendijk AJ, de Boer OJ, Roelofs JJ, van der Poll T. Protease activated receptor 4 limits bacterial growth and lung pathology during late stage Streptococcus pneumoniae induced pneumonia in mice. Thromb Haemost. 2013;110:582–92.
107. Holzhausen M, Spolidorio LC, Ellen RP, Jobin MC, Steinhoff M, Andrade-Gordon P, Vergnolle N. Protease-activated receptor-2 activation: a major role in the pathogenesis of Porphyromonas gingivalis infection. Am J Pathol. 2006;168:1189–99.
108. Francis N, Ayodele BA, O'Brien-Simpson NM, Birchmeier W, Pike RN, Pagel CN, Mackie EJ. Keratinocyte-specific ablation of protease-activated receptor-2 prevents gingival inflammation and bone loss in a mouse model of periodontal disease. Cell Microbiol. 2018;20(11):e12891.
109. Lourbakos A, Chinni C, Thompson P, Potempa J, Travis J, Mackie EJ, Pike RN. Cleavage and activation of proteinase-activated receptor-2 on human neutrophils by gingipain-R from Porphyromonas gingivalis. FEBS Lett. 1998;435:45–8.
110. Lourbakos A, Potempa J, Travis J, D'Andrea MR, Andrade-Gordon P, Santulli R, Mackie EJ, Pike RN. Arginine-specific protease from Porphyromonas gingivalis activates protease-activated receptors on human oral epithelial cells and induces interleukin-6 secretion. Infect Immun. 2001;69:5121–30.
111. Lourbakos A, Yuan YP, Jenkins AL, Travis J, Andrade-Gordon P, Santulli R, Potempa J, Pike RN. Activation of protease-activated receptors by gingipains from Porphyromonas gingivalis leads to platelet aggregation: a new trait in microbial pathogenicity. Blood. 2001;97:3790–7.
112. Papapanagiotou D, Nicu EA, Bizzarro S, Gerdes VE, Meijers JC, Nieuwland R, van der Velden U, Loos BG. Periodontitis is associated with platelet activation. Atherosclerosis. 2009;202:605–11.
113. Ikawa K, Nishioka T, Yu Z, Sugawara Y, Kawagoe J, Takizawa T, Primo V, Nikolic B, Kuroishi T, Sasano T, et al. Involvement of neutrophil recruitment and protease-activated receptor 2 activation in the induction of IL-18 in mice. J Leukoc Biol. 2005;78:1118–26.
114. Kida Y, Inoue H, Shimizu T, Kuwano K. Serratia marcescens serralysin induces inflammatory responses through protease-activated receptor 2. Infect Immun. 2007;75:164–74.
115. Chen X, Earley K, Luo W, Lin SH, Schilling WP. Functional expression of a human thrombin receptor in Sf9 insect cells: evidence for an active tethered ligand. Biochem J. 1996;314(Pt 2):603–11.

116. Ubl JJ, Sergeeva M, Reiser G. Desensitisation of protease-activated receptor-1 (PAR-1) in rat astrocytes: evidence for a novel mechanism for terminating Ca2+ signalling evoked by the tethered ligand. J Physiol. 2000;525(Pt 2): 319–30.
117. Tripathi T, Abdi M, Alizadeh H. Protease-activated receptor 2 (PAR2) is upregulated by Acanthamoeba plasminogen activator (aPA) and induces proinflammatory cytokine in human corneal epithelial cells. Invest Ophthalmol Vis Sci. 2014;55:3912–21.
118. Laing GD, Compton SJ, Ramachandran R, Fuller GL, Wilkinson MC, Wagstaff SC, Watson SP, Kamiguti AS, Theakston RD, Senis YA. Characterization of a novel protein from Proatheris superciliaris venom: proatherocytin, a 34-kDa platelet receptor PAR1 agonist. Toxicon. 2005;46:490–9.
119. Santos BF, Serrano SM, Kuliopulos A, Niewiarowski S. Interaction of viper venom serine peptidases with thrombin receptors on human platelets. FEBS Lett. 2000;477:199–202.
120. Sun G, Stacey MA, Schmidt M, Mori L, Mattoli S. Interaction of mite allergens Der p3 and Der p9 with protease-activated receptor-2 expressed by lung epithelial cells. J Immunol. 2001;167:1014–21.
121. Asokananthan N, Graham PT, Fink J, Knight DA, Bakker AJ, McWilliam AS, Thompson PJ, Stewart GA. Activation of protease-activated receptor (PAR)-1, PAR-2, and PAR-4 stimulates IL-6, IL-8, and prostaglandin E2 release from human respiratory epithelial cells. J Immunol. 2002;168:3577–85.
122. Lin YP, Nelson C, Kramer H, Parekh AB. The allergen Der p3 from house dust mite stimulates store-operated ca(2+) channels and mast cell migration through PAR4 receptors. Mol Cell. 2018;70:228–241.e225.
123. Kondo S, Helin H, Shichijo M, Bacon KB. Cockroach allergen extract stimulates protease-activated receptor-2 (PAR-2) expressed in mouse lung fibroblast. Inflamm Res. 2004;53:489–96.
124. Asaduzzaman M, Nadeem A, Arizmendi N, Davidson C, Nichols HL, Abel M, Ionescu LI, Puttagunta L, Thebaud B, Gordon J, et al. Functional inhibition of PAR2 alleviates allergen-induced airway hyperresponsiveness and inflammation. Clin Exp Allergy. 2015;45:1844–55.
125. Polley DJ, Mihara K, Ramachandran R, Vliagoftis H, Renaux B, Saifeddine M, Daines MO, Boitano S, Hollenberg MD. Cockroach allergen serine proteinases: isolation, sequencing and signalling via proteinase-activated receptor-2. Clin Exp Allergy. 2017;47:946–60.
126. Chiu LL, Perng DW, Yu CH, Su SN, Chow LP. Mold allergen, pen C 13, induces IL-8 expression in human airway epithelial cells by activating protease-activated receptor 1 and 2. J Immunol. 2007;178:5237–44.
127. Kauffman HF, Tomee JF, van de Riet MA, Timmerman AJ, Borger P. Protease-dependent activation of epithelial cells by fungal allergens leads to morphologic changes and cytokine production. J Allergy Clin Immunol. 2000;105:1185–93.
128. Boitano S, Flynn AN, Schulz SM, Hoffman J, Price TJ, Vagner J. Potent agonists of the protease activated receptor 2 (PAR2). J Med Chem. 2011;54: 1308–13.
129. Boitano S, Flynn AN, Sherwood CL, Schulz SM, Hoffman J, Gruzinova I, Daines MO. Alternaria alternata serine proteases induce lung inflammation and airway epithelial cell activation via PAR2. Am J Physiol Lung Cell Mol Physiol. 2011;300:L605–14.
130. Snelgrove RJ, Gregory LG, Peiro T, Akthar S, Campbell GA, Walker SA, Lloyd CM. Alternaria-derived serine protease activity drives IL-33-mediated asthma exacerbations. J Allergy Clin Immunol. 2014;134:583–592.e586.
131. Rathnavelu V, Alitheen NB, Sohila S, Kanagesan S, Ramesh R. Potential role of bromelain in clinical and therapeutic applications. Biomed Rep. 2016;5:283–8.
132. Borrelli F, Capasso R, Severino B, Fiorino F, Aviello G, De Rosa G, Mazzella M, Romano B, Capasso F, Fasolino I, Izzo AA. Inhibitory effects of bromelain, a cysteine protease derived from pineapple stem (Ananas comosus), on intestinal motility in mice. Neurogastroenterol Motil. 2011;23:745–e331.
133. Reddy VB, Lerner EA. Plant cysteine proteases that evoke itch activate protease-activated receptors. Br J Dermatol. 2010;163:532–5.
134. Lerner DJ, Chen M, Tram T, Coughlin SR. Agonist recognition by proteinase-activated receptor 2 and thrombin receptor. Importance of extracellular loop interactions for receptor function. J Biol Chem. 1996;271:13943–7.
135. Hollenberg MD, Yang SG, Laniyonu AA, Moore GJ, Saifeddine M. Action of thrombin receptor polypeptide in gastric smooth muscle: identification of a core pentapeptide retaining full thrombin-mimetic intrinsic activity. Mol Pharmacol. 1992;42:186–91.
136. Blackhart BD, Emilsson K, Nguyen D, Teng W, Martelli AJ, Nystedt S, Sundelin J, Scarborough RM. Ligand cross-reactivity within the protease-activated receptor family. J Biol Chem. 1996;271:16466–71.
137. Hollenberg MD, Saifeddine M, al-Ani B, Kawabata A. Proteinase-activated receptors: structural requirements for activity, receptor cross-reactivity, and receptor selectivity of receptor-activating peptides. Can J Physiol Pharmacol. 1997;75:832–41.
138. Vassallo RR Jr, Kieber-Emmons T, Cichowski K, Brass LF. Structure-function relationships in the activation of platelet thrombin receptors by receptor-derived peptides. J Biol Chem. 1992;267:6081–5.
139. al-Ani B, Saifeddine M, Hollenberg MD. Detection of functional receptors for the proteinase-activated-receptor-2-activating polypeptide, SLIGRL-NH2, in rat vascular and gastric smooth muscle. Can J Physiol Pharmacol. 1995;73: 1203–7.
140. McGuire JJ, Saifeddine M, Triggle CR, Sun K, Hollenberg MD. 2-furoyl-LIGRLO-amide: a potent and selective proteinase-activated receptor 2 agonist. J Pharmacol Exp Ther. 2004;309:1124–31.
141. Al-Ani B, Saifeddine M, Kawabata A, Hollenberg MD. Proteinase activated receptor 2: role of extracellular loop 2 for ligand-mediated activation. Br J Pharmacol. 1999;128:1105–13.
142. Al-Ani B, Saifeddine M, Kawabata A, Renaux B, Mokashi S, Hollenberg MD. Proteinase-activated receptor 2 (PAR(2)): development of a ligand-binding assay correlating with activation of PAR(2) by PAR(1)- and PAR(2)-derived peptide ligands. J Pharmacol Exp Ther. 1999;290:753–60.
143. Nakanishi-Matsui M, Zheng YW, Sulciner DJ, Weiss EJ, Ludeman MJ, Coughlin SR. PAR3 is a cofactor for PAR4 activation by thrombin. Nature. 2000;404:609–13.
144. Hollenberg MD, Saifeddine M, Al-Ani B, Gui Y. Proteinase-activated receptor 4 (PAR4): action of PAR4-activating peptides in vascular and gastric tissue and lack of cross-reactivity with PAR1 and PAR2. Can J Physiol Pharmacol. 1999;77:458–64.
145. Faruqi TR, Weiss EJ, Shapiro MJ, Huang W, Coughlin SR. Structure-function analysis of protease-activated receptor 4 tethered ligand peptides. Determinants of specificity and utility in assays of receptor function [In process citation]. J Biol Chem. 2000;275:19728–34.
146. Lin H, Trejo J. Transactivation of the PAR1-PAR2 heterodimer by thrombin elicits beta-arrestin-mediated endosomal signaling. J Biol Chem. 2013;288: 11203–15.
147. O'Brien PJ, Prevost N, Molino M, Hollinger MK, Woolkalis MJ, Woulfe DS, Brass LF. Thrombin responses in human endothelial cells. Contributions from receptors other than PAR1 include the transactivation of PAR2 by thrombin-cleaved PAR1. J Biol Chem. 2000;275:13502–9.
148. Shi X, Gangadharan B, Brass LF, Ruf W, Mueller BM. Protease-activated receptors (PAR1 and PAR2) contribute to tumor cell motility and metastasis. Mol Cancer Res. 2004;2:395–402.
149. Jaber M, Maoz M, Kancharla A, Agranovich D, Peretz T, Grisaru-Granovsky S, Uziely B, Bar-Shavit R. Protease-activated-receptor-2 affects protease-activated-receptor-1-driven breast cancer. Cell Mol Life Sci. 2014;71:2517–33.
150. Leger AJ, Jacques SL, Badar J, Kaneider NC, Derian CK, Andrade-Gordon P, Covic L, Kuliopulos A. Blocking the protease-activated receptor 1-4 heterodimer in platelet-mediated thrombosis. Circulation. 2006;113:1244–54.
151. Sveshnikova AN, Balatskiy AV, Demianova AS, Shepelyuk TO, Shakhidzhanov SS, Balatskaya MN, Pichugin AV, Ataullakhanov FI, Panteleev MA. Systems biology insights into the meaning of the platelet's dual-receptor thrombin signaling. J Thromb Haemost. 2016;14:2045–57.
152. Hansen KK, Saifeddine M, Hollenberg MD. Tethered ligand-derived peptides of proteinase-activated receptor 3 (PAR3) activate PAR1 and PAR2 in Jurkat T cells. Immunology. 2004;112:183–90.
153. Cunningham MR, McIntosh KA, Pediani JD, Robben J, Cooke AE, Nilsson M, Gould GW, Mundell S, Milligan G, Plevin R. Novel role for proteinase-activated receptor 2 (PAR2) in membrane trafficking of proteinase-activated receptor 4 (PAR4). J Biol Chem. 2012;287:16656–69.
154. Gieseler F, Ungefroren H, Settmacher U, Hollenberg MD, Kaufmann R. Proteinase-activated receptors (PARs) - focus on receptor-receptor-interactions and their physiological and pathophysiological impact. Cell Commun Signal. 2013;11:86.
155. Amadesi S, Cottrell GS, Divino L, Chapman K, Grady EF, Bautista F, Karanjia R, Barajas-Lopez C, Vanner S, Vergnolle N, Bunnett NW. Protease-activated receptor 2 sensitizes TRPV1 by protein kinase Cepsilon- and A-dependent mechanisms in rats and mice. J Physiol. 2006;575:555–71.
156. Grant AD, Cottrell GS, Amadesi S, Trevisani M, Nicoletti P, Materazzi S, Altier C, Cenac N, Zamponi GW, Bautista-Cruz F, et al. Protease-activated receptor 2 sensitizes the transient receptor potential vanilloid 4 ion channel to cause mechanical hyperalgesia in mice. J Physiol. 2007;578:715–33.

157. Zhao P, Lieu T, Barlow N, Sostegni S, Haerteis S, Korbmacher C, Liedtke W, Jimenez-Vargas NN, Vanner SJ, Bunnett NW. Neutrophil elastase activates protease-activated Receptor-2 (PAR2) and transient receptor potential Vanilloid 4 (TRPV4) to cause inflammation and pain. J Biol Chem. 2015;290: 13875–87.
158. Saifeddine M, El-Daly M, Mihara K, Bunnett NW, McIntyre P, Altier C, Hollenberg MD, Ramachandran R. GPCR-mediated EGF receptor transactivation regulates TRPV4 action in the vasculature. Br J Pharmacol. 2015;172:2493–506.
159. Tsubota M, Ozaki T, Hayashi Y, Okawa Y, Fujimura A, Sekiguchi F, Nishikawa H, Kawabata A. Prostanoid-dependent bladder pain caused by proteinase-activated receptor-2 activation in mice: involvement of TRPV1 and T-type ca(2+) channels. J Pharmacol Sci. 2018;136:46–9.
160. Chung H, Ramachandran R, Hollenberg MD, Muruve DA. Proteinase-activated receptor-2 transactivation of epidermal growth factor receptor and transforming growth factor-beta receptor signaling pathways contributes to renal fibrosis. J Biol Chem. 2013;288:37319–31.
161. Sabri A, Guo J, Elouardighi H, Darrow AL, Andrade-Gordon P, Steinberg SF. Mechanisms of protease-activated receptor-4 actions in cardiomyocytes. Role of Src tyrosine kinase. J Biol Chem. 2003;278:11714–20.
162. Fasano A. Zonulin, regulation of tight junctions, and autoimmune diseases. Ann N Y Acad Sci. 2012;1258:25–33.
163. Al-Ani B, Hewett PW, Cudmore MJ, Fujisawa T, Saifeddine M, Williams H, Ramma W, Sissaoui S, Jayaraman PS, Ohba M, et al. Activation of proteinase-activated receptor 2 stimulates soluble vascular endothelial growth factor receptor 1 release via epidermal growth factor receptor transactivation in endothelial cells. Hypertension. 2010;55:689–97.
164. Palm E, Demirel I, Bengtsson T, Khalaf H. The role of toll-like and protease-activated receptors in the expression of cytokines by gingival fibroblasts stimulated with the periodontal pathogen Porphyromonas gingivalis. Cytokine. 2015;76:424–32.
165. Damien P, Cognasse F, Payrastre B, Spinelli SL, Blumberg N, Arthaud CA, Eyraud MA, Phipps RP, McNicol A, Pozzetto B, et al. NF-kappaB links TLR2 and PAR1 to soluble Immunomodulator factor secretion in human platelets. Front Immunol. 2017;8:85.
166. Rallabhandi P, Nhu QM, Toshchakov VY, Piao W, Medvedev AE, Hollenberg MD, Fasano A, Vogel SN. Analysis of proteinase-activated receptor 2 and TLR4 signal transduction: a novel paradigm for receptor cooperativity. J Biol Chem. 2008;283:24314–25.
167. Nhu QM, Shirey K, Teijaro JR, Farber DL, Netzel-Arnett S, Antalis TM, Fasano A, Vogel SN. Novel signaling interactions between proteinase-activated receptor 2 and toll-like receptors in vitro and in vivo. Mucosal Immunol. 2010;3:29–39.
168. Williams JC, Lee RD, Doerschuk CM, Mackman N. Effect of PAR-2 deficiency in mice on KC expression after Intratracheal LPS administration. J Signal Transduct. 2011;2011:415195.
169. Zhou B, Zhou H, Ling S, Guo D, Yan Y, Zhou F, Wu Y. Activation of PAR2 or/ and TLR4 promotes SW620 cell proliferation and migration via phosphorylation of ERK1/2. Oncol Rep. 2011;25:503–11.
170. Bucci M, Vellecco V, Harrington L, Brancaleone V, Roviezzo F, Mattace Raso G, Ianaro A, Lungarella G, De Palma R, Meli R, Cirino G. Cross-talk between toll-like receptor 4 (TLR4) and proteinase-activated receptor 2 (PAR(2)) is involved in vascular function. Br J Pharmacol. 2013;168:411–20.
171. Weithauser A, Bobbert P, Antoniak S, Bohm A, Rauch BH, Klingel K, Savvatis K, Kroemer HK, Tschope C, Stroux A, et al. Protease-activated receptor-2 regulates the innate immune response to viral infection in a coxsackievirus B3-induced myocarditis. J Am Coll Cardiol. 2013;62:1737–45.
172. Oldham WM, Hamm HE. How do receptors activate G proteins? Adv Protein Chem. 2007;74:67–93.
173. Oldham WM, Hamm HE. Heterotrimeric G protein activation by G-protein-coupled receptors. Nat Rev Mol Cell Biol. 2008;9:60–71.
174. Feistritzer C, Riewald M. Endothelial barrier protection by activated protein C through PAR1-dependent sphingosine 1-phosphate receptor-1 crossactivation. Blood. 2005;105:3178–84.
175. Finigan JH, Dudek SM, Singleton PA, Chiang ET, Jacobson JR, Camp SM, Ye SQ, Garcia JG. Activated protein C mediates novel lung endothelial barrier enhancement: role of sphingosine 1-phosphate receptor transactivation. J Biol Chem. 2005;280:17286–93.
176. Mosnier LO, Sinha RK, Burnier L, Bouwens EA, Griffin JH. Biased agonism of protease-activated receptor 1 by activated protein C caused by noncanonical cleavage at Arg46. Blood. 2012;120:5237–46.
177. Hung DT, Wong YH, Vu TK, Coughlin SR. The cloned platelet thrombin receptor couples to at least two distinct effectors to stimulate phosphoinositide hydrolysis and inhibit adenylyl cyclase. J Biol Chem. 1992; 267:20831–4.
178. Hung DT, Vu TK, Wheaton VI, Ishii K, Coughlin SR. Cloned platelet thrombin receptor is necessary for thrombin-induced platelet activation. J Clin Invest. 1992;89:1350–3.
179. Santulli RJ, Derian CK, Darrow AL, Tomko KA, Eckardt AJ, Seiberg M, Scarborough RM, Andrade-Gordon P. Evidence for the presence of a protease-activated receptor distinct from the thrombin receptor in human keratinocytes. Proc Natl Acad Sci U S A. 1995;92:9151–5.
180. McLaughlin JN, Shen L, Holinstat M, Brooks JD, Dibenedetto E, Hamm HE. Functional selectivity of G protein signaling by agonist peptides and thrombin for the protease-activated receptor-1. J Biol Chem. 2005;280: 25048–59.
181. Offermanns S, Laugwitz KL, Spicher K, Schultz G. G proteins of the G12 family are activated via thromboxane A2 and thrombin receptors in human platelets. Proc Natl Acad Sci U S A. 1994;91:504–8.
182. Klages B, Brandt U, Simon MI, Schultz G, Offermanns S. Activation of G12/ G13 results in shape change and rho/rho-kinase- mediated myosin light chain phosphorylation in mouse platelets. J Cell Biol. 1999;144:745–54.
183. Moers A, Nieswandt B, Massberg S, Wettschureck N, Gruner S, Konrad I, Schulte V, Aktas B, Gratacap MP, Simon MI, et al. G13 is an essential mediator of platelet activation in hemostasis and thrombosis. Nat Med. 2003;9:1418–22.
184. Klarenbach SW, Chipiuk A, Nelson RC, Hollenberg MD, Murray AG. Differential actions of PAR2 and PAR1 in stimulating human endothelial cell exocytosis and permeability: the role of Rho-GTPases. Circ Res. 2003;92:272–8.
185. Fortunato TM, Vara DS, Wheeler-Jones CP, Pula G. Expression of protease-activated receptor 1 and 2 and anti-tubulogenic activity of protease-activated receptor 1 in human endothelial colony-forming cells. PLoS One. 2014;9:e109375.
186. Kim YV, Di Cello F, Hillaire CS, Kim KS. Differential Ca2+ signaling by thrombin and protease-activated receptor-1-activating peptide in human brain microvascular endothelial cells. Am J Physiol Cell Physiol. 2004;286: C31–42.
187. Soh UJ, Trejo J. Activated protein C promotes protease-activated receptor-1 cytoprotective signaling through beta-arrestin and dishevelled-2 scaffolds. Proc Natl Acad Sci U S A. 2011;108:E1372–80.
188. Roy RV, Ardeshirylajimi A, Dinarvand P, Yang L, Rezaie AR. Occupancy of human EPCR by protein C induces beta-arrestin-2 biased PAR1 signaling by both APC and thrombin. Blood. 2016;128:1884–93.
189. DeFea KA, Zalevsky J, Thoma MS, Dery O, Mullins RD, Bunnett NW. Beta-arrestin-dependent endocytosis of proteinase-activated receptor 2 is required for intracellular targeting of activated ERK1/2. J Cell Biol. 2000;148: 1267–81.
190. Ge L, Shenoy SK, Lefkowitz RJ, DeFea K. Constitutive protease-activated receptor-2-mediated migration of MDA MB-231 breast cancer cells requires both beta-arrestin-1 and -2. J Biol Chem. 2004;279:55419–24.
191. Stalheim L, Ding Y, Gullapalli A, Paing MM, Wolfe BL, Morris DR, Trejo J. Multiple independent functions of arrestins in the regulation of protease-activated receptor-2 signaling and trafficking. Mol Pharmacol. 2005;67:78–87.
192. Hein L, Ishii K, Coughlin SR, Kobilka BK. Intracellular targeting and trafficking of thrombin receptors. A novel mechanism for resensitization of a G protein-coupled receptor. J Biol Chem. 1994;269:27719–26.
193. Takafuta T, Wu G, Murphy GF, Shapiro SS. Human beta-filamin is a new protein that interacts with the cytoplasmic tail of glycoprotein Ibalpha. J Biol Chem. 1998;273:17531–8.
194. Vouret-Craviari V, Grall D, Chambard JC, Rasmussen UB, Pouyssegur J, Van Obberghen-Schilling E. Post-translational and activation-dependent modifications of the G protein-coupled thrombin receptor. J Biol Chem. 1995;270:8367–72.
195. Schuepbach RA, Feistritzer C, Brass LF, Riewald M. Activated protein C-cleaved protease activated receptor-1 is retained on the endothelial cell surface even in the presence of thrombin. Blood. 2008;111:2667–73.
196. Trejo J, Hammes SR, Coughlin SR. Termination of signaling by protease-activated receptor-1 is linked to lysosomal sorting. Proc Natl Acad Sci U S A. 1998;95:13698–702.
197. Paing MM, Johnston CA, Siderovski DP, Trejo J. Clathrin adaptor AP2 regulates thrombin receptor constitutive internalization and endothelial cell resensitization. Mol Cell Biol. 2006;26:3231–42.

198. Dores MR, Chen B, Lin H, Soh UJ, Paing MM, Montagne WA, Meerloo T, Trejo J. ALIX binds a YPX(3)L motif of the GPCR PAR1 and mediates ubiquitin-independent ESCRT-III/MVB sorting. J Cell Biol. 2012;197:407–19.
199. Grimsey NJ, Coronel LJ, Cordova IC, Trejo J. Recycling and endosomal sorting of protease-activated Receptor-1 is distinctly regulated by Rab11A and Rab11B proteins. J Biol Chem. 2016;291:2223–36.
200. Trejo J, Coughlin SR. The cytoplasmic tails of protease-activated receptor-1 and substance P receptor specify sorting to lysosomes versus recycling. J Biol Chem. 1999;274:2216–24.
201. Ishii K, Chen J, Ishii M, Koch WJ, Freedman NJ, Lefkowitz RJ, Coughlin SR. Inhibition of thrombin receptor signaling by a G-protein coupled receptor kinase. J Biol Chem. 1994;269:1125–30.
202. Paing MM, Stutts AB, Kohout TA, Lefkowitz RJ, Trejo J. Beta -Arrestins regulate protease-activated receptor-1 desensitization but not internalization or down-regulation. J Biol Chem. 2002;277:1292–300.
203. Chen J, De S, Damron DS, Chen WS, Hay N, Byzova TV. Impaired platelet responses to thrombin and collagen in AKT-1-deficient mice. Blood. 2004; 104:1703–10.
204. Wolfe BL, Marchese A, Trejo J. Ubiquitination differentially regulates clathrin-dependent internalization of protease-activated receptor-1. J Cell Biol. 2007; 177:905–16.
205. Soh UJ, Dores MR, Chen B, Trejo J. Signal transduction by protease-activated receptors. Br J Pharmacol. 2010;160:191–203.
206. Jacob C, Cottrell GS, Gehringer D, Schmidlin F, Grady EF, Bunnett NW. c-Cbl mediates ubiquitination, degradation, and down-regulation of human protease-activated receptor 2. J Biol Chem. 2005;280:16076–87.
207. Dery O, Thoma MS, Wong H, Grady EF, Bunnett NW. Trafficking of proteinase-activated receptor-2 and beta-arrestin-1 tagged with green fluorescent protein. Beta-Arrestin-dependent endocytosis of a proteinase receptor. J Biol Chem. 1999;274:18524–35.
208. Ricks TK, Trejo J. Phosphorylation of protease-activated receptor-2 differentially regulates desensitization and internalization. J Biol Chem. 2009; 284:34444–57.
209. Jimenez-Vargas NN, Pattison LA, Zhao P, Lieu T, Latorre R, Jensen DD, Castro J, Aurelio L, Le GT, Flynn B, et al. Protease-activated receptor-2 in endosomes signals persistent pain of irritable bowel syndrome. Proc Natl Acad Sci U S A. 2018;115:E7438–47.
210. Steinhoff M, Buddenkotte J, Shpacovitch V, Rattenholl A, Moormann C, Vergnolle N, Luger TA, Hollenberg MD. Proteinase-activated receptors: transducers of proteinase-mediated signaling in inflammation and immune response. Endocr Rev. 2005;26:1–43.
211. Smith TH, Coronel LJ, Li JG, Dores MR, Nieman MT, Trejo J. Protease-activated receptor-4 signaling and trafficking is regulated by the Clathrin adaptor protein complex-2 independent of beta-Arrestins. J Biol Chem. 2016;291:18453–64.
212. Shapiro MJ, Weiss EJ, Faruqi TR, Coughlin SR. Protease-activated receptors 1 and 4 are shut off with distinct kinetics after activation by thrombin. J Biol Chem. 2000;275:25216–21.
213. Smith TH, Li JG, Dores MR, Trejo J. Protease-activated receptor-4 and purinergic receptor P2Y12 dimerize, co-internalize, and activate Akt signaling via endosomal recruitment of beta-arrestin. J Biol Chem. 2017;292: 13867–78.
214. Kahn ML, Nakanishi-Matsui M, Shapiro MJ, Ishihara H, Coughlin SR. Protease-activated receptors 1 and 4 mediate activation of human platelets by thrombin. J Clin Invest. 1999;103:879–87.
215. Andrade-Gordon P, Derian CK, Maryanoff BE, Zhang HC, Addo MF, Cheung W, Damiano BP, D'Andrea MR, Darrow AL, de Garavilla L, et al. Administration of a potent antagonist of protease-activated receptor-1 (PAR-1) attenuates vascular restenosis following balloon angioplasty in rats. J Pharmacol Exp Ther. 2001;298:34–42.
216. Bae JS, Rezaie AR. Thrombin inhibits nuclear factor kappaB and RhoA pathways in cytokine-stimulated vascular endothelial cells when EPCR is occupied by protein C. Thromb Haemost. 2009;101:513–20.
217. Buddenkotte J, Stroh C, Engels IH, Moormann C, Shpacovitch VM, Seeliger S, Vergnolle N, Vestweber D, Luger TA, Schulze-Osthoff K, Steinhoff M. Agonists of proteinase-activated receptor-2 stimulate upregulation of intercellular cell adhesion molecule-1 in primary human keratinocytes via activation of NF-kappa B. J Invest Dermatol. 2005;124:38–45.
218. Howells GL, Macey MG, Chinni C, Hou L, Fox MT, Harriott P, Stone SR. Proteinase-activated receptor-2: expression by human neutrophils. J Cell Sci. 1997;110(Pt 7):881–7.
219. Borbiev T, Birukova A, Liu F, Nurmukhambetova S, Gerthoffer WT, Garcia JG, Verin AD. p38 MAP kinase-dependent regulation of endothelial cell permeability. Am J Physiol Lung Cell Mol Physiol. 2004;287:L911–8.
220. Schuepbach RA, Feistritzer C, Fernandez JA, Griffin JH, Riewald M. Protection of vascular barrier integrity by activated protein C in murine models depends on protease-activated receptor-1. Thromb Haemost. 2009;101:724–33.
221. Birukova AA, Alekseeva E, Mikaelyan A, Birukov KG. HGF attenuates thrombin-induced endothelial permeability by Tiam1-mediated activation of the Rac pathway and by Tiam1/Rac-dependent inhibition of the rho pathway. FASEB J. 2007;21:2776–86.
222. Grimsey NJ, Aguilar B, Smith TH, Le P, Soohoo AL, Puthenveedu MA, Nizet V, Trejo J. Ubiquitin plays an atypical role in GPCR-induced p38 MAP kinase activation on endosomes. J Cell Biol. 2015;210:1117–31.
223. Hamilton JR, Frauman AG, Cocks TM. Increased expression of protease-activated receptor-2 (PAR2) and PAR4 in human coronary artery by inflammatory stimuli unveils endothelium-dependent relaxations to PAR2 and PAR4 agonists. Circ Res. 2001;89:92–8.
224. Saifeddine M, al-Ani B, Cheng CH, Wang L, Hollenberg MD. Rat proteinase-activated receptor-2 (PAR-2): cDNA sequence and activity of receptor-derived peptides in gastric and vascular tissue. Br J Pharmacol. 1996;118: 521–30.
225. Moffatt JD, Cocks TM. Endothelium-dependent and -independent responses to protease-activated receptor-2 (PAR-2) activation in mouse isolated renal arteries. Br J Pharmacol. 1998;125:591–4.
226. Sobey CG, Moffatt JD, Cocks TM. Evidence for selective effects of chronic hypertension on cerebral artery vasodilatation to protease-activated receptor-2 activation. Stroke. 1999;30:1933–40 discussion 1941.
227. McGuire JJ, Hollenberg MD, Andrade-Gordon P, Triggle CR. Multiple mechanisms of vascular smooth muscle relaxation by the activation of proteinase-activated receptor 2 in mouse mesenteric arterioles. Br J Pharmacol. 2002;135:155–69.
228. Villari A, Giurdanella G, Bucolo C, Drago F, Salomone S. Apixaban enhances vasodilatation mediated by protease-activated receptor 2 in isolated rat arteries. Front Pharmacol. 2017;8:480.
229. Roy SS, Saifeddine M, Loutzenhiser R, Triggle CR, Hollenberg MD. Dual endothelium-dependent vascular activities of proteinase-activated receptor-2-activating peptides: evidence for receptor heterogeneity. Br J Pharmacol. 1998;123:1434–40.
230. Arisato T, Sarker KP, Kawahara K, Nakata M, Hashiguchi T, Osame M, Kitajima I, Maruyama I. The agonist of the protease-activated receptor-1 (PAR) but not PAR3 mimics thrombin-induced vascular endothelial growth factor release in human smooth muscle cells. Cell Mol Life Sci. 2003;60:1716–24.
231. Seymour ML, Binion DG, Compton SJ, Hollenberg MD, MacNaughton WK. Expression of proteinase-activated receptor 2 on human primary gastrointestinal myofibroblasts and stimulation of prostaglandin synthesis. Can J Physiol Pharmacol. 2005;83:605–16.
232. Iablokov V, Hirota CL, Peplowski MA, Ramachandran R, Mihara K, Hollenberg MD, MacNaughton WK. Proteinase-activated receptor 2 (PAR2) decreases apoptosis in colonic epithelial cells. J Biol Chem. 2014;289:34366–77.
233. Rolland-Fourcade C, Denadai-Souza A, Cirillo C, Lopez C, Jaramillo JO, Desormeaux C, Cenac N, Motta JP, Larauche M, Tache Y, et al. Epithelial expression and function of trypsin-3 in irritable bowel syndrome. Gut. 2017;66:1767–78.
234. Vergnolle N, Macnaughton WK, Al-Ani B, Saifeddine M, Wallace JL, Hollenberg MD. Proteinase-activated receptor 2 (PAR2)-activating peptides: identification of a receptor distinct from PAR2 that regulates intestinal transport. Proc Natl Acad Sci U S A. 1998;95:7766–71.
235. Buresi MC, Schleihauf E, Vergnolle N, Buret A, Wallace JL, Hollenberg MD, MacNaughton WK. Protease-activated receptor-1 stimulates Ca(2 +)-dependent Cl(-) secretion in human intestinal epithelial cells. Am J Physiol Gastrointest Liver Physiol. 2001;281:G323–32.
236. Chin AC, Vergnolle N, MacNaughton WK, Wallace JL, Hollenberg MD, Buret AG. Proteinase-activated receptor 1 activation induces epithelial apoptosis and increases intestinal permeability. Proc Natl Acad Sci U S A. 2003;100:11104–9.
237. Idell S, Gonzalez K, Bradford H, Macarthur CK, Fein AM, Maunder RJ, Garcia JGN, Griffith DE, Weiland J, Martin TR, et al. Procoagulant activity in Bronchoalveolar lavage in the adult respiratory-distress syndrome - contribution of tissue factor associated with factor-vii. Am Rev Respir Dis. 1987;136:1466–74.
238. Jarjour NN, Calhoun WJ, Schwartz LB, Busse WW. Elevated bronchoalveolar lavage fluid histamine levels in allergic asthmatics are associated with increased airway obstruction. Am Rev Respir Dis. 1991;144:83–7.

239. Clark JM, Abraham WM, Fishman CE, Forteza R, Ahmed A, Cortes A, Warne RL, Moore WR, Tanaka RD. Tryptase inhibitors block allergen-induced airway and inflammatory responses in allergic sheep. Am J Respir Crit Care Med. 1995;152:2076–83.
240. Rice KD, Tanaka RD, Katz BA, Numerof RP, Moore WR. Inhibitors of tryptase for the treatment of mast cell-mediated diseases. Curr Pharm Des. 1998;4: 381–96.
241. Hauck RW, Schulz C, Schomig A, Hoffman RK, Panettieri RA Jr. Alpha-thrombin stimulates contraction of human bronchial rings by activation of protease-activated receptors. Am J Phys. 1999;277:L22–9.
242. Cocks TM, Fong B, Chow JM, Anderson GP, Frauman AG, Goldie RG, Henry PJ, Carr MJ, Hamilton JR, Moffatt JD. A protective role for protease-activated receptors in the airways. Nature. 1999;398:156–60.
243. Cocks TM, Moffatt JD. Protease-activated receptor-2 (PAR2) in the airways. Pulm Pharmacol Ther. 2001;14:183–91.
244. Schmidlin F, Amadesi S, Vidil R, Trevisani M, Martinet N, Caughey G, Tognetto M, Cavallesco G, Mapp C, Geppetti P, Bunnett NW. Expression and function of proteinase-activated receptor 2 in human bronchial smooth muscle. Am J Respir Crit Care Med. 2001;164:1276–81.
245. Schmidlin F, Amadesi S, Dabbagh K, Lewis DE, Knott P, Bunnett NW, Gater PR, Geppetti P, Bertrand C, Stevens ME. Protease-activated receptor 2 mediates eosinophil infiltration and hyperreactivity in allergic inflammation of the airway. J Immunol. 2002;169:5315–21.
246. Knight DA, Lim S, Scaffidi AK, Roche N, Chung KF, Stewart GA, Thompson PJ. Protease-activated receptors in human airways: upregulation of PAR-2 in respiratory epithelium from patients with asthma. J Allergy Clin Immunol. 2001;108:797–803.
247. Miotto D, Hollenberg MD, Bunnett NW, Papi A, Braccioni F, Boschetto P, Rea F, Zuin A, Geppetti P, Saetta M, et al. Expression of protease activated receptor-2 (PAR-2) in central airways of smokers and non-smokers. Thorax. 2002;57:146–51.
248. Ricciardolo FL, Steinhoff M, Amadesi S, Guerrini R, Tognetto M, Trevisani M, Creminon C, Bertrand C, Bunnett NW, Fabbri LM, et al. Presence and bronchomotor activity of protease-activated receptor-2 in Guinea pig airways. Am J Respir Crit Care Med. 2000;161:1672–80.
249. Carr MJ, Schechter NM, Undem BJ. Trypsin-induced, neurokinin-mediated contraction of Guinea pig bronchus. Am J Respir Crit Care Med. 2000;162: 1662–7.
250. Lin C, von der Thusen J, Daalhuisen J, ten Brink M, Crestani B, van der Poll T, Borensztajn K, Spek CA. Protease-activated receptor (PAR)-2 is required for PAR-1 signalling in pulmonary fibrosis. J Cell Mol Med. 2015;19:1346–56.
251. Cork MJ, Danby SG, Vasilopoulos Y, Hadgraft J, Lane ME, Moustafa M, Guy RH, Macgowan AL, Tazi-Ahnini R, Ward SJ. Epidermal barrier dysfunction in atopic dermatitis. J Invest Dermatol. 2009;129:1892–908.
252. Derian CK, Eckardt AJ, Andrade-Gordon P. Differential regulation of human keratinocyte growth and differentiation by a novel family of protease-activated receptors. Cell Growth Differ. 1997;8:743–9.
253. D'Andrea MR, Derian CK, Santulli RJ, Andrade-Gordon P. Differential expression of protease-activated receptors-1 and -2 in stromal fibroblasts of normal, benign, and malignant human tissues. Am J Pathol. 2001;158:2031–41.
254. Wakita H, Furukawa F, Takigawa M. Thrombin and trypsin induce granulocyte-macrophage colony-stimulating factor and interleukin-6 gene expression in cultured normal human keratinocytes. Proc Assoc Am Physicians. 1997;109:190–207.
255. Kanke T, Macfarlane SR, Seatter MJ, Davenport E, Paul A, McKenzie RC, Plevin R. Proteinase-activated receptor-2-mediated activation of stress-activated protein kinases and inhibitory kappa B kinases in NCTC 2544 keratinocytes. J Biol Chem. 2001;276:31657–66.
256. Kawagoe J, Takizawa T, Matsumoto J, Tamiya M, Meek SE, Smith AJ, Hunter GD, Plevin R, Saito N, Kanke T, et al. Effect of protease-activated receptor-2 deficiency on allergic dermatitis in the mouse ear. Jpn J Pharmacol. 2002;88: 77–84.
257. Lee SE, Jeong SK, Lee SH. Protease and protease-activated receptor-2 signaling in the pathogenesis of atopic dermatitis. Yonsei Med J. 2010;51: 808–22.
258. Steinhoff M, Neisius U, Ikoma A, Fartasch M, Heyer G, Skov PS, Luger TA, Schmelz M. Proteinase-activated receptor-2 mediates itch: a novel pathway for pruritus in human skin. J Neurosci. 2003;23:6176–80.
259. Calabresi E, Petrelli F, Bonifacio AF, Puxeddu I, Alunno A. One year in review 2018: pathogenesis of rheumatoid arthritis. Clin Exp Rheumatol. 2018;36: 175–84.
260. Kelso EB, Lockhart JC, Hembrough T, Dunning L, Plevin R, Hollenberg MD, Sommerhoff CP, McLean JS, Ferrell WR. Therapeutic promise of proteinase-activated receptor-2 antagonism in joint inflammation. J Pharmacol Exp Ther. 2006;316:1017–24.
261. Ferrell WR, Lockhart JC, Kelso EB, Dunning L, Plevin R, Meek SE, Smith AJ, Hunter GD, McLean JS, McGarry F, et al. Essential role for proteinase-activated receptor-2 in arthritis. J Clin Invest. 2003;111:35–41.
262. McCulloch K, McGrath S, Huesa C, Dunning L, Litherland G, Crilly A, Hultin L, Ferrell WR, Lockhart JC, Goodyear CS. Rheumatic disease: protease-activated Receptor-2 in synovial joint pathobiology. Front Endocrinol (Lausanne). 2018;9:257.
263. Crilly A, Palmer H, Nickdel MB, Dunning L, Lockhart JC, Plevin R, McInnes IB, Ferrell WR. Immunomodulatory role of proteinase-activated receptor-2. Ann Rheum Dis. 2012;71:1559–66.
264. Andrade-Gordon P, Maryanoff BE, Derian CK, Zhang HC, Addo MF, Darrow AL, Eckardt AJ, Hoekstra WJ, McComsey DF, Oksenberg D, et al. Design, synthesis, and biological characterization of a peptide-mimetic antagonist for a tethered-ligand receptor. Proc Natl Acad Sci U S A. 1999;96:12257–62.
265. Brass LF, Vassallo RR Jr, Belmonte E, Ahuja M, Cichowski K, Hoxie JA. Structure and function of the human platelet thrombin receptor. Studies using monoclonal antibodies directed against a defined domain within the receptor N terminus. J Biol Chem. 1992;267:13795–8.
266. Chackalamannil S, Wang Y, Greenlee WJ, Hu Z, Xia Y, Ahn HS, Boykow G, Hsieh Y, Palamanda J, Agans-Fantuzzi J, et al. Discovery of a novel, orally active himbacine-based thrombin receptor antagonist (SCH 530348) with potent antiplatelet activity. J Med Chem. 2008;51:3061–4.
267. Kogushi M, Matsuoka T, Kawata T, Kuramochi H, Kawaguchi S, Murakami K, Hiyoshi H, Suzuki S, Kawahara T, Kajiwara A, Hishinuma I. The novel and orally active thrombin receptor antagonist E5555 (Atopaxar) inhibits arterial thrombosis without affecting bleeding time in guinea pigs. Eur J Pharmacol. 2011;657:131–7.
268. Covic L, Gresser AL, Talavera J, Swift S, Kuliopulos A. Activation and inhibition of G protein-coupled receptors by cell-penetrating membrane-tethered peptides. Proc Natl Acad Sci U S A. 2002;99:643–8.
269. Aisiku O, Peters CG, De Ceunynck K, Ghosh CC, Dilks JR, Fustolo-Gunnink SF, Huang M, Dockendorff C, Parikh SM, Flaumenhaft R. Parmodulins inhibit thrombus formation without inducing endothelial injury caused by vorapaxar. Blood. 2015;125:1976–85.
270. Becker RC, Moliterno DJ, Jennings LK, Pieper KS, Pei J, Niederman A, Ziada KM, Berman G, Strony J, Joseph D, et al. Safety and tolerability of SCH 530348 in patients undergoing non-urgent percutaneous coronary intervention: a randomised, double-blind, placebo-controlled phase II study. Lancet. 2009;373:919–28.
271. Morrow DA, Braunwald E, Bonaca MP, Ameriso SF, Dalby AJ, Fish MP, Fox KA, Lipka LJ, Liu X, Nicolau JC, et al. Vorapaxar in the secondary prevention of atherothrombotic events. N Engl J Med. 2012;366:1404–13.
272. Tricoci P, Huang Z, Held C, Moliterno DJ, Armstrong PW, Van de Werf F, White HD, Aylward PE, Wallentin L, Chen E, et al. Thrombin-receptor antagonist vorapaxar in acute coronary syndromes. N Engl J Med. 2012;366:20–33.
273. Goto S, Ogawa H, Takeuchi M, Flather MD, Bhatt DL, Investigators JL. Double-blind, placebo-controlled phase II studies of the protease-activated receptor 1 antagonist E5555 (atopaxar) in Japanese patients with acute coronary syndrome or high-risk coronary artery disease. Eur Heart J. 2010;31: 2601–13.
274. Wiviott SD, Flather MD, O'Donoghue ML, Goto S, Fitzgerald DJ, Cura F, Aylward P, Guetta V, Dudek D, Contant CF, et al. Randomized trial of atopaxar in the treatment of patients with coronary artery disease: the lessons from antagonizing the cellular effect of thrombin-coronary artery disease trial. Circulation. 2011;123:1854–63.
275. O'Donoghue ML, Bhatt DL, Wiviott SD, Goodman SG, Fitzgerald DJ, Angiolillo DJ, Goto S, Montalescot G, Zeymer U, Aylward PE, et al. Safety and tolerability of atopaxar in the treatment of patients with acute coronary syndromes: the lessons from antagonizing the cellular effects of thrombin-acute coronary syndromes trial. Circulation. 2011;123:1843–53.
276. Gurbel PA, Bliden KP, Turner SE, Tantry US, Gesheff MG, Barr TP, Covic L, Kuliopulos A. Cell-penetrating Pepducin therapy targeting PAR1 in subjects with coronary artery disease. Arterioscler Thromb Vasc Biol. 2016;36:189–97.
277. Wilson SJ, Ismat FA, Wang Z, Cerra M, Narayan H, Raftis J, Gray TJ, Connell S, Garonzik S, Ma X, et al. PAR4 (protease-activated receptor 4) antagonism with BMS-986120 inhibits human ex vivo Thrombus formation. Arterioscler Thromb Vasc Biol. 2018;38:448–56.

278. Rasmussen UB, Gachet C, Schlesinger Y, Hanau D, Ohlmann P, Van Obberghen-Schilling E, Pouyssegur J, Cazenave JP, Pavirani A. A peptide ligand of the human thrombin receptor antagonizes alpha- thrombin and partially activates platelets. J Biol Chem. 1993;268:14322–8.
279. Seiler SM, Peluso M, Tuttle JG, Pryor K, Klimas C, Matsueda GR, Bernatowicz MS. Thrombin receptor activation by thrombin and receptor-derived peptides in platelet and CHRF-288 cell membranes: receptor-stimulated GTPase and evaluation of agonists and partial agonists. Mol Pharmacol. 1996;49:190–7.
280. Cisowski J, O'Callaghan K, Kuliopulos A, Yang J, Nguyen N, Deng Q, Yang E, Fogel M, Tressel S, Foley C, et al. Targeting protease-activated receptor-1 with cell-penetrating pepducins in lung cancer. Am J Pathol. 2011;179:513–23.
281. Cheung WM, D'Andrea MR, Andrade-Gordon P, Damiano BP. Altered vascular injury responses in mice deficient in protease-activated receptor-1. Arterioscler Thromb Vasc Biol. 1999;19:3014–24.
282. Young SE, Duvernay MT, Schulte ML, Lindsley CW, Hamm HE. Synthesis of indole derived protease-activated receptor 4 antagonists and characterization in human platelets. PLoS One. 2013;8:e65528.
283. Zhang HC, Derian CK, Andrade-Gordon P, Hoekstra WJ, McComsey DF, White KB, Poulter BL, Addo MF, Cheung WM, Damiano BP, et al. Discovery and optimization of a novel series of thrombin receptor (par-1) antagonists: potent, selective peptide mimetics based on indole and indazole templates. J Med Chem. 2001;44:1021–4.
284. Xu QL, Guo XH, Liu JX, Chen B, Liu ZF, Su L. Blockage of protease-activated receptor 1 ameliorates heat-stress induced intestinal high permeability and bacterial translocation. Cell Biol Int. 2015;39:411–7.
285. Nawata S, Murakami A, Hirabayashi K, Sakaguchi Y, Ogata H, Suminami Y, Numa F, Nakamura K, Kato H. Identification of squamous cell carcinoma antigen-2 in tumor tissue by two-dimensional electrophoresis [in process citation]. Electrophoresis. 1999;20:614–7.
286. Kato Y, Kita Y, Nishio M, Hirasawa Y, Ito K, Yamanaka T, Motoyama Y, Seki J. In vitro antiplatelet profile of FR171113, a novel non-peptide thrombin receptor antagonist. Eur J Pharmacol. 1999;384:197–202.
287. Tanaka M, Arai H, Liu N, Nogaki F, Nomura K, Kasuno K, Oida E, Kita T, Ono T. Role of coagulation factor Xa and protease-activated receptor 2 in human mesangial cell proliferation. Kidney Int. 2005;67:2123–33.
288. Kogushi M, Matsuoka T, Kobayashi H, Sato N, Suzuki S, Kawahara T, Kajiwara A, Hishinuma I. Biological characterization of ER129614-06, a novel, non-peptide protease-activated receptor-1 (PAR-1) antagonist. Atheroscler Suppl. 2003;4:245–6.
289. Kawahara T, Suzuki S, Kogushi M, Matsuoka T, Kobayashi H, Kajiwara A, Hishinuma I. Discovery and optimization of potent orally active small molecular thrombin receptor PAR-1 antagonists. Abstr Pap Am Chem Soc. 2003;225:U215.
290. Perez M, Lamothe M, Maraval C, Mirabel E, Loubat C, Planty B, Horn C, Michaux J, Marrot S, Letienne R, et al. Discovery of novel protease activated receptors 1 antagonists with potent antithrombotic activity in vivo. J Med Chem. 2009;52:5826–36.
291. Monjotin N, Gillespie J, Farrie M, Le Grand B, Junquero D, Vergnolle N. F16357, a novel protease-activated receptor 1 antagonist, improves urodynamic parameters in a rat model of interstitial cystitis. Br J Pharmacol. 2016;173:2224–36.
292. Ahn HS, Foster C, Boykow G, Stamford A, Manna M, Graziano M. Inhibition of cellular action of thrombin by N3-cyclopropyl-7-[[4-(1-methylethyl)phenyl]methyl]-7H-pyrrolo[3, 2-f]quinazoline-1,3-diamine (SCH 79797), a nonpeptide thrombin receptor antagonist. Biochem Pharmacol. 2000;60:1425–34.
293. Gupta N, Liu R, Shin S, Sinha R, Pogliano J, Pogliano K, Griffin JH, Nizet V, Corriden R. SCH79797 improves outcomes in experimental bacterial pneumonia by boosting neutrophil killing and direct antibiotic activity. J Antimicrob Chemother. 2018;73:1586–94.
294. Kogushi M, Kobayashi H, Matsuoka T, Suzuki S, Kawahara T, Kajiwara A, Hishinuma L. Anti-thrombotic and bleeding time effects of E5555, an orally active protease-activated receptor-1 antagonist, in Guinea pigs. Circulation. 2003;108:280.
295. Asteriti S, Daniele S, Porchia F, Dell'anno MT, Fazzini A, Pugliesi I, Trincavelli ML, Taliani S, Martini C, Mazzoni MR, Gilchrist A. Modulation of PAR(1) signaling by benzimidazole compounds. Br J Pharmacol. 2012;167(1):80–94.
296. Deng X, Mercer PF, Scotton CJ, Gilchrist A, Chambers RC. Thrombin induces fibroblast CCL2/JE production and release via coupling of PAR1 to Galphaq and cooperation between ERK1/2 and rho kinase signaling pathways. Mol Biol Cell. 2008;19:2520–33.
297. Tressel SL, Kaneider NC, Kasuda S, Foley C, Koukos G, Austin K, Agarwal A, Covic L, Opal SM, Kuliopulos A. A matrix metalloprotease-PAR1 system regulates vascular integrity, systemic inflammation and death in sepsis. EMBO Mol Med. 2011;3:370–84.
298. Lin C, Duitman J, Daalhuisen J, Ten Brink M, von der Thusen J, van der Poll T, Borensztajn K, Spek CA. Targeting protease activated receptor-1 with P1pal-12 limits bleomycin-induced pulmonary fibrosis. Thorax. 2014;69:152–60.
299. Dockendorff C, Aisiku O, Verplank L, Dilks JR, Smith DA, Gunnink SF, Dowal L, Negri J, Palmer M, Macpherson L, et al. Discovery of 1,3-diaminobenzenes as selective inhibitors of platelet activation at the PAR1 receptor. ACS Med Chem Lett. 2012;3:232–7.
300. De Ceunynck K, Peters CG, Jain A, Higgins SJ, Aisiku O, Fitch-Tewfik JL, Chaudhry SA, Dockendorff C, Parikh SM, Ingber DE, Flaumenhaft R. PAR1 agonists stimulate APC-like endothelial cytoprotection and confer resistance to thromboinflammatory injury. Proc Natl Acad Sci U S A. 2018;115:E982–91.
301. Zhong W, Chen S, Zhang Q, Xiao T, Qin Y, Gu J, Sun B, Liu Y, Jing X, Hu X, et al. Doxycycline directly targets PAR1 to suppress tumor progression. Oncotarget. 2017;8:16829–42.
302. Zhong W, Chen S, Qin Y, Zhang H, Wang H, Meng J, Huai L, Zhang Q, Yin T, Lei Y, et al. Doxycycline inhibits breast cancer EMT and metastasis through PAR-1/NF-kappaB/miR-17/E-cadherin pathway. Oncotarget. 2017;8:104855–66.
303. Bendeck MP, Conte M, Zhang M, Nili N, Strauss BH, Farwell SM. Doxycycline modulates smooth muscle cell growth, migration, and matrix remodeling after arterial injury. Am J Pathol. 2002;160:1089–95.
304. Wei H, Wei Y, Tian F, Niu T, Yi G. Blocking proteinase-activated receptor 2 alleviated neuropathic pain evoked by spinal cord injury. Physiol Res. 2016; 65:145–53.
305. Al-Ani B, Saifeddine M, Wijesuriya SJ, Hollenberg MD. Modified proteinase-activated receptor-1 and -2 derived peptides inhibit proteinase-activated receptor-2 activation by trypsin. J Pharmacol Exp Ther. 2002;300:702–8.
306. Kanke T, Kabeya M, Kubo S, Kondo S, Yasuoka K, Tagashira J, Ishiwata H, Saka M, Furuyama T, Nishiyama T, et al. Novel antagonists for proteinase-activated receptor 2: inhibition of cellular and vascular responses in vitro and in vivo. Br J Pharmacol. 2009;158:361–71.
307. Goh FG, Ng PY, Nilsson M, Kanke T, Plevin R. Dual effect of the novel peptide antagonist K-14585 on proteinase-activated receptor-2-mediated signalling. Br J Pharmacol. 2009;158:1695–704.
308. Boitano S, Hoffman J, Flynn AN, Asiedu MN, Tillu DV, Zhang Z, Sherwood CL, Rivas CM, DeFea KA, Vagner J, Price TJ. The novel PAR2 ligand C391 blocks multiple PAR2 signalling pathways in vitro and in vivo. Br J Pharmacol. 2015;172:4535–45.
309. Suen JY, Barry GD, Lohman RJ, Halili MA, Cotterell AJ, Le GT, Fairlie DP. Modulating human proteinase activated receptor 2 with a novel antagonist (GB88) and agonist (GB110). Br J Pharmacol. 2012;165:1413–23.
310. Gardell LR, Ma JN, Seitzberg JG, Knapp AE, Schiffer HH, Tabatabaei A, Davis CN, Owens M, Clemons B, Wong KK, et al. Identification and characterization of novel small-molecule protease-activated receptor 2 agonists. J Pharmacol Exp Ther. 2008;327:799–808.
311. Chanakira A, Westmark PR, Ong IM, Sheehan JP. Tissue factor-factor VIIa complex triggers protease activated receptor 2-dependent growth factor release and migration in ovarian cancer. Gynecol Oncol. 2017;145:167–75.
312. Sun Q, Wang Y, Zhang J, Lu J. ENMD-1068 inhibits liver fibrosis through attenuation of TGF-beta1/Smad2/3 signaling in mice. Sci Rep. 2017;7:5498.
313. Barry GD, Suen JY, Le GT, Cotterell A, Reid RC, Fairlie DP. Novel agonists and antagonists for human protease activated receptor 2. J Med Chem. 2010;53: 7428–40.
314. Lieu T, Savage E, Zhao P, Edgington-Mitchell L, Barlow N, Bron R, Poole DP, McLean P, Lohman RJ, Fairlie DP, Bunnett NW. Antagonism of the proinflammatory and pronociceptive actions of canonical and biased agonists of protease-activated receptor-2. Br J Pharmacol. 2016;173:2752–65.
315. Cheng RKY, Fiez-Vandal C, Schlenker O, Edman K, Aggeler B, Brown DG, Brown GA, Cooke RM, Dumelin CE, Dore AS, et al. Structural insight into allosteric modulation of protease-activated receptor 2. Nature. 2017;545:112–5.
316. Sevigny LM, Austin KM, Zhang P, Kasuda S, Koukos G, Sharifi S, Covic L, Kuliopulos A. Protease-activated receptor-2 modulates protease-activated receptor-1-driven neointimal hyperplasia. Arterioscler Thromb Vasc Biol. 2011;31:e100–6.

317. Michael ES, Kuliopulos A, Covic L, Steer ML, Perides G. Pharmacological inhibition of PAR2 with the pepducin P2pal-18S protects mice against acute experimental biliary pancreatitis. Am J Physiol Gastrointest Liver Physiol. 2013;304:G516–26.
318. Ishikawa C, Tsuda T, Konishi H, Nakagawa N, Yamanishi K. Tetracyclines modulate protease-activated receptor 2-mediated proinflammatory reactions in epidermal keratinocytes. Antimicrob Agents Chemother. 2009; 53:1760–5.
319. Liu XJ, Mu ZL, Zhao Y, Zhang JZ. Topical tetracycline improves MC903-induced atopic dermatitis in mice through inhibition of inflammatory cytokines and Thymic stromal Lymphopoietin expression. Chin Med J. 2016; 129:1483–90.
320. Castro ML, Franco GC, Branco-de-Almeida LS, Anbinder AL, Cogo-Muller K, Cortelli SC, Duarte S, Saxena D, Rosalen PL. Downregulation of proteinase-activated Receptor-2, Interleukin-17, and other Proinflammatory genes by subantimicrobial doxycycline dose in a rat periodontitis model. J Periodontol. 2016;87:203–10.
321. Hollenberg MD, Saifeddine M. Proteinase-activated receptor 4 (PAR4): activation and inhibition of rat platelet aggregation by PAR4-derived peptides. Can J Physiol Pharmacol. 2001;79:439–42.
322. Kim HY, Goo JH, Joo YA, Lee HY, Lee SM, Oh CT, Ahn SM, Kim NH, Hwang JS. Impact on inflammation and recovery of skin barrier by nordihydroguaiaretic acid as a protease-activated receptor 2 antagonist. Biomol Ther (Seoul). 2012;20:463–9.
323. Yang J, Wu J, Kowalska MA, Dalvi A, Prevost N, O'Brien PJ, Manning D, Poncz M, Lucki I, Blendy JA, Brass LF. Loss of signaling through the G protein, Gz, results in abnormal platelet activation and altered responses to psychoactive drugs. Proc Natl Acad Sci U S A. 2000;97:9984–9.
324. Wu CC, Huang SW, Hwang TL, Kuo SC, Lee FY, Teng CM. YD-3, a novel inhibitor of protease-induced platelet activation. Br J Pharmacol. 2000;130: 1289–96.
325. Wu CC, Hwang TL, Liao CH, Kuo SC, Lee FY, Lee CY, Teng CM. Selective inhibition of protease-activated receptor 4-dependent platelet activation by YD-3. Thromb Haemost. 2002;87:1026–33.
326. Young SE, Duvernay MT, Schulte ML, Nance KD, Melancon BJ, Engers J, Wood MR, Hamm HE, Lindsley CW. A novel and selective PAR4 antagonist: ML354. In: Probe reports from the NIH molecular libraries program. Bethesda: National Center for Biotechnology Information; 2015.
327. Wen W, Young SE, Duvernay MT, Schulte ML, Nance KD, Melancon BJ, Engers J, Locuson CW 2nd, Wood MR, Daniels JS, et al. Substituted indoles as selective protease activated receptor 4 (PAR-4) antagonists: discovery and SAR of ML354. Bioorg Med Chem Lett. 2014;24:4708–13.
328. Peek GJ, Elbourne D, Mugford M, Tiruvoipati R, Wilson A, Allen E, Clemens F, Firmin R, Hardy P, Hibbert C, et al. Randomised controlled trial and parallel economic evaluation of conventional ventilatory support versus extracorporeal membrane oxygenation for severe adult respiratory failure (CESAR). Health Technol Assess. 2010;14:1–46.
329. Wong PC, Seiffert D, Bird JE, Watson CA, Bostwick JS, Giancarli M, Allegretto N, Hua J, Harden D, Guay J, et al. Blockade of protease-activated receptor-4 (PAR4) provides robust antithrombotic activity with low bleeding. Sci Transl Med. 2017;9. https://doi.org/10.1126/scitranslmed.aaf5294.
330. Hollenberg MD, Saifeddine M, Sandhu S, Houle S, Vergnolle N. Proteinase-activated receptor-4: evaluation of tethered ligand-derived peptides as probes for receptor function and as inflammatory agonists in vivo. Br J Pharmacol. 2004;143:443–54.

18

Successful use of eculizumab to treat atypical hemolytic uremic syndrome in patients with inflammatory bowel disease

Ramy M. Hanna[1,2*], Noah Merin[3], Richard M. Burwick[4], Lama Abdelnour[1], Umut Selamet[1], Beshoy Yanny[5], Patrick Bui[6], Mary Fouad[6] and Ira Kurtz[1,7]

Abstract

Background: Atypical hemolytic uremic syndrome is a rare group of disorders that have in common underlying complement amplifying conditions. These conditions can accelerate complement activation that results in a positive feedback cycle. The known triggers for complement activation can be diverse and include, infection, autoimmune disease, and malignancy. Recent reports suggest that certain autoimmune and rheumatological triggers of complement activation may result in atypical hemolytic uremic syndrome that does not resolve despite treating the underlying disorder. Specifically, patients with systemic lupus erythematosus and microangiopathic hemolysis may not respond to treatment of their underlying rheumatological trigger but responded to complement blockade.

Case presentations: We report two patients with inflammatory bowel disease complicated by development of atypical hemolytic uremic syndrome. In both cases, patients were on treatment for inflammatory bowel disease, that was not well controlled/flaring at the time. The first patient is a male who developed Crohn's disease and microangiopathic hemolysis at age 5 and was treated with eculizumab successfully. Discontinuation of the medication led to multiple relapses, and the patient currently is being treated with eculizumab and has normal hematological and stable renal parameters. The second patient is a 49-year-old female with Ulcerative Colitis treated with 6-Mercaptopurine. She developed acute kidney injury and microangiopathic hemolysis. Prompt diagnosis and treatment with eculizumab resulted in the recovery of kidney injury along with a complete hematological response.

Conclusions: These two cases are the fifth and sixth patients to be published in the literature with atypical hemolytic uremic syndrome and inflammatory bowel disease treated with complement blockade. This confirms that C5 complement blockade is effective in treating complement mediated thrombotic microangiopathy/atypical hemolytic uremic syndrome when it is triggered in patients with inflammatory bowel disease.

Keywords: Atypical hemolytic uremic syndrome, Complement, Complement blockade, Thrombotic Microangiopathy, Inflammatory bowel disease, Crohn's colitis/Crohn's disease, Ulcerative colitis

* Correspondence: rhannamd81@yahoo.com; rmhanna@mednet.ucla.edu
[1]Department of Medicine, Division of Nephrology, UCLA David Geffen School of Medicine, Room 7-155, Factor Bldg. 700 Tiverton Ave, Los Angeles, CA 90095, USA
[2]Department of Medicine, Division of Nephrology, University of California Irvine School of Medicine, Irvine CA, USA
Full list of author information is available at the end of the article

Key points

1) IBD can trigger aHUS, and in both cases, we report patients received eculizumab and other biologic agents for IBD concurrently.
2) We report the 5th and 6th cases of complement mediated TMA/aHUS in IBD patients (1 with UC and 1 with CD), with successful resolution of aHUS with complement blockade.
3) We report prevention of relapse with ongoing complement blockade in one case.

Background

Atypical Hemolytic Uremic Syndrome (aHUS), is a syndrome resulting from complement activation. The causes of complement activation are numerous, and the genetics are diverse. In affected patients, a latent susceptibility becomes unmasked by a complement amplifying condition resulting in Thrombotic Microangiopathy (TMA) [1]. New research is revealing complement biology as having a role in microangiopathic hemolysis and endothelial dysfunction in many rheumatological diseases [1]. Recent publications by Park et al. disclose that ongoing microangiopathic hemolysis stemming from rheumatological conditions like systemic lupus erythematosus can be treated with eculizumab when more traditional immunosuppression has failed [2, 3]. Microangiopathic hemolysis due to Antiphospholipid antibody syndrome and catastrophic antiphospholipid antibody syndrome (CAPS) variant has also been successfully treated with complement blockade [4].

The final common pathway of a positive feedback complement activation loop is the etiologic feature of aHUS, despite the various triggers [1]. In most cases, a strong genetic susceptibility is unmasked by the presence of a strong temporary, or a milder but lingering complement amplifying condition [1]. Indeed, some cases of aHUS can be self-limiting (due to hypertension, post-partum aHUS), and others can be due to slow ongoing complement activation (C3 glomerulopathy, and chronic hypertension due to occult TMA) [1].

Inflammatory bowel disease (IBD), resulting in Crohn's Disease (CD) and Ulcerative Colitis (UC) as clinical phenotypes, is one such group of complement amplifying conditions [5]. As with other autoimmune diseases, first-line treatment has always focused on suppression of the immune system to eliminate inflammation and clinical symptoms. In the current literature, there are three published case reports and one abstract detailing patients with IBD and aHUS treated with eculizumab [6–10].

We report two additional cases of patients with IBD and aHUS treated with eculizumab. Patient 1 is a 19-year-old male with CD who was treated with eculizumab at age 5 and had multiple relapses off the drug. He was maintained on eculizumab with near-normal renal and hematological parameters for 14 years. Patient 2 is a 49-year-old female with UC who developed severe aHUS and was promptly treated with eculizumab resulting in normalization of renal and hematological parameters.

Case presentations

Patient 1

The first patient is a 19-year-old Caucasian male who presented for the continuation of complement blockade therapy with eculizumab. He was diagnosed with CD at age 5 and started on per rectum steroid therapy with budesonide and mesalamine therapy. He did require continuous infusions of adalimumab then infliximab for control of his disease. During the initial episode of colitis/CD in 2005, he was noted to have thrombocytopenia, hemolytic anemia, schistocytes with elevations in lactate dehydrogenase. The ADAMTS13 level was not available early on in the time of his diagnosis. Though Shiga toxin was not found in stool, vitamin B12 levels was 781 nanogram (ng)/mL, red blood cell (RBC) folate levels were within normal limits 981 nanomoles/L excluding other known TMA causes at that time. Rheumatolgoic serologies like anti nuclear antibody (ANA), anti double stranded DNA (dsDNA), anti ribonucleic protein smith (RNP-smith), anti cardiolipin (anti CL), dilute russell viper venom time assay (DRVVT), and anti phospholipid antibody (anti APL Ab) studies were all negative during initial workup.

As the scientific understanding of aHUS developed, it was decided that eculizumab would be initiated on an experimental basis with institutional review board approval (in a health system in Florida). He underwent a several-month course of eculizumab with clinical response. Following discontinuation of the medication, the patient suffered a relapse a few months later. Therapy was reinitiated and was subsequently weaned off successfully. Complement testing found a high level of C5 convertase function with high soluble membrane attack complex that improved with eculizumab use.

In April 2010 the patient had a severe flare of his underlying CD despite ongoing use of adalimumab, his disease remained poorly controlled clinically. He had a concomitant serum creatinine elevation peaking at 3.13 mg/dL in 04/2010 (baseline of 0.57 mg/dL 08/2009) with a drop in hemoglobin from 12.9 g/L baseline in 08/2009 to 7.6 g/L in 03/2010. The total bilirubin concentration was not elevated but increased from a baseline of 0.1 (mg/dL) in 03/2010 to 0.8 (mg/dL) in 04/2010 during the flare. Lactate dehydrogenase (LDH) was significantly elevated at 2223 Units/L in 4/2010 and decreased after that. The platelets had decreased by ~ 50% from 366, 000/uL at baseline in 02/2010 to 187,000/uL in 4/2010. ADAMTS 13 activity level was within normal limits at

44% ruling out ADAMTS13 deficiency and thrombotic thrombocytopenic purpura (TTP). Prothrombin time (PT) was 11 s with an International Normalized Ratio (INR) was 1, and activated partial thromboplastin time (aPTT) was 39 s both within normal. Fibrinogen levels were high at 471 mg/dL, and a D-dimer was high at 6.17 microgram/mL as expected in an inflammatory condition. Absolute reticulocyte count was in the upper limit of normal / slightly elevated at 2.92%.

Since the patient had an established aHUS diagnosis, eculizumab was promptly reintroduced at 900 mg (infusion dose) each week with normalization of the platelet count and LDH levels, and improvement of the serum Cr to 1 mg/dL. The patient received eculizumab (1200 mg infusions) every two weeks for 8.5 years continually after this aHUS relapse. The patient had already received appropriate meningococcal vaccinations from prior courses and was aware of the risks and benefits of pharmacological complement blockade. He has not had any issues with meningitis or other infections linked to eculizumab. Clinically, his IBD symptoms slowly improved with use of ongoing adalimumab and steroid therapy for CD/IBD along with eculizumab use for aHUS.

The effect of this regimen was the maintenance on his aHUS was the maintenance of normal hematological parameters and renal function with current serum Cr of 1 mg/dL, hemoglobin of 10–12 g/L, and LDH within normal limits. The patient has agreed to continue lifelong eculizumab in this case given his two prior relapses. No antibodies against complement factors were found, and no mutations were found. Only 60% of patients with aHUS have detectable mutations, and in some cases, the mutations can be present but are currently unknown [11]. Figure 1 details laboratory findings in patient 1 and his course during his last flare in 2010.

Patient 2

The second patient is a 49-year-old Caucasian female with a history of UC treated with six mercaptopurine (6MP), and steroids per rectum [budesonide]. She developed a flare of ulcerative colitis symptoms, along with increased platelet count from 322,000 platelets/uL (09/2018) to 622,000 platelets/uL (10/2018). This was a reactive thrombocytosis that occurred initially, and was clinically concomitant with a flare despite treatment with budesonide and 6MP.Her serum Cr at baseline was 0.51 mg/dL, and her hemoglobin was 13.1 g/L. She presented to the emergency department one month after the start of this flare with weakness and was found to have a decreased hemoglobin concentration of 4.3 g/L, serum Cr of 3.2 mg/dL, LDH of 1000 units/L, and schistocytes on the peripheral smear. Her platelets had droped from baseline of 322, 000 platelets/ul and more recent peak of 622,000 platelets/ul to 83,000 platelets/ul; this level dropped further to 43,000 plateltes/ul at a nadir 11/17/2018. These findings and the hemolysis suggested thrombotic microangiopathy like aHUS as the likely diagnosis over 6-mercaptopurine bone marrow toxicity which was another possibility. Vitamin B12 level was normal at 769 ng/mL, serum folate level was also

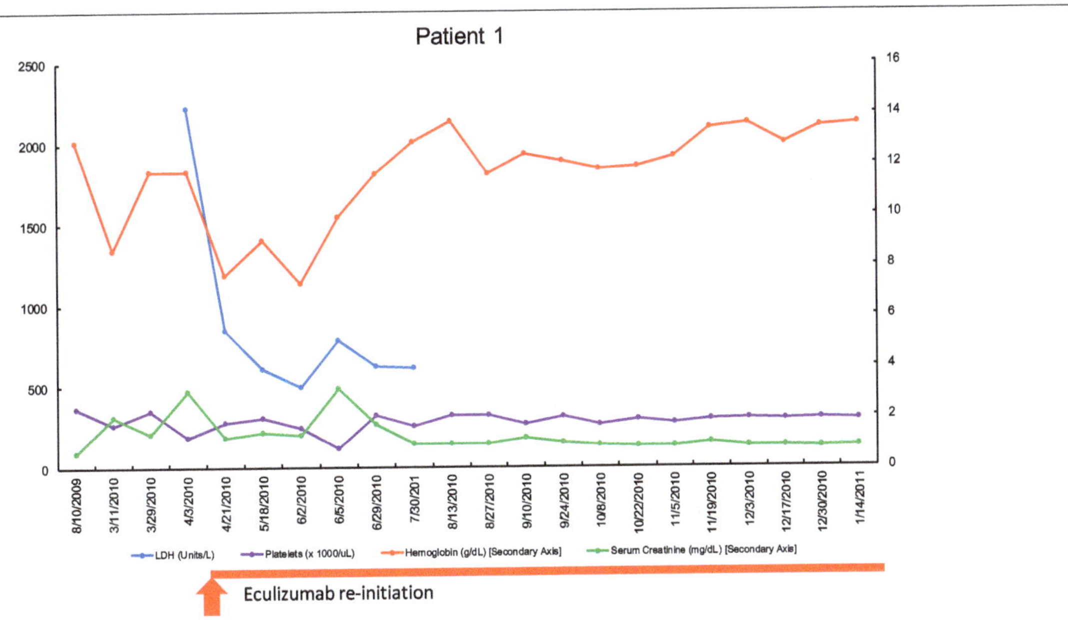

Fig. 1 Laboratory tests for patient one as a function of time. Abbreviations: dL, deciliter; g, gram; L, liter; LDH, lactate dehydrogenase; mg, milligram; uL, microliter

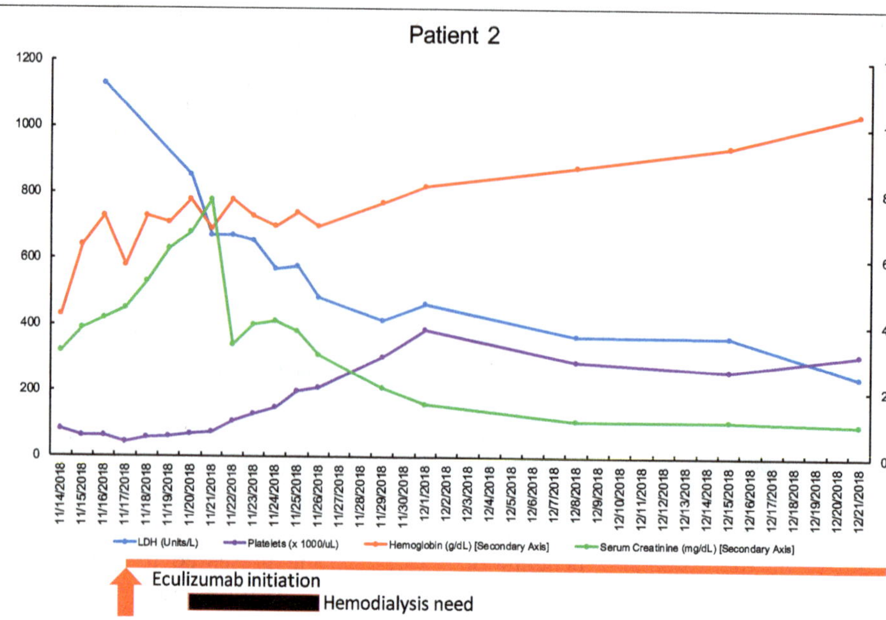

Fig. 2 Laboratory results for patient two as a function of time. Abbreviations: dL, deciliter; g, gram; L, liter; LDH, lactate dehydrogenase; mg, milligram; uL, microliter

normal at 12.9 ng/mL. PT was 13 s with an INR of 1.3, aPTT was slightly elevated at 39 s. Fibrinogen levels were high normal range at 313 mg/dL, with an elevated D-dimer 3.28 microgram/mL. Rheumatologic serologies (ANA, anti dsDNA, anti RNP sm, anti CL, DRVVT, APL Ab) were all negative. Reticulocyte counts were obtained at 2% (normal), and one result was 3% which is slightly elevated, and not expected in bone marrow toxicity.

The finding of an ADAMTS13 level of 80.8% suggested that the cause of TMA was not TTP. The patient was transferred back to a quaternary care center, and then eculizumab was started after meningococcal vaccination. The patient agreed to the risks and benefits of complement blockade in the absence of waiting the requisite two weeks for development of meningococcal antibody titers after immunization, and antibiotic prophylaxis with penicillin-based antibiotics was provided.

The serum Cr, platelets, hemoglobin, LDH all improved quickly and normalized (Fig. 2). The patient required hemodialysis for ~ 1 week and was weaned off. Eculizumab was administered at a given dose of 900 mg weekly infusions. The patient was discharged on 1200 mg infusions of eculizumab every other week with normal lab parameters. She is currently undergoing genetic testing at the University of Iowa laboratory to look for autoantibodies and genetic mutations. The response to eculizumab and the clinical course are typical for aHUS and diagnostic.

Table 1 aHUS Cases Associated with Inflammatory Bowel Disease

Reference #	n	Pt age	Pt gender	UC/CD	Creatinine improvement (Y/N)	TMA resolution (Y/N)	Ab/Mutation found?
Case Reports							
Beers [6] [abstract]	1	19	F	IBD**	Y	Y	CFH Auto-Ab & HZ CF1/CFH3 Del
Webb [7]	1	16	M	UC	Y	Y	*HZ MTHFR mutation****
Viada Bris [8, 9]*	1	15	F	UC	NT	NT	HZ CFH, HZ MCP
Green H [10]	1	27	F	UC	Y	Y	CFH Auto-Ab
Hanna et.al. Case 1	(1/2)	19	M	CD	Y	Y	No clear mutation found****
Hanna et.al. Case 2	(2/2)	49	F	UC	Y	Y	Testing not completed

Ab antibody, *CD* Crohn's disease, *CF1* complement factor 1, *CF3* complement factor 3, *CFH* complement factor H, *HZ* heterozygous, *MCP* membrane cofactor protein, *MTHFR* methylenetetrahydrofolate reductase gene/protein product, *n* number, *N* no, *NT* not treated, *Pt* patient, *UC* ulcerative colitis, *Y* yes. *8, 9 describe same case. ** IBD is only stated, it is not otherwise specified whether UC/CD.***Heterozygous MTHFR mutations are not generally thought to cause aHUS, therefore a known causative mutation was not found in Webb et.al. [7]. ****Early on in Case 1 a borderline positive C3 nephritic factor antibody was found and complement testing revealed a high level of soluble membrane attack complex

Discussion and conclusions

We report a series of two patients with two variants of IBD presented with a very rare association with aHUS. The pathophysiology of aberrant complement activation does not require a specific trigger, and like rheumatological diseases, gastroenterological autoimmune disease like IBD can also trigger complement activation in susceptible individuals [1]. Interestingly, both patients had poorly controlled IBD at the time of aHUS diagnosis. The patients' IBD (CD case 1, UC case 2), was eventually well controlled with intensive IBD therapy, alongside complement blockade for aHUS.

The lack of finding a complement mutation in patient 1 who had a relapsing course is suggestive of either: 1) a new complement mutation that has not been previously reported; 2) background gene effect. The two known classes of complement mutations are complement factor mutations resulting in auto-activation or increased activation of complement (factor B, factor I), and loss of function mutations resulting in decreased ability of self-cells to disarm complement proteins (like membrane co-factor protein, etc.) [11]. Other complement proteins interact with the complement cascade that when mutated can also affect complement function. For example, vitronectin is the newest genetic defect described that has been linked to aHUS [11, 12].

In addition to four other published case, we report the fifth and sixth cases of IBD associated aHUS successfully treated with eculizumab. Table 1 summarizes previous case reports in the literature and our current cases that highlight the infrequent association between IBD (CD and UC) and aHUS [6–10].

Abbreviations

6MP: Six mercaptopurine; ADAMTS13: A Disintegrin and Metalloproteinase Thrombospondin Motif # 13, member 1; aHUS: Atypical Hemolytic Uremic Syndrome; ANA: Anti Nuclear Antibody; anti APL Ab: Anti Phospholipid Antibody; anti CL: Anti Cardiolipin; aPTT: Activated Partial Thromboplastin Time; CAPS: Catastrophic Antiphospholipid Antibody Syndrome; CD: Crohn's Disease; Cr: Creatinine; dL: Deciliter; DRVVT: Dilute Russell Viper Venom Time Assay; dsDNA: Anti Double Stranded DNA; IBD: Inflammatory bowel disease; INR : International Normalized Ratio; L: Liter; LDH: Lactate Dehydrogenase; mg: Milligram; microliter: uL; mL: Milliliter; ng: Nanogram; PT: Prothrombin Time; RBC: Red blood cell; RNP-smith,: Anti Ribonucleic protein Smith; TMA: Thrombotic Microangiopathy; TTP.: Thrombotic Thrombocytopenic Purpura; UC: Ulcerative Colitis

Acknowledgements

IK is supported in part by funds from the NIH (R01-DK077162), the Allan Smidt Charitable Fund, the Factor Family Foundation, and the Ralph Block Family Foundation.

Authors' contributions

RMH, directed manuscript writing as first author. NM, assisted in writing case 2 and revised text of manuscript. RB, assisted in the background and with case 2. LA, assisted in writing case 1. US, helped with case 1 and figures. BY, edited case with oversight of gastrointestinal portions of introduction and conclusion. PB, assisted in making figures. MF, edited text and compiled references. IK, edited text of manuscript as senior author. All authors read and approved the final manuscript.

Authors' information

Ramy M, Hanna MD FASN FACP Is an assistant Professor and nephrology consultant at UCLA David Geffen School of Medicine with expertise in complement mediated renal disorders.
Noah Merin MD PhD is an assistant professor and hematology / oncology consultant at Cedars Sinai Medical Center with expertise in complement mediated diseases.
Richard Burwick MD MPH is an associate professor and maternal fetal medicine consultant at Cedars Sinai Medical Center with key expertise in pre-eclampsia and complement mediated disorders during pregnancy.
Ira Kurtz MD is a long term-NIH funded investigator, and the division chief of nephrology at UCLA.

Author details

[1]Department of Medicine, Division of Nephrology, UCLA David Geffen School of Medicine, Room 7-155, Factor Bldg. 700 Tiverton Ave, Los Angeles, CA 90095, USA. [2]Department of Medicine, Division of Nephrology, University of California Irvine School of Medicine, Irvine CA, USA. [3]Department of Medicine, Division of Hematology Oncology, Cedars Sinai Medical Center, Los Angeles, USA. [4]Department of Obstetrics and Gynecology, Division of Maternal Fetal Medicine, Cedars Sinai Medical Center, Los Angeles, USA. [5]Department of Medicine, Division of Digestive Diseases, UCLA David Geffen School of Medicine, Los Angeles, USA. [6]Department of Medicine, UCLA David Geffen School of Medicine, Los Angeles, USA. [7]UCLA Brain Research Institute, Los Angeles, USA.

References

1. Hanna R, Marina B, Vandross A et.al. Atypical hemolytic uremic syndrome and complement blockade: established and emerging uses of complement activation. Curr Opin Nephrol Hypertens 2019, 28:000–000 DOI: https://doi.org/10.1097/MHN.0000000000000499.
2. Park MHCN, Ulmer S, Weitz IC. Complement-mediated thrombotic microangiopathy associated with lupus nephritis. Blood Advances. 2018;2:2090–4.
3. Park MHWI. Effective treatment of thrombotic Microangiopathy associated with lupus nephritis with Eculizumab: a series of 8 cases. Blood. 2017;130:1043.
4. Hanna RMHM, Larson B, Lopez EA, Wilson J, Hendifar A. Thrombotic Microangiopathy due to catastrophic antiphospholipid antibody syndrome confirmed on skin biopsy and treated with. Eculizumab J Onco Nephrol. 2017;1:2.
5. Bianchi L, Gaiani F, Vincenzi F, et.al. Hemolytic uremic syndrome: differential diagnosis with the onset of inflammatory bowel diseases. Acta Biomed 2018;89(9-S):153–157. (Review).
6. Beers K, Zarrabi K, Yezina N, et al. Atypical HUS in a patient with inflammatory bowel disease. Am J Kidney Dis. 2017;69(4):A1–A105 Am J Kidney Dis. 2017;69(4): A1-A105.
7. Webb TN, Griffiths H, Miyashita Y. Atypical hemolytic uremic syndrome and chronic ulcerative colitis treated with Eculizumab. Int J Med Pharm Case Reports Int J Med Pharm Case Reports. 2015;4(5):105–12.
8. Viada Bris JF, Velasco Rodríguez-Belvís M, de Lucas Collantes C. Et.al. Ulcerative colitis and atypical hemolytic-uremic syndrome: an unusual but potentially life-threating life complication. Inflamm Bowel Dis. 2018 Jun 20.
9. Viada Bris JF, Velasco Rodríguez-Belvís M, de Lucas Collantes C, et.al. Ulcerative Colitis and Atypical Hemolytic-Uremic Syndrome: An Unusual But Potentially Life-threatening Complication Inflamm Bowel Dis • Volume 00, Number 00, Month 2018 (letter to the editor).
10. Green H, Harari E, Davidovits M, et al. Atypical HUS due to factor H antibodies in an adult patient successfully treated with eculizumab. Ren Fail. 2014;36(7):1119–21.
11. Rodriguez de Cordoba S, Hidalgo MS, Pinto S. Et.al. Genetics of atypical hemolytic uremic syndrome (aHUS). Semin Thromb Hemost. 2014;40(4):422–30.
12. Bu F, Zhang Y, Wang K, et.al. Genetic analysis of 400 patients refines understanding and implicates a new gene in atypical hemolytic uremic syndrome. J Am Soc Nephrol 2018;29(12):2809–2819.

19

Genotypic and phenotypic character of Chinese neonates with congenital protein C deficiency

Xiaoying Li[1], Xiaoyan Li[2], Xiao Li[1], Yuanhua Zhuang[1], Lili Kang[1] and Xiuli Ju[3*]

Abstract

Background: Our objective was to study the phenotype of and molecular genetic mechanisms underlying congenital protein C (PC) deficiency in Chinese neonates. We report the case of a neonate who presented 4 h after birth with purpura fulminans of the skin and thrombosis in the kidney. We also carried out a through literature review to study the genotype and phenotype, relevance, diagnosis, management, and prognosis of neonates with congenital PC deficiency in China.

Case presentation and literature review: Following a septic work-up and check of PC and protein S (PS) levels that showed PC deficiency, we investigated the patient's and her parents' genotypes. Our patient was found to have a plasma PC level of 0.8%. Molecular testing revealed a compound heterozygous mutation of the *PROC* gene: From the father, a c._262 G > T p. ASP88Tyr mutation in exon 4; from the mother, a C. 400 + 5G mutation in intron 5 that had been previously reported as likely pathogenic. Both parents were found to have heterozygous mutations for PC deficiency. In China, 5 other cases of congenital PC deficiency in the neonatal period were reported in the literature. In those cases, purpura fulminans and thrombosis were the main symptoms, and homozygous or compound heterozygous mutations of the *PROC* gene were identified.

Conclusion: Congenital PC deficiency should be ruled out for neonates presenting with purpura fulminans and thrombosis.

Keywords: Neonate, *PROC* gene, Protein C, Purpura fulminans, Thrombosis

Background

Congenital protein C (PC) deficiency is a rare genetic disorder associated with an elevated risk of venous thromboembolism (VTE) and an incidence of 1/40,000–1/250,000 [1–3]. Patients with a mild form of this disorder may be asymptomatic but remain susceptible to VTE at any age [4]. Severe congenital PC deficiency is a life-threatening coagulopathy associated with significant clinical signs of thrombophilia, including purpura fulminans and substantial skin necrosis during the early neonatal period. Protein C is encoded by the *PROC* gene on chromosome 2q13–q14, which consists of nine exons (1790 bp) [5]. Congenital PC deficiency is typically inherited in an autosomal dominant manner from heterozygous carriers. Severe PC deficiency has been associated with homozygosity or compound heterozygosity for *PROC* gene mutations that affect the normal sequence of mRNA and the protein synthesis [4]. Here, we report a neonate who presented with purpura fulminans on the first day of life. We discuss the clinical phenotype and the localization of the genetic mutations. We also present a comprehensive review of all cases of neonatal congenital PC deficiency reported in China during the past 40 years.

* Correspondence: jxlqlyy@163.com
[3]Qilu Hospital of Shandong University, No107, Cultural west Road, Lixia District, Jinan, Shandong, China
Full list of author information is available at the end of the article

Case presentation

A one-day-old female neonate was transferred to our neonatal intensive care unit from a local hospital with a main complaint of progressive necrotic lesions from toe to heel on the right foot beginning at four hours after birth (Fig. 1). The patient was a G2P2, small-for-gestational-age neonate with birth weight 2.36 Kg and gestational age 38 weeks. She had a complicated perinatal history of fetal distress and premature rupture of membranes (PROM) with blood-stained amniotic fluid. Before birth, fetal ultrasound revealed a left kidney enlargement. At birth, the patient experienced an episode of asphyxia and required intensive resuscitation. She had an Apgar score of 1′-4, 5′-7. The patient's physical examination was generally unremarkable except for the black, necrotic skin lesions covering the right foot. The maternal history was unrevealing, and the mother had one other child: a healthy, four-year-old boy.

Antibiotics were started initially for suspected sepsis and vitamin K was given at the first admission day. After a clinical diagnosis of congenital PC deficiency was made, fresh frozen plasma (FFP) was given at 10 mL/kg per day following injection of low-molecular-weight heparin (LMWH). A laboratory work-up was carried out and PC activity levels were measured with a chromogenic assay. Results showed the following: red blood cell count (RBC): 3.3×10^{12}/L; hemoglobin (Hb): 11.2 g/dL; Hct: 32.7%; MCV: 99.2 fl; WBC: 13.00×10^{9}/L; platelet: 299×10^{9}/L; PT: 17 s; PTT: 49.1 s; PT-INR: 1.38; D-dimer: 24.04 mg/dL; fibrinogen: 1.29 mg/dL. The patient's plasma PC activity level was severely low at 0.8% (reference range 70–140%). Protein S (PS) level was 90.7% (reference range 60–130%), and lupus anticoagulant (LA) factor ratio LA1/LA2 was 0.96 (reference range 0.8–1.2). An ultrasound showed a left kidney infarction. Both the patient's parents were found to have low PC activity (mother: 48.7%; father: 64.7%) and normal PS activity (mother: 63.6%; father: 80.7%) (Table 1). We carried out a genetic evaluation of our patient and both her parents. This investigation revealed two heterozygous mutations of the *PROC* gene (NM-000312.3) inherited from the father and mother individually. From the father, a c._262 G > T mutation in exon 4 that had previously been reported as pathogenic. From the mother, a c.400 + 5G > A micromutation in intron 5 that we believe was pathogenic and caused the low PC activity level in the mother, and had previously been reported in PC deficiency patients as likely pathogenic. Both parents were found to be heterozygous carriers for the mutation individually (Fig. 2).

Our patient remained stable during her hospitalization except for the necrotic skin lesions. Her vital signs also remained stable. The skin lesions on her foot resolved under the FFP and LMWH SC treatment. The patient was not treated with activated PC concentrate because this product is currently not available in China. She experienced another two episodes of purpura fulminans on her left arm and abdomen wall in the hospital. Repeat assessments showed her plasma PC activity level remained stable at 0.8% with no improvement. Our patient's total hospital admission time was 50 days. During this time, her skin lesions resolved. Subsequently, her enlarged kidney reduced in size and she recovered from the left kidney infarction. Our patient passed away two months after presentation.

Discussion and conclusions

Protein C is a vitamin K-dependent plasma glycoprotein zymogen synthesized in hepatocytes [6] and activated by the thrombin-thrombomodulin complex. It exerts anticoagulant activity through inactivation of Factor V and VIII via cleavage of bonds of the active forms of these factors [6, 7]. Protein S, Factor V, calcium, and phospholipids are co-factors involved during this process [8]. Low plasma PC levels increase the risk of thromboembolic events. Congenital PC deficiency is a rare cause

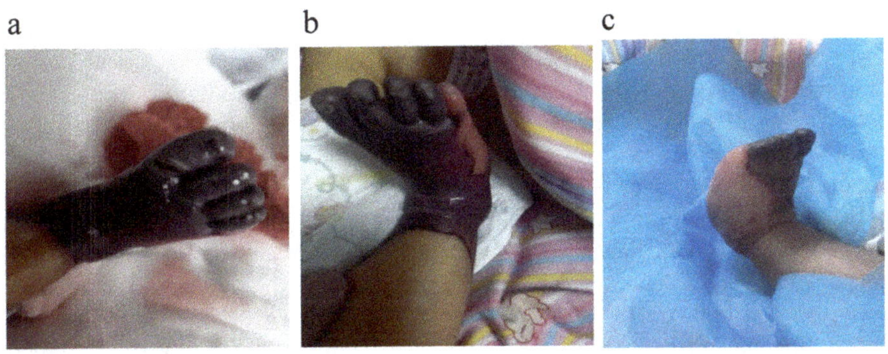

Fig. 1 Phenotype of a One-day-old Female Presenting with Purpura Fulminans. A one-day-old female neonate was transferred to our neonatal intensive care unit from a local hospital presenting with purpura fulminans that began four hours after birth. Note the progressive necrotic lesions from toe to heel on the right foot (**a**–**c**)

Table 1 Selected Laboratory Results for the Patient and Parents

Measure (units) (Reference range)	PC (%) (70–140%)	PS (%) (60–130%)	PLT ($\times 10^9$/L)	D-dimer (mg/L) (0-1 mg/L)	Fibrinogen (g/L) (2-4 g/L)
Patient	0.8	90.7	299	24.04	1.29
Mother	48.7	63.6	287	0.5	3.2
Father	64.7	80.7	319	0.4	3.5

Abbreviations: *PC* protein C, *PLT* platelet count, *PS* protein S

of hereditary thrombophilia and is implicated in only about 3–5% of all thromboembolic events [1]. In the adult Asian population, studies from Taiwan, Hong Kong, Japan, and Thailand concluded that thrombophilia is associated with PC system dysfunction via various rare mutations [9–12]. Recently, Tang L et al. completed a case-control study of 310 adult VTE patients showing that the most common genetic risk factor for VTE in the Chinese population, with a genetic prevalence of approximately 2.36%, is a PROC c.574_576 del variant associated with both decreased PC anticoagulant activity and increased risk of VTE [13]. Thrombotic manifestations are relatively rare (less than 5%) in children with inherited thrombophilia [14]. Fewer reports are published in this area.

PC deficiency was first described in 1962 as a suspected inherited lesion identified in three siblings with similar skin lesions [5]. In 1976, Stenflo [4] first described a vitamin K-dependent anticoagulant glycoprotein synthesized by the liver as a proenzyme for PC deficiency. PC deficiency is very rare, with a frequency of 1 case in 40,000–250,000 individuals [15]. The frequency of homozygous PC deficiency is estimated at 1 in 500,000–750,000 neonates, with an equal distribution of males and females [16]. The typical clinical symptoms of PC deficiency are purpura fulminans, disseminated intravascular coagulation, and thrombotic complications or subsequent hemorrhage during the neonatal period. The plasma PC level of the neonate is physiologically low and reaches adult levels by six months of age [17]. A physiologically low vitamin K status after birth further increases the risk of mortality related to low PC activity during the neonatal period. Thus, neonates with congenital PC deficiency typically exhibit severe symptoms and high mortality rates. A search of the literature reveals that almost 30 cases of neonatal congenital PC deficiency were reported globally. In mainland China, 5 cases of neonatal congenital PC deficiency were reported during the past 40 years.

There are two types of congenital PC deficiency: type I (quantitative) is characterized by reduced PC antigenic level and function; type II (qualitative) [18] is characterized by decreased PC function relative to the antigenic level. The majority of PC deficiency cases are of type I [7].. In our patient's case, the type of deficiency (I or II) was unknown since PC activity levels, but not antigen levels, were measured. Congenital PC deficiency is encoded by the *PROC* gene on the long arm of chromosome 2q13–q14, which consists of nine exons (1790 bp). Mutations in the *PROC* gene cause PC deficiency, which is inherited in an autosomal dominant manner (leading to the heterozygous form) or an autosomal recessive manner (leading to the homozygous form). More than 360 unique mutations in the *PROC* gene have been identified to-date [7]. Most of these are missense or nonsense mutations that cause either quantitative PC deficiency (type I) or qualitative PC deficiency (type II). In adults, heterozygous PC deficiency is associated with a 10-fold increase in the risk of developing thromboembolic events compared to the general population [19]. Although homozygous or compound heterozygous PC deficiencies usually lead to life-threatening thrombosis and increased mortality rates in neonates [20], previous reports indicate that some homozygous PC-deficient patients may have a milder course with low but measurable PC levels [21]. One such patient who had a PC level < 10% was reported to be asymptomatic at the age of 38 years [22]. This suggests that the course of this disorder may vary in patients with relatively higher PC levels. Of note, other reports showed family histories of third-degree consanguineous marriage for some patients with PC deficiency [23, 24].

Our patient was compound heterozygous for PC deficiency, with one *PROC* gene mutation on exon 4 inherited from the father and the other mutation on intron 5 inherited from the mother. The compound heterozygous *PROC* exon gene mutation caused the low plasma PC level of 0.8%. Her high D-dimer level and low fibrinogen level revealed thrombosis was ongoing. The severe clinical consequences of the patient's extremely low PC level (0.8%) caused thrombosis in the left kidney during the fetal period and purpura fulminans with hallmark necrotic skin lesion soon after birth. Her parents both had differing low levels of PC and normal levels of PS, Di-dimer, and fibrinogen without clinical symptoms. We hypothesize that both the severely low fetal PC level and the moderately low maternal PC level related to the maternal heterozygous *PROC* gene intron mutation may have resulted in the PROM and blood-stained amniotic fluid recorded in the perinatal history by causing abnormal coagulation leading to thrombosis and hemorrhage in the placenta.

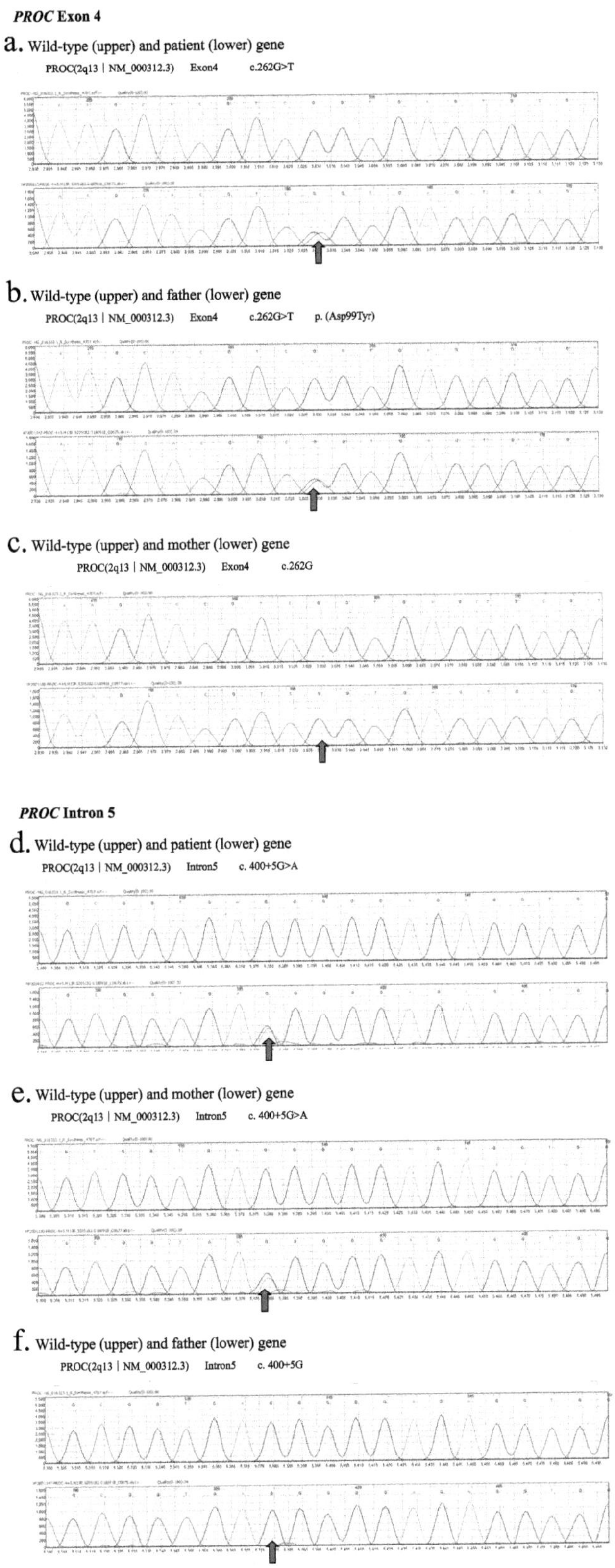

Fig. 2 (See legend on next page.)

(See figure on previous page.)
Fig. 2 Sequence Diagrams for the Patient and Parents. A genetic evaluation revealed the patient had compound heterozygous mutations of the *PROC* gene (NM-000312.3) inherited from the father and mother individually. From the father, a c._262 G > T mutation in exon 4 (red arrows in **a**–**c**). From the mother, a c.400 + 5G > A micromutation in intron 5 (red arrows in **d**–**f**). Both parents were heterozygous carriers for the mutation individually

Table 2 presents a summary of key baseline, clinical, physiological (PC), and genetic characteristics of the 6 neonatal congenital PC deficiency cases reported in China during the past 40 years [8, 25–27]. These patients include our current case and 5 others reported in the literature. Among reported cases, the ratio of males:females was even at 3:3, even though some reports suggest morbidity is higher males than in females. One patient was late preterm and 5 were term at birth, including 1 intrauterine growth restriction. All showed typical symptoms within 2 days after birth. All had significant symptoms that included purpura fulminans, cranial thrombosis (6 patients), eye injures (2 patients), intracranial hemorrhage (2 patients), or renal thrombosis (2 patients). The earliest-reported case was from 1988. That case was based on a clinical diagnosis without plasma PC level or genotype. In all 5 cases with PC levels available, patients had PC levels ranging from < 1–22% (normal range: 60% or 70–140%). Four cases also had severely low PC levels ≤8%. The lowest PC level was 0.8% (in our patient), and one case had a PC level of 22% (almost 1/3 of normal) and presented with skin lesions. M. Ichiyama and colleagues reported a Japanese deep vein thrombosis patient with a similar PC level of 21% who presented with fetal hydrocephalus and neonatal stroke [28]. The authors detected a heterozygous c.574_576delAAG, p.Lys193del mutation in exon 7 of the *PROC* gene. The 3-base deletion raises the thromboembolic risk in Japanese and Chinese populations [13]. Two cases reported before 2016 were diagnosed based on PC level alone (without genotype data). Of the 3 cases with genotype available, all had compound heterozygous *PROC* mutations. Where PC levels of the parents were measured, all were mildly lower than the normal range and not symptomatic.

For the parents of neonates with congenital PC deficiency, molecular diagnosis offers an option for a reliable prenatal diagnosis during the next pregnancy. Early diagnosis can be especially important considering the poor prognosis for this serious condition and the associated need for multidisciplinary care to address the ophthalmological and neurological sequelae in affected individuals.

The suggested treatment approach for PC deficiency is PC replacement using FFP, activated PC concentrate, or human PC concentrate. After bleeding is controlled, anticoagulants can be initiated; LMWH is recommended. The American College of Chest Physicians (ACCP) guidelines for antithrombotic therapy in symptomatic neonates and children recommend treatment with either 10 to 20 mL/kg FFP every 12 h or PC concentrates at 20 to 60 IU/kg until resolution of clinical lesions. Prophylactic treatment is initiated after stabilization of the clinical symptoms. PC concentrate may be individually reduced to a prophylactic level: Dose regimens of 24 to 90 IU/kg once a day, 250 to 350 IU/kg every other day, or 90 IU/kg three times a week have been reported [19]. A curative therapy for severe congenital PC deficiency is liver transplantation. We treated our patient initially with FFP replacement, and then with

Table 2 Summary of Clinical Characteristics and Laboratory Results for Neonatal Congenital Protein C Deficiency Cases Reported during the Past 40 Years in China

Case	Year	Sex	Age	GA (wk)	BW	Clinical Symptom			PC Level (%) (ref: 70~140%)			Mutation
						Skin	Thromb	Hemorrhage	Patient	Mother	Father	
1	1988	male	19 h	term	–	yes	–	cranial	no	no	no	not investigated
2	2012	male	10 h	39	–	yes	–	no	22	61	53	not investigated
3	2015	female	38 h	35^{+6}	2300	yes	cerebral eyes	cranial	8.0[a]	51[a]	60[a]	not investigated
4	2015	female	1 day	39^{+6}	2780	yes	cerebral	no	< 1	49	53	c.755C > T(P. 252.A > V)[c] intron5 + 5G > A[d]
5	2017	male	1 day	term	–	yes	yes	no	4	43	50	exon8(c.795_796insA)[c] exon9(c.1206_1207insG)[d]
6[b]	2018	female	43 h	34	2300	yes	yes	no	0.8	48.7	64.7	exon4 c.262G > T p. (Asp99Tyr)[c] intron5c.400 + 5G > A P?[d]

[a]Reference range 60–140%
[b] Denotes our current case
[c] Mutation inherited from father
[d]Mutation inherited from mother
Abbreviations: *BW* birth weight, *GA* gestational age, *PC* protein C, *ref.,* reference range, *Thromb* thrombosis, *wk.* weeks

LMWH by subcutaneous injection. We did not give activated PC concentrate because the PC product is currently not available in China. At discharge, she was receiving only LMWH injections.

For neonates with PC deficiency, the prognosis is poor, and the mortality rate is higher than for adults with the condition. Neonates who survive infancy are at risk for poor neurological outcomes due to clinical symptoms such as intracranial thrombosis and hemorrhage. In all 6 congenital PC deficiency cases reported in China, the infants received no comprehensive follow-up care and subsequently died. In conclusion, congenital PC deficiency, although rare, should be considered in neonates presenting with purpura fulminans and thrombosis. Early diagnosis and replacement of PC deficiency may prevent or reduce the mortality and severe morbidities typically associated with this condition. In all cases we reviewed (our own case and the other 5 cases reported in China), the neonates presented with extremely low PC levels and severe clinical signs, and experienced early death. We identified that compound heterozygous *PROC* gene mutations inherited separately from the father and mother can be responsible for the severe phenotypes of these patients. We recommend further family pedigree study be done for the family health consultation when a neonate presents with possible PC deficiency.

Abbreviations

ACCP: American College of Chest Physicians; FFP: Fresh frozen plasma; LMWH: Low-molecular-weight heparin; PC: Protein C; PROM: Premature rupture of membranes; PS: Protein S; VTE: Venous thromboembolism

Acknowledgements

The authors thank Heather McDonald Kinkaid, PhD, from the Maternal-Infant Care Research Centre at Mount Sinai Hospital, Toronto, Ontario, Canada for editorial assistance in the preparation of this manuscript.

Authors' contributions

Xiaoying Li: designing the study and drafting the manuscript. Xiaoyan Li: acquisition, analysis, and interpretation of the data and substantively revising the work. Xiao Li: acquisition, analysis, and interpretation of the data. YZ: acquisition, analysis, and interpretation of the data. LK: acquisition, analysis, and interpretation of the data. XJ: contributing to the study concept and design. All authors read and approved the final manuscript as submitted and agreed to be accountable for both their own contributions and the accuracy and integrity of the entire work.

Author details

[1]Qilu Children's Hospital of Shandong University, Jinan, Shandong, China. [2]People's Hospital of Rongcheng, Weihai, Shandong, China. [3]Qilu Hospital of Shandong University, No107, Cultural west Road, Lixia District, Jinan, Shandong, China.

References

1. Seligsohn U, Lubetsky A. Genetic susceptibility to venous thrombosis. N Engl J Med. 2001;344:1222–31.
2. Williams MD, Chalmers EA, Gibson BE. Haemostasis and thrombosis task force, British Committee for Standards in Haematology. The investigation and management of neonatal haemostasis and thrombosis. Br J Haematol. 2002;119:295–309.
3. Alhenc-Gelas M, Gandrille S, Aubry ML, Aiach M. Thirty three novel mutations in the protein C gene. French INSERM network on molecular abnormalities responsible for protein C and protein S. Thromb Haemost. 2000;83:86–92.
4. Marlar RA, Montgomery RR, Broekmans AW. Diagnosis and treatment of homozygous protein C deficiency: report of the working party on homozygous protein C deficiency of the subcommittee on protein C and protein S, international committee on thrombosis and haemostasis. J Pediatr. 1989;114:528–34.
5. Van Der Horst RL. Purpura fulminans in a newborn baby. Arch Dis Child. 1962;37:436–41.
6. Ohga S, Kang D, Kinjo T, Ochiai M, Doi T, Ishimura M, et al. Paediatric presentation and outcome of congenital protein C deficiency in Japan. Haemophilia. 2013;19:378–84.
7. Castoldi E, Rosing J. APC resistance: biological basis and acquired influences. J Thromb Haemost. 2010;8:445–53.
8. Zhang H, Bi X, Su Z, Tu X, Wang L, Shen B. A novel compound heterozygous mutations in protein C gene causing neonatal purpura fulminans. Blood Coagul Fibrinolysis. 2018;29:216–9.
9. Shen MC, Lin JS, Tsay W. High prevalence of antithrombin III, protein C and protein S deficiency, but no factor V Leidenmutation in venous thrombophilic Chinese patients in Taiwan. Thromb Res. 1997;87:377–85.
10. Tsuda H, Hattori S, Tanabe S, Iida H, Nakahara M, Nishioka S, et al. Screening for aetiology of thrombophilia: a high prevalence of protein S abnormality. Ann Clin Biochem. 1999;36:423–32.
11. Kinoshita S, Iida H, Inoue S, Watanabe K, Kurihara M, Wada Y, et al. Protein S and protein C gene mutations in Japanese deep vein thrombosis patients. Clin Biochem. 2005;38:908–15.
12. Chumpia W, Peerapittayamongkol C, Angchaisuksiri P, Komanasin N, Muta K, Kuaha K, et al. Single nucleotide polymorphisms and haplotypes of protein C and protein S genes in the Thai population. Blood Coagul Fibrinolysis. 2006;17:13–8.
13. Tang L, Lu X, Yu JM, Wang QY, Yang R, Guo T, et al. PROC c.574_576del polymorphism: a common genetic risk factor for venous thrombosis in the Chinese population. J Thromb Haemost. 2012;10:2019–26.
14. Nowak-Gottl U, Auberger K, Gobel U, Kreuz W, Schneppenheim R, Vielhaber H, et al. Inherited defects of the protein C anticoagulant system in childhood thrombo-embolism. Eur J Pediatr. 1996;155(11):921–7.
15. Kazi GI, Siddiqui E, Tabassum S, Afzal B, Khan IQ. Neonatal purpura fulminans, a rare genetic disorder due to protein C deficiency: a case report. J Pak Med Assoc. 2018;68(3):463–5.
16. Marlar RA, Mastovich S. Hereditary protein C deficiency: a review of the genetics, clinical presentation, diagnosis and treatment. Blood Coagul Fibrinolysis. 1990;1:319–30.
17. Goldenberg NA, Manco-Johnson MJ. Protein C deficiency. Haemophilia. 2008;14:1214–21.
18. Al-Hamed MH, AlBatniji F, AlDakheel GA, El-Faraidi H, Al-Zahrani A, Al-Abbass F, et al. Molecular characterization of novel splice site mutation causing protein C deficiency. Blood Coagul Fibrinolysis. 2016;27:585–8.
19. Kroiss S, Albisetti M. Use of human protein C concentrates in the treatment of patients with severe congenital protein C deficiency. Biologics. 2010;4:51–60.
20. Kovacs KB, Pataki I, Bardos H, Fekete A, Pfliegler G, Haramura G, et al. Molecular characterization of p.Asp77Gly and the novel p nongenital protein C deficiency and venous thromboembolism. Throm Res. 2015;135:718–26.
21. Cooper PC, Hill M, Maclean RM. The phenotypic and genetic assessment of protein C deficiency. Int J Lab Hematol. 2012;34:336–46.

22. Tripodi A, Franchi F, Krachmalnicoff A, Mannucci PM. Asymptomatic homozygous protein C deficiency. Acta Haematol. 1990;83:152–5.
23. Unal S, Gumruk F, Yigit S, Tuncer M, Tavil B, Cil O, et al. A novel mutation in protein C gene (PROC) causing severe phenotype in neonatal period. Pediatr Blood Cancer. 2014;61(4):763–4.
24. Monagle K, Ignjatovic V, Hardikar W, Newall F, Monagle P. Long-term follow-up of homozygote protein C deficiency after multimodal therapy. J Pediatr Hematol Oncol. 2014;36(7):e452–5.
25. Wei Y, Dong S. A neonatal purpura fulminans due to inheritied protein C deficiency. Chinese J Pediatr. 2016;54(12):950.
26. Yang F, Li H, Chen Q. A case report of neonatal protein C deficiency with purpura fulminans. Chinese J Neonatalogy. 2015;30(3):230–1.
27. Bao Y, Shi L, Wu X, Du L. A case report of neonatal inherited protein C deficiency. Chinese J Pediatr. 2011;49(5):390–1.
28. Ichiyama M, Ohga S, et al. Fetal hydrocephalus and neonatal stroke as the first presentation of protein C deficiency. Brain Dev. 2016;38:253–6.

20

Usefulness of the APTT waveform for the diagnosis of DIC and prediction of the outcome or bleeding risk

Kei Suzuki[1], Hideo Wada[2*], Takeshi Matsumoto[3], Makoto Ikejiri[4], Kohshi Ohishi[3], Yoshiki Yamashita[5], Hiroshi Imai[1], Toshiaki Iba[6] and Naoyuki Katayama[5]

Abstract

Background: The usefulness of the activated partial thromboplastin time (APTT) waveform has been reported in hemophilia, acquired hemophilia and monitoring for anticoagulants.

Material and methods: The APTT waveform was examined in patients suspected of having disseminated intravascular coagulation (DIC) to analyze its usefulness for the diagnosis of DIC or the prediction of the outcome or bleeding risk.

Results: DIC with fibrinogen < 2 g/L was frequently associated with infectious diseases (43.3%). The heights of the first derivative peak (1stDP) and second DP (2ndDP) were extremely low in DIC, especially DIC with hypofibrinogenemia, but high in infectious patients without DIC. The peak time and width of the 1stDP and 2ndDP were prolonged in patients with DIC. The heights of the 1stDP and 2ndDP were markedly low in patients with a poor outcome or those with hemoglobin < 8.0 g/dl.

Discussion and conclusion: As bleeding type DIC was observed in infectious DIC, DIC without hypofibrinogenemia might switch to DIC with hypofibrinogenemia by the progression of DIC. The height of the 1stDP and 2ndDP is useful for the diagnosis of DIC and prediction of the bleeding risk or outcome.

Keywords: APTT waveform, Hypofibrinogenemia, Bleeding, Outcome

Introduction

Disseminated intravascular coagulation (DIC) [1, 2] is a serious and fatal disease that causes microvascular thrombosis associated with thrombocytopenia, a bleeding tendency with hyperfibrinolysis and organ failure. The basic mechanism underlying the onset for DIC is the marked activation and consumption of the coagulation system followed by the activation of secondary fibrinolysis [3]. DIC also has several clinical subtypes, including asymptomatic type, marked bleeding type, organ failure type and complication types such as thrombotic microangiopathy (TMA) [4, 5]. Marked bleeding type DIC is associated with hypofibrinogenemia and is generally observed in patients with leukemia, trauma or aneurysm, while organ failure type DIC is associated with an elevated fibrinogen level and generally observed in infectious DIC such as sepsis [5, 6]. Asymptomatic DIC is considered pre-DIC.

The activated partial thromboplastin time (APTT) was previously considered to be an end-point clotting time assay which was useful for diagnosing an intrinsic pathway of coagulation factor deficiency as hemophilia and acquired hemophilia [7, 8], antiphospholipid antibody (aPL), such as lupus anticoagulant (LA) [9], or for monitoring unfractionated heparin (UFH) treatment [10]. Optical coagulation analyzers can visualize the clotting curve, and abnormal biphasic clotting curves have been reported to be associated with DIC [11, 12].

The ACL TOP analyzer for the using APTT-synthetic phospholipids (SPs) recently showed the associated first- and second-derivative peaks (1st and 2nd DPs, respectively) [13] It has been reported that the evaluation of the 1st and 2nd DPs in the APTT is valuable for diagnosing

* Correspondence: wadahide@clin.medic.mie-u.ac.jp
[2]Departments of Molecular and Laboratory Medicine, Mie University Graduate School of Medicine, Tsu, Mie 514-8507, Japan
Full list of author information is available at the end of the article

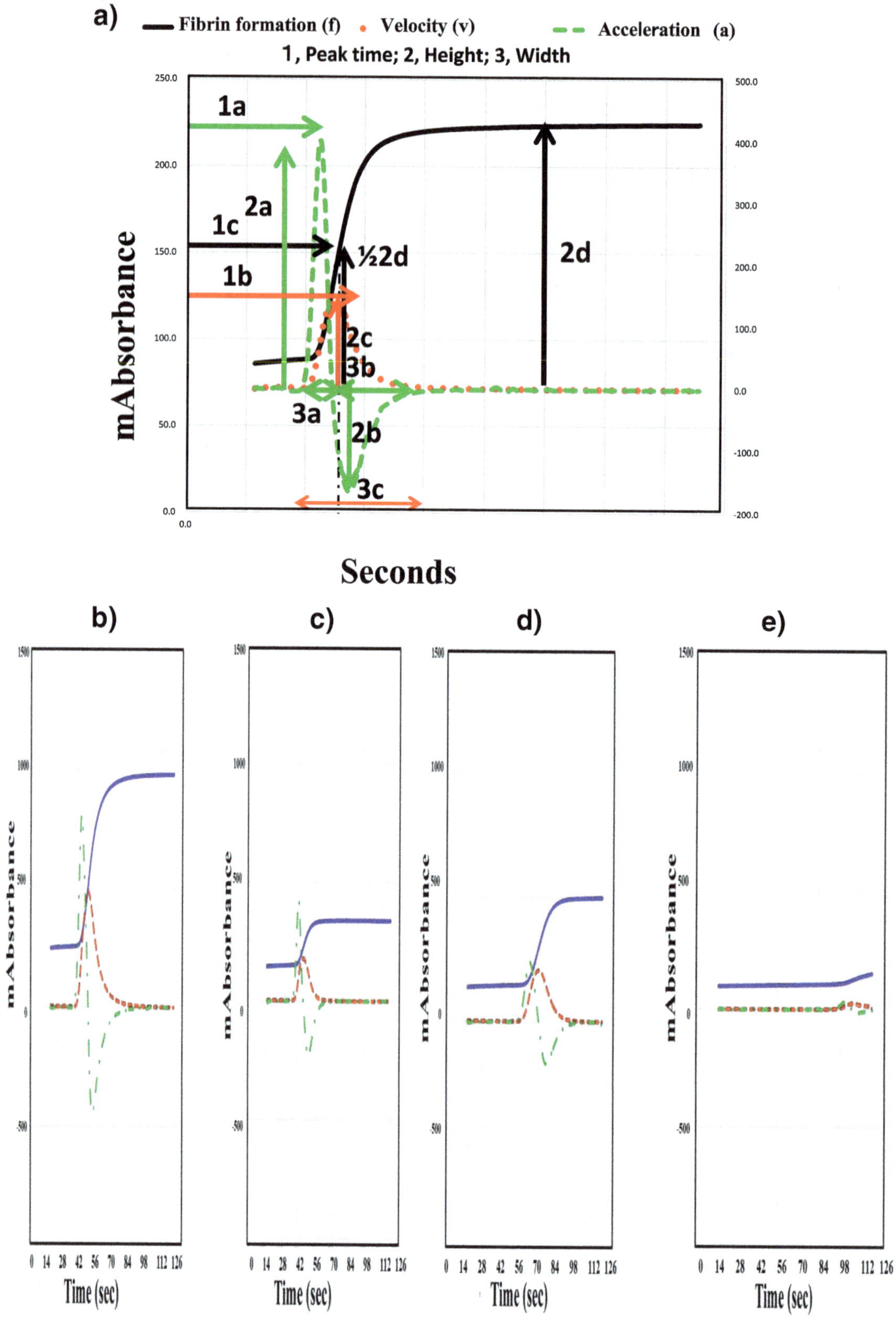

Fig. 1 The APTT waveform in a healthy volunteer (**a**), a patient without DIC (**b**), a patient with pre-DIC (**c**), a patients with DIC without hypofibrinogenemia (**d**) and a patient with DIC with hypofibrinogenemia (**e**). a, 2nd derivative; b, negative acceleration, c, 1st derivative; d, coagulation curve, 1, time for peak; 2, height of peak; 3, width of peak

Table 1 Non-DIC, Pre-DIC, DIC without hypofibrinogenemia and DIC with hypofibrinogenemia

	Non-DIC	Pre-DIC	DIC without hypofibrinogenemia	DIC with hypofibrinogenemia
Number	121	25	35	30
Age (years)	61 (37–75)	71 (67–80)*	71 (58–76)	63 (54–70)+
Sex (F:M)	60: 61	8: 17	13: 22	13: 17
Platelet counts (10^4/μl)	15.8 (9.7–26.7)	7.5 (5.0–11.4)***	4.8 (3.3–6.7)***, ##	4.4 (2.7–5.9)***, ###
PT-ratio	1.1 (1.0–1.2)	1.3 (1.1–1.5)**	1.3 (1.1–1.5)***,	1.4 (1.3–1.7)***, #,+
FDP (μg/ml)	13.6 (7.5–29.5)	28.4 (15.0–40.7) **	44.4 (23.2–71.8)***	29.9 (13.8–57.8)***
Fibrinogen (mg/dl)	443 (320–598)	365 (293–449)*	345 (219–404)**,	143 (103–169)***, ###, +++
DIC score	3.0 (2.0–4.0)	6.0 (6.0–6.0)***	7.0 (7.0–8.0)***, ###	8.0 (7.0–9.0)***, ###

DIC disseminated intravascular coagulation, *PT* prothrombin time, *FDP* fibrinogen and fibrin degradation products

*, $P < 0.05$; **, $P < 0.01$; ***, $P < 0.001$ compared with non-DIC

#, $P < 0.05$; ##, $P < 0.01$; ###, $P < 0.001$ compared with pre-DIC

+, $P < 0.05$; +++, $P < 0.001$ compared with DIC without hypofibrinogen

coagulation factor abnormalities and monitoring anti-Xa inhibitor [14, 15].

In this study, we measured and analyzed the APTT waveform in patients suspected of having DIC and examined the relationship between the bleeding risk or outcome and the parameters of the APTT waveform.

Materials and methods

Patients

A total of 211 patients suspected of having DIC, who had hemostatic abnormalities such as platelet count < 120,000/μl, fibrinogen and fibrin degradation products (FDP) > 10.0 μg/ml, prothrombin time (PT) ratio > 1.27 or fibrinogen < 150 mg/dl, at Mie University Hospital from June 1, 2011 to December 31, 2017, were enrolled in the study. DIC was diagnosed according to the Japanese Ministry of Health, Labor and Welfare (JMHLW) diagnostic criteria for DIC [16]. A JMHLW DIC score of < 6 points, 6 points or ≧7 points was defined as non-DIC, pre-DIC and DIC, respectively. In our definition of this study, fibrinogen levels of < 2 and ≧2 g/L were considered to indicate hypofibrinogen and no hypofibrinogen, respectively.

The study protocol was approved by the Human Ethics Review Committee of the Mie University School of Medicine and a signed consent form was obtained from each subject. This study was faithfully carried out in accordance with the principles of the Declaration of Helsinki.

The routine assay for the diagnosis of DIC

PT, FDP, fibrinogen and platelet counts were measured as previously reported methods [17, 18].

The APTT waveform assay

APTT waveform assays were carried out in 30 healthy volunteers (HVs: 10 females and 20 males; median age, 21 years; 25th–75th percentile, 20–24 years) and 211 patients suspected of having DIC. The APTT was measured using the APTT-SP® including silica as an activator of FXII and synthetic phospholipids (Instrumentation Laboratory, Bedford, MA, USA) using an ACL-TOP® system (Instrumentation Laboratory). After measuring the APTT, we performed a waveform analysis based on the results of each APTT assay [15, 19]. Three types of curves are shown on the monitor of the ACL-TOP® system (Fig. 1a); a curve showing the changes in the absorbance observed while measuring the APTT, corresponding to the fibrin formation; a curve showing the first derivative of the absorbance, corresponding to the coagulation velocity; and a curve showing the second derivative of the absorbance, corresponding to the coagulation acceleration. For the waveform analysis, we first checked the presence of an abnormal curve showing a biphasic

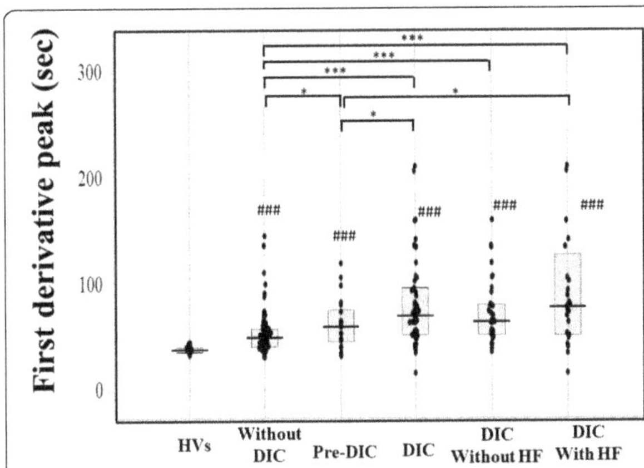

Fig. 2 Peak time of 1st derivative (b) of the APTT waveform in healthy volunteers, patients without DIC, patients with pre-DIC, patients with DIC without hypofibrinogenemia and patients with DIC with hypofibrinogenemia. HVs, healthy volunteers; DIC, disseminated intravascular coagulation; HF, hypofibrinogenemia. ###, $p < 0.001$ in comparison with HVs. ***, $p < 0.001$, **, $p < 0.01$; *, $p < 0.05$. HVs, healthy volunteers; DIC, disseminated intravascular coagulation; HF, hypofibrinogenemia. ###, $p < 0.001$ in comparison with HVs. ***, $p < 0.001$, **, $p < 0.01$; *, $p < 0.05$

Table 2 Parameters of APTT waveform in patients with and without DIC and healthy volunteers

		HV	Without DIC	Pre-DIC	DIC	Without HF	With HF
2ndDP	Time (seconds)	35.5*** (32.6–37.4)	43.9 (36.6–52.7)	54.6* (41.0–71.3)	63.7*** (47.4–84.9)	58.8*** (46.2–73.9)	72.5*** (49.7–110)
	Height (mabs)	457** (411–535)	648 (379–973))	460* (191–772)	151*** (83.2–326)	286*** (144–427)	107*** (26.5–148)
	NH (mabs)	234** (170–255)	316 (164–499)	176** (93.2–308)	66.4*** (39.4–150)	134*** (61.6–188)	55.0*** (14.5–64.5)
	Width (sec)	7.45*** (6.80–8.00)	12.9 (11.2–16.8)	17.7** (12.5–24.3)	15.1* (12.4–20.5)	15.1* (12.4–20.5)	21.4*** (12.7–38.2)
1stDP	Time (sec)	38.7*** (36.3–40.8)	50.2 (41.2–58.6)	60.5* (45.8–76.4)	70.7*** (52.1–96.8)	64.9*** (52.2–81.0)	78.6*** (51.8–128.2)
	Height (mabs)	183*** (144–200))	389 (241–544)	251** (154–358)	103*** (64.2–208)	200*** (116–257)	61.2*** (36.8–93.9)
	Width (sec)	19.5*** (18.0–20.3)	65,2 (54.8–72.1)	65.2 (54.0–195)	66.5 (56.0–200)	64.6 (57.6–200)	75.1 (55.0–200)
1/2FFP	Time (sec)	40.9*** (37.8–42.7)	52.5 (43.1–61.5)	61.6* (47.1–78.9)	75.0*** (54.8–99.3)	69.1*** (54.8–87.9)	80.9*** (54.7–142)
	Height (mabs)	242*** (207–305)	462 (338–586)	328* (271–490)	265*** (178–371)	319*** (240–399)	195*** (170–286)

2ndDP, second derivative peak; 1stDP, first derivative peak, 1/2FFP, fibrin formation peak; NH, negative height; DIC, disseminated intravascular coagulation; HF, hypofibrinogenemia
Data was shown as median (25–75%tile)
***, $p < 0.001$, **, $p < 0.01$ or: *$p < 0.05$ compared with patients without DIC

waveform on the 1stDP and/or 2ndDP. Furthermore, as shown in Fig. 1a, we calculated the following 10 parameters on the first- or second- derivative curve by manual handling of the mouse: 1, peak time, 1stDP, 2ndDP1 and 1/2 fibrin formation (1/2FF); 2, width, 1stDP, 2ndDP1 and 2ndDP2; 3, height; 1stDP, 2ndDP1, 2ndDP2 and 1/2FF.

Statistical analyses

The data are expressed as the medians and 25th–75th percentiles. The differences in factors among HVs, non-DIC patients, patients with pre-DIC and DIC patients with and without hypofibrinogenemia between survivor and non-survivor groups or groups of Hb < 8.0 and ≥8.0 g/L were examined using the Mann-Whitney U-test. A p-value of < 0.05 was considered to be statistically significant. All statistical analyses were performed using the Stat Flex, version 6, software package (Artec Co. Ltd., Osaka, Japan).

Results

The underlying diseases of patients suspected of having DIC was sepsis in 58 (causes for sepsis; pneumonia 30, peritonitis 7, pyelonephritis 5, pleurisy 5, catheter infection 3, necrotizing fasciitis 3, meningitis 2 and others 3), pneumonia in 43, trauma in 41, aneurysm in 15 and hemangioma, living donor liver transplantation in 13, hematopoietic malignancy in 10, cardiopulmonary arrest 10, gynecological complication in 5, autoimmune disease in 4, solid cancer 3 and other infection in 9. Of the 211 patients suspected of having DIC, 65 were diagnosed with DIC (DIC score ≥7), and 25 were diagnosed with pre-DIC (DIC score = 6). Thirty patients showed low levels of fibrinogen (< 2.0 g/L, DIC with hypofibrinogenemia) among 65 patients with DIC (Table 1). Figure 1 shows the APTT waveform in patients without DIC and those with pre-DIC and DIC with and without hypofibrinogenemia. Infectious diseases such as sepsis and pneumonia were 28/35 (80.0%) in DIC without hypofibrinogenemia and 13/30 (43.3%) in DIC with hypofibrinogenemia. The 1stDP and 2ndDP was high in patient without DIC and the width of the 1stDP and 2ndDP were enlarged in patients suspected of

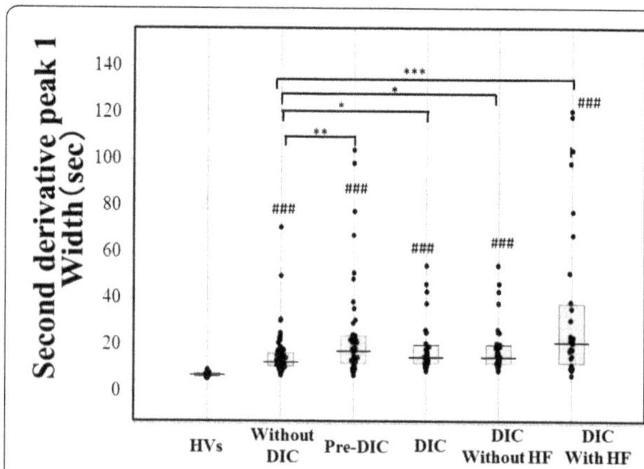

Fig. 3 Width of the 2nd derivative-1 of the APTT waveform in healthy volunteers, patients without DIC, patients with pre-DIC, patients with DIC without hypofibrinogenemia and patients with DIC with hypofibrinogenemia. HVs, healthy volunteers; DIC, disseminated intravascular coagulation; HF, hypofibrinogenemia. ###, $p < 0.001$ in comparison with HVs. ***, $p < 0.001$, **, $p < 0.01$; *, $p < 0.05$

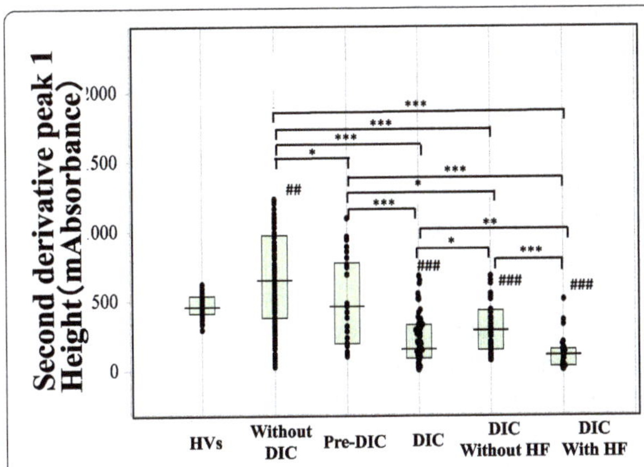

Fig. 4 Height of the 2nd derivative-1 of the APTT waveform in healthy volunteers, patients without DIC, patients with pre-DIC, patients with DIC without hypofibrinogenemia and patients with DIC with hypofibrinogenemia. HVs, healthy volunteers; DIC, disseminated intravascular coagulation; HF, hypofibrinogenemia. ###, $p < 0.001$, ##, $p < 0.01$ in comparison with HVs. ***, $p < 0.001$, **, $p < 0.01$; *, $p < 0.05$

having DIC. The height of the 1[st]DP and 2ndDP was reduced in DIC patients, especially in DIC patients with hypofibrinogenemia.

The biphasic pattern of 2ndDP was often observed in patients with DIC. The peak time of the 1[st]DP and 2ndDP and 1/2FF were significantly prolonged in patients suspected of having DIC in comparison with HVs ($p < 0.001$) and those were significantly longer in patients with pre-DIC ($p < 0.05$) and in patients with DIC ($p < 0.001$) than in patients without DIC (Fig. 2 and Table 2). There were no significant differences in the peak time of the 1[st]DP and 2ndDP and 1/2FF between DIC patients with and without hypofibrinogenemia. The widths of the 1[st]DP, 2ndDP1 and 2ndDP2 were significantly longer in patients suspected of having DIC ($p < 0.001$) than in HVs (Fig. 3). The width of the 2[nd]DP1 was larger in patients with pre-DIC ($p < 0.01$) and DIC ($p < 0.05$) than in HVs and that of the 1stDP was longer in patients with DIC than in patients without DIC. There were no significant differences in the 2[nd]DP2 among patients suspected of having DIC. The heights of the 2[nd]DP1 and 2[nd]DP2 were significantly higher in patients without DIC and significantly lower in those with DIC than in HVs (Fig. 4). The height of the 1stDP was significantly higher in patients without DIC and in those with DIC and significantly lower in patients

Table 3 The Correlation between the DIC score and the parameters of the APTT waveform

Parameters	r	P	
2nd derivative peak-1 time	0.2768	$P < 0.001$	Y = 4.74 X + 36.57
2nd derivative peak-1 wide	0.1870	$P < 0.01$	Y = 1.44 X + 12.21
2nd derivative peak-1 height	−0.5342	$P < 0.001$	Y = − 94.56 X + 1002.41
2nd derivative peak-2 wide	0.2991	$P < 0.001$	Y = 6.07 X + 43.55
2nd derivative peak-2 height	−0.5605	$P < 0.001$	Y = −51.87X + 514.24
1st derivative peak time	0.2950	$P < 0.001$	Y = 5.05 X + 40.83
1st derivative peak height	−0.5714	$P < 0.001$	Y = −50.39 X + 553.33
1st derivative peak wide	0.2944	$P < 0.0001$	Y = 7.98 X + 54.30
1/2 fibrin formation time	0.3098	$P < 0.0001$	Y = 5.89 X + 40.64
1/2 fibrin formation height	−0.3593	$P < 0.0001$	Y = −30.37 X + 563.74

DIC disseminated intravascular coagulation, *APTT* activated partial thromboplastin time

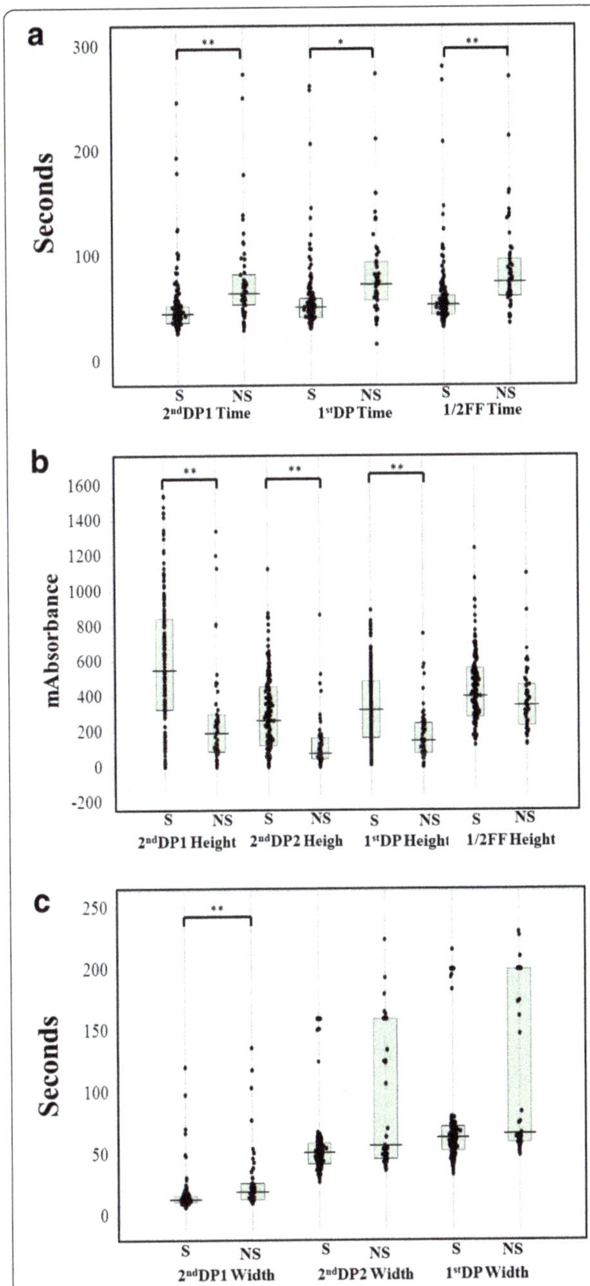

Fig. 5 Relationships between the outcome and APTT waveform. Peak time (**a**), peak height (**b**) and peak width (**c**) of the APTT waveform. APTT, activated partial thromboplastine time; DP, derivative peak; FF, fibrin formation; S, survivor; NS, non-survivor. ***, $p < 0.001$, **, $p < 0.01$; *, $p < 0.05$

with DIC than in HVs. The height of the 1stDP, 2^nd^DP and 1/2FF were significantly lower ($p < 0.001$) in patients with pre-DIC and those with DIC than in those without DIC. The height of the 1stDP, 2ndDP1 and 2ndDP2 were significantly lower in DIC patients with hypofibrinogenemia than in those without hypofibrinogenemia. The height of the 1stDP and 2ndDP1 and 2ndDP2 were well correlated with DIC score in comparison with the time of the 1stDP and 2ndDP (Table 3). Fifty five patients died within 28 days form the APTT wave analysis. The time of 2ndDP1 ($p < 0.01$), 1stDP ($p < 0.05$) and 1/2FF ($p < 0.01$) was significantly longer in non-survivors ($n = 55$) than in survivors ($n = 158$) (Fig. 5a). The height of 2ndDP1 and 2ndDP2, and 1stDP was significantly lower ($p < 0.01$, respectively) in non-survivors than in survivors (Fig. 5b). The width of the 2ndDP1 was significantly larger ($p < 0.01$) in non-survivors than in survivors (Fig. 5c). Although the time of the 2ndDP was significantly longer ($p < 0.05$) in patients with Hb < 8 g/dl ($n = 56$) than in those with Hb ≧8 g/dl ($n = 157$), there were no significant differences in the time of the 1stDP and 1/2FF between patients with Hb ≧8.0 g/dl and those with Hb < 8.0 g/dl (Fig. 6). The heights of the 2ndDP1, 2ndDP2 and 1stDP were significantly lower ($p < 0.01$, respectively) in patients with Hb < 8.0 g/dl than in those with Hb ≧8.0 g/dl. The width of the 2ndDP1 was significantly larger ($p < 0.05$) in those with Hb < 8.0 g/dl than in those with Hb ≧8.0 g/dl.

Discussion

The APTT waveform analysis indicated that the heights of the 1stDP and 2ndDP were increased in infectious patient without DIC, and that after the onset of DIC, the widths of the 1stDP and 2ndDP were enlarged, the heights of the 1stDP and 2ndDP were reduced, and the biphasic pattern of the 2ndDP was often observed. The reduced heights of the 1stDP and 2ndDP were more significant in patients with DIC with hypofibrinogenemia (bleeding type DIC) than in those with DIC without hypofibrinogenemia. Although septic patients with DIC are usually not associated with bleeding type DIC or hypofibrinogenemia [2, 3], in this study, 43.3% of cases of DIC with fibrinogen level < 2.0 g/L were associated with infectious DIC, indicating that bleeding type DIC with hypofibrinogenemia often occurs in septic patients with DIC. Given that a shift from DIC without hypofibrinogenemia to DIC with hypofibrinogenemia was observed, DIC with hypofibrinogenemia might be a more severe type of DIC than DIC without hypofibrinogenemia. Indeed, the heights of the 1stDP, 2ndDP1 and 2ndDP2 were well correlated with the DIC score, while the heights of the 1stDP, 2ndDP1 and 2ndDP2 were not well correlated with the fibrinogen levels in cases of hemophilia [8].

The peak time of the 1stDP and 2ndDP and 1/2FF were significantly longer in patients with DIC than in those without DIC, but there were no significant differences in the peak time of the 1stDP, 2ndDP and 1/2FF between DIC patients with and without hypofibrinogenemia, suggesting that a prolonged peak time of the 1stDP and 2ndDP and 1/2FF might indicate a diagnostic ability for DIC which is similar to that of a routine APTT assay. A prolonged peak time of the 1stDP and 2ndDP and 1/2FF are therefore considered to be less useful for the diagnosis of DIC than the height of 1stDP and 2ndDP. The widths of the 1stDP, 2ndDP1and 2ndDP2 were significant larger

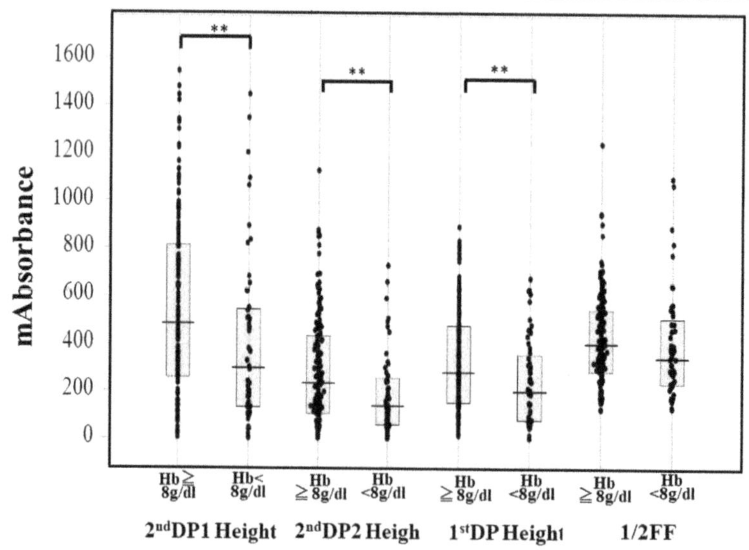

Fig. 6 Relationships between hemoglobin level and APTT waveform. The peak height of the APTT waveform. APTT, activated partial thromboplastine time; DP, derivative peak; FF, fibrin formation; S, survivor; NS, non-survivor; Hb, hemoglobin. ***, $p < 0.001$, **, $p < 0.01$; *, $p < 0.05$

in patients suspected of having DIC including both pre-DIC and DIC, than in HVs, suggesting that the width of the APTT waveform indicates the presence of underlying diseases of DIC. The heights of the 2^{nd}DP1, 2^{nd}DP2 and 1^{st}DP were significantly higher in patients without DIC than in those with DIC, suggesting that the heights of the 1^{st}DP and 2^{nd}DP were markedly high in non-DIC patients suspected of having DIC but reduced in patients with DIC, especially DIC patients with hypofibrinogenemia. As DIC with hypofibrinogenemia is considered bleeding-type DIC, a reduced 1^{st}DP and 2^{nd}DP might suggest a bleeding risk. The heights of the 2^{nd}DP1, 2^{nd}DP2 and 1^{st}DP were significantly lower in patients with Hb < 8 g/dl than in those with Hb ≥8 g/dl, suggesting that a reducing 1^{st}DP and 2^{nd}DP might indicate an increased risk for severe bleeding. In orthopedic patients treated with edoxaban, a reduced height of the 1^{st}DP was reported to be risk factor for bleeding [15]. Taken together, these findings suggest that anticoagulant therapy should begin in the Pre-DIC state, and strong anticoagulant therapy is not recommended in patients with DIC with hypofibrinogenemia. However, there was no significant difference in the time of the 1^{st}DP between the two groups. The height of the 2^{nd}DP1 was reported to be well correlated with the FVIII activity [8]. The heights of the 2^{nd}DP1, 2^{nd}DP2, and 1^{st}DP were also observed in bleeding-type DIC. The times of the 2^{nd}DP1, 1^{st}DP and 1/2FF were significantly longer; the heights of the 2^{nd}DP1, 2^{nd}DP2 and 1^{st}DP were significantly lower; and the width of the 2^{nd}DP1 was significantly larger in non-survivors than in survivors. These findings suggest that the APTT waveform might be useful for predicting the outcome.

Conclusion

As bleeding type DIC was observed in infectious DIC, DIC without hypofibrinogenemia might switch to DIC with hypofibrinogenemia by the progression of DIC. The height of the 1^{st}DP and 2^{nd}DP is useful for the diagnosis of DIC and prediction of the bleeding risk or outcome in patients with DIC.

Abbreviations

1/2FF: 1/2 fibrin formation; aPL: Antiphospholipid antibody; APTT: Activated partial thromboplastin time; DIC: Disseminated intravascular coagulation; DP: Derivative peak; FDP: Fibrinogen and fibrin degradation products; Hb: Haemoglobin; HV: Healthy volunteer; JMHLW: Japanese Ministry of Health, Labor and Welfare; LA: Lupus anticoagulant; PT: Prothrombin time; SP: Synthetic phospholipids; TMA: Thrombotic microangiopathy; UFH: Unfractionated heparin

Acknowledgements

We thank Mrs. Yumi Sakano and Mrs. Hiroko Nishii to measure the APTT.

Authors' contributions

SK, MT HI and YY obtained samples, WH, IM and OK measured and analyzed APTT waveform, and IT and KN reviewed this manuscript. All authors read and approved the final manuscript.

Author details

[1]Emergency Critical Care Center, Mie University Graduate School of Medicine, Tsu, Mie, Japan. [2]Departments of Molecular and Laboratory Medicine, Mie University Graduate School of Medicine, Tsu, Mie 514-8507, Japan. [3]Division of Blood Transfusion Medicine and Cell Therapy, Mie University Graduate School of Medicine, Tsu, Mie, Japan. [4]Central laboratory, Mie University Graduate School of Medicine, Tsu, Mie, Japan. [5]Department of Hematology and Oncology, Mie University Graduate School of Medicine, Tsu, Mie, Japan. [6]Department of Emergency and Disaster Medicine, Juntendo University Graduate School of Medicine, Tokyo, Japan.

References

1. Taylor FB Jr, Toh CH, Hoots WK, Wada H, Levi M. Scientific subcommittee on disseminated intravascular coagulation (DIC) of the international society on thrombosis and Haemostasis (ISTH): towards definition, clinical and laboratory criteria, and a scoring system for disseminated intravascular coagulation. Thromb Haemost. 2001;86:1327–30.
2. Wada H, Matsumoto T, Yamashita Y, Hatada T. Disseminated intravascular coagulation: testing and diagnosis. Clin Chim Acta. 2014;436C:130–4.
3. Wada H, Asakura H, Okamoto K, Iba T, Uchiyama T, Kawasugi K, Koga S, Mayumi T, Koike K, Gando S, Kushimoto S, Seki Y, Madoiwa S, Maruyama I, Yoshioka A. Japanese Society of Thrombosis Hemostasis/DIC subcommittee: expert consensus for the treatment of disseminated intravascular coagulation in Japan. Thromb Res. 2010;125:6–11.
4. Wada H, Matsumoto T, Yamashita Y. Natural history of thrombotic thrombocytopenic Purpura and hemolytic uremic syndrome. Semin Thromb Hemost. 2014;40:866–73.
5. Wada H. Disseminated intravascular coagulation. Clin Chim Acta. 2004; 344:13–21.
6. Wada H, Thachil J, Di Nisio M, Mathew P, Kurosawa S, Gando S, Kim HK, Nielsen JD, Dempfle CE, Levi M, Toh CH. The scientific standardization committee on DIC of the international society on thrombosis Haemostasis.: guidance for diagnosis and treatment of DIC from harmonization of the recommendations from three guidelines. J Thromb Haemost. 2013;11:761–7.
7. Matsumoto T, Nogami K, Shima M. A combined approach using global coagulation assays quickly differentiates coagulation disorders with prolonged aPTT and low levels of FVIII activity. Int J Hematol. 2017;105: 174–83.
8. Katayama H, Matsumoto T, Wada H, Fujimoto N, Toyoda J, Abe Y, Ohishi K, Yamashita Y, Ikejiri M, Habe K, Katayama N. An evaluation of hemostatic abnormalities in patients with hemophilia according to the activated partial thromboplastin time waveform. Clin Appl Thromb Hemost. 2018 Jan 1.
9. Tokutake T, Baba H, Shimada Y, Takeda W, Sato K, Hiroshima Y, et al. Exogenous magnesium chloride reduces the activated partial thromboplastin times of lupus anticoagulant-positive patients. PLoS One. 2016;11:e0157835.
10. Byun JH, Jang IS, Kim JW, Koh EH. Establishing the heparin therapeutic range using aPTT and anti-Xa measurements for monitoring unfractionated heparin therapy. Blood Res. 2016;51:171–4.
11. Toh CH, Giles AR. Waveform analysis of clotting test optical profiles in the diagnosis and management of disseminated intravascular coagulation (DIC). Clin Lab Haematol. 2002;24:321–7.
12. Matsumoto T, Wada H, Nishioka Y, Nishio M, Abe Y, Nishioka J, et al. Frequency of abnormal biphasic aPTT clot waveforms in patients with underlying disorders associated with disseminated intravascular coagulation. Clin Appl Thromb Hemost. 2006;12:185–92.
13. Solano C, Zerafa P, Bird R. A study of atypical APTT derivative curves on the ACL TOP coagulation analyser. Int J Lab Hematol. 2011;33:67–78.
14. Tokunaga N, Inoue C, Sakata T, Kagawa K, Abe M, Takamatsu N, et al. Usefulness of the second-derivative curve of activated partial thromboplastin time on the ACL-TOP coagulation analyzer for detecting factor deficiencies. Blood Coagul Fibrinolysis. 2016;27:474–6.
15. Hasegawa M, Wada H, Tone S, Yamaguchi T, Wakabayashi H, Ikejiri M, Watanabe M, Fujimoto N, Matsumoto T, Ohishi K, Yamashita Y, Katayama N, Sudo A. Monitoring of hemostatic abnormalities in major orthopedic surgery patients treated with edoxaban by APTT waveform. Int J Lab Hematol. 2018;40:49–55.

16. Kobayashi N, Maekawa T, Takada M, Tanaka H, Gonmori H. Criteria for diagnosis of DIC based on the analysis of clinical and laboratory findings in 345 DIC patients collected by the research committee on DIC in Japan. Bibl Haematol. 1983;49:265–75.
17. Takemitsu T, Wada H, Hatada T, Ohmori Y, Ishikura K, Takeda T, Sugiyama T, Yamada N, Maruyama K, Katayama N, Isaji S, Shimpo H, Kusunoki M, Nobori T. Prospective evaluation of three different diagnostic criteria for disseminated intravascular coagulation. Thromb Haemost. 2011;105:40–4.
18. Hatada T, Wada H, Kawasugi K, Okamoto K, Uchiyama T, Kushimoto S, Seki Y, Okamura T, Imai H, Kaneko T, Nobori T. Analysis of the cutoff values in fibrin-related markers for the diagnosis of overt DIC. Clin Appl Thromb Hemost. 2012;18(5):495–500.
19. Matsumoto T, Wada H, Fujimoto N, Toyoda J, Abe Y, Ohishi K, Yamashita Y, Ikejiri M, Hasegawa K, Suzuki K, Imai H, Nakatani K, Katayama N. An evaluation of the activated partial thromboplastin time waveform. Clin Appl Thromb Hemost. 2017.

21

Use of laboratory testing for prediction of postoperative bleeding volume in cardiovascular surgery

Yoshie Kawahara[1,2], Kohei Ohtsuka[1], Kimine Tanaka[1], Mayumi Yamanaka[1], Hiroyuki Kamiya[3], Takayuki Kunisawa[4] and Satoshi Fujii[1*]

Abstract

Background: Coagulopathy and following massive bleeding are complications of cardiovascular surgery, particularly occurring after procedures requiring prolonged cardiopulmonary bypass (CPB). Reliable and rapid tests for coagulopathy are desirable for guiding transfusion. Measuring multiple coagulation parameters may prove useful. The purpose of this study is to determine the laboratory parameters predicting massive bleeding.

Methods: In a prospectively collected cohort of 48 patients undergoing cardiovascular surgery, markers of coagulation and fibrinolysis were measured using automated analyzer and their correlations with bleeding volume were determined.

Results: Operation time was 318 (107–654) min. CPB time was 181 (58–501) min. Bleeding volume during surgery was 2269 (174–10,607) ml. Number of transfusion units during surgery were packed red blood cells 12 (0–30) units, fresh frozen plasma 12 (0–44) units, platelets 20 (0–60) units and intraoperative autologous blood collection 669 (0–4439) ml. Post-surgery activities of coagulation factors II (FII), FV, FVII, FVIII, FIX, FX, FXI and FXII were decreased. Values of fibrinogen, antithrombin, α2 plasmin inhibitor (α2PI) and FXIII were also decreased. Values of thrombin-antithrombin complex (TAT) were increased. Values of FII, FIX, FXI and α2PI before surgery were negatively correlated with bleeding volume (FII, $r = -0.506$: FIX, $r = -0.504$: FXI, $r = -0.580$; α2PI, $r = -0.418$). Level of FIX after surgery was negatively correlated with bleeding volume ($r = -0.445$) and level of TAT after surgery was positively correlated with bleeding volume ($r = 0.443$).

Conclusions: These results suggest that several clinical and routine laboratory parameters of coagulation were individually associated with bleeding volume during cardiovascular surgery. Determining the patterns of coagulopathy may potentially help guide transfusion during cardiovascular surgery.

Keywords: Cardiovascular surgery, Diluted coagulopathy, Hemorrhage, Laboratory tests

* Correspondence: sfujii@asahikawa-med.ac.jp
[1]Department of Medical Laboratory and Blood Center, Asahikawa Medical University Hospital, Midorigaoka-Higashi 2-1-1-1, Asahikawa 078-8510, Japan
Full list of author information is available at the end of the article

Background

Coagulopathy and massive bleeding are severe complications of cardiovascular surgery, particularly occurring after procedures requiring prolonged cardiopulmonary bypass (CPB) [1, 2]. During surgery massive transfusion of blood products may become necessary. Transfusion practices may vary among different countries [3]. To predict and prevent massive bleeding thoracic surgeons and cardiovascular anesthesiologists have made substantial efforts [4]. Reliable and predictive laboratory tests on coagulation function may help guide clinicians to appropriate transfusion and may decrease variabilities in transfusion practices. Appropriate coagulation parameters would help to predict bleeding after cardiovascular surgery. Such knowledge could potentially enable clinicians to effectively treat coagulopathy after surgery requiring CPB. However, risk scores for blood transfusion [5] and hemostatic laboratory tests have not been fully evaluated for hemorrhagic risk prediction in Asian patients. The purpose of this study is to determine the markers predicting massive bleeding in Japanese patients. Markers of coagulation and fibrinolysis were measured during cardiovascular surgery and their correlations with bleeding volume were determined.

Methods

Study design and subjects

In a prospectively collected cohort of patients undergoing cardiovascular surgery at Asahikawa Medical University Hospital (Asahikawa, Japan) from November 2015 to March 2016, we examined the association between coagulation parameters and bleeding volume. The characteristics of the subjects are shown in Table 1. This study was approved by the institutional review board (approval number 14127). Informed consent was obtained from individual patient for the use of blood normally discarded after routine laboratory coagulation testing and medical records. Clinicians caring for patients were not provided with assay results. Institutional clinical or transfusion practices were unchanged during the study period. Heparin (300 units/kg body weight) was used in each case during CPB and great care was taken so that blood samples for analysis would avoid heparin contamination.

Data and sample collection

Patients' medical records were reviewed for surgical procedures, general laboratory test results and blood product transfusions. One milliliter of citrated whole blood was obtained from specimens collected as part of routine clinical care. For each case blood was obtained at 2 points. Time of sampling included (point 1) pre-CPB after induction of anesthesia, or at the time immediately before surgery in patients without CPB, and (point 2) post-CPB, or at the time right after surgery procedure in patients without CPB.

Table 1 Demographic and operative patient characteristics

Characteristic	Median (IQR) or n (%)	
Age (years)	69	(39–85)
Gender		
Male	29	(60%)
Female	19	(40%)
Race		
Asian	48	(100%)
Operative Data		
Operation time (min)	318	(107–654)
CPB time (min)	181	(58–501)
Bleeding volume (ml)	2269	(174–10,607)
Blood Transfusion		
RBC (units)	12	(0–30)
FFP (units)	12	(0–44)
PC (units)	20	(0–60)
Intraoperative autologous		
Blood collection (ml)	669	(0–4439)
Type of operation		
CABG only	6	(13%)
Valve surgery only	22	(46%)
CABG and valve surgery	3	(6%)
CABG and aortic surgery	11	(23%)
Valve and aortic surgery	2	(4%)
Other procedures	4	(8%)

CPB cardiopulmonary bypass, *RBC* red blood cell, *FFP* fresh frozen plasma, *PC* platelet concentrate, *CABG* coronary artery bypass grafting

Analysis

After centrifuge blood samples were stored below −30 °C. PT, APTT and fibrinogen (Fib) were measured with automated analyzer (coagulation one-stage assay, Coapresta2000, Sekisui Medical Company, Tokyo, Japan). Activities of coagulation factor II (FII), FV, FVII, FVIII, FIX, FX, FXI and FXII were measured with Coapresta3000 (coagulation one-stage assay, Sekisui Medical). α2 Plasmin inhibitor (α2PI, synthetic substrate assay), antithrombin (AT, synthetic substrate assay), thrombin-antithrombin complex (TAT, chemiluminescent enzyme immunoassay) and FXIII activity (synthetic substrate assay) were measured using automated analyzer (STACIA, LSI Medience Corporation, Tokyo, Japan).

Statistical methods

Data are reported as box plot with lower extreme, lower quartile, median, upper quartile and upper extreme

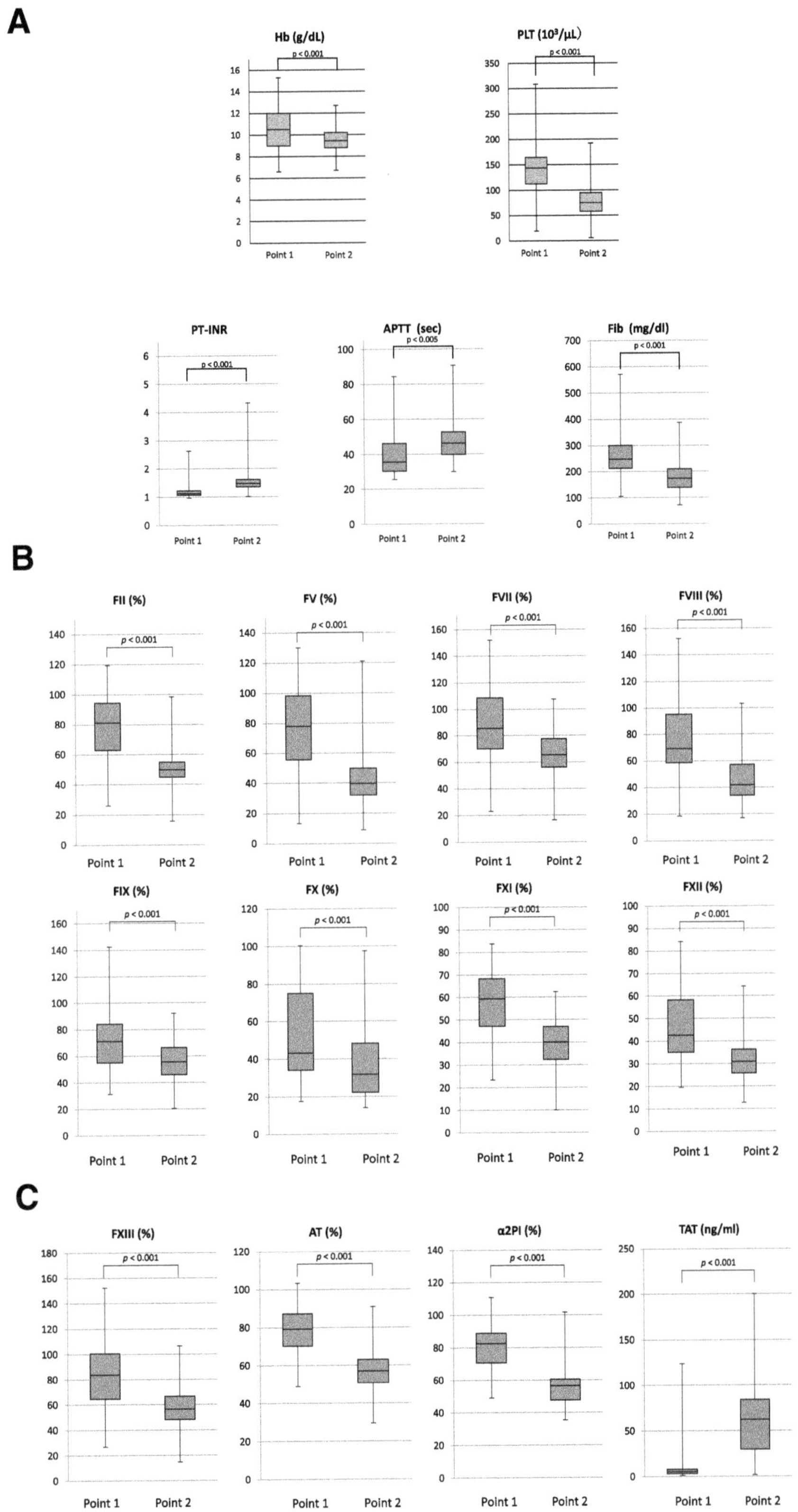

Fig. 1 (See legend on next page.)

(See figure on previous page.)
Fig. 1 (A) Values of hemoglobin (Hb), platelet (PLT), PT-INR, APTT and fibrinogen (Fib) at point 1 (pre-CPB after induction of anesthesia, or at the time immediately before surgery in patients without CPB) and at point 2 (post-CPB, or at the time right after surgery procedure in patients without CPB). (B) Values of coagulation factor II (FII), FV, FVII, FVIII, FIX, FX, FXI and FXII at point 1 (pre-CPB after induction of anesthesia, or at the time immediately before surgery in patients without CPB) and at point 2 (post-CPB, or at the time right after surgery procedure in patients without CPB). (C) Values of FXIII, antithrombin (AT), α2 plasmin inhibitor (α2PI) and thrombin-antithrombin complex (TAT) at point 1 (pre-CPB after induction of anesthesia, or at the time immediately before surgery in patients without CPB) and at point 2 (post-CPB, or at the time right after surgery procedure in patients without CPB)

values. Student or Mann-Whitney rank sum tests for quantitative variables were used to determine differences in unadjusted preoperative and postoperative characteristics. A p value less than 0.05 was considered as statistically significant. The correlation of coagulation and fibrinolysis markers and bleeding volume were analyzed using the Pearson's correlation coefficient.

Results

No differences in bleeding volume or number of transfusion units with or without CPB

Among 48 cases, 42 cases were on CPB. No differences in bleeding volume or number of transfusion units were observed between the cases with CPB and the cases without CPB. There was also no significant difference in bleeding volume among patients with blood type-A, -B, -O and -AB (median value type-A 1402 ml, type-B 1372 ml, type-O 1777 ml, type-AB 884 ml, respectively). The correlation coefficient between dose (unit) of FFP infused and bleeding volume was 0.73.

Laboratory parameters predicting massive bleeding

When the laboratory assay values obtained at point 1 were compared with values obtained at point 2, values of hemoglobin, platelet, PT-INR, APTT and Fib at point 2 were changed (Fig. 1A). Activities of coagulation factors (Fig. 1B) and values of FXIII, ATIII and α2PI were also decreased and values of TAT were increased at point 2 (Fig. 1C). Values of FII, FIX, FXI andα2PI at point 1 were negatively correlated with bleeding volume (FII, $r = -0.506$: FIX, $r = -0.504$: FXI $r = -0.580$; α2PI, $r = -0.418$) (Fig. 2). Correlations between the assay values at point 2 and bleeding volume are shown in Fig. 3. Values of FIX were negatively correlated ($r = -0.445$) and values of TAT were positively correlated with bleeding volume ($r = 0.443$).

Discussion

To evaluate the potential predictive value of hemostatic laboratory tests for postoperative bleeding in cardiovascular surgery laboratory markers of coagulation and fibrinolysis were measured and their correlations with bleeding volume were determined. Levels of coagulation factors were uniformly decreased at point 2 as compared to the values at point 1, suggesting that use of anticoagulants and consumption due to operative stress can decrease the levels of coagulation factors. Markers of increased fibrinolysis, such as fibrin-fibrinogen degradation products, D-dimer and plasmin-α2 plasmin inhibitor complex exhibited relatively more volatility than coagulation factors (results not shown). Increase in TAT

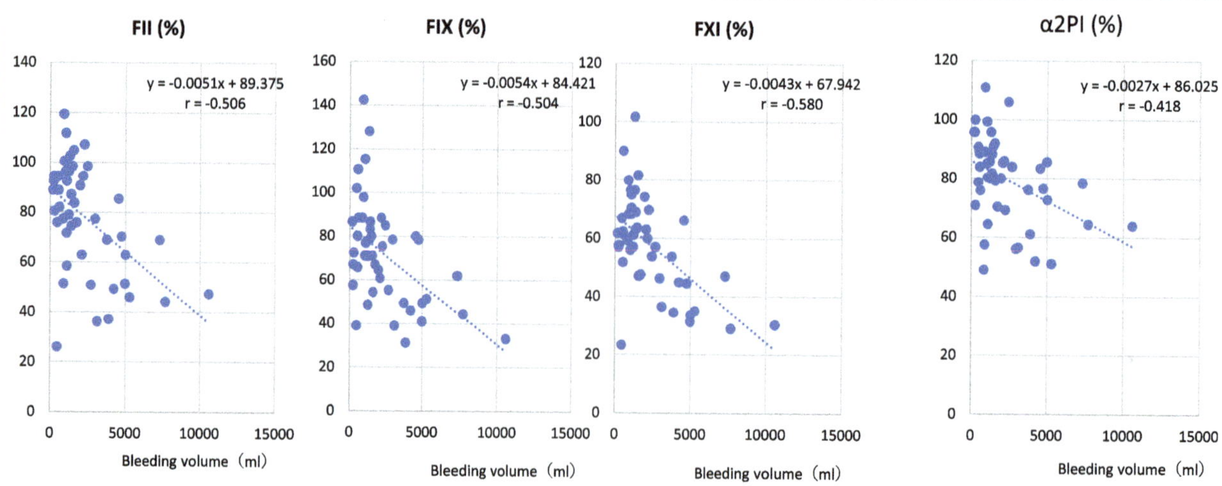

Fig. 2 Correlation of bleeding volume and values of FII, FIX, FXI and α2PI at point 1 (pre-CPB after induction of anesthesia, or at the time immediately before surgery in patients without CPB)

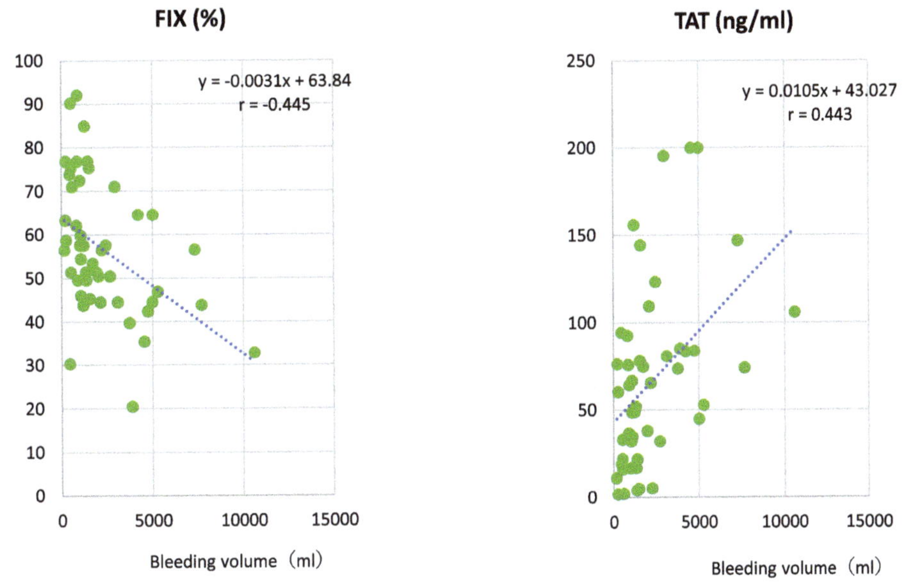

Fig. 3 Correlation of bleeding volume and values of FIX and TAT at point 2 (post-CPB, or at the time right after surgery procedure in patients without CPB)

at point 2 may suggest tissue damage due to operative stress.

Activities of FII, FIX and FXI at point 1 were negatively correlated with bleeding volume. These activities were not below the critical levels necessary for hemostasis even at point 2, suggesting that coagulopathy is not due to a decrease in one single factor but due to the decrease in multiple coagulation factors. Furthermore, ATIII with anticoagulant activity, FXIII with thrombus stabilizing activity andα2PI, hypofibrinolytic marker, were negatively correlated with bleeding volume. In particular, α2PI was closely correlated with bleeding volume. At point 2 TAT was positively correlated with bleeding volume, suggesting that TAT reflected operative stress. Taken together, coagulopathy seen during cardiovascular surgery reflects combined effects of decrease in coagulation factors and factors related to thrombus formation.

This is a single center observational study. The number of patients studied was relatively low and did not allow us to conduct multivariate analysis. The relatively low ability of laboratory markers to predict the volume of postoperative bleeding might be due to the differences in the in vitro testing of coagulation and fibrinolysis and the local hemostatic environment. There might be a local high fibrinolytic activity in the traumatized tissue that is not represented by circulating blood samples [6]. Fibrinogen levels were reported to be moderately associated with postoperative bleeding [7]. Fibrinogen is a key molecule for hemostasis and is relatively easily administered using fibrinogen concentrate [8]. Preoperative fibrinogen administration might reduce postoperative bleeding [9]. Thromboelastometry is also reported to predict major bleeding in pediatric cardiac surgery [10].

There was no increased blood loss in patients with particular blood group. This observation in cardiovascular surgery was in contrast to the previous reports in postpartum blood loss [11, 12]. Prospective further studies taking into account coagulation profile are necessary to determine whether ABO system can be considered a heritable risk of perioperative period hemorrhage.

This is a single center study with relatively small number of patients and the antigen levels of coagulation factors were not measured. Further study involving more patients would clarify factor(s) independently predicting postoperative bleeding.

Conclusions

Our results suggested that FII, FIX, FXI, FXIII and α2PI are associated with postoperative bleeding volume. Diluted coagulopathy, decrease in stabilized fibrin due to decrease in FXIII and hyperfibrinolysis due to decrease in α2PI may contribute to hemostatic defect. Thrombin burst due to operative stress as represented by increase in TAT values may also contribute to hemostatic defect. Further larger scale clinical studies will improve our understanding of the pathogenesis of coagulopathy and provide good predicting markers of bleeding.

Abbreviations

CPB: Cardiopulmonary bypass; RBC: Red blood cells; FFP: Fresh frozen plasma; PC: Platelet concentrate; CABG: Coronary artery bypass grafting; AT: Antithrombin; α2PI: α2 plasmin inhibitor; TAT: Thrombin-antithrombin complex

Acknowledgements
The secretarial assistance of Ms. Noriko Saito is greatly appreciated.

Authors' contributions
YK, KO, KT and MY conducted assays, analyzed the data and wrote the manuscript. TK and HK managed and treated the enrolled patients and collected samples. SF provided major input into the conceptual development of the studies and supervised all of the investigations. All of the authors read and approved the final manuscript.

Author details
[1]Department of Medical Laboratory and Blood Center, Asahikawa Medical University Hospital, Midorigaoka-Higashi 2-1-1-1, Asahikawa 078-8510, Japan. [2]Present address: Japanese Red Cross Hokkaido Block Blood Center, Sapporo, Japan. [3]Department of Cardiac Surgery, Asahikawa Medical University, Asahikawa, Japan. [4]Department of Anesthesiology, Asahikawa Medical University, Asahikawa, Japan.

References
1. Besser MW, Klein AA. The coagulopathy of cardiopulmonary bypass. Crit Rev Clin Lab Sci. 2010;47:197–212.
2. Colson PH, Gaudard P, Fellahi J-L, Bertet H, Faucanie M, Amour J, et al. Active bleeding after cardiac surgery: a prospective observational multicenter study. PLoS One. 2016. https://doi.org/10.1371/journal.pone.0162396.
3. Bennett-Guerrero E, Zhao Y, O'Brien SM, Ferguson TB Jr, Peterson ED, Gammie JS, et al. Variation in use of blood transfusion in coronary artery bypass graft surgery. JAMA. 2010;304:1568–75.
4. The Society of Thoracic Surgeons Blood Conservation Guideline Task Force, The Society of Cardiovascular Anesthesiologists Special Task Force on Blood Transfusion, The International Consortium for Evidence Based Perfusion. 2011 Update to The Society of Thoracic Surgeons and the Society of Cardiovascular Anesthesiologists Blood Conservation Clinical Practice Guidelines. Ann Thorac Surg. 2011;91:944–82.
5. Goudie R, Sterne JAC, Verheyden V, Bhabra M, Ranucci M, Murphy GJ. Risk scores to facilitate preoperative prediction of transfusion and large volume blood transfusion associated with adult cardiac surgery. Br J Anaesth. 2015; 114:757–66.
6. Pelletier MP, Solymoss S, Lee A, Chiu RC. Negative reexploration for cardiac postoperative bleeding: can it be therapeutic? Ann Thorac Surg. 1998;65: 999–1002.
7. Gielen C, Dekkers O, Stijnen T, Schoones J, Brand A, Klautz R, et al. The effects of pre- and postoperative fibrinogen levels on blood loss after cardiac surgery: a systematic review and meta-analysis. Interact Cardiovasc Thorac Surg. 2014;18:292–9.
8. Yamamoto K, Yamaguchi A, Sawano M, Matsuda M, Anan M, Inokuchi K, et al. Pre-emptive administration of fibrinogen concentrate contributes to improved prognosis in patients with severe trauma. Trauma Surg Acute Care Open. 2016;1:e000037.
9. Jeppsson A, Waldén K, Roman-Emanuel C, Thimour-Bergström L, Karlsson M. Preoperative supplementation with fibrinogen concentrate in cardiac surgery: a randomized controlled study. Br J Anaesth. 2016;116:208–14.
10. Nakayama Y, Nakajima Y, Tanaka KA, Sessler DI, Maeda S, Iida J, et al. Thromboelastometry-guided intraoperative haemostatic management reduces bleeding and red cell transfusion after paediatric cardiac surgery. Br J Anaesth. 2015;114:91–102.
11. Kahr MK, Franke D, Brun R, Wisser J, Zimmermann R, Haslinger C. Blood group O: a novel risk factor for increased postpartum blood loss? Haemophilia. 2018;24:e207–12.
12. Bade NA, Kazma JM, Amdur RL, Ellis-Kahana J, Ahmadzia HK. Blood type association with bleeding outcomes at delivery in a large multi-center study. J Thromb Thrombolysis. 2020;50:439–45.

22

Factor V Leiden 1691G > A mutation and the risk of recurrent pregnancy loss (RPL)

Mohammad Masoud Eslami[1], Majid khalili[2,3], Mina Soufizomorrod[1], Saeid Abroun[1] and Bahman Razi[1*]

Abstract

Background: Although numerous replication case-control studies have attempted to determine the association between Factor V Leiden (FVL) 1691G > A mutation and susceptibility to Recurrent pregnancy loss (RPL), there have been confliction among the results of various ethnic groups. To address this limitation, here we implemented first meta-analysis to provide with consistent conclusion of the association between FVL 1691G > A mutation and RPL risk.

Methods: After a systematic literature search, pooled odds ratio (OR) and their corresponding 95% confidence interval (CI) were used to evaluate the strength of the association. Additionally, meta-regression analyses were performed to find potential source of heterogeneity.

Results: In this meta-analysis, 62 studies, containing 10,410 cases and 9406 controls, were included in quantitative analysis. Overall population analysis revealed a significant positive association in the dominant (OR = 2.15, 95% CI = 1.84–2.50, $P < 0.001$), over-dominant (OR = 1.88, 95% CI = 1.61–2.19, $P < 0.001$), allelic (OR = 2.05, 95% CI = 1.79–2.35, $P < 0.001$), and heterozygote (OR = 1.97, 95% CI = 1.68–2.30, $P < 0.001$) models. Moreover, a significant association of dominant (OR = 3.04, 95% CI = 2.04–4.54, $P < 0.001$), over-dominant (OR = 2.65, 95% CI = 1.74–4.05, $P < 0.001$), and heterozygote (OR = 2.67, 95% CI = 1.81–4.22, $P < 0.001$) models was found in the Iranian population. The subgroup analysis indicated strong significant association in Asian, European, Africa population, and case-control studies but not in South Americans and cohort studies.

Conclusion: The FVL 1691G > A mutation and the risk of RPL confers a genetic contributing factor in increasing the risk of RPL, particularly in Iranians, except for South Americans.

Keywords: Recurrent pregnancy loss, Factor V Leiden, 1691G > A mutation, Meta-analysis, Meta-regression

Introduction

Recurrent pregnancy loss (RPL) is a heterogeneous disorder which affects women of reproductive age. Recently, The American Society of Reproductive Medicine has defined RPL as two or more than two failed pregnancies before the 20th week of pregnancy [1–3]. Overall, 1–5% of women during reproductive ages could be affected [4]. From pathophysiological point of view, RLP might be influenced by various items, such as genetic factors (chromosomal aberrations, genetic polymorphisms), infectious diseases, structural abnormalities of the uterus, coagulative disorders (thrombophilia), endocrinological problems (thyroid disease and diabetes), and immunological disease (autoimmune disorder and inflammatory diseases) [5–7]. With considering these factors, still approximately 40 to 50% of cases remained idiopathic [8].

Although pregnancy as a physiological condition is associated with a hypercoagulable state, and the contact

* Correspondence: b.razi@modares.ac.ir
[1]Department of Hematology, Faculty of Medical Sciences, Tarbiat Modares University, North Kargar Av, Tehran 14117, Iran
Full list of author information is available at the end of the article

between placenta and maternal circulation is crucial for the establishment of a successful pregnancy, but any abnormality in this circulation, especially abnormal blood clotting in the small placental blood vessels, may results in RPL [9, 10]. During last decades, thrombophilia attracted a lot of attention as a risk factor for RLP. Thrombophilia is characterized as a hemostatic disorder which leads to an increased tendency of thromboembolic processes. Classically, thrombophilia could be classified into acquired and inherited forms [11, 12]. In this regards, antiphospholipid syndrome is an established acquired thrombophilia factor which increase the risk of RPL. Among inherited factors, mutation in Factor V Leiden (FVL) of the FV gene, G20210A of the FII (prothrombin) gene, and C677T of the methylenetetrahydrofolate reductase (MTHFR) gene are believed to play a key role in pathogenesis of RPL [13, 14].

FVL mutation shows an autosomal dominant pattern which occurs by substitution of guanine by adenine (CGA--- > CAA) at the nucleotide 1691 in the exon 10. As a result of this missense mutation, arginine (Arg) at amino acid 506 is substituted with glutamine (Gln), leading to generation of FVL resistant to the activated protein C (APC). APC is a natural anticoagulant which in normal situation cleaves activated factor V at amino acid 506 and makes it inactive [15–20].

Studies have shown that FVL mutation increases the risk of venous thrombosis 7 times in heterozygote and 80 times in homozygote carriers. In addition, it has been reported that this mutation increases the risk of preeclampsia in FVL carriers [21, 22]. The exact mechanism that FVL mutation influence the etiology of RPL is a controversial issue and has not yet been divulged thoroughly, but several studies suggested that production of micro thrombosis could sediment in delicate placental blood vessels and cause placental infarction and subsequent maternal and fetal complications [23, 24].

In spite of all findings, still the exact association between FVL mutation and the risk RPL is unclear and several investigators worldwide try to clarify this question. Therefore, here we conducted the first and the most comprehensive meta-analysis on the association between FVL 1691G > A mutation and risk of RPL by exerting 62 studies encompassing 10,410 cases and 9406 health control to achieve more reliable conclusion.

Methods

Ethical approval is not necessary for this meta-analysis. The current meta-analysis was conducted according to the Preferred Reporting Items for Systematic reviews and Meta-Analyses (PRISMA) statement [25], including publication search, study selection, inclusion and exclusion criteria, data extraction, quality assessment, and statistical analysis.

Publication search

A comprehensive systematic search in the ISI Web of Science, Scopus, and PubMed/Medline databases was conducted to retrieve all publications evaluating the associations between FVL 1691G > A mutation and susceptibility to RPL prior to May 2020. The following combinations of key words were used: ("Miscarriage" OR "abortion" OR "pregnancy loss" OR "habitual abortion" OR "fetal loss" OR "Recurrent Pregnancy Loss") AND ("Factor V Leiden" OR "FV Leiden" OR "1691G > A" OR "rs6025") AND ("polymorphism*" OR "variant" OR "mutation" OR "genotype" OR "allele" OR "single nucleotide polymorphism" OR "SNP"). In spite of detailed search, a manual cross-check of eligible studies and reviews was carried out to include other potential studies. Original data in English language and human population studies were collected.

Study selection

Primary search strategy generates 1266 studies that were exported into Endnote X8 software. The duplicated studies were removed and title & abstract of remaining studies were reviewed by two investigators and irrelevant studies were excluded. Full-text verification was performed if we could not classify studies based on title & abstract. Any disagreements during study selection were discussed and resolved by consensus.

Inclusion and exclusion criteria

Studies considered eligible if they met the following inclusion criteria: a) Studies concerning the association between FVL 1691G > A mutation and susceptibility to recurrent pregnancy loss as the main outcome; b) Studies that their case group have recurrent pregnancy loss (two or more times of abortion); c) Studies with case-control and cohort design; d) Studies reporting sufficient data of genotype or allele frequency that could confer feasibility of calculating the odds ratios (ORs) and 95% confidence intervals (CIs). On the other hand, duplicates, case reports, book chapters, reviews, letter to editor, studies with insufficient data, and abstracts were all excluded.

Data extraction and quality assessment

According to a standardized extraction form, the following data were independently extracted by two investigators: the first author's last name, journal and year of publication, country of origin, ethnicity, allele and genotype frequency in cases and controls, mean or range of age, genotyping method, and total sample size of cases and controls. The third investigator finalized the extracted data, and potential discrepancies were resolved by consensus. For quality assessment of the included publications, the Newcastle-Ottawa Scale (NOS) was applied [26]. In this respect, studies with 0–3, 4–6 or 7–9 scores were of, respectively, low, moderate, and high-quality.

Statistical analysis

Deviation from Hardy–Weinberg equilibrium (HWE) for distribution of the genotype frequencies was analyzed by χ2-test in the control group. The strength of the association between FVL 1691G > A mutation and RPL risk was evaluated by the pooled OR and its corresponding 95% CI. Different comparison models for FVL 1691G > A mutation were as follow: dominant model (AA+GA vs. GG), overdominant model (GA vs. GG + AA), allelic model (A vs. G), and heterozygote (GA vs. GG). It should be noted that due to the AA genotype frequency of zero in both cases and controls, the recessive and homozygote models were not calculable. Presence of heterogeneity between included studies was estimated by Cochran's Q-statistic (P value< 0.10 was considered as statistically significant) [27]. Besides, to report quantitative heterogeneity I-squared (I^2) tests was used. The fixed-effected model (FEM) was used if $P_{Q\text{-statistic}} > 0.10$ or I^2 was< 50%; otherwise, the random-effected model (REM) was applied. In order to assessed the predefined sources of heterogeneity among included studies, subgroup analysis and meta-regression analysis based on year of population, the continent of the study population, and genotyping method were performed. Additionally, sensitivity analysis was conducted in presence of heterogeneity [28, 29]. Publication bias was estimated by Begg's funnel plots and Egger's regression test (P value< 0.05 was considered as statistically significant) [30, 31]. The funnel plot asymmetry was assessed with the Egger's test. Practically, in case of no evidence of publication bias, studies with high precision (large study effects) will be located near the average line, and studies with low precision (small study effects) will be spread equally on both sides of the average line; any deviation from this shape can indicate publication bias. The data analyses were carried out using STATA (version 14.0; Stata Corporation, College Station, TX) and SPSS (version 23.0; SPSS, Inc. Chicago, IL) software.

Results

Study characteristics

The four-phase search and screening process of the literatures based on the PRISMA statement is depicted in the Fig. 1. According to the aforementioned keywords, a

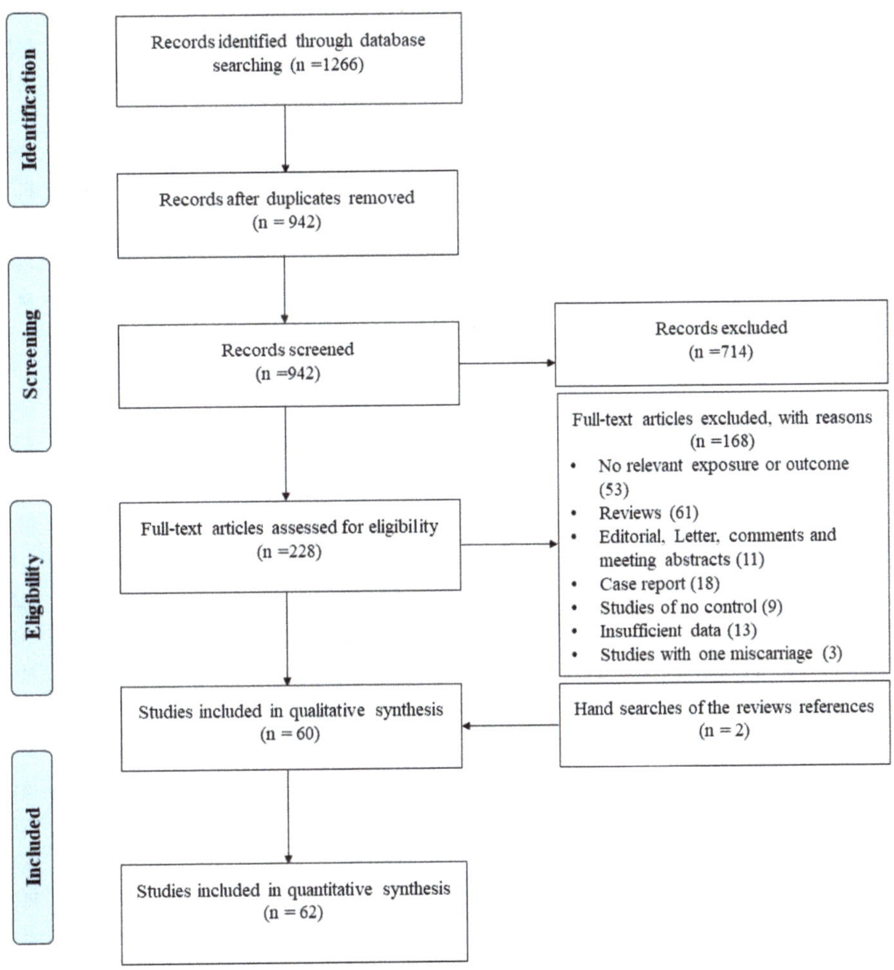

Fig. 1 Flow diagram of study selection process

Table 1 Characteristics of studies included in meta-analysis

Study author	Year	Country	Study design	Ethnicity	Total cases/ controls	Age case/control (Mean)	Genotyping method	Quality score
Souza et al. [34]	1999	Brazil	case-control	South America	56/384	29.6 / 24.3	RLFP-PCR	7
Brenner et al. [35]	1999	Israel	case-control	Asia	76/106	31 ± 5 / 31 ± 6	RLFP-PCR	6
Wramsby et al. [36]	2000	Sweden	case-control	Europe	62/69	21–39 / 21–39	RLFP-PCR	7
Murphy et al. [37]	2000	Ireland	case-control	Europe	41/540	32 ± 0.74 / NR	RLFP-PCR	6
Pihusch et al. [33]	2000	Germany	case-control	Europe	102/128	35 / 32	RLFP-PCR	6
Younis et al.	2000	Israel	case-control	Asia	78/139	30.0 ± 4.4 / 30.7 ± 4.2	RLFP-PCR	6
Foka et al. [14]	2000	Greece	case-control	Europe	80/100	33 / 35	RLFP-PCR	6
Rai et al.	2001	London	cohort	Europe	1111/150	33.5 / 33	RLFP-PCR	8
Carp et al.	2002	Israel	case-control	Asia	108/82	31 / 36	RLFP-PCR	6
Finan et al. [38]	2002	Lebanon	case-control	Asia	110/67	32.3 ± 5.3 / 33.9 ± 7.3	RLFP-PCR	6
Hohlagschwandtner et al.	2003	Australia	case-control	Oceania	145/101	32 / 56	Multiplex PCR	7
Pauer et al. [39]	2003	German	case-control	Europe	30/122	31.3 / NR	RLFP-PCR	6
Mtiraoui et al	2004	Tunisia	case-control	Africa	146/99	29.0 ± 6.1 / 28.9 ± 5.3	RLFP-PCR	6
Aksoy et al.	2005	Turkey	case-control	Europe	41/50	32 ± 5.54 / 29 ± 4.66	PCR	5
Mahjoub et al. [40]	2005	Tunisia	case-control	Africa	200/200	28.68 ± 5.61 / 28.24 ± 5.51	RLFP-PCR	8
Ulukus et al.	2006	Turkey	case-control	Europe	10/53	29.1 ± 5.2 / 28.0 ± 4.8	PCR	5
Sotiriadis et al.	2006	Greece	case-control	Europe	99/102	32.2 / 32.2	RLFP-PCR	6
Mohammad et al. [21]	2007	Syrian	case-control	Asia	35/45	29.6 ± 6.3 / 28.8 ± 6.8	Q-PCR	5
Altintas et al. [41]	2007	Turkey	case-control	Europe	114/185	30.6 ± 4.4 / 30.5 ± 4.3	Q-PCR	7
Toth et al. [42]	2008	Germany	case-control	Europe	151/157	33.2 ± 4.6 / 45.2 ± 12.6	RLFP-PCR	7
Pasquier et al	2008	France	case-control	Europe	311/599	32.8 / 34.3	Q-PCR	8
Biswas et al. [43]	2008	India	case-control	Asia	85/31	27.9 ± 0.3 / 26 ± 0.5	RLFP-PCR	6
Lvanov et al.	2009	Bulgaria	case-control	Europe	153/100	29.7 / 31.0	RLFP-PCR	7
Mukhopadhyay et al. [44]	2009	India	case-control	Asia	84/80	24.9 ± 3.3 / 24.9 ± 3.3	RLFP-PCR	6
Ciacci et al. [45]	2009	Italy	case-control	Europe	39/72	36.24 ± 8.26 / 30.10 ± 8.60	Multiplex PCR	6
Mohamed et al. [46]	2010	Egypt	case-control	Africa	20/20	29.0 ± 4.80 / 31.4 ± 6.82	PCR	5
Hussein et al. [47]	2010	Palestine	case-control	Asia	145/205	31.9 / 32	ARMS-PCR	7
Serrano et al. [17]	2011	Portugal	case-control	Europe	100/100	32 ± 4.25 / 30.9 ± 5.19	PCR	7
Settin et al.	2011	Egypt	case-control	Africa	72/70	19 to 38 / 19 to 38	PCR	6
Dissanayake et al. [32]	2012	Sri Lanka	case-control	Asia	200/200	32.1 ± 5.6 / 32.4 ± 4.6	RLFP-PCR	8
Gazi et al.	2012	Turkey	case-control	Europe	57/47	30.12 ± 7.32 / 27.80 ± 6.36	PCR	6
Karata et al.	2012	Turkey	case-control	Europe	84/84	31.6 ± 3.7 / 32.2 ± 3.9	Q-PCR	6
Mierla et al. [48]	2012	Romania	case-control	Europe	283/100	33.76 / 32.8	RLFP-PCR	7
Ozdemir et al. [49]	2012	Turkey	case-control	Europe	543/106	27.8 ± 2.1 / 28.9 ± 2.2	Q-PCR	7
Torabi et al. [50]	2012	Iran	case-control	Asia	100/100	NR / NR	RLFP-PCR	6
Kaur et al.	2012	India	case-control	Asia	107/588	24.89 / 25.32	RLFP-PCR	7
Parveen et al.	2012	India	case-control	Asia	1000/500	28.4 ± 5.9 / 31.9 ± 7.3	ARMS-PCR	8
Ardestani et al.	2012	Iran	case-control	Asia	80/80	28.8 / 23.6	RLFP-PCR	6
Cardona et al. [51]	2012	Colombia	case-control	South America	93/206	34.1 ± 0.9 / 41.6 ± 0.7	RLFP-PCR	7
Kazerooni et al. [52]	2013	Iran	case-control	Asia	60/ 60	24.8 ± 3.9 / 24.6 ± 4.7	PCR	5
Baumann et al.	2013	Germany	cohort	Europe	641/157	32.95 ± 4.94 / 33.16 ± 6.24	RLFP-PCR	8
Parand et al. [53]	2013	Iran	case-control	Asia	90/44	29.21 ± 5.9 / 28.75 ± 5.2	RLFP-PCR	6
Zonouzi et al. [54]	2013	Iran	case-control	Asia	89/50	30.18 ± 4.95 / 31.54 ± 4.81	ARMS-PCR	6

Table 1 Characteristics of studies included in meta-analysis *(Continued)*

Study author	Year	Country	Study design	Ethnicity	Total cases/ controls	Age case/control (Mean)	Genotyping method	Quality score
Dutra et al.	2013	Brazil	case-control	South America	145/135	31.72 / 29.86	Q-PCR	6
Isaoglu et al.	2013	Turkey	case-control	Europe	60/40	29.14 ± 6.18 / 30.50 ± 6.77	NR	6
Pietropolli et al. [55]	2014	Italy	case-control	Europe	186/129	35.2 ± 5.1 / 40.4 ± 5.3	Rapid-cycle PCR	7
Lino et al.	2014	Brazil	case-control	South America	83/98	30.3 / 40.2	Q-PCR	6
Sharma et al. [56]	2015	India	case-control	Asia	78/78	28.6 ± 3.32 / 30.5 ± 2.57	RLFP-PCR	6
Farahmand et al.	2015	Iran	case-control	Asia	330/350	30.37 / 29.88	PCR	8
Kashif et al. [57]	2015	Pakistan	case-control	Asia	56/56	28.55 ± 4.69 / 28.61 ± 4.38	PCR	6
Gonçalves et al. [58]	2016	Brazil	case-control	South America	137/100	32.1 / 25.8	RLFP-PCR	7
Khaniani et al. [59]	2016	Iran	case-control	Asia	210/160	less than 40 / NR	RLFP-PCR	7
Eldeen et al.	2017	Arabia	case-control	Asia	96/96	37.7 ± 4.6 / 36.5 ± 5.8	PCR	6
Wolski et al. [60]	2017	Poland	case-control	Europe	359/400	30.99 ± 4.50 / 30.05 ± 3.81	RLFP-PCR	8
Elgari et al. [61]	2017	Arabia	case-control	Asia	60/80	38 ± 12 / 38 ± 12	Multiplex PCR	6
Mahmutbegović et al. [62]	2017	Bosnia	case-control	Europe	51/154	32.9 ± 5.1 / 31.7 ± 6.6	Q-PCR	6
Wingeyer et al.	2017	Argentina	case-control	South America	247/107	32 / NR	Q-PCR	7
Jusić et al.	2018	Bosnia	case-control	Europe	60/80	33.05 / 34.08	RLFP-PCR	6
Taghi Kardi et al.	2018	Iran	case-control	Asia	250/116	29.7 ± 3.4 / 30.4 ± 3.2	Multiplex PCR	7
Xu et al.	2018	China	case-control	Asia	426/444	29.26 ± 4.294 / 34.50 ± 4.895	Multiplex PCR	8
Bigdeli et al. [63]	2018	Iran	case-control	Asia	200/200	23.0 ± 3.8 / 25.1 ± 4.4	RLFP-PCR	8
Reddy et al. [64]	2019	India	case-control	Asia	50/28	26.8 / 27.6	RLFP-PCR	5
Yengel et al.	2019	turkey	case-control	Europe	145/105	30.5 ± 6.5 /30.5 ± 6.7	real-time PCR	6

total of 1266 studies were retrieved (PubMed: 254, Scopus: 512, and ISI Web of Science: 500). Subsequently, application of inclusion/exclusion criteria resulted in the exclusion of 1206 studies (324 duplicates studies, 714 and 168 studies excluded according to title & abstract and full-text examination, respectively). Eventually, 62 qualified studies were included in the quantitative analysis, of which two studies were detected by cross-check of eligible studies and reviews [32, 33]. All eligible studies were published between 1999 to 2019 and had an overall good methodological quality with NOS scores ranging from 5 to 8. The Restriction fragment length polymorphism (RFLP)-PCR was the most genotyping methods which used in the included studies. Except two studies which had cohort design, other **60** studies had case-control design. **Tables 1** and **2** summarize the characteristics and allele/genotype frequency of the included studies.

Meta-analysis of FVL 1691G > A mutation and the risk of RPL

Overall, 62 studies with 10,**410** cases and 9**406** controls included in quantitative analysis of the association between FVL 1691G > A mutation and the risk of RPL. Of those, 25 studies were in Asian countries [21, 22, 32, 35, 38, 43, 44, 47, 50, 52–54, 56, 57, 59, 61, 63–71], **26** studies were conducted in European countries [17, 33, 36, 37, 39, 41, 42, 45, 48, 49, 55, 60, 62, 72–82], 6 studies in South American countries [34, 51, 58, 83–85], 4 studies in African countries [40, 46, 86, 87] and one study in Oceania. The analysis of overall population revealed a significant positive association between FVL 1691G > A mutation and the risk of RPL across all possible genotype models, including dominant model (OR = 2.1**5**, 95% CI = 1.84–2.**50**, $P < 0.001$, FEM), over-dominant model (OR = 1.88, 95% CI = 1.61–2.19, $P < 0.001$, FEM), allelic model (OR = 2.05, 95% CI = 1.**79**–2.**35**, $P < 0.001$, REM), and heterozygote model (OR = 1.97, 95% CI = 1.68–2.30, $P < 0.001$, FEM) (Table 3 and Fig. 2).

Meta-analysis of FVL 1691G > A mutation and the risk of RPL in Iranian population

Among the included studies, studies performed in Iran with 9 publications (1409 cases and 1160 controls) were in the first rank with respect to sample size and the number of studies, therefore we performed separate analysis. Our results found a significant association between FVL 1691G > A mutation and increased risk of RPL in this population under dominant model (OR = 3.04, 95% CI = 2.04–4.54, $P < 0.001$, FEM), over-dominant model (OR = 2.65, 95% CI = 1.74–4.05, $P < 0.001$, FEM), and heterozygote model (OR = 2.67, 95% CI = 1.81–4.22, $P <$

Table 2 Distribution of genotype and allele among RPL patients and controls

Study author	RPL cases					Healthy control					P-HWE	MAF
	GG	GA	AA	G	A	GG	GA	AA	G	A		
Souza et al. [34]	52	4	0	108	4	378	6	0	762	6	0/87	0/007
Brenner et al. [35]	52	19	5	123	29	95	11	0	201	11	0/57	0/051
Wramsby et al. [36]	51	10	1	112	12	67	2	0	136	2	0/9	0/014
Murphy et al. [52]	39	2	0	80	2	527	13	0	1067	13	0/77	0/012
Pihusch et al. [33]	94	8	0	196	8	117	11	0	245	11	0/61	0/042
Younis et al.	63	12	3	138	18	131	8	0	270	8	0/72	0/028
Foka et al. [14]	65	15	0	145	15	96	4	0	196	4	0/83	0/02
Rai et al.	1037	72	2	2146	76	138	12	0	288	12	0/6	0/04
Carp et al.	104	4	0	212	4	77	5	0	159	5	0/77	0/03
Finan et al. [38]	65	38	7	168	52	56	11	0	123	11	0/46	0/082
Hohlagschwandtner et al.	130	15	0	275	15	97	4	0	198	4	0/83	0/019
Pauer et al. [39]	28	2	0	58	2	113	9	0	235	9	0/67	0/036
Mtiraoui et al.	116	24	6	256	36	93	6	0	192	6	0/75	0/03
Aksoy et al.	31	9	1	71	11	45	5	0	95	5	0/7	0/05
Mahjoub et al. [40]	152	40	8	344	56	189	11	0	389	11	0/68	0/027
Ulukus et al.	7	3	0	17	3	49	3	1	101	5	≤0.001	0/047
Sotiriadis et al.	94	5	0	193	5	99	3	0	201	3	0/88	0/014
Mohammad et al. [21]	25	10	0	60	10	41	4	0	86	4	0/75	0/044
Altintas et al. [41]	105	9	0	219	9	172	13	0	357	13	0/62	0/035
Toth et al. [42]	138	13	0	289	13	145	12	0	302	12	0/61	0/038
Pasquier et al.	296	15	0	607	15	574	25	0	1173	25	0/6	0/02
Biswas et al. [43]	83	2	0	168	2	31	0	0	62	0	≤0.001	0
Lvanov et al.	133	19	1	285	21	93	7	0	193	7	0/71	0/035
Mukhopadhyay et al. [44]	80	4	0	164	4	80	0	0	160	0	≤0.001	0
Ciacci et al. [45]	38	1	0	77	1	70	2	0	142	2	0/9	0/013
Mohamed et al. [46]	6	12	2	24	16	19	1	0	39	1	0/9	0/025
Hussein et al. [47]	104	36	5	244	46	181	24	0	386	24	0/37	0/058
Serrano et al. [17]	95	5	0	195	5	95	5	0	195	5	0/79	0/025
Settin et al.	54	17	1	125	19	69	1	0	139	1	0/95	0/007
Dissanayake et al. [32]	196	4	0	396	4	195	5	0	395	5	0/85	0/012
Gazi et al.	50	6	1	106	8	43	4	0	90	4	0/76	0/042
Karata et al.	66	16	2	148	20	66	18	0	150	18	0/27	0/107
Mierla et al. [48]	260	21	2	541	25	95	5	0	195	5	0/79	0/025
Ozdemir et al. [49]	433	109	1	975	111	104	2	0	210	2	0/92	0/009
Torabi et al. [50]	87	12	1	186	14	96	4	0	196	4	0/83	0/02
Kaur et al.	102	4	1	208	6	573	15	0	1161	15	0/75	0/012
Parveen et al.	950	50	0	1950	50	488	12	0	988	12	0/78	0/012
Ardestani et al.	78	2	0	158	2	79	1	0	159	1	0/95	0/006
Cardona et al. [51]	92	1	0	185	1	205	1	0	411	1	0/97	0/002
Kazerooni et al. [52]	43	12	5	98	22	54	4	2	112	8	0.48	0.734
Baumann et al.	592	49	0	1233	49	145	12	0	302	12	0/61	0/038
Parand et al. [53]	72	15	3	159	21	38	6	0	82	6	0/62	0/068
Zonouzi et al. [54]	87	2	0	176	2	50	0	0	100	0	≤0.001	0

Table 2 Distribution of genotype and allele among RPL patients and controls *(Continued)*

Study author	RPL cases					Healthy control					P-HWE	MAF
	GG	GA	AA	G	A	GG	GA	AA	G	A		
Dutra et al.	142	3	0	287	3	131	4	0	266	4	0/86	0/014
Isaoglu et al.	47	13	0	107	13	39	1	0	79	1	0/93	0/012
Pietropolli et al. [55]	168	18	0	354	18	125	4	0	254	4	0/85	0/015
Lino et al.	79	4	0	162	4	96	2	0	194	2	0/91	0/01
Sharma et al. [56]	36	40	2	112	44	77	1	0	155	1	0/95	0/006
Farahmand et al.	302	28	0	632	28	340	10	0	690	10	0/78	0/014
Kashif et al. [57]	53	3	0	109	3	56	0	0	112	0	≤0.001	0
Gonçalves et al. [58]	133	4	0	270	4	98	2	0	198	2	0/91	0/01
Khaniani et al. [59]	202	8	0	412	8	158	2	0	318	2	0/93	0/006
Eldeen et al.	0	72	24	72	120	0	94	2	94	98	≤0.001	0/51
Wolski et al. [60]	333	26	0	692	26	378	21	1	777	23	0/23	0/028
Elgari et al. [61]	56	4	0	116	4	74	6	0	154	6	0/72	0/037
Mahmutbegović et al. [62]	44	7	0	95	7	142	12	0	296	12	0/61	0/038
Wingeyer et al.	239	8	0	486	8	105	2	0	212	2	0/92	0/009
Jusić et al.	51	9	0	111	9	77	3	0	157	3	0/86	0/018
Taghi Kardi et al.	236	12	2	484	16	109	5	2	223	9	≤0.001	0/038
Xu et al.	426	0	0	852	0	443	1	0	887	1	0/98	0/001
Bigdeli et al. [63]	150	30	20	330	70	192	8	0	392	8	0/77	0/02
Yengel et al.	130	1	14	261	29	102	0	3	204	6	0/65	0/394

P-HWE p-value for Hardy–Weinberg equilibrium; *MAF* Minor allele frequency of control group

0.001, FEM) but not allelic model (OR = 2.09, 95% CI = 0.88–4.94, $P = 0.09$, REM) (Table 3).

Subgroup analysis by continent

The included studies were performed in Asia (25 studies), Europe (26 studies), South America (6 studies), Africa (4 studies) and Oceania (1 article). Since there was only one study for Oceania, we exclude it from the subgroup analysis. The final results revealed strong significant association between FVL 1691G > A mutation and the risk of RPL in Asian, European, and Africa population, but not in South Americans (Fig. 3). The results of pooled ORs, heterogeneity tests, and publication bias tests in different analysis models are shown in the Table 3.

Subgroup analysis by study design

The stratification of studies based on study design caused to the inclusion of two studies with 1752 cases and 307 controls in cohort group, and 60 studies with 8658 cases and 9099 controls in case-control group. The findings demonstrated a statistical significant association between FVL 1691G > A mutation and the risk of RPL in case-control studies across dominant model (OR = 2.33, 95% CI = 1.99–2.74, $P < 0.001$, FEM), over-dominant model (OR = 2.05, 95% CI = 1.74–2.41, $P < 0.001$, FEM), allelic model (OR = 2.18, 95% CI = 1.8–2.52, $P < 0.001$, FEM), and heterozygote model (OR = 2.16, 95% CI = 1.83–2.55, $P < 0.001$, FEM). However, no significant association was observed in cohort studies (Table 3).

Heterogeneity and publication bias

To check existence of publication bias, Egger's linear regression and Begg's funnel plot test were used. The shape of the funnel plots did not disclose obvious asymmetry under all the genotype model of the *FVL* 1691G > A mutation (Fig. 4). Additionally, some degree of heterogeneity was detected in overall population. Therefore, we stratified study by continent and study design to find its potential source.

Meta-regression analyses

Meta-regression analyses were performed to explore potential sources of heterogeneity among included studies (Table 4). The findings indicated that none of the expected heterogeneity parameter were the source of heterogeneity (Fig. 5).

Sensitivity analysis

The impact of individual study on pooled OR was evaluated by sequential omission of each studies. The analysis results showed that no individual study significantly affected the pooled ORs under any genotype models of the FVL 1691G > A mutation (Fig. 6).

Table 3 Main results of pooled ORs in meta-analysis of FVL 1691G > A mutation

Subgroup		Sample size	Test of association		Test of heterogeneity		Test of publication bias (Begg's test)		Test of publication bias (Egger's test)	
	Genetic model	**Case/Control**	**OR**	**95% CI (*P*-value)**	**OR**	**P**	**Z**	**P**	**T**	**P**
Overall	Dominant	10,410 / 9406	**2.15**	**1.84–2.50 (< 0.001)**	38.3	0.002	1.49	0.13	1.64	0.11
	Over-Dominant	10,410 / 9406	**1.88**	**1.61–2.19 (< 0.001)**	35.8	0.005	1.33	0.17	1.45	0.14
	Allelic model	10,410 / 9406	**2.05**	**1.79–2.35 (< 0.001)**	48.6	≤0.001	1.45	0.16	1.59	0.13
	GA vs. GG	10,410 / 9406	**1.97**	**1.68–2.30 (< 0.001)**	28.3	0.03	1.51	0.11	2.01	0.04
Iranian population	Dominant	1409 / 1160	**3.04**	**2.04–4.54 (< 0.001)**	37.3	0.13	−0.45	0.65	−0.53	0.61
	Over-Dominant	1409 / 1160	**2.65**	**1.74–4.05 (< 0.001)**	0	0.66	−1.05	0.29	−0.64	0.55
	Allelic model	1409 / 1160	2.09	0.88–4.94 (< 0.09)	76.8	0.008	−0.45	0.65	−0.33	0.75
	GA vs. GG	1409 / 1160	**2.67**	**1.81–4.22 (< 0.001)**	0	0.59	−1.05	0.29	−0.67	0.53
Subgroup (continent)										
Asia	Dominant	4153 / 3957	**2.80**	**2.20–3.56(< 0.001)**	35.4	0.06	−0.80	0.42	− 0.44	0.64
	Over-Dominant	4153 / 3957	**2.22**	**1.73–2.85 (< 0.001)**	47.2	0.01	−1.43	0.15	−0.98	0.34
	Allelic model	4153 / 3957	**2.35**	**1.92–2.87 (< 0.001)**	62.6	0.003	− 0.45	0.64	0.39	0.7
	GA vs. GG	4153 / 3957	**2.51**	**1.95–3.21 (< 0.001)**	11.9	0.31	−1.10	0.27	−0.35	0.73
Europe	Dominant	4913 / 3929	**1.49**	**1.20–1.84 (0.001)**	14.7	0.25	3.06	0.002	3.79	0.001
	Over-Dominant	4913 / 3929	**1.43**	**1.15–1.79 (0.002)**	16	0.23	2.99	0.003	3.65	0.001
	Allelic model	4913 / 3929	**1.48**	**1.19–1.81 (0.001)**	9.9	0.32	1.37	0.16	1.58	0.13
	GA vs. GG	4913 / 3929	**1.44**	**1.15–1.80 (0.001)**	16	0.23	2.96	0.003	3.65	0.001
South America	Dominant	761 / 1030	2.04	0.88–4.74 (0.09)	0	0.76	0.19	0.85	−0.53	0.62
	Over-Dominant	761 / 1030	2.04	0.88–4.74 (0.09)	0	0.76	0.19	0.85	−0.53	0.62
	Allelic model	761 / 1030	2	0.87–4.60 (0.1)	0	0.76	−0.19	0.85	−0.47	0.66
	GA vs. GG	761 / 1030	2.04	0.88–4.74 (0.09)	0	0.76	0.19	0.85	−0.53	0.62
Africa	Dominant	438 / 389	**5.65**	**3.15–10.14 (< 0.001)**	3.9	0.37	1.36	0.17	2.55	0.12
	Over-Dominant	438 / 389	**4.44**	**2.45–8.03 (< 0.001)**	3.2	0.37	1.36	0.17	2.41	0.13
	Allelic model	438 / 389	**5.93**	**3.38–10.40 (< 0.001)**	0	0.55	1.36	0.17	2.59	0.12
	GA vs. GG	438 / 389	**4.70**	**2.59–8.53 (< 0.001)**	12.3	0.33	1.36	0.17	2.47	0.13
Subgroup (Study design)										
Case-Control	Dominant	8658 / 9099	**2.33**	**1.99–2.74 (< 0.001)**	31.5	0.01	0.28	0.78	0.79	0.43
	Over-Dominant	8658 / 9099	**2.05**	**1.74–2.41 (< 0.001)**	29.3	0.02	1.43	0.15	1.28	0.21
	Allelic model	8658 / 9099	**2.18**	**1.8–2.52 (< 0.001)**	44.9	0.003	0.71	0.47	0.87	0.39
	GA vs. GG	8658 / 9099	**2.16**	**1.83–2.55 (< 0.001)**	18.2	0.13	1.70	0.09	1.78	0.08
Cohort	Dominant	1752 / 307	**0.90**	**0.55–1.49 (0.68)**	0	0.69	−1.23	0.47	−1.88	0.11
	Over-Dominant	1752 / 307	**0.88**	**0.54–1.46 (0.63)**	0	0.64	−1.23	0.47	−0.95	0.38
	Allelic model	1752 / 307	**0.91**	**0.56–1.49 (0.71)**	0	0.72	−1.23	0.47	− 1.27	0.25
	GA vs. GG	1752 / 307	**0.88**	**0.54–1.46 (0.63)**	0	0.64	−1.23	0.47	−1.68	0.14

Discussion

RPL has been one of the most prevalent obstetric complications, that affect more than 30% of gestations. A remarkable amount of pregnancy losses has been attributed to genetic variations, of which over 50% have been related to chromosomal abnormalities. Several investigations have reported the association of FVL 1691G > A mutation with RPL; that notwithstanding, there have been conflicting results among various ethnicities. The inconsistent results have been attributed to variety in the race of included subjects, different diagnostic criteria of patients, little statistical power, small sample sizes, and the linkage disequilibrium (LD) between various genes and variations [88]. However, meta-analysis strategy provides a pertinent tool to settle the problem of confliction by resolving the limitations of single replication studies, such as limited statistical power and little sample size. Thus, here we conducted

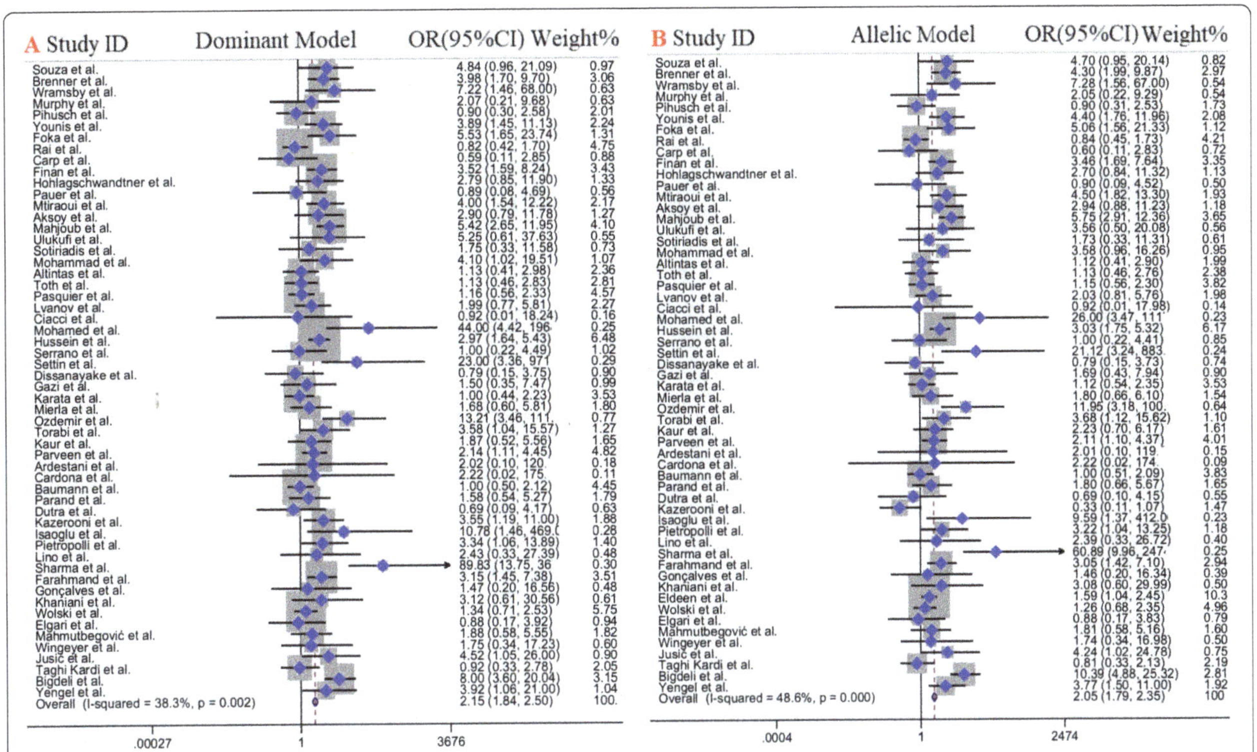

Fig. 2 Pooled odds OR and 95% confidence interval of individual studies and pooled data for the association between FVL 1691G > A mutation and the risk of RPL in overall populations for **a**; Dominant Model, **b**; Allelic Model

the first meta-analysis to find a valid estimation of the association between FVL 1691G > A mutation and risk of RPL.

The FVL 1691G > A mutation is a G-to-A point mutation at nucleotide 1691 in the factor V gene, that results in the single amino-acid replacement Arg506Gln, leading to resistance to be cleaved and, therefore, inactivation by APC and promoted susceptibility to clotting [89, 90]. This mutation enhances the risk of venous thrombosis up to 50–100 times in homozygote carriers [22].

In this meta-analysis, 62 studies, containing 10,410 cases and 9406 controls, were included in quantitative analysis. The analysis of overall population indicated that all genetic comparisons of the FVL 1691G > A mutation, including dominant model (OR = 2.15), over-dominant model (OR = 1.88), allelic model (OR = 2.05), and heterozygote model (OR = 1.97) significantly increased the risk of RPL susceptibility. In 2015, Sergi et al. [91] by including nine studies, containing a total of 2147 women for the FVL mutation, 1305 women with early RPL, and 842 women with no gestational complications, indicated higher carrier frequency of FVL mutation in women with early RPL (OR = 1.68). Moreover, Marcelo and colleagues [92] in 2019 revealed that there was no association between recurrent miscarriage and inherited thrombophilias in patients with polycystic ovarian syndrome, with respect to FVL (OR = 0.74; 95% CI = 0.38–1.45; $P = 0.38$), among others. On the other hand, a comprehensive systematic review and meta-analysis in 2016 [93], by exerting 369 articles evaluating 124 polymorphisms of 73 genes, to explore the potential genetic biomarkers for recurrent miscarriage identified increased risk of the disease in the recessive and over-dominant models, but a decreased risk in the dominant and allelic models for FVL 1691G > A mutation, both in overall analysis and subgroup analysis in Caucasians. Our analysis is unique of its type, as it included only patients having RPL diagnosis. Moreover, our subgroup analysis based on the continent of the study population divulged a strong association between FVL 1691G > A mutation and the risk of RPL in Asian, European, and Africa populations, but not in South Americans. It should be noted that among the 62 case-control studies included, 25 studies were in Asia, 26 studies in Europe, 6 studies in South America, 4 studies in Africa, and 1 study in Oceania. Although the subgroup analysis of 6 studies in South America indicated an OR < 1 (which was not significant across all genetic models), all other populations (which made large portion of the studies included) had OR > 1, imply that the South America data had little effect on the pooled effect estimation. The other parameter for subgroup analysis was study design. In this regard, a significant positive association between FVL 1691G > A mutation and the risk of

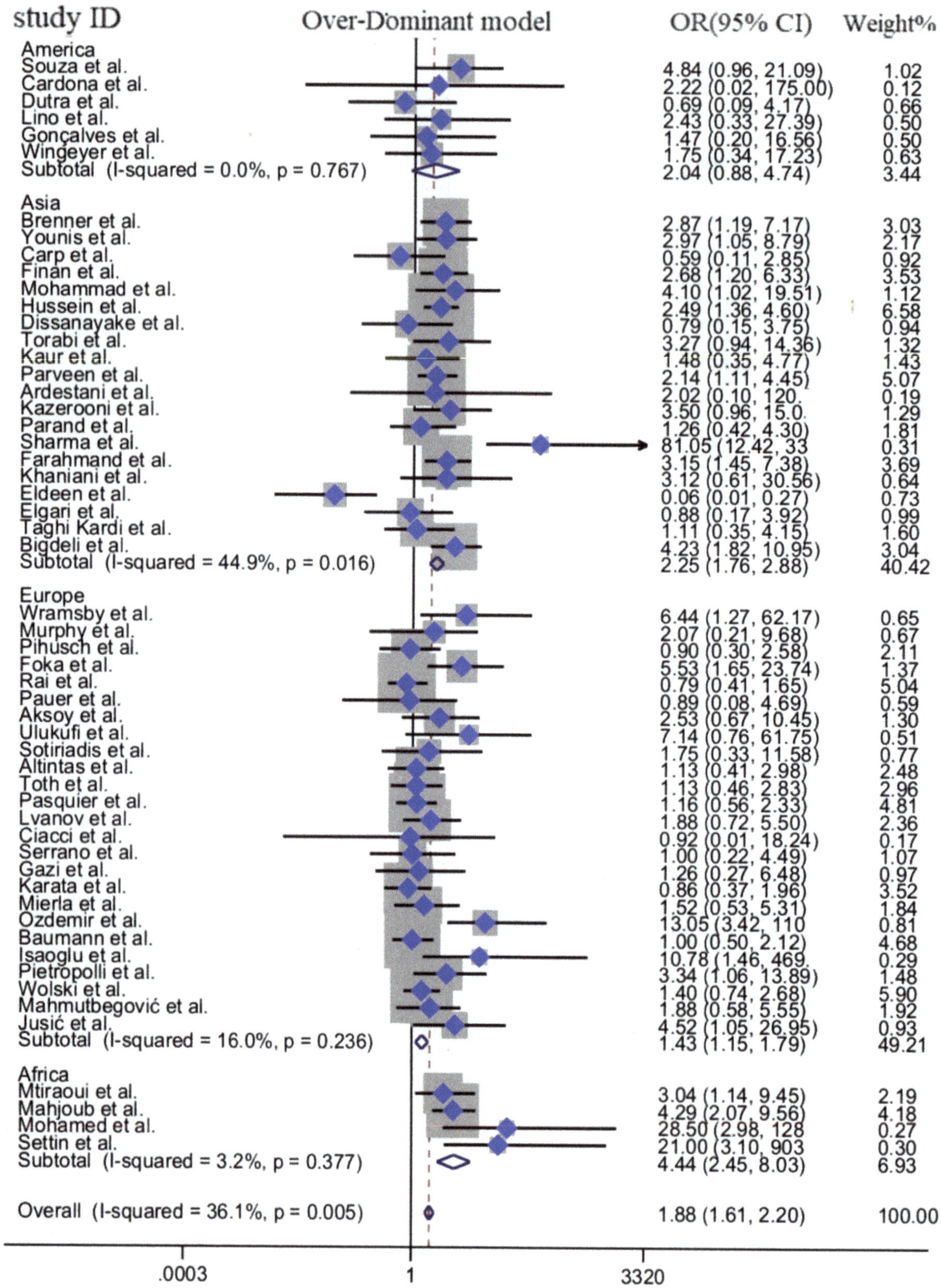

Fig. 3 Pooled OR and 95% CI of individual studies and pooled data for the association between FVL 1691G > A mutation and the risk of RPL in different continents based on subgroup analysis for Over-Dominant model

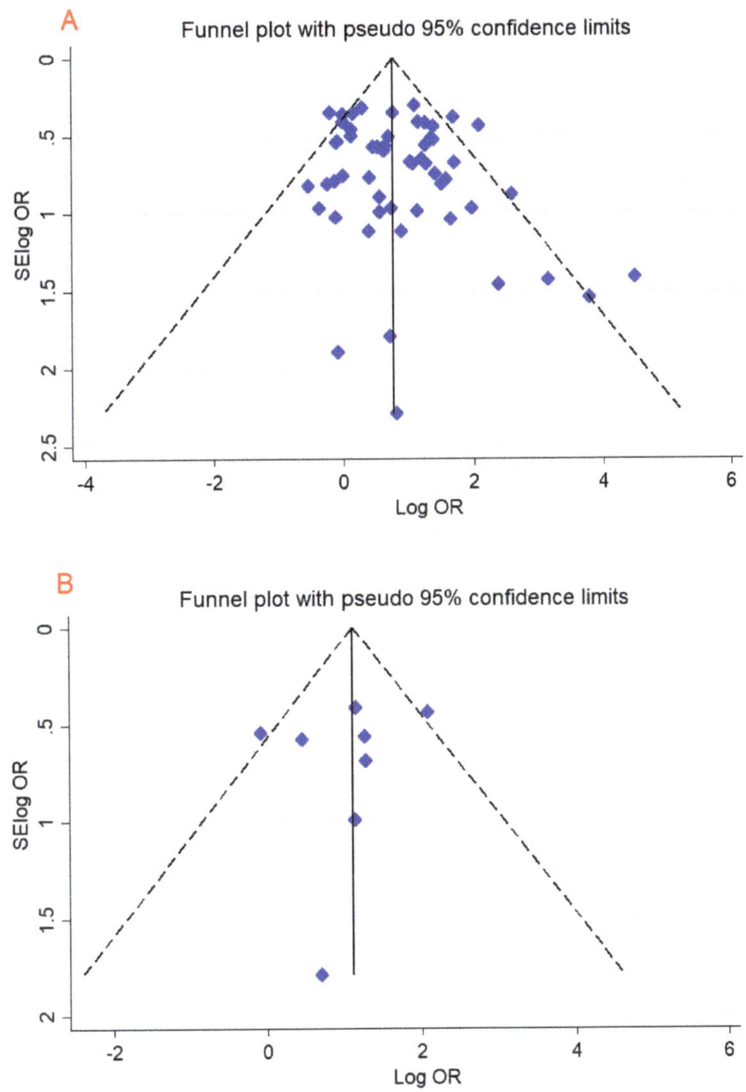

Fig. 4 Begg's funnel plot for publication bias test for the association between FVL 1691G > A mutation and the risk of RPL in the dominant model; **a**:overall population, **b**: Iranian studies . Each point represents a separate study for the indicated association

RPL was observed in case-control studies, while cohort studies revealed no such association. The result of this subgroup should interpret with caution because of imbalance between included studies in each group (60 vs. 2).

On the other side, the analysis was also performed in the Iranian population, containing 9 publications with 1409 cases and 1160 controls. The previous meta-analysis in Iranian population by Kamali et al. [94] in 2018, by employing 7 studies, indicated significant increased risk of RPL only in the allelic (OR = 2.252) and dominant models (OR = 2.217). However, our analysis indicated that the measured genetic models, including dominant model (OR = 2.97), over-dominant model (OR = 2.58), and heterozygote model (OR = 2.67, 95%) increased the risk of RPL. The difference between our analysis and the previous one was that we included two more study with higher sample size.

There was a degree of heterogeneity during the overall analysis. From statistical perspective, this heterogeneity describes the variability between included studies and may originate from clinical or methodological heterogeneity, from other unreported, unknown study characteristics, or may be due to chance. Therefore, for finding any sources of heterogeneity and attenuating their effects, we conducted subgroup analysis and weighted meta-regression. Collectively, the results of meta-regression showed that none of the parameters, including publication year, the continent of the study population, and genotyping methods were the expected source of heterogeneity. However, subgroup analysis reduced heterogeneity in all groups and explained part of the

Table 4 Meta-regression analyses of potential source of heterogeneity

Heterogeneity Factor		Coefficient	SE	T-test	P-value	95% CI	
						UL	LL
Publication Year	Dominant	0.296	0.31	0.85	0.39	−0.365	0.905
	Over-Dominant	0.211	0.26	0.79	0.43	−0.325	0.747
	Allelic model	0.159	0.20	0.77	0.44	−0.257	0.576
	GA vs. GG	0.253	0.29	0.86	0.39	−0.341	0.848
Continent	Dominant	0.879	1.92	0.46	0.65	−2.99	4.74
	Over-Dominant	0.498	1.63	0.30	0.76	−2.79	3.78
	Allelic model	0.650	1.27	0.51	0.61	−1.90	3.20
	GA vs. GG	0.72	1.80	0.40	0.69	−2.90	4.35
Genotyping Methods	Dominant	−0.04	1.35	−0.04	0.97	−2.76	2.66
	Over-Dominant	0.028	1.15	0.02	0.98	−2.29	2.35
	Allelic model	− 0.115	0.89	−0.13	0.89	−1.92	1.68
	GA vs. GG	0.016	1.26	0.01	0.98	−2.52	2.55

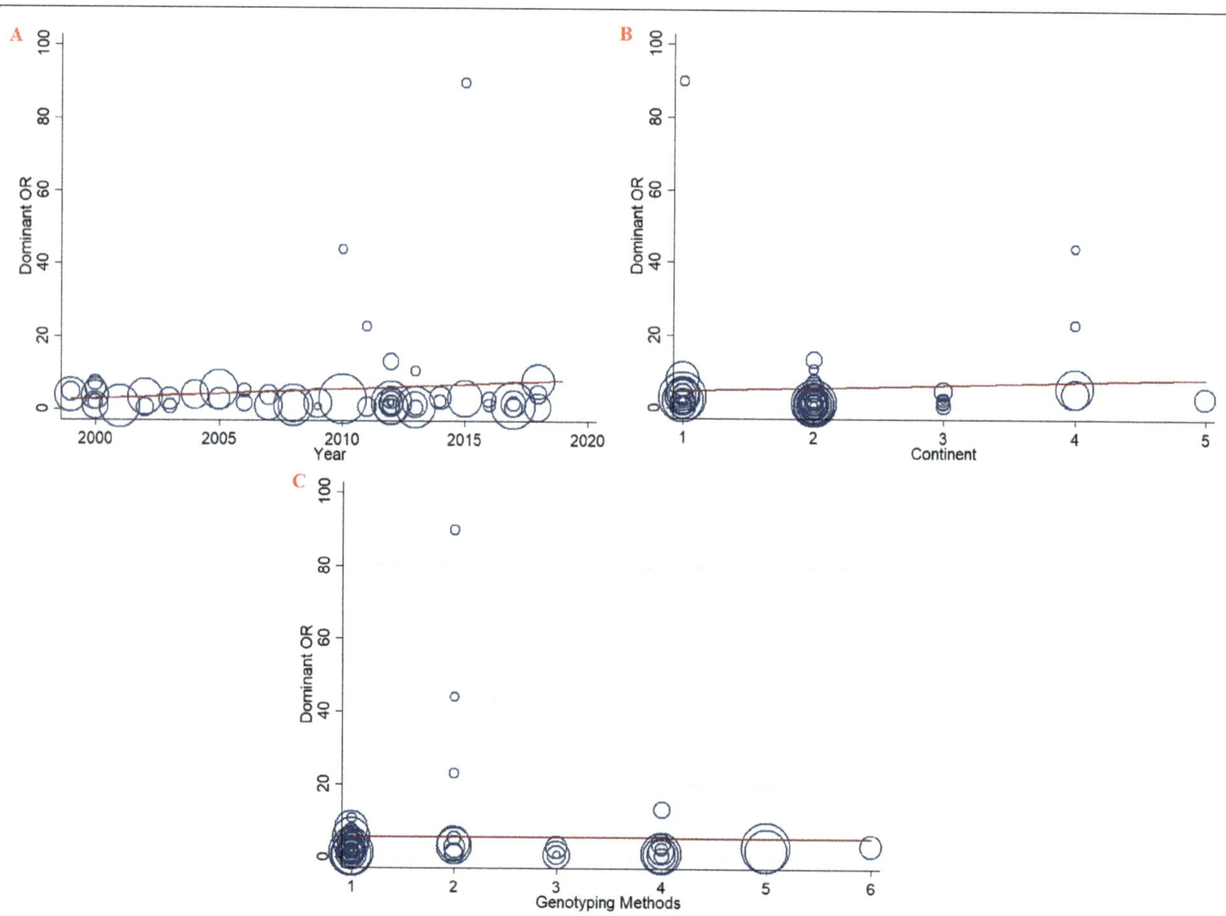

Fig. 5 Meta-regression plots of the association between FVL 1691G > A mutation and risk of RPL (Dominant model) based on; **a**: Publication year, **b**: Continent, **c**: Genotyping methods

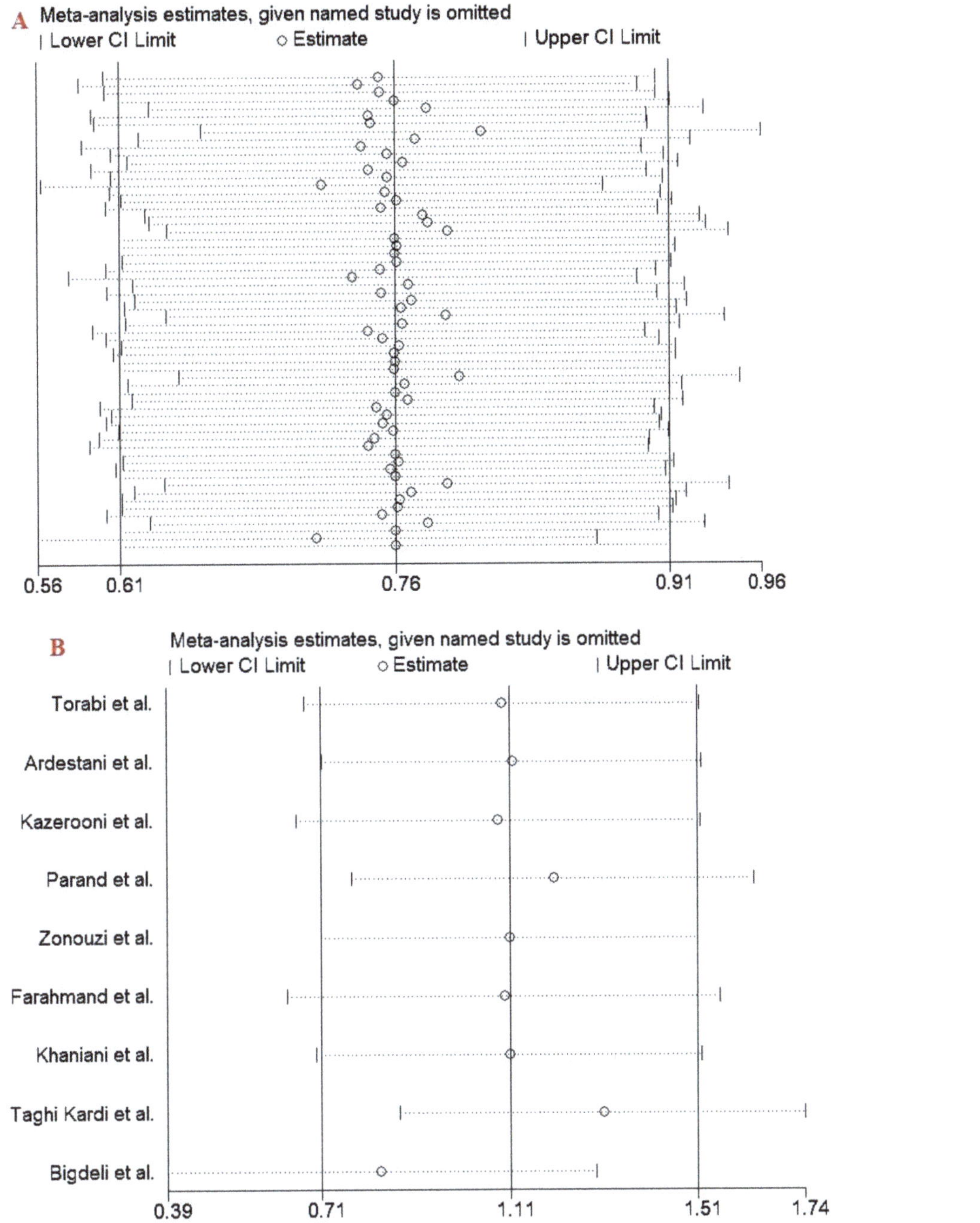

Fig. 6 Sensitivity analysis in the present meta-analysis investigates the association of FVL 1691G > A mutation an risk of RPL; **a**: overall population, **b**: Iranian studies

observed heterogeneity expect Asians and studies with cohort design. Furthermore, the other way of dealing with statistical heterogeneity, which we used in our analysis, was to incorporate "Random" term to account for it in a random-effects. Random effect model typically produces more conservative estimates of the significance of a result (a wider confidence interval). As it gives proportionately higher weights to smaller studies and lower weights to larger studies than fixed effect analysis.

To address the limitations in the current meta-analysis, it should be stated that, first our literature search was limited to only studies published in English language. Second, there was a degree of heterogeneity during the overall analysis. But not in all subgroup analyses, indicating the role of genetic diversity and other confounders in susceptibility to RPL. Third, as this meta-analysis a crude estimation of the association between FVL 1691G > A mutation and the risk of RPL, thus the roles of age, paternal genetic impression, environmental factors, and the effect of gene-gene interactions in conferring the susceptibility risk to RPL were neglected.

Considering all the facts, this meta-analysis, the first one of its type to our best knowledge, retrieved 62 studies, encompassing 10,410 cases and 9406 health controls, to find a consistent result of the association between FVL 1691G > A mutation and risk of RPL. Our results indicated statistically significant increased risk of RPL in the overall analysis. The increased susceptibility to RPL was also observed in Iranian, Asian, European, Africa populations, and studies with case-control design, but not in South Americans and studies with cohort design. Further experiments, alongside with inclusion of additional studies with large sample sizes, should consider the role cofounders in susceptibility to RPL.

Abbreviations

FVL: Factor V leiden; RPL: Recurrent pregnancy loss; MTHFR: Methylenetetrahydrofolate reductase; CI: Confidence interval; OR: Odds ratio; SNP: Single-nucleotide polymorphism; PRISMA: Preferred reporting items for systematic reviews and meta-analyses; NOS: Newcastle–Ottawa scale; HWE: Hardy–Weinberg equilibrium

Acknowledgments

We would like to thank Dr. Saeed Aslani for his valuable comments that greatly improved the manuscript.

Authors' contributions

ME and BR originated the study, acquired data. BR and MK performed statistical analysis, interpreted data, drafted the manuscript. SA revised the manuscript. SA and MS approved the manuscript. All authors read and approved the final manuscript.

Author details

[1]Department of Hematology, Faculty of Medical Sciences, Tarbiat Modares University, North Kargar Av, Tehran 14117, Iran. [2]Department of Basic sciences, Maragheh University of medical sciences, Maragheh, Iran. [3]Rahat Breach and Sleep Research Center, Tabriz University of Medical Sciences, Tabriz, Iran.

References

1. Reddy UM. Management of pregnancy after stillbirth. Clin Obstet Gynecol. 2010;53(3):700–9.
2. Ford HB, Schust DJ. Recurrent pregnancy loss: etiology, diagnosis, and therapy. Rev Obstet Gynecol. 2009;2(2):76.
3. von Eye Corleta H. It is time to respect the American Society for Reproductive Medicine definition of recurrent pregnancy loss. Fertil Steril. 2010;94(4):e61.
4. Rai R, Regan L. Recurrent miscarriage. Lancet. 2006;368(9535):601–11.
5. Li T, et al. Recurrent miscarriage: aetiology, management and prognosis. Hum Reprod Update. 2002;8(5):463–81.
6. Allison JL, Schust DJ. Recurrent first trimester pregnancy loss: revised definitions and novel causes. Curr Opin Endocrinol, Diabetes Obes. 2009; 16(6):446–50.
7. McNamee K, Dawood F, Farquharson R. Recurrent miscarriage and thrombophilia: an update. Curr Opin Obstet Gynecol. 2012;24(4):229–34.
8. Obstetricians, A.C.o. and Gynecologists, Frequently asked questions 100, pregnancy: repeated miscarriages [Internet]. Washington, DC: The American College of Obstetricians and Gynecologists; c2017 [cited 2017 Oct 25].
9. Brenner B. Haemostatic changes in pregnancy. Thromb Res. 2004;114(5–6): 409–14.
10. Uchikova EH, Ledjev II. Changes in haemostasis during normal pregnancy. Eur J Obstet Gynecol Reprod Biol. 2005;119(2):185–8.
11. Buchholz T, Thaler CJ. Inherited thrombophilia: impact on human reproduction. Am J Reprod Immunol. 2003;50(1):20–32.
12. Di Micco P, et al. Recurrent pregnancy loss and thrombophilia. Clin Lab. 2007;53(5/6):309.
13. Coulam CB, et al. Multiple thrombophilic gene mutations rather than specific gene mutations are risk factors for recurrent miscarriage. Am J Reprod Immunol. 2006;55(5):360–8.
14. Foka Z, et al. Factor V Leiden and prothrombin G20210A mutations, but not methylenetetrahydrofolate reductase C677T, are associated with recurrent miscarriages. Hum Reprod. 2000;15(2):458–62.
15. Grandone E, et al. Factor V Leiden is associated with repeated and recurrent unexplained fetal losses. Thromb Haemost. 1997;77(05):0822–4.
16. Reznikoff-Etiévant M, et al. Factor V Leiden and G20210A prothrombin mutations are risk factors for very early recurrent miscarriage. Br J Obstet Gynaecol. 2001;108(12):1251–4.
17. Serrano F, et al. Factor V Leiden and prothrombin G20210A in Portuguese women with recurrent miscarriage: is it worthwhile to investigate? Arch Gynecol Obstet. 2011;284(5):1127–32.
18. Kujovich JL. Factor v Leiden thrombophilia. Genetics in Medicine. 2011;13(1): 1.
19. Lindqvist PG, et al. Activated protein C resistance (FV: Q506) and pregnancy. Thromb Haemost. 1999;81(04):532–7.
20. Svensson PJ, Dahlback B. Resistance to activated protein C as a basis for venous thrombosis. N Engl J Med. 1994;330(8):517–22.
21. Mohammadi, M.M.A., M.G. Al-Halabi, and F.M.S. Monem, Prevalence of factor V Leiden mutation and its relation with recurrent spontaneous pregnancy loss in a group of Syrian women. 2007.
22. Rodeghiero F, Tosetto A. Activated protein C resistance and factor V Leiden mutation are independent risk factors for venous thromboembolism. Ann Intern Med. 1999;130(8):643–50.
23. Baek K-H, Lee E-J, Kim Y-S. Recurrent pregnancy loss: the key potential mechanisms. Trends Mol Med. 2007;13(7):310–7.
24. Jivraj S, et al. Pregnancy outcome in women with factor V Leiden and recurrent miscarriage. BJOG Int J Obstet Gynaecol. 2009;116(7):995–8.
25. Moher D, et al. Preferred reporting items for systematic reviews and meta-analyses: the PRISMA statement. Ann Intern Med. 2009;151(4):264–9.
26. Stang A. Critical evaluation of the Newcastle-Ottawa scale for the assessment of the quality of nonrandomized studies in meta-analyses. Eur J Epidemiol. 2010;25(9):603–5.
27. Huedo-Medina TB, et al. Assessing heterogeneity in meta-analysis: Q statistic or I^2 index? Psychol Methods. 2006;11(2):193.
28. Mantel N, Haenszel W. Statistical aspects of the analysis of data from retrospective studies of disease. J Natl Cancer Inst. 1959;22(4):719–48.
29. DerSimonian R, Laird N. Meta-analysis in clinical trials. Control Clin Trials. 1986;7(3):177–88.
30. Begg CB, Mazumdar M. Operating characteristics of a rank correlation test for publication bias. Biometrics. 1994;50(4):1088–101.
31. Egger M, et al. Bias in meta-analysis detected by a simple, graphical test. Bmj. 1997;315(7109):629–34.
32. Dissanayake VH, et al. Candidate gene study of genetic thrombophilic polymorphisms in pre-eclampsia and recurrent pregnancy loss in Sinhalese women. J Obstet Gynaecol Res. 2012;38(9):1168–76.

33. Pihusch R, et al. Thrombophilic gene mutations and recurrent spontaneous abortion: prothrombin mutation increases the risk in the first trimester. Am J Reprod Immunol. 2001;46(2):124–31.
34. Souza S, et al. Factor V Leiden and factor II G20210A mutations in patients with recurrent abortion. Hum Reprod. 1999;14(10):2448–50.
35. Brenner B, et al. Thrombophilic polymorphisms are common in women with fetal loss without apparent cause. Thromb Haemost. 1999;82(07):6–9.
36. Wramsby M, Sten-Linder M, Bremme K. Primary habitual abortions are associated with high frequency of factor V Leiden mutation. Fertil Steril. 2000;74(5):987–91.
37. Murphy RP, et al. Prospective evaluation of the risk conferred by factor V Leiden and thermolabile methylenetetrahydrofolate reductase polymorphisms in pregnancy. Arterioscler Thromb Vasc Biol. 2000;20(1):266–70.
38. Finan RR, et al. Prevalence of factor V G1691A (factor V-Leiden) and prothrombin G20210A gene mutations in a recurrent miscarriage population. Am J Hematol. 2002;71(4):300–5.
39. Pauer HU, et al. Analyzes of three common thrombophilic gene mutations in German women with recurrent abortions. Acta Obstet Gynecol Scand. 2003;82(10):942–7.
40. Mahjoub T, et al. Association between adverse pregnancy outcomes and maternal factor V G1691A (Leiden) and prothrombin G20210A genotypes in women with a history of recurrent idiopathic miscarriages. Am J Hematol. 2005;80(1):12–9.
41. Altintas A, et al. Factor V Leiden and G20210A prothrombin mutations in patients with recurrent pregnancy loss: data from the southeast of Turkey. Ann Hematol. 2007;86(10):727–31.
42. Toth B, et al. Paternal thrombophilic gene mutations are not associated with recurrent miscarriage. Am J Reprod Immunol. 2008;60(4):325–32.
43. Biswas A, et al. Recurrent abortions in Asian Indians: no role of factor V Leiden Hong Kong/Cambridge mutation and MTHFR polymorphism. Clin Appl Thromb Hemost. 2008;14(1):102–4.
44. Mukhopadhyay R, Saraswathy KN, Ghosh PK. MTHFR C677T and factor V Leiden in recurrent pregnancy loss: a study among an endogamous group in North India. Genet Test Mole Biomark. 2009;13(6):861–5.
45. Ciacci C, et al. Early pregnancy loss in celiac women: the role of genetic markers of thrombophilia. Dig Liver Dis. 2009;41(10):717–20.
46. Mohamed MA, et al. Thrombophilic gene mutations in women with repeated spontaneous miscarriage. Genet Test Mole biomark. 2010;14(5): 593–7.
47. Hussein AS, Darwish H, Shelbayeh K. Association between factor V Leiden mutation and poor pregnancy outcomes among Palestinian women. Thromb Res. 2010;126(2):e78–82.
48. Mierla D, et al. Association of prothrombin (A20210G) and factor V Leiden (A506G) with recurrent pregnancy loss. Maedica. 2012;7(3):222.
49. Ozdemir O, et al. Recurrent pregnancy loss and its relation to combined parental thrombophilic gene mutations. Genet Test Mole Biomark. 2012;16(4):279–86.
50. Torabi R, et al. Combination of thrombophilic gene polymorphisms as a cause of increased the risk of recurrent pregnancy loss. J Reprod Infertility. 2012;13(2):89.
51. Cardona H, et al. Lack of association between recurrent pregnancy loss and inherited thrombophilia in a group of Colombian patients. Thrombosis. 2012;2012.
52. Kazerooni T, et al. Correlation between thrombophilia and recurrent pregnancy loss in patients with polycystic ovary syndrome: a comparative study. J Chin Med Assoc. 2013;76(5):282–8.
53. Parand A, et al. Inherited thrombophilia and recurrent pregnancy loss. Iran Red Crescent Med J. 2013;15(12):e13708. (PMID: 24693393).
54. Zonouzi AP, et al. The association between thrombophilic gene mutations and recurrent pregnancy loss. J Assist Reprod Genet. 2013;30(10):1353–9.
55. Pietropolli A, et al. Plasminogen activator inhibitor-1, factor V, factor II and methylenetetrahydrofolate reductase polymorphisms in women with recurrent miscarriage. J Obstet Gynaecol. 2014;34(3):229–34.
56. Sharma A, et al. Polymorphisms in factor V and antithrombin III gene in recurrent pregnancy loss: a case–control study in Indian population. J Thromb Thrombolysis. 2015;39(4):481–8.
57. Kashif S, Kashif MA, Saeed A. The association of factor V Leiden mutation with recurrent pregnancy loss. J Pak Med Assoc. 2015;65(11):1169–72.
58. Gonçalves, R.O., et al., Association between the thrombophilic polymorphisms MTHFR C677T, factor V Leiden, and prothrombin G20210A and recurrent miscarriage in Brazilian women. 2016.
59. Khaniani MS, et al. Evaluation of Thrombophilic genes in recurrent pregnancy loss: a case-control study in Iranian women. Int J Hum Genet. 2016;16(1–2):48–52.
60. Wolski H, et al. Contribution of inherited thrombophilia to recurrent miscarriage in the polish population. Ginekol Pol. 2017;88(7):385–92.
61. Elgari MM, et al. Frequency of Thrombophilic gene mutations in patients with deep vein thrombosis and in women with recurrent pregnancy loss. Open Life Sci. 2017;12(1):162–6.
62. Mahmutbegović E, et al. Prevalence of F5 1691G> a, F2 20210G> a, and MTHFR 677C> T polymorphisms in Bosnian women with pregnancy loss. Bosn J Basic Med Sci. 2017;17(4):309.
63. Bigdeli R, et al. Association between thrombophilia gene polymorphisms and recurrent pregnancy loss risk in the Iranian population. Syst Biol Reprod Med. 2018;64(4):274–82.
64. Reddy R, et al. Recurrent pregnancy loss: can factor V Leiden mutations be a cause. Obstet Gynecol Sci. 2019;62(3):179–82.
65. Xu Z, et al. Polymorphisms of F2, PROC, PROZ, and F13A1 genes are associated with recurrent spontaneous abortion in Chinese Han women. Clin Appl Thromb Hemost. 2018;24(6):894–900.
66. Kardi MT, et al. Association of factor V Leiden and prothrombin G20210A polymorphisms in women with recurrent pregnancy loss in Isfahan province, Iran. Int J Prev Med. 2018;9(13):8–14. (PMID: 29541428).
67. Badawy A, AlSel BA, Fawzy M. Factor V Leiden G1691A and Prothrombin G20210A mutations are associated with repeated spontaneous miscarriage in Northern area of Saudi Arabia. Genet Mol Res. 2017;16(4):gmr16039810.
68. Farahmand K, et al. Thrombophilic genes alterations as risk factor for recurrent pregnancy loss. J Matern Fetal Neonatal Med. 2016;29(8):1269–73.
69. Ardestani MT, et al. Case control study of the factor V Leiden and factor II G20210A mutation frequency in women with recurrent pregnancy loss. Iranian journal of reproductive medicine. 2013;11(1):61.
70. Parveen F, Shukla A, Agrawal S. Should factor V Leiden mutation and prothrombin gene polymorphism testing be done in women with recurrent miscarriage from North India? Arch Gynecol Obstet. 2013;287(2):375–81.
71. Kaur L, et al. Genetic thromobophilia in pregnancy: a case–control study among north Indian women. J Thromb Thrombolysis. 2013;35(2):250–6.
72. Ivanov PD, et al. Association of inherited thrombophilia with embryonic and postembryonic recurrent pregnancy loss. Blood Coagul Fibrinolysis. 2009; 20(2):134–40.
73. Sotiriadis A, et al. Combined thrombophilic mutations in women with unexplained recurrent miscarriage. Am J Reprod Immunol. 2007;57(2):133–41.
74. Eroglu Z, et al. Frequency of factor V Leiden (G1691A), prothrombin (G20210A) and methylenetetrahydrofolate reductase (C677T) genes mutations in woman with adverse pregnancy outcome. J Turk German Gynecol Assoc. 2006;7:195–201.
75. Aksoy M, et al. The role of thrombofilia related to factor V Leiden and factor II G20210A mutations in recurrent abortions. J Pak Med Assoc. 2005;55(3): 104–8.
76. Hohlagschwandtner M, et al. Combined thrombophilic polymorphisms in women with idiopathic recurrent miscarriage. Fertil Steril. 2003;79(5):1141–8.
77. Jusić A, et al. The association of factor V G1961A (factor V Leiden), prothrombin G20210A, MTHFR C677T and PAI-1 4G/5G polymorphisms with recurrent pregnancy loss in Bosnian women. Med Glas (Zenica). 2018;15(2): 158–63.
78. Isaoglu U, et al. The association between inherited thrombophilia and recurrent pregnancy loss in Turkish women. Clin Exp Obstet Gynecol. 2014; 41(2):177–81.
79. Baumann K, et al. Maternal factor V Leiden and prothrombin mutations do not seem to contribute to the occurrence of two or more than two consecutive miscarriages in Caucasian patients. Am J Reprod Immunol. 2013;70(6):518–21.
80. Karata S, et al. Hereditary thrombophilia, anti-beta2 glycoprotein 1 IgM, and anti-annexin V antibodies in recurrent pregnancy loss. Am J Reprod Immunol. 2012;67(3):251–5.
81. Suer N, et al. Inherited thrombophilia with recurrent pregnancy loss in Turkish women–a real phenomenon? Ginekol Pol. 2012;83(8).
82. Yengel I, Yorulmaz T, Api M. Association between FVL G1691A, FII G20210A, and MTHFR C677T and A1298C polymorphisms and Turkish women with recurrent pregnancy loss. Medicinski Glasnik. 2020;17(1).
83. Wingeyer, S.P., et al., Inherited thrombophilia and pregnancy loss. Study of an Argentinian cohort. Medicina Clínica (English Edition), 2019. 152(7): p. 249–254.

84. Lino FL, et al. Thrombophilic mutations and polymorphisms, alone or in combination, and recurrent spontaneous abortion. Clin Appl Thromb Hemost. 2015;21(4):365–72.
85. Dutra CG, et al. Lack of association between thrombophilic gene variants and recurrent pregnancy loss. Hum Fertil. 2014;17(2):99–105.
86. Mtiraoui N, et al. Prevalence of antiphospholipid antibodies, factor V G1691A (Leiden) and prothrombin G20210A mutations in early and late recurrent pregnancy loss. Eur J Obstet Gynecol Reprod Biol. 2005;119(2): 164–70.
87. Settin A, et al. Factor V Leiden and prothrombin gene mutations in Egyptian cases with unexplained recurrent pregnancy loss. Hematology. 2011;16(1):59–63.
88. Page JM, Silver RM. Genetic causes of recurrent pregnancy loss. Clin Obstet Gynecol. 2016;59(3):498–508.
89. Rosendaal F, et al. High risk of thrombosis in patients homozygous for factor V Leiden (activated protein C resistance)[see comments]. Blood. 1995; 85(6):1504–8.
90. Ridker PM, et al. Mutation in the gene coding for coagulation factor V and the risk of myocardial infarction, stroke, and venous thrombosis in apparently healthy men. N Engl J Med. 1995;332(14):912–7.
91. Sergi C, Al Jishi T, Walker M. Factor V Leiden mutation in women with early recurrent pregnancy loss: a meta-analysis and systematic review of the causal association. Arch Gynecol Obstet. 2015;291(3):671–9.
92. Cavalcante MB, et al. Coagulation biomarkers in women with recurrent miscarriage and polycystic ovarian syndrome: systematic review and meta-analysis. Geburtshilfe Frauenheilkd. 2019;79(07):697–704.
93. Shi X, et al. Maternal genetic polymorphisms and unexplained recurrent miscarriage: a systematic review and meta-analysis. Clin Genet. 2017;91(2): 265–84.
94. Kamali M, et al. Association between thrombophilic genes polymorphisms and recurrent pregnancy loss susceptibility in the Iranian population: a systematic review and meta-analysis. Iran Biomed J. 2018;22(2):78.

23

Validation of the Khorana score in acute myeloid leukemia patients: A single-institution experience

Abu-Sayeef Mirza[1*], Seongseok Yun[1,2], Najla Al Ali[2], Hannah Shin[3], Joseph Luke O'Neil[3], Maher Elharake[1], Daniel Schwartz[3], Katherine Robinson[1], Ethan Nowell[3], Grace Engle[3], Ibraahim Badat[4], Thomas Brimer[1], Amra Kuc[3], Ashton Sequeira[1], Sabbir Mirza[3], Dhiraj Sikaria[3], Jesus Diaz Vera[3], Noah Hackney[3], Sammy Abusrur[3], Jose Jesurajan[3], Jameson Kuang[3], Shreyans Patel[3], Sabrina Khalil[3], Sonya Bhaskar[3], Alexander Beard[3], Toaa Abuelenen[3], Kevin Ratnasamy[3], Nathan Visweshwar[1], Rami Komrokji[2] and Michael Jaglal[2]

Abstract

Background: Although patients with acute myeloid leukemia (AML) were shown to have an increased risk of thrombosis, no thrombosis risk assessment scoring system has been developed for AML patients. The Khorana Risk Score (KRS), which has been widely used for thrombosis risk assessment in the clinical setting, was developed on the basis of solid tumor data and has not been validated among AML patients. This study aims to validate the use of the KRS as a thrombosis risk-scoring system among patients with AML.

Methods: Using data from H. Lee Moffitt Cancer Center and Research Institution's Total Cancer Care Research Study, we retrospectively identified patients who were histologically confirmed with AML from 2000 to 2018. Clinical and laboratory variables at the time of AML diagnosis were characterized and analyzed. The thrombotic event rate was estimated with the Kaplan-Meier method and compared using the log-rank test.

Results: A total of 867 AML patients were included in the analysis. The median age at AML diagnosis was 75 years (range, 51–96), and the majority were male (65%, $n = 565$). A total of 22% ($n = 191$), 51% ($n = 445$), 24% ($n = 207$), and 3% ($n = 24$) of patients had a KRS of 0, 1, 2, and 3, respectively. A total of 42 thrombotic events (3% [$n = 6/191$] with a KRS of 1; 5% [$n = 23/445$] with a KRS of 2; 6.3% [$n = 13/207$] with a KRS of 3) were observed, with a median follow-up of 3 months (range, 0.1–307). There was no statistical difference in the risk of thrombosis between these groups ($P = .1949$).

Conclusions: Although there was an increased risk of thrombosis associated with a higher KRS among AML patients with a KRS of 1 to 3, the difference was not statistically significant. Furthermore, only a few patients were found to have a KRS > 3, and this was largely due to pancytopenia, which is commonly associated with AML. These results indicate the need for a better thrombotic risk-scoring system for AML patients.

Keywords: Acute myeloid leukemia, Khorana score, Thrombosis, Risk prediction

* Correspondence: mirzaa@mail.usf.edu
[1]Department of Internal Medicine, H. Lee Moffitt Cancer Center and Research Institution, University of South Florida, 17 Davis Blvd., Suite 308, Tampa, FL 33606, USA
Full list of author information is available at the end of the article

Background

Both solid and hematologic malignancies are associated with an increased risk of thromboembolic events [1]. One of the most commonly used risk-scoring systems for thrombosis, the Khorana Risk Score (KRS), was developed by studying patients with non-hematologic cancers receiving outpatient chemotherapy treatments. The KRS is based on body mass index (BMI) and white blood cell, hemoglobin, and platelet counts [2]. The KRS has proven to be the best validated risk assessment model for predicting thrombotic events among patients with solid tumors, given the hypercoagulable state of solid tumor malignancies [2, 3]. Despite its initial application in the ambulatory setting, its utility in inpatient settings also became evident [4].

However, the risk of developing thrombosis varies depending on the type of malignancy, and over the last decade, researchers have tried to identify the varying risks of thrombosis [3]. Recent research has shown that the KRS cannot be used to predict thrombosis among patients with certain solid organ malignancies [5, 6]. Among patients with lymphoid malignancies, such as diffuse large B cell lymphoma, the KRS was found to be inadequate for stratifying thrombosis risk, and other factors were found to be more predictive of thrombosis [7]. Therefore, a more specific prognostic scoring system was required to predict thrombosis in lymphomatous malignancies [8, 9].

Unlike patients with solid tumors, patients with acute leukemia commonly present with pancytopenia at the time of diagnosis [10]. Paradoxically, there is a risk of thrombosis in thrombocytopenic patients with hematological malignancies, but management with anticoagulation is often curtailed [11]. According to a 2017 meta-analysis, the incidence rate of thromboembolic events among patients with acute myeloid leukemia (AML) was approximately 6% [1]. According to current literature, the KRS has not been validated among patients with AML [12]. Furthermore, prophylactic and therapeutic management of thrombosis has not been standardized for patients with AML [13–15]. Thus, the goal of our current study was to validate the use of the KRS as a risk-scoring system for thrombosis among patients with AML.

Methods and materials

Patients and sample acquisition

Using data from H. Lee Moffitt Cancer Center and Research Institution's Total Cancer Care Research Study, we retrospectively identified patients who were histologically confirmed with AML from 2000 to 2018. Patients with primary and secondary AML were included. Clinical and laboratory variables at the time of AML diagnosis (including age, sex, previous cancer history, use of growth factors, underlying coagulopathy, and white blood cell, hemoglobin, and platelet counts) were characterized and annotated using descriptive statistics. Information regarding venous thrombosis after AML diagnosis and prior thrombotic events before AML diagnosis were collected from individual patient chart reviews. Additional clinical information, including cytogenetics, AML risk stratification, treatment regimens, and Eastern Cooperative Oncology Group performance statuses, were also obtained. This study was approved by the H. Lee Moffitt Cancer Center and Research Institution Scientific Review Committee (MCC #18648) and the University of South Florida Institutional Review Board (**Pro00025683**). Study data were collected and managed by using REDCap electronic data capture tools hosted at the University of South Florida [16].

Primary end point and outcome measure

The primary end point of our study was the rate of venous thromboembolic event according to the KRS and arterial embolic events were excluded from our analysis. We included both superficial and deep vein thrombotic events as our end point. Thrombosis was counted as an event when VTE was confirmed by imaging modalities including venous duplex ultrasound and CT angiogram. Also, we counted VTE as an event when there is a clear documentation on the medical chart in patients who were transferred from another medical center.

KRS calculation and statistical analyses

To calculate the KRS, we used complete blood count data from immediately before induction chemotherapy. Each of the following categories constituted a single point: pre-chemotherapy platelet count $> 350 \times 10^9$/L, leukocyte count $> 11 \times 10^9$/L, hemoglobin count < 10 g/dL, the use of erythropoiesis-stimulating agents, and a BMI > 35 kg/m^2 [2]. Intermediate-risk patients had 1 to 2 points, and high-risk patients had > 3 points. A cancer type was designated as "other" with a score of zero indicating that it was not associated with thrombosis per KRS. The VTE rate was estimated with the Kaplan-Meier method, and univariate comparisons were completed using the log-rank test. A P value $< .05$ was regarded as statistically significant. We performed additional multivariate analysis utilizing clinical and laboratory parameters such as age, gender, race, BMI, prior history of cancer, and albumin. All statistical analyses were performed using SPSS v24.0 (SAS Institute Inc., Cary, NC, USA) and GraphPad Prism 7.00 (GraphPad Software, La Jolla, California, USA).

Results

Patient characteristics

A total of 867 AML patients were included in this study, of whom 44% (n = 383) and 56% (n = 483) had de novo and secondary AML, respectively. The median age at AML diagnosis was 75 years (range, 51–96), and the majority of patients were male (65%, $n = 565$). A total of 28% ($n = 241$) of patients had a prior history of cancer (hematologic malignancies, $n = 34$; solid tumors, $n = 207$). A total of 14% ($n = 126$), had a prior history of thrombosis, however information about whether it was venous or provoked was incomplete. A total of 12% ($n = 101$), 7% ($n = 58$), and 1% ($n = 8$) of patients were treated with erythropoietin-stimulating agents, granulocyte colony-stimulating factor, or granulocyte-macrophage colony-stimulating factor, respectively, before AML diagnosis. Eighty-four percent ($n = 728$) of patients were treated with growth factors of any type after AML diagnosis. The median BMI of patients was 27.1 (range, 14.8–103), and 5% ($n = 47$) of patients had a BMI > 35. A total of 26% ($n = 229$) of patients had a white blood cell count ≥ 11 × 10^9/L; 67% ($n = 584$) of patients had a hemoglobin count < 10 g/L; and 1% ($n = 10$) of patients had a platelet count > 350 × 109/L. Additional demographic profiles and clinical parameters are described in Table 1.

Table 1 Additional clinical characteristics and demographic profiles of patients

Patient Characteristics	Value ($N = 867$)
Patient age	
median (range), y	75 (51–96)
Sex, No. (%)	
Male	565 (65)
Female	302 (35)
Race, No. (%)	
Caucasian	801 (92)
African American	18 (2)
Hispanic	25 (3)
Other	19 (2)
AML classification, No. (%)	
De novo AML	383 (44)
Secondary AML	483 (56)
AML risk stratification, No. (%)	
Favorable-risk group	21 (2)
Intermediate-risk group	485 (56)
Poor-risk group	266 (31)
History of prior or concurrent cancer, No. (%)	
Hematologic malignancies	34 (4)
Solid tumors	207 (24)
Prior thrombosis (arterial/venous)	126 (14)
Growth factors, No. (%)	
Before AML diagnosis	167 (19)
EPO	101 (12)
G-CSF	58 (7)
GM-CSF	8 (1)
Post AML diagnosis	728 (84)
Complete blood counts at diagnosis, No. (range)	
WBC (× 10^9/L)	3.165 (0.08–413.74)
Hemoglobin (g/dL)	9.3 (5.6–15.2)
Platelet (× 10^9/L)	46 (1–800)
Treatment regimen, No. (%)	
Chemotherapy	250 (29)
Hypomethylating agent	240 (28)
Best supportive care	225 (26)
BMI, No. (%)	
≤ 35 kg/m^2	820 (65)
> 35 kg/m^2	47 (5)

Abbreviations: *AML* acute myeloid leukemia, *BMI* body mass index, *EPO* erythropoietin, *G-CSF* granulocyte colony-stimulating factor, *GM-CSF* granulocyte-macrophage colony-stimulating factor, *WBC* white blood cells

KRS and thrombosis risk analyses

A total of 22% ($n = 191$), 51% ($n = 445$), 24% ($n = 207$), and 3% ($n = 24$) of patients had a KRS of 0, 1, 2, and 3, respectively (Table 2). A total of 42 thrombotic events were observed, with a median follow-up of 3 months (range, 0.1–307). A total of 3% ($n = 6/191$), 5% ($n = 23/445$), 6.3% ($n = 13/207$), and 0% ($n = 0/24$) of these thrombotic events occurred among patients with a KRS of 0, 1, 2, and 3, respectively (Fig. 1). A log-rank (Mantel-Cox) test showed no statistical difference of thrombosis risk between individual subgroups ($P = 0.1949$). In the comparison between patients with KRS 0 vs. KRS 1–3, there was no statistical difference in the rates of VTE between these two groups (3% vs. 5%, $P = 0.2555$) (Table 2). In an additional multivariate analysis including age (HR 1.056, 95%CI 0.983–1.134, $P = 0.135$), gender (HR 1.118, 95%CI 0.466–2.678, $P = 0.803$), race (HR 0.998, 95%CI 0.977–1.021, $P = 0.890$), BMI (HR 1.030, 95%CI 0.984–1.079, $P = 0.207$), prior history of cancer (HR 0.422, 95%CI 0.129–1.387, $P = 0.155$), and albumin (HR 0.808, 95%CI 0.473–1.379, $P = 0.434$), there was no clinical or laboratory parameter that was significantly associated with increased risk of VTE.

Discussion

Patients with AML were shown to have an increased risk of thrombosis, and thrombotic episodes represent significant morbidity and mortality among AML patients [17–19]. Although the KRS is the most widely used validated risk assessment model for predicting thrombotic events, it was not developed specifically for AML patients. Current predictive and prognostic models that are widely used in clinical settings do not take into account several risk factors that are relevant to AML patients [20, 21]. Therefore, we performed a single-institution retrospective study to validate the use of KRS in assessing the risk of thrombosis among AML patients.

Table 2 Khorana Risk Score and Thrombosis Event Correlation. Fisher's Exact test showed that there is no statistical VTE difference between AML patients with KRS 0 vs. KRS 1–3 ($P = 0.2555$). Also, a log-rank (Mantel-Cox) test showed no statistical difference of VTE risk between individual subgroups ($P = 0.1949$)

Khorana Score	No. (%) of patients, $n = 867$	No. (%) of thrombosis events
0	191 (22)	6 (3)
1	445 (51)	23 (5)
2	207 (24)	13 (6.3)
3	24 (3)	0 (0)
1, 2, 3, and 4 (all patients)	867 (100)	42 (5)

We observed an increased number of VTE among AML patients with a KRS of 1 to 3 compared to patients with KRS 0. However, there was no statistical difference of thrombosis risk between these groups. Furthermore, only a few patients had a KRS ≥ 3, indicating that the current KRS system is not optimal to categorize the thrombosis risk group. This limitation of the KRS for AML patients is largely due to unique clinical characteristics associated with this patient group (such as pancytopenia at the time of diagnosis). Moreover, the KRS system does not include other clinical parameters—such as female sex, older age, number of chronic comorbidities, and presence of a central venous catheter—that were previously shown to be associated with an increased risk of thrombosis among AML patients [22]. Hematopoietic growth factors (erythropoietin, granulocyte colony-stimulating factor, and granulocyte-macrophage colony-stimulating factor) play a role in activating coagulation factors and cytokines that alter coagulation, modulate hemostasis, and cause platelet aggregation—all of which contribute to thrombotic events [23–27]. In addition, the KRS does not stratify by type of thrombosis, and arterial thromboembolism may be associated with greater morbidity and mortality than venous thromboembolism [27, 28].

The pathophysiology of arterial and venous thrombosis both follow the principles of Virchow's Triad: alterations in flow, abnormal cell and molecular properties, and altered blood vessel walls [27]. The multifactorial pathogenesis of thrombosis in leukemia involves blast cells secreting prothrombotic tissue factor, cancer procoagulants, and cytokines. When combined with the release of growth factors, cytokines activate platelets [11]. These interactions are multiplied when chemotherapy induces massive cell death, which may explain why thrombosis can occur while leukemic patients are thrombocytopenic [22, 29]. According to one study, the majority (75%) of thrombotic events consist of deep vein thrombosis/pulmonary embolism, cardiac events, and cerebrovascular accidents, > 80% of which occur before or during induction chemotherapy [30]. High-risk patients who are taking chemotherapy have been a highlighted sub-population in recent studies who may benefit from routine thromboprophylaxis (https://www.nejm.org/doi/full/10.1056/NEJMoa1814630, https://www.nejm.org/doi/full/10.1056/NEJMoa1814468).

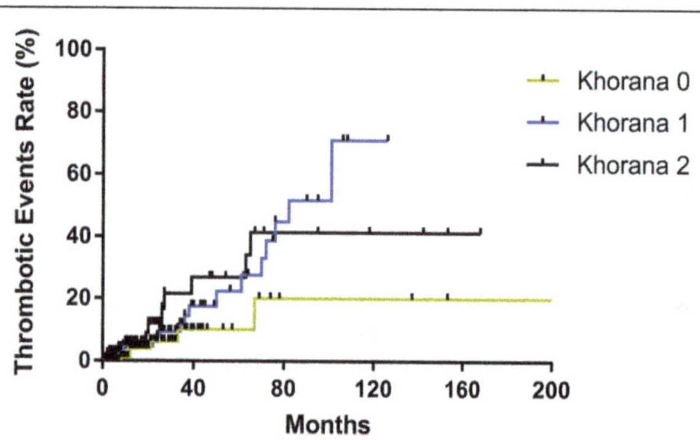

Fig. 1 Risk of Thrombosis. VTE rates are plotted based on Khorana Score over time. AML patients with KRS 3 and 4 are not included in the analysis since there was no observed VTE in these groups

In acute promyelocytic leukemia, several factors were related to a higher incidence of thrombosis: leukocytes > 10 × 109/L (9% vs 4%, $P < .01$), M3-variant subtype (11% vs 4%, $P = .02$), fibrinogen < 170 mg/dl (7% vs 3%, $P = .02$), and hemoglobin > 10 g/dl (8% vs 4%, $P = .03$) [31]. In addition to previously known prognostic factors, independent risk factors for thrombosis also include increasing age and cytogenetic risk [32, 33]. Recently, disseminated intravascular coagulation was found to be significantly predictive of venous and arterial thrombosis, with thrombosis incidence being approximately 10% among patients who were treated with multiple rounds of chemotherapy [34]. Recent data suggest that malignant leukocytes mediate both disseminated intravascular coagulation and primary hyperfibrinolysis in acute promyelocytic leukemia [35]. Additionally, both hyperleukocytosis and platelet aggregation seem to contribute to arterial thrombosis in acute leukemia [36, 37]. Collectively, these results suggest that further studies are warranted to develop an optimal thrombosis risk-scoring system that reflects the unique clinical and pathogenic features of AML patients.

Although limited by its retrospective nature, this study involved the experience of a single institution from 2000 to 2018, during which time the institution used the same diagnostic and treatment protocols and maintained follow-up data in a large comprehensive database. A large group of trained personnel conducted the extensive chart review but any threats to data quality was mitigated by using a centralized, organized extraction tool via REDCap. Another limitation of this study was the lack of routine screening for thrombotic events. Only symptomatic events were included, which may have falsely decreased observed frequency rates.

Conclusions

Among AML patients with a KRS of 1 to 3, we found a higher incidence of thrombotic events; however, the difference between these groups was not statistically significant. The proportion of patients with a KRS ≥ 3 was relatively low, which is largely due to pancytopenia, a common presentation among AML patients. These results suggest that the development of a better thrombotic risk-scoring system is warranted for AML patients.

Abbreviations

AML: Acute Myeloid Leukemia; BMI: Body Mass Index; KRS: Khorana Risk Score

Acknowledgments

We thank Paul Fletcher and Daley Drucker of H. Lee Moffitt Cancer Center and Research Institution for editorial support. They were not compensated beyond their regular salary. We thank the Morsani College of Medicine's Office of Research, Innovation, and Scholarly Endeavors (RISE) for collaborating with H. Lee Moffitt Cancer Center and Research Institution in completing this extensive chart review.

Authors' contributions

Extensive manual chart review: JLO, SA, KRo, SP, EN, TA, SB, JK, DSc, HS, AB, AK, JJ, SK, TB, ME, AS, KRa, IB, GE, SM, JDV, NH, DSi. Study design: ASM, SY, NV, RK, MJ. Organized Chart Review: ASM, MJ. Manuscript Contribution: ASM, SY, NV, MJ. Data set contribution: NAA, RK. Study supervision: MJ. All authors read and approved the final manuscript.

Authors' information

ASM - Resident Physician, Department of Internal Medicine, University of South Florida.

SY - Fellow, Department of Hematology and Oncology, H. Lee Moffitt Cancer Center and Research Institution.

NAA - Data Scientist, Malignant Hematology Department, H. Lee Moffitt Cancer Center and Research Institution.

NV - Assistant Professor, Hematology Department, University of South Florida.

RK - Senior member of the Malignant Hematology and Experimental Therapeutics Program at H. Lee Moffitt Cancer Center and Research Institution and a Professor in Medicine and Oncologic Sciences at the University of South Florida Morsani College of Medicine.

MJ - Assistant Professor, College of Medicine Internal Medicine, Department of Oncologic Sciences and a member of the Moffitt Cancer Center Medical Group faculty based at Tampa General Hospital.

Author details

[1]Department of Internal Medicine, H. Lee Moffitt Cancer Center and Research Institution, University of South Florida, 17 Davis Blvd., Suite 308, Tampa, FL 33606, USA. [2]Department of Malignant Hematology, H. Lee Moffitt Cancer Center and Research Institution, Tampa, FL, USA. [3]Morsani College of Medicine, University of South Florida, Tampa, FL, USA. [4]College of Arts and Sciences, University of South Florida, Tampa, FL, USA.

References

1. Wu YY, Tang L, Wang MH. Leukemia and risk of venous thromboembolism: a meta-analysis and systematic review of 144 studies comprising 162,126 patients. Sci Rep. 2017;7:1167.
2. Khorana AA, Kuderer NM, Culakova E, Lyman GH, Francis CW. Development and validation of a predictive model for chemotherapy-associated thrombosis. Blood. 2008;111:4902–7.
3. Khorana AA, Francis CW. Risk prediction of cancer-associated thrombosis: appraising the first decade and developing the future. Thromb Res. 2018; 164(Suppl 1):S70–s76.
4. Parker A, Peterson E, Lee AYY, de Wit C, Carrier M, Polley G, Tien J, Wu C. Risk stratification for the development of venous thromboembolism in hospitalized patients with cancer. J Thromb Haemost. 2018;16:1321–6.
5. van Es N, Franke VF, Middeldorp S, Wilmink JW, Buller HR. The Khorana score for the prediction of venous thromboembolism in patients with pancreatic cancer. Thromb Res. 2017;150:30–2.
6. Wang Y, Attar BM, Fuentes HE, Yu J, Zhang H, Tafur AJ. Performance of Khorana risk score for prediction of venous thromboembolism in patients with hepatocellular carcinoma. Clin Appl Thromb Hemost. 2018;24:471–6.
7. Rupa-Matysek J, Gil L, Kazmierczak M, Baranska M, Komarnicki M. Prediction of venous thromboembolism in newly diagnosed patients treated for lymphoid malignancies: validation of the Khorana risk score. Med Oncol. 2017;35:5.
8. Antic D, Biljana M, Milic N, Cheson BD, Narkhede M, Panovska I, Trajkova S, Popova M, Llamas P, Garcías Raso A, et al. Development and external validation of thrombosis lymphoma /Throly/ score–interim analysis. Blood. 2017;130:558.
9. Antic D, Biljana M, Milic N, Cheson BD, Narkhede M, Panovska I, Trajkova S, Popova M, Llamas P, Raso AG, et al. Internal and external validation of THROLY (thrombosis lymphoma) score. Thromb Res. 2018;164:S187.
10. Swords R, Santini V. In elderly patients with AML, which patients should be considered fit or unfit for standard induction therapy? Hematology Am Soc Hematol Educ Program. 2012;2012:74–5.
11. Del Principe MI, Del Principe D, Venditti A. Thrombosis in adult patients with acute leukemia. Curr Opin Oncol. 2017;29:448–54.
12. Sud R, Khorana AA. Cancer-associated thrombosis: risk factors, candidate biomarkers and a risk model. Thromb Res. 2009;123(Suppl 4):S18–21.
13. Zimran E, Hoffman R, Kremyanskaya M. Current approaches to challenging scenarios in myeloproliferative neoplasms. Expert Rev Anticancer Ther. 2018; 18:567–78.
14. Khan M, Cox TM, Nassif M, Alzubaidi MA, Garg N, Qiao W, Aung FM, Oo TH, Rojas-Hernandez CM. Comparative outcomes of thrombocytopenic acute leukemic patients with venous thromboembolism at a Comprehensive Cancer Center. J Thromb Thrombolysis. 2018;45:377–85.
15. Farge D, Debourdeau P, Beckers M, Baglin C, Bauersachs RM, Brenner B, Brilhante D, Falanga A, Gerotzafias GT, Haim N, et al. International clinical practice guidelines for the treatment and prophylaxis of venous thromboembolism in patients with cancer. J Thromb Haemost. 2013;11: 56–70.
16. Harris PA, Taylor R, Thielke R, Payne J, Gonzalez N, Conde JG. Research electronic data capture (REDCap)—a metadata-driven methodology and

workflow process for providing translational research informatics support. J Biomed Inform. 2009;42:377–81.
17. Stone RM. Thrombosis in AML? Yes, but when to worry? Blood. 2016;128: 1784–5.
18. Streiff MB, McCrae KR, Kuderer NM, Milentijevic D, Germain G, Laliberté F, Le N, Lefebvre P, Lyman GH, Khorana AA. Healthcare costs in patients with Cancer increase with increasing risk of venous thromboembolism. Blood. 2018;132:3799.
19. Shrotriya S, Dhakal P, Sharma M, Gardiner J, Al-Janadi A, Lupi A, Delshad A, Rayamajhi S. CAN WE predict risk factors of venous thromboembolism among Cancer inpatients? Blood. 2018;132:5871.
20. Al-Ani F, Wang YP, Lazo-Langner A. Development of a clinical prediction rule for the risk of venous thromboembolism in patients with acute leukemia. Blood. 2018;132:2223.
21. Khorana AA, Francis CW, Culakova E, Kuderer NM, Lyman GH. Thromboembolism is a leading cause of death in cancer patients receiving outpatient chemotherapy. J Thromb Haemost. 2007;5:632–4.
22. Vu K, Luong NV, Hubbard J, Zalpour A, Faderl S, Thomas DA, Yang D, Kantarjian H, Kroll MH. A retrospective study of venous thromboembolism in acute leukemia patients treated at the University of Texas MD Anderson Cancer Center. Cancer Med. 2015;4:27–35.
23. Bussolino F, Ziche M, Wang JM, Alessi D, Morbidelli L, Cremona O, Bosia A, Marchisio PC, Mantovani A. In vitro and in vivo activation of endothelial cells by colony-stimulating factors. J Clin Invest. 1991;87:986–95.
24. Kawachi Y, Watanabe A, Uchida T, Yoshizawa K, Kurooka N, Setsu K. Acute arterial thrombosis due to platelet aggregation in a patient receiving granulocyte colony-stimulating factor. Br J Haematol. 1996;94:413–6.
25. Nadir Y, Hoffman R, Brenner B. Drug-related thrombosis in hematologic malignancies. Rev Clin Exp Hematol. 2004;8:E4.
26. Sisson SD, Dinarello CA. Production of interleukin-1 alpha, interleukin-1 beta and tumor necrosis factor by human mononuclear cells stimulated with granulocyte-macrophage colony-stimulating factor. Blood. 1988;72:1368–74.
27. Visweshwar N, Jaglal M, Sokol L, et al. Hematological malignancies and arterial thromboembolism. Indian J Hematol Blood Transfus. 2019;1–14. https://doi.org/10.1007/s12288-019-01085-x, https://link.springer.com/article/10.1007%2Fs12288-019-01085-x.
28. Khorana AA, Francis CW, Culakova E, Fisher RI, Kuderer NM, Lyman GH. Thromboembolism in hospitalized neutropenic cancer patients. J Clin Oncol. 2006;24:484–90.
29. Lebois M, Josefsson EC. Regulation of platelet lifespan by apoptosis. Platelets. 2016;27:497–504.
30. Rashidi A, Silverberg ML, Conkling PR, Fisher SI. Thrombosis in acute promyelocytic leukemia. Thromb Res. 2013;131:281–9.
31. Montesinos P, de la Serna J, Vellenga E, Rayon C, Bergua J, Parody R, Esteve J, Gonzalez M, Brunet S, Sanz M. Incidence and risk factors for thrombosis in patients with acute Promyelocytic leukemia. Experience of the PETHEMA LPA96 and LPA99 protocols. Blood. 2006;108:1503.
32. Lee YG, Kim I, Kwon JH, Yoon SS, Park S, Song L, Yoon JH, Shin SH, Min WS, Kim HJ. Implications of cytogenetics for venous thromboembolism in acute myeloid leukaemia. Thromb Haemost. 2015;113:201–8.
33. O'Donnell MR, Tallman MS, Abboud CN, Altman JK, Appelbaum FR, Arber DA, Bhatt V, Bixby D, Blum W, Coutre SE, et al. Acute Myeloid Leukemia, Version 3.2017, NCCN clinical practice guidelines in oncology. J Natl Compr Cancer Netw. 2017;15:926–57.
34. Libourel EJ, Klerk CPW, van Norden Y, de Maat MPM, Kruip MJ, Sonneveld P, Löwenberg B, Leebeek FWG. Disseminated intravascular coagulation at diagnosis is a strong predictor for thrombosis in acute myeloid leukemia. Blood. 2016;128:1854–61.
35. Mantha S, Tallman MS, Soff GA. What's new in the pathogenesis of the coagulopathy in acute promyelocytic leukemia? Curr Opin Hematol. 2016; 23:121–6.
36. Fass R, Haddad M, Zaizov R, Sandbank Y, Yaniv I, Cohen IJ, Stark B, Zelikovski A. Recurrent peripheral arterial occlusion by leukemic cells sedimentation in acute promyelocytic leukemia. J Pediatr Surg. 1992;27:665–7.
37. Kafetzakis A, Foundoulakis A, Ioannou CV, Stavroulaki E, Koutsopoulos A, Katsamouris AN. Acute lower limb ischemia as the initial symptom of acute myeloid leukemia. Vasc Med. 2007;12:199–202.

Sensory evaluation of edoxaban orally disintegrating tablets: An open-label interventional study (secondary publication)

Takeshi Yamashita[1*], Joji Hagii[2], Yoshiyuki Morishima[3], Takaaki Akasaka[3], Takuyuki Matsumoto[3] and Tetsuya Kimura[3]

Abstract

Background: This study involved a sensory evaluation of edoxaban orally disintegrating (OD) tablets in patients with nonvalvular atrial fibrillation who had been receiving the existing edoxaban film-coated tablets before the study.

Methods: Edoxaban OD tablets 30 or 60 mg were prescribed for patients who had been receiving the existing 30- or 60-mg edoxaban film-coated tablets before the study. Each dose group was randomized into groups taking the tablets with or without water. After ingestion of the edoxaban OD tablet, each patient was asked to complete a sensory evaluation questionnaire (12 items).

Results: In the evaluation of satisfaction with edoxaban OD tablets, 52.8% of the patients perceived "no difference" from the existing edoxaban film-coated tablets and 34.9% indicated that they were more satisfied with the OD tablets, thus demonstrating a relatively high degree of satisfaction. When asked about convenience and reliability in using edoxaban OD tablets, about half of the patients perceived "no difference" from the existing edoxaban film-coated tablets and the remaining half indicated preference for the OD tablets. Responses about taste, flavor, ease of ingestion, and motivation to continue taking edoxaban indicated the overall acceptance of the OD tablets. Recognition of edoxaban OD tablets was rated as "easy" by about half of the patients and "difficult" by the remaining half. Among all patients, 49.5% preferred a change to edoxaban OD tablets. The degree of satisfaction with taste, flavor, and ease of ingestion, as well as overall satisfaction, tended to be greater when the OD tablets were taken with rather than without water, and the percentage of patients who preferred a change was higher in the group taking the OD tablets with water.

Conclusions: This study indicated that the degree of satisfaction with taste, flavor, ease of ingestion, and convenience, as well as overall satisfaction, in addition to motivation to continue drug intake and sense of confidence were greater for OD tablets than for the existing edoxaban film-coated tablets. Edoxaban OD tablet is a promising formulation for inducing greater patient adherence to medication and therefore ensures better treatment response.

Trial registration: UMIN-CTR UMIN000028788, registered 23-Aug-2017.

Keywords: Edoxaban, Nonvalvular atrial fibrillation, Orally disintegrating (OD) tablet, Sensory evaluation

* Correspondence: yamt-tky@umin.ac.jp
[1]The Cardiovascular Institute, 3-2-19, Nishiazabu, Minato-ku, Tokyo 106-0031, Japan
Full list of author information is available at the end of the article

Background

Some patients with diseases requiring anticoagulant therapy such as atrial fibrillation have impaired swallowing function due to reasons including advanced age and history of stroke [1]. Decreased medication adherence in these patients could be attributable to difficulty of taking standard tablets. Improvement of medication adherence is essential for a safe and effective anticoagulant therapy, as anticoagulants require long-term use. Use of orally disintegrating (OD) tablets is one way of improving medication adherence. OD tablets are easy to be taken by patients with restricted water ingestion, the elderly, and those with difficulty swallowing because they readily disintegrate in the oral cavity and can be taken anywhere without water. Thus, the use of OD tablets is expected to improve medication adherence, which results in improved therapeutic performance. Patients who changed from standard tablets to OD tablets have been reported to show improvement of medication adherence and therapeutic efficacy [2, 3].

Edoxaban (product name in Japan: Lixiana®) is a direct oral anticoagulant (DOAC) with indications for prevention of stroke and systemic embolism in patients with nonvalvular atrial fibrillation and treatment and prevention of recurrent venous thromboembolism, among others. Since 2017, edoxaban OD tablet, which is the first OD tablet formulation of DOACs, has been marketed in Japan.

This questionnaire survey was conducted for sensory evaluation of edoxaban OD tablets (a single dose of 30 or 60 mg) to assess the impression upon ingestion, sense of convenience, ease of recognition as edoxaban OD tablets, and desire to switch to edoxaban OD tablets in patients with nonvalvular atrial fibrillation who had been receiving existing edoxaban film-coated tablets (30 or 60 mg) before the study.

Methods

Patients

This study included 108 patients aged ≥20 years who had nonvalvular atrial fibrillation and visited the Cardiovascular Institute Hospital or Hirosaki Stroke and Rehabilitation Center between August 2017 and January 2018. All patients had been receiving 30- or 60-mg edoxaban film-coated tablets for ≥2 weeks, and provided written consent to participate in this study.

The following patients were excluded: those receiving dual antiplatelet therapy; those with a cardiovascular event (stroke, myocardial infarction, cardiac intervention other than myocardial infarction, or heart failure requiring hospitalization) or those hospitalized for bleeding within 2 weeks before enrollment; or those who were pregnant, lactating, or possibly pregnant.

Study design

This was an open-label interventional study, with patients given a single dose of one edoxaban OD tablet (30 or 60 mg).

On the day of the visit, each patient underwent blood sampling for measuring prothrombin time to confirm that no edoxaban tablet had been taken that day. A corresponding 30- or 60-mg edoxaban OD tablet was prescribed for each patient who had been receiving 30- or 60-mg edoxaban film-coated tablets before the start of the study. Each of these 2 groups was subdivided randomly into 2 groups (taking an OD tablet with or without water) in an allocation ratio of 1:1 (4 groups in total). A dynamic allocation was performed using study site, age at enrollment (≥75 or < 75 years), and sex as adjustment factors to avoid their influence on outcome. Simultaneous ingestion of edoxaban OD tablet with any other drugs was prohibited. The target total number of participants was 100 patients in 4 groups (set at 20 or 30 patients/group), in accordance with previous clinical studies on sensory evaluation of OD tablets [4–6].

After ingestion of the 30- or 60-mg edoxaban OD tablet, each patient was asked to answer all 12 questions in the sensory evaluation questionnaire (Table 1), and the answers to each question were analyzed. An open-ended answer was permitted for the "Reason for the selected answer."

Safety was assessed by recording any adverse events (AEs) from immediately after ingestion of edoxaban OD tablets to the following day.

The protocol was registered with the open database UMIN Clinical Trials Registry (Trial ID: UMIN000028788) prior to the start of the study, and the summarized results were also registered after completion of the study.

Statistical analyses

For patient background characteristics and answers to the sensory evaluation questionnaire (12 items), summary statistics for continuous data and numbers and percentages for categorical data were calculated. For safety evaluation, the number of patients with AEs or adverse drug reactions (ADRs) was counted.

These analyses were conducted for the subgroups (age < 75 or ≥ 75 years; presence or absence of difficulty taking the current medication). These statistical analyses were performed using SAS Ver. 9.4.

Results

In total, 108 patients were enrolled and allocated for this study. The 66 patients who were receiving 30-mg edoxaban film-coated tablets before the start of the study were allocated to group 1 (30-mg edoxaban OD tablet without water; $n = 33$) and group 3 (30-mg edoxaban OD tablet with water; $n = 33$). The 42 patients

Table 1 Sensory evaluation questionnaire

	Question
Please answer the questions about the drugs you are routinely taking.	
Question 1	Have you ever found difficulty taking any of the tablets you are routinely taking?
	1. Yes 2. No
Question 2	Have you ever been unable to identify the drug taken out of the sheet, causing you to stop taking the drug or any other difficulty?
	1. Yes 2. No
Question 3	How many types of drugs are you routinely taking every morning, including the edoxaban film-coated tablet?
	[types]
Next, please answer the questions about the edoxaban orally disintegrating (OD) tablet you have just taken.	
Question 4	What is the extent of your satisfaction with the edoxaban OD tablet you have just taken, if your satisfaction with the current medication (edoxaban film-coated tablet) rates 5?
	Quite unsatisfactory No difference Quite satisfactory
	0 1 2 3 4 5 6 7 8 9 10
Question 5	The product name and dose level are printed on this tablet. Was it easy for you to identify that the drug was edoxaban OD tablet 30 mg or 60 mg?
	1. Quite easy to identify
	2. Easier to identify than the current edoxaban film-coated tablet
	3. More difficult to identify than the current edoxaban film-coated tablet
	4. Quite difficult to identify
Question 6	What was your impression about the size of this tablet when you placed it in your mouth?
	1. Very large
	2. Slightly large
	3. Slightly small
	4. Very small
Question 7	What was your impression of the taste and flavor of this drug when it disintegrated in your mouth?
	1. Very good
	2. Relatively good
	3. Relatively bad (i: Tolerable, ii: Not tolerable)
	4. Very bad (i: Tolerable, ii: Not tolerable)
	→Please give the reason for selecting 1 through 4, if possible: [Reason:]
Question 8	Have you found this drug easier to take than the current medication (edoxaban film-coated tablet)?
	1. Quite easy to take
	2. Slightly easier to take
	3. No difference
	4. Slightly more difficult to take
	5. Quite difficult to take
	→Please give the reason for selecting 1 through 5, if possible: [Reason:]
This drug can be taken both with or without water, and simultaneously with other drugs. The product name and dose level are printed on this tablet.	
Question 9	Do you think that this drug is more convenient to take than the current medication (edoxaban film-coated tablet)?
	1. Much more convenient
	2. Slightly more convenient
	3. No difference
	4. Slightly less convenient
	5. Much less convenient

Table 1 Sensory evaluation questionnaire *(Continued)*

	Question
	→Please give the reason for selecting 1 through 5, if possible: [Reason:]
Question 10	Do you think this drug is easier to continue than the current medication (edoxaban film-coated tablet)?
	1. Much easier to continue
	2. Slightly easier to continue
	3. No difference
	4. Slightly less easy to continue
	5. Much less easy to continue
	→Please give the reason for selecting 1 through 5, if possible: [Reason:]
Question 11	Do you think that the printed product name and dose level on the tablet make this drug more reliable than the current medication (edoxaban film-coated tablet)?
	1. Much more reliable
	2. Slightly more reliable
	3. No difference
	4. Slightly less reliable
	5. Much less reliable
Question 12	This drug will be sold at the same price as the current medication (edoxaban film-coated tablet). Will you want to change to this drug after marketed?
	1. Desire to change
	2. No desire to change
	→Please give the reason for selecting 1 or 2, if possible: [Reason:]

who had been receiving 60-mg edoxaban film-coated tablets before the start of the study were allocated to group 2 (60-mg edoxaban OD tablet without water; $n = 21$) and group 4 (60-mg edoxaban OD tablet with water; $n = 21$).

Nine patients (group 1, $n = 4$; group 3, $n = 3$; and group 4, $n = 2$) discontinued the study because of violation of the inclusion/exclusion criteria ($n = 8$) and cancellation of the study drug administration at the discretion of the physician ($n = 1$). The efficacy analysis set included 106 of 108 patients, excluding 1 patient with a serious violation of the study protocol and 1 patient in whom the administration was cancelled at the discretion of the physician. The safety analysis set included 107 patients, excluding 1 patient in whom the administration was cancelled at the discretion of the physician.

Background characteristics

Table 2 shows the background characteristics of the efficacy analysis set ($n = 106$). The patients' mean (± standard deviation) age was 73.0 ± 8.6 years. Of the patients, 49.1% (52/106) were aged ≥75 years and 60.4% (64/106) were men. For the 79 patients who had data on height, the mean height was 161.6 ± 10.0 cm. The mean body weight was 61.1 ± 13.5 kg.

Sensory evaluation questionnaire

For the efficacy analysis set ($n = 106$), the responses to the sensory evaluation questionnaire for the total and subgroups are shown in Fig. 1 and Table 3, respectively. Questions 1–3 pertain to the current medication, and questions 4–12 pertain to edoxaban OD tablets.

Current medication

For question 1 (Have you ever found difficulty taking any of the tablets you are routinely taking?), 13.2% (14/106) of the respondents answered "Yes."

For question 2 (Have you ever been unable to identify the drug taken out of the sheet, causing you to stop taking the drug or any other difficulty?), 6.6% (7/106) selected "Yes."

For question 3 (How many types of drugs are you routinely taking every morning, including the edoxaban film-coated tablet?), "three types or more" was selected most frequently (78.3%, 83/106), followed by "2 types" (12.3%, 13/106) and "1 type (edoxaban film-coated tablet alone)" (9.4%, 10/106).

Edoxaban OD tablets

For question 4 (What is the extent of your satisfaction with the edoxaban OD tablet you have just taken, if your

Table 2 Background characteristics (efficacy analysis set)

		n	Mean ± SD or %
Age, y, mean ± SD	Total	106	73.0 ± 8.6
	Group 1 (30 mg without water)[a]	32	77.0 ± 6.2
	Group 3 (30 mg with water)[b]	32	76.7 ± 6.3
	Group 2 (60 mg without water)[c]	21	67.7 ± 8.9
	Group 4 (60 mg with water)[d]	21	66.9 ± 8.3
Age (≥75 years old), %	Total	52	49.1
	Group 1 (30 mg without water)[a]	21	65.6
	Group 3 (30 mg with water)[b]	20	62.5
	Group 2 (60 mg without water)[c]	6	28.6
	Group 4 (60 mg with water)[d]	5	23.8
Male, %	Total	64	60.4
	Group 1 (30 mg without water)[a]	11	34.4
	Group 3 (30 mg with water)[b]	12	37.5
	Group 2 (60 mg without water)[c]	21	100.0
	Group 4 (60 mg with water)[d]	20	95.2
Height, cm, mean ± SD	Total	79	161.64 ± 9.97
	Group 1 (30 mg without water)[a]	24	156.79 ± 8.43
	Group 3 (30 mg with water)[b]	24	156.25 ± 7.19
	Group 2 (60 mg without water)[c]	17	170.69 ± 7.93
	Group 4 (60 mg with water)[d]	14	168.19 ± 7.22
Body weight, kg, mean ± SD	Total	106	61.12 ± 13.49
	Group 1 (30 mg without water)[a]	32	52.69 ± 7.38
	Group 3 (30 mg with water)[b]	32	52.80 ± 6.71
	Group 2 (60 mg without water)[c]	21	71.85 ± 11.13
	Group 4 (60 mg with water)[d]	21	75.91 ± 10.09
Prior cerebral infarction/transient ischemic attack, %	Total	46	43.4
	Group 1 (30 mg without water)[a]	17	53.1
	Group 3 (30 mg with water)[b]	14	43.8
	Group 2 (60 mg without water)[c]	8	38.1
	Group 4 (60 mg with water)[d]	7	33.3

[a]Edoxaban OD tablet 30 mg without water
[b]edoxaban OD tablet 30 mg with water
[c]edoxaban OD tablet 60 mg without water
[d]edoxaban OD tablet 60 mg with water
OD orally disintegrating, *SD* standard deviation

satisfaction with the current medication [edoxaban film-coated tablet] rates 5?), the answers on the 11-grade scale are shown in Fig. 2. "No difference (rating: 5)" was selected most frequently (52.8%, 56/106), followed by "satisfactory (rating: 6–10)" (34.9%, 37/106) and "unsatisfactory (rating: 0–4)" (12.3%, 13/106) (Fig. 1).

For question 5 (The product name and dose level are printed on this tablet. Was it easy for you to identify that the drug was edoxaban OD tablet 30 mg or 60 mg?), "quite difficult to identify" was selected most frequently (30.5%, 32/105), followed by "easier to identify than the current edoxaban film-coated tablet" (27.6%, 29/105), "quite easy to identify" (26.7%, 28/105), and "more difficult to identify than the current edoxaban film-coated tablet" (15.2%, 16/105).

For question 6 (What was your impression about the size of this tablet when you placed it in your mouth?), "slightly large" was selected most frequently (59.2%, 58/98), followed by "slightly small" (37.8%, 37/98), "very large" (2.0%, 2/98), and "very small" (1.0%, 1/98).

For question 7 (What was your impression of the taste and flavor of this drug when it disintegrated in your mouth?), "relatively good" was selected most frequently (64.1%, 59/92), followed by "relatively bad" (18.5%, 17/92)

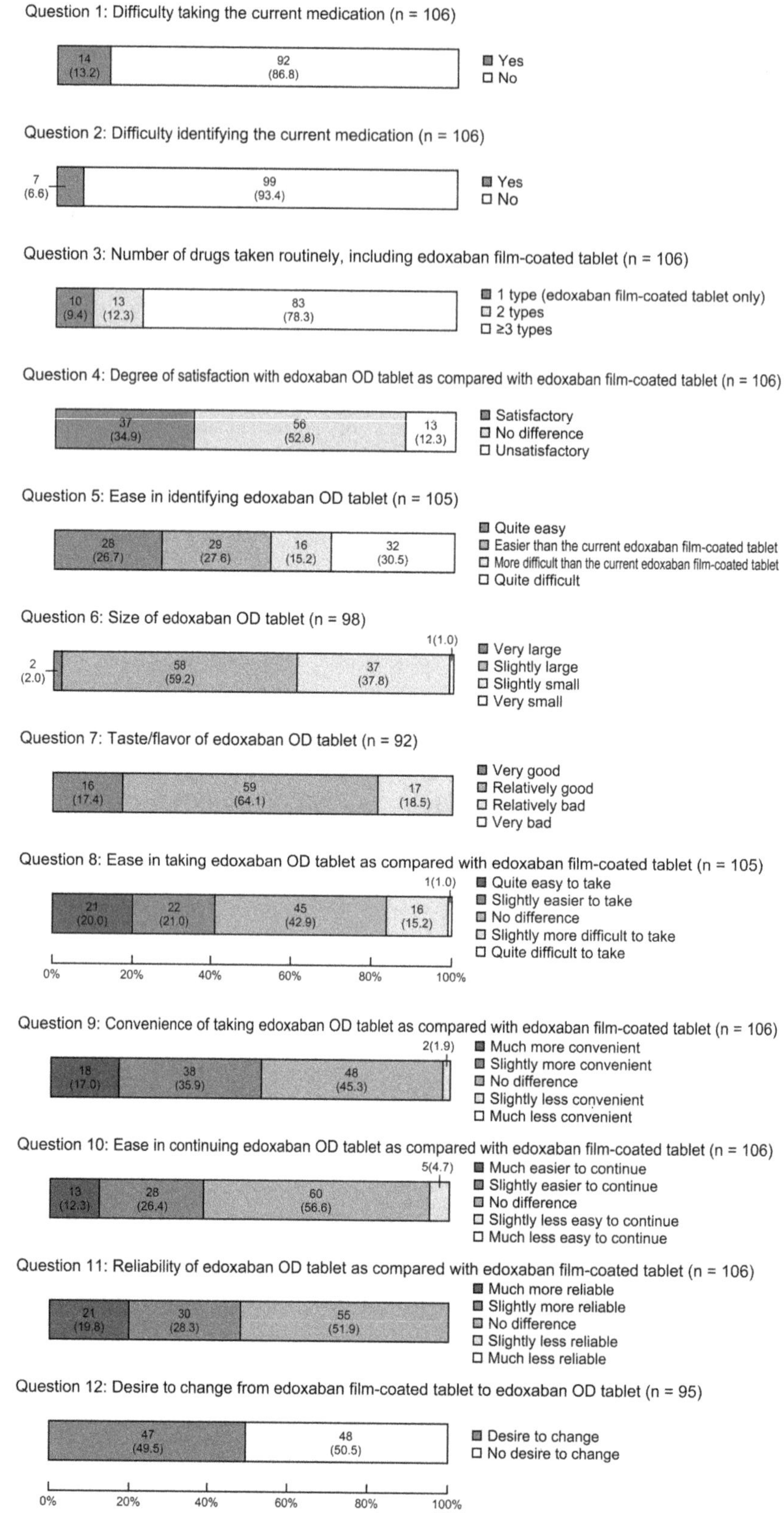

Fig. 1 Results of the sensory evaluation questionnaire survey on edoxaban OD tablets. Values in the bar graph represents the number of patients (%). *OD* orally disintegrating

Table 3 Results of the sensory evaluation questionnaire about edoxaban OD tablets shown by the group

		Group 1 (30 mg without water)[c] $N = 32$	Group 3 (30 mg with water)[d] $N = 32$	Group 2 (60 mg without water)[e] $N = 21$	Group 4 (60 mg with water)[f] $N = 21$
		n (%)	n (%)	n (%)	n (%)
Question 1: Difficulty in ingestion (current)	Yes	6 (18.8)	5 (15.6)	2 (9.5)	1 (4.8)
	No	26 (81.3)	27 (84.4)	19 (90.5)	20 (95.2)
Question 2: Difficulty in identifying (current)	Yes	2 (6.3)	3 (9.4)	0 (0.0)	2 (9.5)
	No	30 (93.8)	29 (90.6)	21 (100.0)	19 (90.5)
Question 3: Number of drugs taken simultaneously	1 type[a]	3 (9.4)	3 (9.4)	1 (4.8)	3 (14.3)
	2 types	6 (18.8)	2 (6.3)	4 (19.1)	1 (4.8)
	≥3 types	23 (71.9)	27 (84.4)	16 (76.2)	17 (81.0)
Question 4: Degree of satisfaction	Unsatisfactory	5 (15.6)	1 (3.1)	7 (33.3)	0 (0.0)
	No difference	15 (46.9)	21 (65.6)	6 (28.6)	14 (66.7)
	Satisfactory	12 (37.5)	10 (31.3)	8 (38.1)	7 (33.3)
Question 5: Ease in identifying	Quite easy	7 (21.9)	7 (21.9)	6 (30.0)	8 (38.1)
	Easier[b]	8 (25.0)	11 (34.4)	7 (35.0)	3 (14.3)
	More difficult[b]	5 (15.6)	3 (9.4)	4 (20.0)	4 (19.1)
	Quite difficult	12 (37.5)	11 (34.4)	3 (15.0)	6 (28.6)
Question 6: Size	Very large	1 (3.6)	0 (0.0)	0 (0.0)	1 (5.0)
	Slightly large	10 (35.7)	19 (61.3)	16 (84.2)	13 (65.0)
	Slightly small	16 (57.1)	12 (38.7)	3 (15.8)	6 (30.0)
	Very small	1 (3.6)	0 (0.0)	0 (0.0)	0 (0.0)
Question 7: Taste at the time of disintegration	Very good	5 (16.7)	5 (20.0)	1 (5.3)	5 (27.8)
	Relatively good	16 (53.3)	19 (76.0)	12 (63.2)	12 (66.7)
	Relatively bad	9 (30.0)	1 (4.0)	6 (31.6)	1 (5.6)
	Very bad	0 (0.0)	0 (0.0)	0 (0.0)	0 (0.0)
Question 8: Ease in ingestion (comparison)	Quite easy to take	8 (25.0)	8 (25.8)	2 (9.5)	3 (14.3)
	Slightly easier to take	6 (18.8)	8 (25.8)	3 (14.3)	5 (23.8)
	No difference	9 (28.1)	15 (48.4)	9 (42.9)	12 (57.1)
	Slightly more difficult to take	8 (25.0)	0 (0.0)	7 (33.3)	1 (4.8)
	Quite difficult to take	1 (3.1)	0 (0.0)	0 (0.0)	0 (0.0)
Question 9: Convenience	Much more convenient	7 (21.9)	3 (9.4)	2 (9.5)	6 (28.6)
	Slightly more convenient	9 (28.1)	15 (46.9)	6 (28.6)	8 (38.1)
	No difference	14 (43.8)	14 (43.8)	13 (61.9)	7 (33.3)
	Slightly less convenient	2 (6.3)	0 (0.0)	0 (0.0)	0 (0.0)
	Much less convenient	0 (0.0)	0 (0.0)	0 (0.0)	0 (0.0)
Question 10: Ease of continuation	Much easier to continue	6 (18.8)	2 (6.3)	1 (4.8)	4 (19.1)
	Slightly easier to continue	6 (18.8)	13 (40.6)	6 (28.6)	3 (14.3)
	No difference	17 (53.1)	17 (53.1)	12 (57.1)	14 (66.7)
	Slightly less easy to continue	3 (9.4)	0 (0.0)	2 (9.5)	0 (0.0)
	Much less easy to continue	0 (0.0)	0 (0.0)	0 (0.0)	0 (0.0)
Question 11: Reliability	Much more reliable	2 (6.3)	7 (21.9)	6 (28.6)	6 (28.6)
	Slightly more reliable	11 (34.4)	9 (28.1)	5 (23.8)	5 (23.8)
	No difference	19 (59.4)	16 (50.0)	10 (47.6)	10 (47.6)
	Slightly less reliable	0 (0.0)	0 (0.0)	0 (0.0)	0 (0.0)
	Much less reliable	0 (0.0)	0 (0.0)	0 (0.0)	0 (0.0)

Table 3 Results of the sensory evaluation questionnaire about edoxaban OD tablets shown by the group *(Continued)*

		Group 1 (30 mg without water)[c] N = 32	Group 3 (30 mg with water)[d] N = 32	Group 2 (60 mg without water)[e] N = 21	Group 4 (60 mg with water)[f] N = 21
		n (%)	n (%)	n (%)	n (%)
Question 12: Desire to change	Desire to change	12 (44.4)	19 (63.3)	6 (30.0)	10 (55.6)
	No desire to change	15 (55.6)	11 (36.7)	14 (70.0)	8 (44.4)

[a]Edoxaban film-coated tablet only
[b]compared with the current medication
[c]edoxaban OD tablet 30 mg without water
[d]edoxaban OD tablet 30 mg with water
[e]edoxaban OD tablet 60 mg without water
[f]edoxaban OD tablet 60 mg with water
OD orally disintegrating

and "very good" (17.4%, 16/92). "Very bad" was not selected by any of the patients. Of the patients who selected "relatively bad," 15 stated that it was "tolerable" and 1 stated that it was "not tolerable." Of the patients who selected "relatively bad," 15 took the tablet without water. The major reasons for responding "good" were "no concerns," "easy to swallow," and "slightly sweet and smoothly dissolving," and those for selecting "relatively bad" were "bitter," "sour," "not dissolving smoothly," and "slightly starchy."

For question 8 (Have you found this drug easier to take than the current medication [edoxaban film-coated tablet]?), "no difference" was selected most frequently (42.9%, 45/105), followed by "slightly easier to take" (21.0%, 22/105), "quite easy to take" (20.0%, 21/105), "slightly more difficult to take" (15.2%, 16/105), and "quite difficult to take" (1.0%, 1/105). Of the patients who selected "slightly more difficult to take" or "quite difficult to take," 16 took the tablet without water. The major reasons for selecting "easy to take" were "sweet," "tasty," and "smoothly dissolving," and those for selecting "difficult to take" were "easier to take with water," "bitter," and "not smoothly dissolving."

For question 9 (Do you think that this drug is more convenient to take than the current medication [edoxaban film-coated tablet]?), "no difference" was selected most frequently (45.3%, 48/106), followed by "slightly more convenient" (35.9%, 38/106), "much more convenient" (17.0%, 18/106), and "slightly less convenient" (1.9%, 2/106). "Much less convenient" was not selected by any of the patients. The major reasons for selecting "more convenient" were "can be taken without water" and "does not require effort to swallow," and those for selecting "no difference" was "I take it with water, simultaneously with other drugs."

For question 10 (Do you think this drug is easier to continue than the current medication [edoxaban film-coated tablet]?), "no difference" was selected most frequently (56.6%, 60/106), followed by "slightly easier to continue" (26.4%, 28/106), "much easier to continue" (12.3%, 13/106), and "slightly less easy to continue" (4.7%, 5/106). "Much

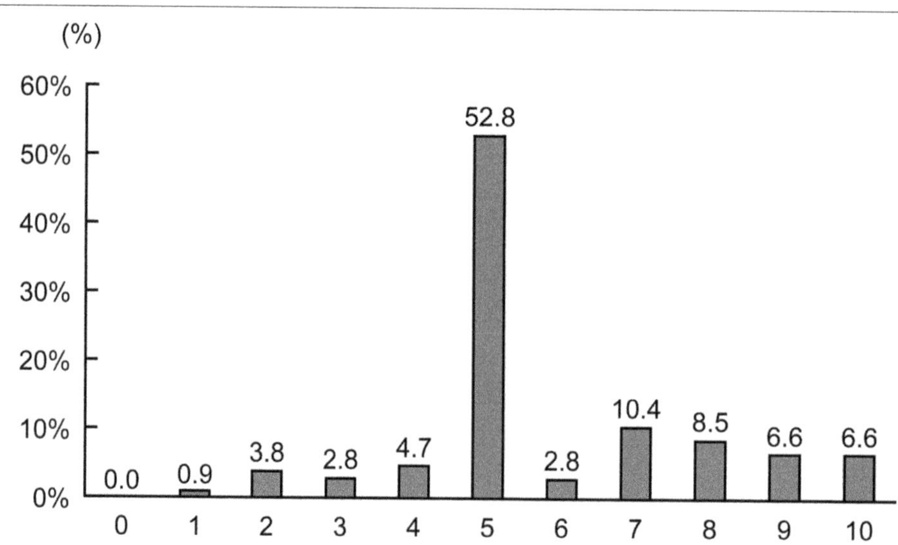

Fig. 2 The histogram of the degree of satisfaction with edoxaban OD tablets (question 4). The graph represents the frequency of patients in each answer (11-grade scale) to the question 4 of the sensory evaluation questionnaire. 0: Quite unsatisfactory, 5: no difference, and 10: quite satisfactory. *OD* orally disintegrating

less easy to continue" was not selected by any of the patients. The major reasons for selecting "easier to continue" were "can be taken without water" and "the smoothly dissolving nature may be advantageous," and those for selecting "no difference" were "both existing film-coated tablet and OD tablet are easy to take" and "finds no difference," and those for selecting "slightly less easy to continue" were "bitter" and "it is better when taken with water."

For question 11 (Do you think that the printed product name and dose level on the tablet make this drug more reliable than the current medication [edoxaban film-coated tablet]?), "no difference" was selected most frequently (51.9%, 55/106), followed by "slightly more reliable" (28.3%, 30/106) and "much more reliable" (19.8%, 21/106). "Slightly less reliable" and "much less reliable" were not selected by any of the patients.

For question 12 (This drug will be sold at the same price as the current medication [edoxaban film-coated tablet]. Will you want to change to this drug after marketed?), "No desire to change" was selected by 50.5% (48/95) and "desire to change" was selected by 49.5% (47/95). The major reasons for "no desire to change" were "water is needed when taking it together with other drugs," "I have some other drugs to take simultaneously," "it does not taste good," and others, and those for selecting "desire to change" were "easier to take," "less likely to incorrectly identify," and others.

Comparison of sensory evaluation between ingestion with and without water

The comparison of sensory evaluation between ingestion without water (groups 1 and 2) and with water (groups 3 and 4) is shown in Fig. 3 and Table S1 (Additional file 1). Figure 3 shows the percentages of patients who answered "5" (no difference) through "10" (quite satisfactory) in question 4; "1" (very good) or "2" (relatively good) in questions 5 and 7; "1" (very large) or "2" (slightly large) in question 6; "1" (very good) through "3" (no difference) in questions 8–11; and "1" (desire to change) in question 12. The major items found to differ between ingestion with and without water were the degree of satisfaction (question 4), taste and flavor at the time of edoxaban OD tablet disintegration (question 7), ease of ingestion (question 8), and desire a change to edoxaban OD tablets from the current medication (question 12).

"Unsatisfactory" for question 4 (degree of satisfaction) was selected by 1.9% (1/53) and 22.6% (12/53) of the patients who ingested edoxaban OD tablets with and without water, respectively. "Relatively bad" for question 7 (taste and flavor) was selected by 4.7% (2/43) and 30.6% (15/49) of the patients who ingested edoxaban OD tablets with and without water, respectively. "Slightly more difficult to take" or "quite difficult to take" for question 8 (ease of ingestion compared with edoxaban film-coated tablets) was selected by 1.9% (1/52) and 30.2% (16/53) of the patients who ingested the drug with and without water, respectively. "Desire to change" to the OD tablet for question 12 was selected by 60.4% (29/48) and 38.3% (18/47) of the patients who ingested the drug with and without water, respectively.

Subgroup analysis

Table 4 shows the results of subgroup analysis of sensory evaluation of edoxaban OD tablets by age (< 75 and ≥ 75 years) and response to question 1 (difficulty taking the current medication).

Age < 75 and ≥ 75 years

The major items found to differ between the < 75 and ≥ 75 year age groups were reliability (question 11) and the desire to change to the OD tablets (question 12). "Much more reliable" was selected by 29.6% (16/54) of the patients aged < 75 years and by 9.6% (5/52) of those aged ≥75 years. "Desire to change" to edoxaban the OD tablets for question 12 was selected by 57.1% (28/49) of the patients aged < 75 years and by 41.3% (19/46) of those aged ≥75 years.

Presence/absence of difficulty taking the current medication

The major items found to differ between the patients who perceived difficulty taking the current medication and those who did not perceive difficulty were "ease of ingestion" (question 8), "ease in continuing drug ingestion" (question 10), and "desire change to the OD tablet" (question 12). Regarding the ease of taking edoxaban OD tablets (question 8), "quite easy to take" was selected by 42.9% (6/14) of those who perceived difficulty taking the current medication and 16.5% (15/91) of those who did not. Regarding ease of continuing drug ingestion (question 10), "much easier to continue" was selected by 35.7% (5/14) of those who perceived difficulty taking the current medication and 8.7% (8/92) of those who did not. "Desire to change" to edoxaban OD tablets in question 12 was selected by 76.9% (10/13) of those who perceived difficulty taking the current medication and 45.1% (37/82) of those who did not.

AEs and ADRs

During the study period, 4 AEs occurred in 4 patients, including palpitation, diarrhea, and chest discomfort (1 event each) in the 30-mg edoxaban OD tablet group and excessive gastrointestinal motility (1 event) in the 60-mg edoxaban OD tablet group. No serious AEs or AEs causally related with edoxaban OD tablets were observed. All AEs were confirmed to be recovered after the end of the study.

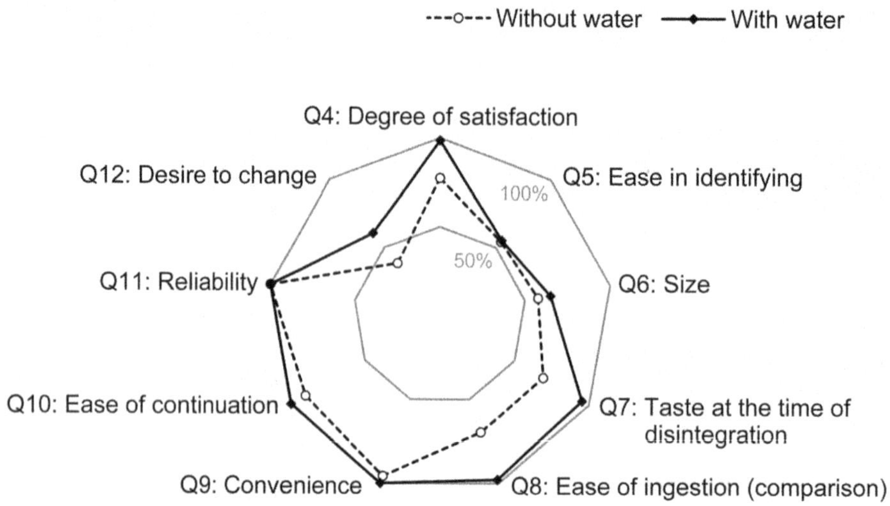

Fig. 3 Comparison of sensory evaluation between groups that ingested edoxaban OD tablets with or without water. The graph shows the percentages of patients who answered "5" (no difference) through "10" (quite satisfactory) in question 4; "1" (very good) or "2" (relatively good) in questions 5 and 7; "1" (very large) or "2" (slightly large) in question 6; "1" (very good) through "3" (no difference) in questions 8–11; and "1" (desire to change) in question 12. *OD* orally disintegrating, *Q* question

Discussion

In this study, the subjects enrolled tended to be older (mean age, 73.0 ± 8.6 years; aged ≥75 years, 49.1%), particularly those who were allocated to the 30-mg edoxaban OD tablet groups (groups 1 and 3).

For the questionnaires regarding the current medication, 86.8% of the subjects reported no difficulty taking the current medication (question 1), 93.4% reported no difficulty identifying the drugs (question 2), and 78.3% reported taking 3 or more drugs (including edoxaban film-coated tablets) every morning (question 3), which indicates that most of the patients in this study perceived no difficulty taking the current medication including edoxaban film-coated tablets and had been taking 3 or more drugs.

In question 4 (degree of satisfaction with edoxaban OD tablets), more than half (52.8%) of the patients selected "no difference." However, considering that "satisfactory" was selected by 34.9% in the group that included many patients who perceived no complaint about taking the current medication, the degree of satisfaction with edoxaban OD tablets was considered to be relatively high. The percentage of patients who selected "unsatisfactory" was higher in the group that ingested the drug without water than in the group with water. This result seems to reflect the fact that the rating of taste/flavor (question 7) and ease of ingestion (question 8) was frequently lower in the group that ingested the drug without water. Furthermore, the responses suggest that the patients who usually ingested drugs with water had not been accustomed to taking drugs without water.

In question 5 (ease in identifying edoxaban OD tablets), "quite difficult to identify" was selected most frequently (30.5%), but "quite easy to identify" or "easier to identify than the current edoxaban film-coated tablet" was selected by 26.7 and 27.6% of the respondents, respectively. Thus, the answers to this question were equally divided among the opposing responses. This response probably reflected the fact that the figures printed on the tablets were considered too small for some patients to read.

In question 6 (size of edoxaban OD tablets), the percentage of those who selected "slightly large" (59.2%) or "slightly small" (37.8%) was 97%, which suggests that the size of these tablets was acceptable on the whole.

In question 7 (taste and flavor at the time of edoxaban OD tablet disintegration), the percentage of selecting "very good" or "relatively good" was 81.5%. Of the 17 patients who selected "relatively bad," 15 answered "tolerable," which suggests no problem with the taste/flavor of this product. The percentage of selecting "relatively bad" was higher in the group that ingested edoxaban OD tablets without water than in the group that ingested it with water. The rating tended to be higher in the group that ingested edoxaban OD tablets with water. The patients who selected "good" for the taste/flavor tended to select "easier to take" in question 8, which suggests that the taste or flavor affected the perceived ease of OD tablet ingestion.

In question 8 (ease of taking edoxaban OD tablets), "no difference" was selected most frequently (42.9%), but the percentage with a favorable impression ("easier to take") was higher than the percentage with a bad impression, which suggests that edoxaban OD tablets were easier to take. Since the percentage of selecting "difficult to take" was higher in the group that ingested edoxaban OD tablets without water than in the group that ingested edoxaban OD tablets with water, the rating tended to be higher in the group that ingested edoxaban OD tablets with water.

Table 4 Results of the sensory evaluation questionnaire about edoxaban OD tablets shown by subgroup

		Age (< 75 years) N = 54	Age (≥75 years) N = 52	Current difficulty in ingestion (Question 1)	
				Yes N = 14	No N = 92
		n (%)	n (%)	n (%)	n (%)
Question 1: Difficulty in ingestion (current)	Yes	6 (11.1)	8 (15.4)	14 (100.0)	0 (0.0)
	No	48 (88.9)	44 (84.6)	0 (0.0)	92 (100.0)
Question 2: Difficulty in identifying (current)	Yes	2 (3.7)	5 (9.6)	2 (14.3)	5 (5.4)
	No	52 (96.3)	47 (90.4)	12 (85.7)	87 (94.6)
Question 3: Number of drugs taken simultaneously	1 type[a]	2 (3.7)	8 (15.4)	1 (7.1)	9 (9.8)
	2 types	7 (13.0)	6 (11.5)	2 (14.3)	11 (12.0)
	≥3 types	45 (83.3)	38 (73.1)	11 (78.6)	72 (78.3)
Question 4: Degree of satisfaction	Unsatisfactory	7 (13.0)	6 (11.5)	1 (7.1)	12 (13.0)
	No difference	27 (50.0)	29 (55.8)	6 (42.9)	50 (54.4)
	Satisfactory	20 (37.0)	17 (32.7)	7 (50.0)	30 (32.6)
Question 5: Ease in identifying	Quite easy	17 (31.5)	11 (21.6)	5 (35.7)	23 (25.3)
	Easier[b]	13 (24.1)	16 (31.4)	5 (35.7)	24 (26.4)
	More difficult[b]	11 (20.4)	5 (9.8)	1 (7.1)	15 (16.5)
	Quite difficult	13 (24.1)	19 (37.3)	3 (21.4)	29 (31.9)
Question 6: Size	Very large	2 (3.8)	0 (0.0)	0 (0.0)	2 (2.4)
	Slightly large	34 (64.2)	24 (53.3)	7 (53.9)	51 (60.0)
	Slightly small	17 (32.1)	20 (44.4)	6 (46.2)	31 (36.5)
	Very small	0 (0.0)	1 (2.2)	0 (0.0)	1 (1.2)
Question 7: Taste at the time of disintegration	Very good	9 (18.4)	7 (16.3)	2 (14.3)	14 (18.0)
	Relatively good	31 (63.3)	28 (65.1)	11 (78.6)	48 (61.5)
	Relatively bad	9 (18.4)	8 (18.6)	1 (7.1)	16 (20.5)
	Very bad	0 (0.0)	0 (0.0)	0 (0.0)	0 (0.0)
Question 8: Ease of ingestion (comparison)	Quite easy to take	10 (18.9)	11 (21.2)	6 (42.9)	15 (16.5)
	Slightly easier to take	14 (26.4)	8 (15.4)	3 (21.4)	19 (20.9)
	No difference	22 (41.5)	23 (44.2)	3 (21.4)	42 (46.2)
	Slightly more difficult to take	6 (11.3)	10 (19.2)	2 (14.3)	14 (15.4)
	Quite difficult to take	1 (1.9)	0 (0.0)	0 (0.0)	1 (1.1)
Question 9: Convenience	Much more convenient	8 (14.8)	10 (19.2)	4 (28.6)	14 (15.2)
	Slightly more convenient	22 (40.7)	16 (30.8)	6 (42.9)	32 (34.8)
	No difference	23 (42.6)	25 (48.1)	4 (28.6)	44 (47.8)
	Slightly less convenient	1 (1.9)	1 (1.9)	0 (0.0)	2 (2.2)
	Much less convenient	0 (0.0)	0 (0.0)	0 (0.0)	0 (0.0)
Question 10: Ease of continuation	Much easier to continue	6 (11.1)	7 (13.5)	5 (35.7)	8 (8.7)
	Slightly easier to continue	16 (29.6)	12 (23.1)	2 (14.3)	26 (28.3)
	No difference	29 (53.7)	31 (59.6)	7 (50.0)	53 (57.6)
	Slightly less easy to continue	3 (5.6)	2 (3.9)	0 (0.0)	5 (5.4)
	Much less easy to continue	0 (0.0)	0 (0.0)	0 (0.0)	0 (0.0)
Question 11: Reliability	Much more reliable	16 (29.6)	5 (9.6)	5 (35.7)	16 (17.4)
	Slightly more reliable	16 (29.6)	14 (26.9)	2 (14.3)	28 (30.4)
	No difference	22 (40.7)	33 (63.5)	7 (50.0)	48 (52.2)
	Slightly less reliable	0 (0.0)	0 (0.0)	0 (0.0)	0 (0.0)
	Much less reliable	0 (0.0)	0 (0.0)	0 (0.0)	0 (0.0)

Table 4 Results of the sensory evaluation questionnaire about edoxaban OD tablets shown by subgroup *(Continued)*

		Age (< 75 years) *N* = 54	Age (≥75 years) *N* = 52	Current difficulty in ingestion (Question 1)	
				Yes *N* = 14	No *N* = 92
		n (%)	n (%)	n (%)	n (%)
Question 12: Desire to change	Desire to change	28 (57.1)	19 (41.3)	10 (76.9)	37 (45.1)
	No desire to change	21 (42.9)	27 (58.7)	3 (23.1)	45 (54.9)

[a]Edoxaban film-coated tablet only
[b]compared with the current medication
OD orally disintegrating

In question 9 (convenience of taking edoxaban OD tablets), "no difference" was selected most frequently (45.3%). However, the percentage of patients selecting "more convenient" was also high (52.8%), which suggests that edoxaban OD tablets were perceived as more convenient than the existing edoxaban film-coated tablets.

In question 10 (ease of continuation to take edoxaban OD tablets), "no difference" was selected most frequently (56.6%). However, the percentage of patients who selected "easier to continue" was relatively high (38.7%), which suggests that edoxaban OD tablets tented to be perceived as easy to continue.

In question 11 (reliability of edoxaban OD tablets), "no difference" was selected most frequently (51.9%). As the only other answer was "more reliable" (indicating a favorable impression), edoxaban OD tablets were considered reliable for reasons such as printing the product name on the tablets.

In question 12 (desire a change to edoxaban OD tablets), 49.5% of the respondents selected "desire to change" and approximately the same percentage selected "no desire to change." Thus, 2 opposing answers were selected by a similar percentage. The desire to change tended to be more common in the group that ingested edoxaban OD tablets with water than in the group that ingested the edoxaban OD tablet without water. The reason for selecting "desire to change" was "easier to take" in most patients, probably reflecting sufficient understanding of the features of OD tablets. The reasons for selecting "no desire to change" were as follows: "water is needed when taking it with other drugs," "I have other drugs to take simultaneously," "it is easier to take with water," and others. These answers suggest that the features of OD tablets (can be taken with water simultaneously with other drugs) had not been sufficiently understood by these patients.

In this study, 90% of the patients had been routinely taking 2 or more tablets, and most of the patients seemed to have been taking edoxaban tablets with water simultaneously with other drugs. For this reason, taking edoxaban OD tablets with water seems to be closer to the actual clinical setting for these patients. This study shows that the degree of satisfaction, taste/flavor, and ease of ingestion were rated higher in the group that ingested the drug with water, which suggests that the advantages of edoxaban OD tablets are better manifested when it is taken with water. As a result, it is inferred that many of the patients desired a change to edoxaban OD tablets in the group taking edoxaban OD tablets with water.

Generally, OD tablets are more advantageous for elderly patients and patients with impaired swallowing function due to cerebral infarction or other reasons. Therefore, we conducted a subgroup analysis according to the ages (< 75 and ≥ 75 years) and presence/absence of difficulty of taking current medications. In patients aged ≥75 years, the rating of reliability and percentage of desiring a change were low, with no item recording a high rating. In patients who perceived difficulty of taking current medications, the percentage of desiring a change to edoxaban OD tablets was as high as 76.9%. These results suggest that the advantages of edoxaban OD tablets were rated high by the patients who perceived difficulty taking current medications.

There are some limitations in this study. The first is that the condition for drug ingestion (designed to take edoxaban OD tablet as a single agent) differed from the actual clinical setting. Most patients were taking 2 or more tablets routinely and probably took edoxaban tablets and other medications simultaneously with water. Irikura et al. [7] investigated whether there are advantages of OD tablets for patients taking multiple drugs simultaneously and reported that 40% of medical professionals (physicians, nurses, and pharmacists) perceived no advantage of OD tablets, while 30% perceived some advantages. They additionally reported that the perceived advantages included easier ingestion if at least one drug is an OD tablet or the ability to take OD tablets without water even if it is at least one of the drugs, among others. We therefore consider it necessary to evaluate the advantages of edoxaban OD tablets under conditions of multiple drug ingestion simultaneously with water identical to that in a practical clinical setting. The second is that only 2 hospitals participated in this study, and the number of patients enrolled was limited (including only 1 woman in the 60-mg edoxaban OD tablet group). Therefore, general extrapolation of the results of this study might be difficult. A further large-scale study would be needed, especially enrolling more women in the 60-mg edoxaban OD tablet group. The third is that we cannot rule out the possibility that the features of edoxaban OD tablets (can be taken with water simultaneously with other drugs) had not been understood sufficiently by the participants. Different results may be obtained if a similar study is conducted in actual clinical practice, in which patients would be adequately informed about the possibility of taking these tablets with water.

Conclusions

A questionnaire survey was conducted to assess edoxaban OD tablets in a study population that included many patients who perceived no difficulty taking the existing film-coated tablets. The degree of satisfaction and rating of taste/flavor, ease of ingestion, convenience, ease of continuation, and reliability were high on the whole, with about half of all the patients reporting a desire to change to edoxaban OD tablets. The rating of these features was higher in the group that ingested the drug with water than in the group without water. The percentage of patients who desired a change to edoxaban OD tablet was higher among those who perceived difficulty taking current medications. Similar results were reported in past studies of sensory evaluation for other OD tablets [4–6]. Moreover, a change to OD tablets from existing standard tablets has been reported to increase medication adherence to dosing instructions, resulting in better treatment responses [2, 3]. We may therefore expect that the use of edoxaban OD tablets will improve the patient adherence to medication and outcomes.

Abbreviations

ADR: Adverse drug reaction; AE: Adverse event; DOAC: Direct oral anticoagulant; OD: Orally disintegrating

Acknowledgments

This article is an English translation of the article originally reported in *Therapeutic Research* in Japanese (Yamashita T, Hagii J, Morishima Y, Akasaka T, Matsumoto T, Kimura T. Clinical study of sensory evaluation using edoxaban orally disintegrating tablet [OD tablet]. 2018; 39:399–417), not including the description of (1) the reasons for desiring or not desiring the change to edoxaban OD tablet from the current medication and (2) the subgroup analysis of sensory evaluation of edoxaban OD tablet by prior cerebral infraction (presence vs. absence) and the number of drug types taken routinely (1 type vs. ≥2 types). The authors are indebted to Sogo Rinsho Medefi Co., Ltd. for their support in drafting and writing the original manuscript written in Japanese. The medical writing assistance in editing and styling the manuscript was performed by inScience Communications, Springer Healthcare, and was funded by Daiichi Sankyo Co., Ltd.

Authors' contributions

TY and JH established the study concept and design, conducted the research process, acquired and interpreted the data, and reviewed the manuscript. YM established the study concept and design, managed and coordinated research activities, interpreted the data, and contributed to drafting of the manuscript. TA and TK established the study concept and design, managed and coordinated the research activity, interpreted the data, and reviewed the manuscript. TM analyzed and interpreted the data. All authors read and approved the final manuscript.

Author details

[1]The Cardiovascular Institute, 3-2-19, Nishiazabu, Minato-ku, Tokyo 106-0031, Japan. [2]Hirosaki Stroke and Rehabilitation Center, 1-2-1, Ougi-machi, Hirosaki, Aomori 036-8104, Japan. [3]Daiichi Sankyo Co., Ltd., 3-5-1, Nihonbashi Honcho, Chuo-ku, Tokyo 103-8426, Japan.

References

1. Taniguchi H, Magara J, Inoue M. Dysphagia in elderly people [in Japanese]. J Jpn Soc Parenter Enteral Nutr. 2013;28:1069–74.
2. Kinon BJ, Hill AL, Liu H, Kollack-Walker S. Olanzapine orally disintegrating tablets in the treatment of acutely ill non-compliant patients with schizophrenia. Int J Neuropsychopharmacol. 2003;6:97–102.
3. Koh N, Sakamoto S, Chino F. Improvement in medication compliance and glycemic control with voglibose oral disintegrating tablet. Tohoku J Exp Med. 2008;216:249–57.
4. Suzuki H, Muraki K, Matsuoka O, Ano R, Kitada M, Koshio Z. Sensory evaluation of Diovan orally integrating tablet (OD tablet) [in Japanese]. J New Remedies Clin. 2013;62:642–8.
5. Ichikawa Y, Matsuoka O, Matsuda A, Nakajima A, Tashiro N, Kitada M. Sensory evaluation of orally disintegrating (OD) Exforge combination tablet [in Japanese]. J N Remedies Clin. 2015;64:16–23.
6. Oda H, Matsumoto T, Harada J. A clinical study of sensor evaluation of olmesartan medoxomil orally disintegrating tablet (OD tablet) in patients with hypertension receiving olmesartan medoxomil tablet [in Japanese]. Ther Res. 2016;37:293–302.
7. Irikura M, Ishizuka Y, Kasano M, Kakuta T, Irie T. Clinical needs of drug preparations easier to take, focusing on orally disintegrating tablets [in Japanese]. Pharmaceuticals Monthly. 2008;50:1669–76.

25

Effects of concomitant use of prasugrel with edoxaban on bleeding time, pharmacodynamics, and pharmacokinetics of edoxaban in healthy elderly Japanese male subjects: A clinical pharmacology study

Ippei Ikushima[1*], Takaaki Akasaka[2], Yoshiyuki Morishima[2], Atsushi Takita[3], Tomoko Motohashi[2] and Tetsuya Kimura[2]

Abstract

Background: Dual therapy with a direct oral anticoagulant (DOAC) plus a $P2Y_{12}$ receptor inhibitor is recommended in patients with nonvalvular atrial fibrillation who undergo percutaneous coronary intervention. Thus, we evaluated the effects of DOAC edoxaban plus $P2Y_{12}$ receptor inhibitor prasugrel on bleeding time (BT), and pharmacodynamics (PD) and pharmacokinetics (PK) of edoxaban in healthy elderly Japanese male subjects.

Methods: A single-center, clinical pharmacology study with randomized, open-label, repeated dosing enrolled 24 participants in two groups of 12 receiving 30 mg edoxaban once daily for 3 days; then 30 mg edoxaban plus 2.5 mg prasugrel (Group 1) or 30 mg edoxaban plus 3.75 mg prasugrel (Group 2) once daily for 5 days. Primary endpoint was BT by the Ivy method. Secondary endpoints were the PD parameters of prothrombin time (PT), activated partial thromboplastin time (aPTT), prothrombin fragment F1 + 2 (F1 + 2), and $P2Y_{12}$ reaction units (PRU), and PK profiles of edoxaban alone and in combination with prasugrel.

Results: Geometric least squares mean of BT ratios (vs. baseline) for 3-day edoxaban treatment were 1.097 (90% confidence interval (CI) 0.911–1.321) in Group 1 and 1.327 (90% CI 1.035–1.703) in Group 2; for 5-day edoxaban plus 2.5 mg and 3.75 mg prasugrel, they were 1.581 (90% CI 1.197–2.087) and 1.996 (90% CI 1.482–2.690), respectively. Contributions of prasugrel to the effects (edoxaban + prasugrel/edoxaban) were 1.442 (90% CI 1.096–1.897) in Group 1 and 1.504 (90% CI 1.172–1.930) in Group 2. Edoxaban prolonged PT and aPTT and decreased F1 + 2. Adding on prasugrel did not appreciably change PT, aPTT, or F1 + 2. Prasugrel reduced PRU, whereas edoxaban had no effect on PRU. We recorded 26 adverse events; 23 were treatment-emergent (positive fecal occult blood test). All participants with adverse events recovered during follow-up.

(Continued on next page)

* Correspondence: ippei-ikushima@lta-med.com
[1]Sumida Hospital, 29-1, Honjo 1-Chome, Sumida-ku, Tokyo 130-0004, Japan
Full list of author information is available at the end of the article

(Continued from previous page)

Conclusions: Coadministration of 2.5 mg and 3.75 mg prasugrel with 30 mg edoxaban prolonged BT in healthy elderly Japanese male subjects. The effect was dependent on the dose of prasugrel. Prasugrel did not affect PD or PK profiles of edoxaban. Edoxaban had no effect on PD of prasugrel.

Trial registration: Japan Registry of Clinical Trials No. jRCTs071190006; registration date, 26-April-2019.

Keywords: Edoxaban, Prasugrel, Healthy male subjects, Elderly Japanese, Bleeding time, Pharmacodynamics, Pharmacokinetics

Background

In patients with nonvalvular atrial fibrillation (NVAF), especially in those of advanced aged, coronary artery disease is common. In the All Nippon AF in the Elderly (ANAFIE) registry of 32,726 Japanese patients ≥75 years of age, 5600 (17.1%) had angina pectoris and 1874 (5.7%), a myocardial infarction [1].

The European Heart Rhythm Association, a branch of the European Society of Cardiology issued a 2018 joint European consensus document on the management of antithrombotic therapy in atrial fibrillation (AF) patients presenting with acute coronary syndrome and undergoing percutaneous coronary intervention (PCI) [2]. The document recommends as short a period as possible for triple therapy with an oral anticoagulant plus dual antiplatelet agents (a $P2Y_{12}$ receptor inhibitor and aspirin), followed by dual therapy with an oral anticoagulant plus a single antiplatelet agent ($P2Y_{12}$ receptor inhibitor).

The direct oral anticoagulant (DOAC) used in our study was edoxaban, a direct inhibitor of activated blood coagulation factor X, indicated to reduce the risk of stroke and systemic embolism in patients with NVAF [3, 4]. Edoxaban is available in a standard dose of 60 mg and a reduced dose of 30 mg for patients with low body weight (≤60 kg), moderate or severe renal impairment (creatinine clearance 15–50 mL/min), or concomitant use of potent P-glycoprotein inhibitors [4, 5]. According to baseline clinical characteristics of patients taking edoxaban enrolled in the ANAFIE registry (all 75 years of age or older), 82.2% of them were taking this reduced 30 mg dose [1].

The $P2Y_{12}$ receptor inhibitor administered with edoxaban in this study was prasugrel, indicated for patients with acute coronary syndrome, stable angina, or old myocardial infarction who are to be managed with PCI in Japan [6, 7]. Compared to clopidogrel, prasugrel is a more potent $P2Y_{12}$ receptor inhibitor [6–9]. In Japan, the doses of prasugrel (loading dose, 20 mg; standard daily maintenance dose, 3.75 mg; reduced daily maintenance dose, 2.5 mg for patients with low body weight ≤ 50 kg) are lower than those approved in western countries (loading dose, 60 mg; daily maintenance dose, 10 mg), and the bleeding risk of prasugrel was shown to be similar to that of clopidogrel [8, 9]. Recently, a retrospective analysis of an observational cohort study reported that in Japanese AF patients undergoing PCI, prasugrel as a part of triple therapy with aspirin and an oral anticoagulant had not increased the risk of bleeding compared to clopidogrel [10]. Besides this observational study, a phase 3b trial [11] was conducted to analyze the safety of edoxaban in combination with a $P2Y_{12}$ receptor inhibitor in AF patients undergoing PCI. The study demonstrated that edoxaban plus a $P2Y_{12}$ receptor inhibitor was non-inferior for bleeding events compared with a vitamin K antagonist in combination with a $P2Y_{12}$ receptor inhibitor and aspirin without significant differences in ischemic events. However, data on the combination of edoxaban and prasugrel are very limited.

Here, we aimed to assess the effects of edoxaban plus prasugrel on bleeding time (BT), and pharmacodynamic (PD) and pharmacokinetic (PK) profiles of edoxaban in healthy elderly Japanese male subjects between the ages of 65 and 80.

Methods

Study population

The target sample size for the study population was 24 healthy elderly Japanese male subjects between the ages of 65 and 80 years, weighing 40 kg–60 kg (body mass index [BMI], 18.5 kg/m^2–25.0 kg/m^2). Subjects were excluded from participation in the study for any of the following reasons: presence of a bleeding disorder (history of intracranial bleeding, gastrointestinal bleeding, vitreous hemorrhage, retinal bleeding, hemoptysis, hematemesis, hematuria, bloody stools, hemorrhoids), prolonged BT of over 9.5 min, and prolonged prothrombin time (PT) and activated partial thromboplastin time (aPTT) exceeding the upper limit of normal values, a positive finding in fecal occult blood test or urine occult blood test, chronic kidney disease (estimated glomerular filtration rate (eGFR) of less than 60 mL/min/1.73 m^2), or use or consumption any of the following within 7 days before the examination: inhibitors or inducers of the drug-metabolizing enzyme CYP3A, inhibitors of P-glycoprotein, and grapefruit or any food products containing the fruit.

Study design

This study was a randomized, open-label, parallel-group, repeated-dose study conducted at a single center.

Subjects who were eligible and agreed to participate in this study were admitted to the hospital for the length of the study and randomly assigned to two groups: Group 1, 30 mg edoxaban (dose for NVAF patients weighing 60 kg or less) plus 2.5 mg prasugrel (reduced maintenance dose in Japan); Group 2, 30 mg edoxaban plus 3.75 mg prasugrel (standard maintenance dose in Japan).

The study design, schedule of procedures, and blood sample collections are summarized in Table 1. This study was conducted at Sumida Hospital in Tokyo, Japan, from May to June 2019.

Bleeding time

We measured BT by the Ivy method [12] using a needle type of BD Microtainer Contact-Activated Lancet device (Becton, Dickinson and Company, USA). The incision was made in the lateral aspect of the forearm by puncturing at one arbitrary site with the device. After the incision, we blotted the blood every 30 s with filter paper until the bleeding reached hemostasis and recorded the time from the puncture to the end of bleeding. When the time to hemostasis exceeded 10 min, it was recorded as such, ending the observation. The schedule of BT evaluations is included in Table 1. Baseline BT was measured on the day of admission at approximately the same time of day as the BT evaluations on Day 3 and Day 8, 3 hours after the administration of study drug(s).

Pharmacodynamics

Effects of edoxaban alone and edoxaban plus prasugrel on the parameters of blood coagulation and platelet aggregation were assessed by determination of PT, aPTT, prothrombin fragment 1 + 2 (F1 + 2), and $P2Y_{12}$ reaction units (PRU). Blood was collected in a vacuum blood collection tube containing 3.2% sodium citrate, and the schedule for the blood sample collection is shown in Table 1. Blood coagulation marker PT was measured by PT reagent Thrombocheck-PT in system CS-200i (Sysmex Corporation, Japan); aPTT by aPTT reagent HemosIL SynthASil in system ACL TOP Base (IL, Japan; Instrumentation Laboratory, USA) and F1 + 2 by ELISA Enzygnost F1 + 2 monoclonal reagent in system Behring ELISA Processor III and DiaSorin (BIT Analytical Instruments GmbH, Germany). These assays were carried out at LSI Medience Corporation (Tokyo, Japan). We measured PRU with VerifyNow system (Instrumentation Laboratory, Bedford, MA, USA) at Sumida Hospital.

Pharmacokinetics

Blood for the determination of plasma edoxaban concentrations was collected from a forearm vein and deposited in a vacuum blood collection tube containing heparin sodium. Blood was collected before the administration and 0.25, 0.5, 1, 2, 3, 4, 8, 12, and 24 h after the drug administration on Day 1, Day 3, and Day 8 (Table 1). Plasma samples were obtained by centrifugation, and the samples were immediately frozen at − 20 °C or lower. Plasma edoxaban concentrations were measured by a liquid chromatography–tandem mass spectrometry system (SCIEX, Framingham, MA, USA) at Shin Nippon Biomedical Laboratories, Ltd. (Wakayama, Japan).

Statistical analysis

The sample size selected was 24 (12 in each group) based on feasibility. According to results of phase 1 studies of edoxaban and prasugrel, inter-individual variations in BT and BT ratio before and after the administration were 20 to 45%, and intra-individual variations were estimated to be 20 to 30%. Under this estimation, effects of drugs on bleeding time could be analyzed with a certain degree of

Table 1 Study design, schedule for BT, PD, and edoxaban PK evaluations

Subjects	Healthy elderly Japanese male subjects (Age: 65–80)	
Study design	Open-label, parallel-group, repeated-dose	
Sample size	24 Subjects randomized into two groups	
	Group 1, $n = 12$ 30 mg Edoxaban + 2.5 mg Prasugrel	Group 2, $n = 12$ 30 mg Edoxaban + 3.75 mg Prasugrel
Treatments	Stage 1: all participants Day 1, 2, 3: 30 mg Edoxaban once daily	
	Stage 2: Day 4, 5, 6, 7, 8 Group 1: 30 mg Edoxaban + 2.5 mg Prasugrel once daily	Stage 2: Day 4, 5, 6, 7, 8 Group 2: 30 mg Edoxaban + 3.75 mg Prasugrel once daily
Schedule of procedures and blood sample collections		
Bleeding time (BT)[a] & PD Evaluation [b]	Day 0: Day of admission Day 3: Three hours after administration of edoxaban Day 8: Three hours after administration of edoxaban + prasugrel	
Edoxaban PK	Day 1: Pre-administration of study drug Day 3: Pre-administration and 0.25, 0.5, 1, 2, 3, 4, 8, 12, 24 h after administration of edoxaban Day 8: Pre-administration and 0.25, 0.5, 1, 2, 3, 4, 8, 12, 24 h after administration of edoxaban + prasugrel	

[a]Ivy Method
[b]PD indices: prothrombin time (PT), activated partial thromboplastin time (aPTT), prothrombin fragment 1 + 2 (F1 + 2), and $P2Y_{12}$ reaction units (PRU)

accuracy. All statistical analyses were performed using SAS Ver. 9.4 (SAS Institute Inc., Cary, NC, USA).

Summary statistics were calculated for baseline and post-dose values for the primary endpoint of BT and secondary endpoints of PD indices (PT, aPTT, F1 + 2, PRU) for each target population. Furthermore, we calculated summary statistics of plasma edoxaban concentrations and estimated pharmacokinetic parameters ($C_{max,ss}$, steady state maximum plasma concentration; $AUC_{tau,ss}$, steady state area under the plasma concentration-time curve; $t_{max,ss}$, steady state time to reach maximum plasma concentration) using a noncompartmental analysis engine of Phoenix WinNonlin Ver. 8.1 based on Model 1200 (Extravascular). $AUC_{tau,ss}$ was calculated by the linear-Log trapezoidal method.

For the primary endpoint analysis, a logarithmically converted value of BT for each sample was the dependent variable in a linear mixed model with fixed and random effects. We calculated geometric least squares means (GLSM) of ratios of BT for the edoxaban alone and the edoxaban plus prasugrel combination, with 90% confidence intervals (CIs) for each treatment group.

For the secondary endpoints analysis, we used the logarithmically converted PK parameters ($C_{max,ss}$, $AUC_{tau,ss}$, $t_{max,ss}$) as dependent variables in a linear mixed model with fixed and variable effects. We calculated GLSM of these values for edoxaban administered alone and edoxaban plus prasugrel combination, with 90% CIs.

Results

Subject disposition and characteristics

A total of 24 healthy Japanese male subjects (age range 66–78 years) participated in this study: 12 administered 30 mg edoxaban plus 2.5 mg prasugrel (Group 1) and 12 administered 30 mg edoxaban plus 3.75 mg prasugrel (Group 2). All participants completed the study as planned. Demographic and clinical characteristics of participants are shown in Table 2.

All participant data was available for the final analyses.

Bleeding time

Mean and median BTs (min) at baseline and after treatments are shown in Table 3.

The increase in BT is expressed as a GLSM of BT ratios (expressed as ratios to baseline) are shown in a box plot (Fig. 1). After 3 days of a once-daily dosing of 30 mg edoxaban, the GLSMs of BT ratios were 1.097 (90% CI

Table 2 Baseline demographic and clinical characteristics

	All subjects n = 24	Group 1 30 mg edoxaban plus 2.5 mg prasugrel n = 12	Group 2 30 mg edoxaban plus 3.75 mg prasugrel n = 12
Age (years)			
Mean (SD)	72 (4)	71 (4)	73 (3)
65 to < 75, n (%)	16 (66.7)	9 (75.0)	7 (58.3)
75 to 80, n (%)	8 (33.3)	3 (25.0)	5 (41.7)
Height (cm)			
Mean (SD)	164.0 (5.7)	162.5 (5.7)	165.5 (5.6)
Weight (kg)			
Mean (SD)	56.4 (2.2)	55.8 (2.4)	57.0 (1.9)
> 50 to 60, n (%)	24 (100.0)	12 (100.0)	12 (100.0)
BMI (kg/m^2)			
Mean (SD)	21.1 (1.6)	21.3 (1.6)	21.0 (1.6)
eGFR (mL/min/1.73m^2)			
Mean (SD)	73.5 (7.0)	75.5 (6.6)	71.6 (7.1)
60 to < 90, n (%)	24 (100.0)	12 (100.0)	12 (100.0)
CL_{cr} (mL/min)			
Mean (SD)	67.2 (7.2)	68.7 (6.1)	65.8 (8.1)
≥50, n (%)	24 (100.0)	12 (100.0)	12 (100.0)
Medical history[a]			
No, n (%)	10 (41.7)	5 (41.7)	5 (41.7)
Yes, n (%)	14 (58.3)	7 (58.3)	7 (58.3)

SD standard deviation, *IQR* interquartile range, *BMI* body mass index, CL_{cr} creatinine clearance

[a]Appendicitis, polyps in the colon, hemorrhoids, cataract, hepatitis A, tuberculosis, and inguinal hernia

Table 3 Mean and median bleeding times (min)

	Group 1 30 mg edoxaban plus 2.5 mg prasugrel $n = 12$	Group 2 30 mg edoxaban plus 3.75 mg prasugrel $n = 12$
Baseline		
Mean (SD)	3.2 (0.8)	3.5 (1.1)
Median (IQR)	3.0 (2.5–3.8)	3.5 (2.5–4.5)
Edoxaban administration		
Mean (SD)	3.5 (0.6)	4.7 (1.6)
Median (IQR)	3.5 (3.0–4.0)	4.5 (3.3–6.0)
Edoxaban plus prasugrel administration		
Mean (SD)	5.4 (2.3)	7.3 (2.8)
Median (IQR)	5.0 (3.5–7.3)	7.5 (5.5–10.0)

SD standard deviation, *IQR* interquartile range

0.911–1.321) in Group 1 and 1.327 (90% CI 1.035–1.703) in Group 2. After coadministration of edoxaban and prasugrel for 5 days, BT was further prolonged. Especially in the edoxaban plus 3.75 mg prasugrel group, bleeding was continued over 10 min in 5 out of 12 subjects. The GLSMs of BT ratios were 1.581 (90% CI 1.197–2.087) in the edoxaban plus 2.5 mg prasugrel group and 1.996 (90% CI 1.482–2.690) in the edoxaban plus 3.75 mg prasugrel group. The GLSM increases in BT by prasugrel (edoxaban plus prasugrel/edoxaban alone) were 1.442 (90% CI 1.096–1.897) in the edoxaban plus 2.5 mg prasugrel group and 1.504 (90% CI 1.172–1.930) in the edoxaban plus 3.75 mg prasugrel group.

Pharmacodynamic parameters of coagulation and platelet aggregation

Figure 2 provides an overview of changes in pharmacodynamic parameters of coagulation and platelet aggregation. The administration of 30 mg edoxaban prolonged PT, in sec, from a mean of 11.8 at baseline to 15.3 (Group 1) and 11.9 at baseline to 15.5 (Group 2). With the addition of 2.5 mg or 3.75 mg prasugrel, the PT values were essentially unchanged for both groups (mean of 15.2 and 15.1). The aPTT, in sec, was also prolonged by edoxaban, from 31.9 at baseline to 38.0 (Group 1) and 34.2 at baseline to 39.9 (Group 2). Again, when 2.5 mg or 3.75 mg prasugrel was added on to the administration, the effect of edoxaban on aPTT, in sec, was unchanged (mean of 37.7 and 39.1). Edoxaban decreased F1 + 2, in pmol/L, from 309 at baseline to 112 (Group 1) and 279 at baseline to 129 (Group 2). The results of F1 + 2, were essentially unchanged when 2.5 mg or 3.75 mg prasugrel was added on to the administration of edoxaban (mean of 123 and 120). In the platelet aggregation assay, edoxaban essentially had no effect on PRU. After 2.5 mg or 3.75 mg prasugrel was added on to edoxaban, the PRU was reduced from a mean of 249 at baseline to 158 and 246 at baseline to 78, respectively.

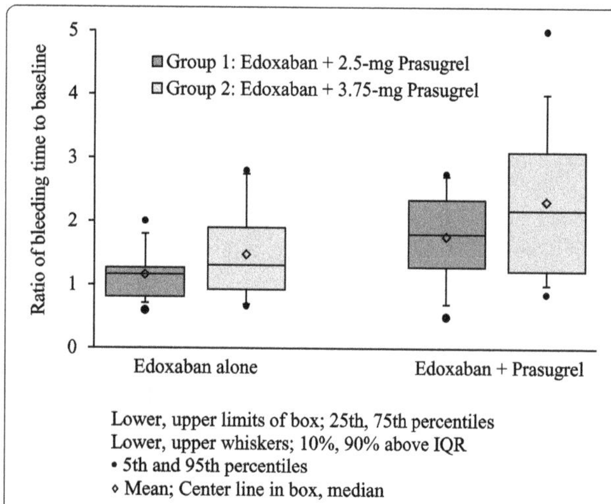

Fig. 1 Changes in bleeding time induced by edoxaban administered alone and with prasugrel

Pharmacokinetics

The plasma edoxaban concentration profiles over 24 h were consistent, whether edoxaban was administered alone or with 2.5 mg or 3.75 mg prasugrel, as shown in Fig. 3. Also, prasugrel did not affect the PK parameters ($C_{max,ss}$, $AUC_{tau,ss}$, $t_{max,ss}$) of edoxaban (Table 4).

Safety

Of the 24 participants, 16 had at least one occurrence of a treatment-emergent adverse event (AE) (16 positive results in the fecal occult blood test without stool blackening; 2, constipation). We observed 26 treatment-emergent AEs: 23 were reported as a treatment-related AE (positive result in the fecal occult blood test), and 3 were an AE unrelated to treatment (constipation) and most likely due to environmental changes at the hospital. We observed four positive results in the fecal occult blood test for edoxaban alone, 10 for edoxaban plus 2.5 mg prasugrel, and 9 for edoxaban plus 3.75 mg prasugrel and confirmed the recovery of all participants during the follow-up period. No serious adverse event was observed in the study. There were also no withdrawals or discontinuation of the study drug due to AEs or any other reason.

Discussion

This study provided additional, preliminary safety results that concur with the 2018 joint European consensus document [2] recommending dual-therapy use of a DOAC plus a $P2Y_{12}$ receptor inhibitor and also found no effect of prasugrel on PK and PD of edoxaban in healthy Japanese

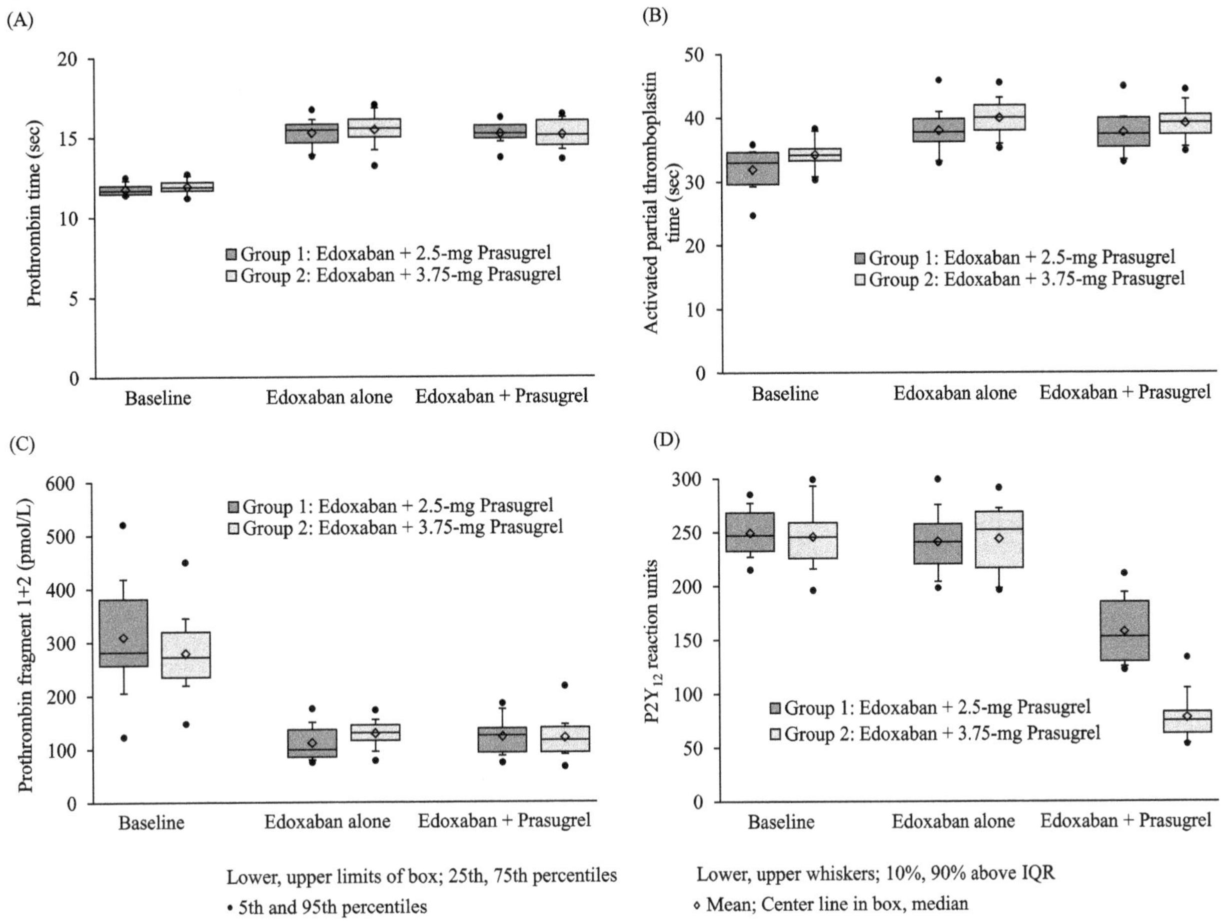

Fig. 2 Changes in pharmacodynamic indices of coagulation and platelet aggregation induced by edoxaban administered alone and with prasugrel (**a**) Prothrombin time; (**b**) Activated partial thromboplastin time; (**c**) Prothrombin fragment 1 + 2; (**d**) $P2Y_{12}$ reaction units

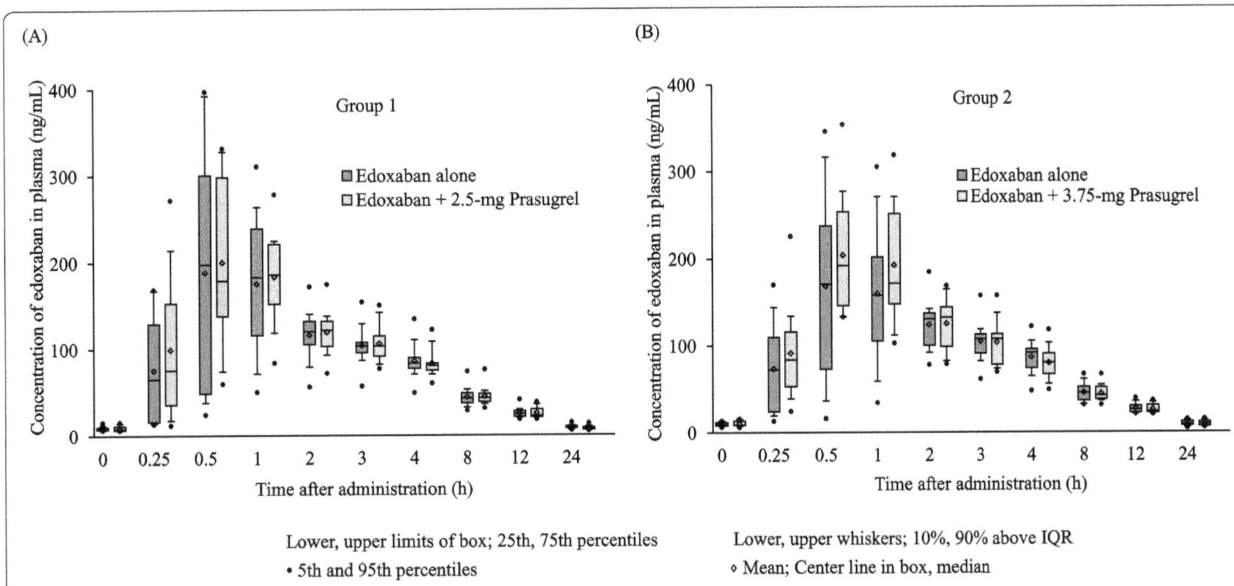

Fig. 3 Effect of prasugrel on edoxaban concentrations in plasma (**a**) Group 1, administration of 30 mg edoxaban alone compared with 30 mg edoxaban plus 2.5 mg prasugrel; (**b**) Group 2, administration of 30 mg edoxaban alone compared with 30 mg edoxaban plus 3.75 mg prasugrel

Table 4 Pharmacokinetic parameters of edoxaban

	Group 1 30 mg edoxaban plus 2.5 mg prasugrel		Group 2 30 mg edoxaban plus 3.75 mg prasugrel	
	$n = 12$		$n = 12$	
	Edoxaban alone	Edoxaban plus prasugrel	Edoxaban alone	Edoxaban plus prasugrel
$C_{max,ss}$ (ng/mL)				
Mean (SD)	216.0 (112.4)	216.0 (80.4)	191.4 (79.7)	216.9 (68.1)
Median (IQR)	199.8 (127.6–300.5)	187.8 (169.0–298.3)	170.5 (134.5–237.3)	216.6 (157.3–269.3)
$AUC_{tau,ss}$ (ng h/mL)				
Mean (SD)	1059.4 (228.0)	1068.6 (189.8)	1070.4 (239.7)	1088.5 (223.2)
Median (IQR)	1025.0 (924.2–1152.4)	1005.8 (912.2–1216.7)	992.5 (953.1–1206.7)	1049.6 (930.4–1240.1)
$t_{max,ss}$ (h)				
Mean (SD)	1.2 (0.96)	0.92 (0.70)	1.0 (0.84)	0.71 (0.26)
Median (IQR)	0.75 (0.50–1.5)	0.75 (0.50–1.0)	0.50 (0.50–1.5)	0.50 (0.50–1.0)

$C_{max,ss}$ steady state maximum plasma concentration, $AUC_{tau,ss}$ steady state area under the plasma concentration-time curve, $t_{max,ss}$ steady state time to reach maximum plasma concentration, *SD* standard deviation, *IQR* interquartile range

male subjects of normal weight (50 to 60 kg; BMI 21.1 ± 1.6 kg/m^2) and advanced age (65 to 80 years).

Results of BT measured by the Ivy method were predictable and confirmatory for coadministration of edoxaban with prasugrel. Edoxaban (30 mg) alone prolonged BT 1.10- or 1.33-fold from baseline, and the addition of 2.5 mg and 3.75 mg prasugrel to edoxaban further prolonged BT 1.58- or 2.00-fold, respectively. The effect of prasugrel on BT was dose-dependent. This two-fold increase was comparable to the coadministration of 60 mg edoxaban with aspirin or naproxen, antiplatelet agents used in the study of Mendell et al. [13] and less than a 3.8-fold increase after the coadministration of 15 mg rivaroxaban with standard doses of clopidogrel [14]. The incidence of bleeding events caused by the concomitant treatment with 60 mg edoxaban plus aspirin has been evaluated in patients with peripheral artery disease who have undergone endovascular treatment [15]. The bleeding risk of 60 mg edoxaban plus aspirin was similar to that of standard dual antiplatelet therapy with clopidogrel plus aspirin [15]. Another clinical study showed that concomitant use of 15 mg rivaroxaban with clopidogrel in NVAF patients undergoing PCI was associated with a lower rate of clinically significant bleeding when compared to standard therapy with a vitamin K antagonist plus dual antiplatelet therapy [16]. Although BT may not always be correlated with bleeding risk in patients receiving antithrombotic agents [17, 18], the present data suggest that bleeding risk by dual-therapy edoxaban and prasugrel might be low.

The results of PD and PK parameters were also predictable and confirmatory. As expected, edoxaban alone prolonged PT and aPTT and decreased F1 + 2. Combined treatment with prasugrel had no additional effect on these parameters. In terms of platelet function, edoxaban alone had no effect on platelet aggregation, and only after administration of prasugrel, platelet aggregation was suppressed. Coadministration of prasugrel with edoxaban did not affect the PK parameters of edoxaban. These results of the combined administration of DOAC edoxaban and $P2Y_{12}$ receptor inhibitor prasugrel are consistent with those reported previously for rivaroxaban and clopidogrel [14]; clopidogrel did not alter the anticoagulant effects and PK of rivaroxaban, whereas platelet aggregation was not affected by rivaroxaban.

This study reports safety results on the coadministration of edoxaban with prasugrel in a small study. All the treatment-emergent AEs were non-serious AEs. Most of the AEs were a positive fecal occult blood test, and all subjects with this AE did not require any treatment during the study period. Additionally, no study withdrawal or discontinuation due to AEs were observed.

It is also important to note that the reduced dose of edoxaban (30 mg) is used in 82% of elderly (≥75 years of age) AF patients in Japan [1]. Thus, for elderly AF patients undergoing PCI, clinicians can consider giving the reduced 30 mg dose of edoxaban with either the 2.5 mg or 3.75 mg dose of prasugrel. Since the effect of prasugrel on BT was dose-dependent, 2.5 mg prasugrel would be an option for patients whose body weight is ≤50 kg or who have other risk factors for bleeding.

There are several limitations of the study. First, this is a small clinical pharmacology study of short-term treatment in healthy elderly Japanese male subjects. Although we have set the duration of treatment for the purpose of reaching the steady states of PK variables for both edoxaban and prasugrel, according to the European consensus document [2] up to 12 months of treatment of dual therapy with an oral anticoagulant plus a $P2Y_{12}$ receptor inhibitor are required for AF patients undergoing PCI. Large clinical studies are required to assess the concomitant use of edoxaban and prasugrel over a longer period in these patients, especially in patients who are at high-risk for bleeding. Second, it would be hard to predict bleeding risk of antithrombotic agents by BT results only [17, 18]. However, BT is currently the only method used to assess the risk of bleeding in clinical pharmacological studies of antithrombotic agents or combined effects of anticoagulant and antiplatelet agents [13, 14]. The increase in BT in a clinical study may indicate a potential for a higher risk of bleeding.

Conclusions

Coadministration of the $P2Y_{12}$ receptor inhibitor 2.5 mg and 3.75 mg prasugrel with the DOAC 30 mg edoxaban prolonged BT in healthy elderly Japanese male subjects. The BT prolongation was dependent on the dose of prasugrel. Prasugrel did not affect PD or PK parameters of edoxaban.

Abbreviations

AE: Adverse event; aPTT: Activated partial thromboplastin time; BT: Bleeding time; CI: Confidence interval; DOAC: Direct oral anticoagulant; eGFR: Estimated glomerular filtration rate; F1 + 2: Prothrombin fragment 1 + 2; GLSM: Geometric least squares mean; NVAF: Nonvalvular atrial fibrillation; PCI: Percutaneous coronary intervention; PD: Pharmacodynamic; PK: Pharmacokinetic; PRU: $P2Y_{12}$ reaction units; PT: Prothrombin time; SAS: Safety analysis set

Acknowledgments

The authors wish to acknowledge Keita Yamaguchi (Daiichi Sankyo Co., Ltd.) and Hideyuki Tomitori (IQVIA Services Japan K.K., Tokyo, Japan) for coordinating the research activity and assisting with the drafting of the manuscript. Additionally, the authors are indebted to Takahiro Horii (IDD Inc., Tokyo, Japan), Kazuhiro Kanmuri (Inter-Professional Inc., Tokyo, Japan) and Mary Y. Nishikawa (Yokohama, Japan) for their medical writing and editorial support.

Authors' contributions

II established the study concept and design, conducted the research process, acquired, and interpreted the data, and reviewed the manuscript. TA and TM established the study concept and design, interpreted the data, and reviewed the manuscript. YM established the study concept and design, managed, and coordinated research activities, interpreted the data, and contributed to the drafting of the manuscript. AT designed the statistical analyses, interpreted the data, and reviewed the manuscript. TK established the study concept and design, managed, coordinated the research activity, interpreted the data, and reviewed the manuscript. All authors read and approved the final manuscript.

Author details

[1]Sumida Hospital, 29-1, Honjo 1-Chome, Sumida-ku, Tokyo 130-0004, Japan. [2]Medical Science Department, Daiichi Sankyo Co., Ltd., 3-5-1 Nihonbashi Honcho, Chuo-ku, Tokyo 103-8426, Japan. [3]Biostatistics & Data Management Department, Daiichi Sankyo Co., Ltd., 1-2-58 Hiromachi, Shinagawa-ku, Tokyo 140-8710, Japan.

References

1. Koretsune Y, Yamashita T, Akao M, Atarashi H, Ikeda T, Okumura K, et al. Baseline demographics and clinical characteristics in the all Nippon AF in the elderly (ANAFIE) registry. Circ J. 2019;83(7):1538–45.
2. Lip GYH, Collet J-P, Haude M, Byrne R, Chung EH, Fauchier L, et al. 2018 Joint European consensus document on the management of antithrombotic therapy in atrial fibrillation patients presenting with acute coronary syndrome and/or undergoing percutaneous cardiovascular interventions: a joint consensus document of the EuropeanHeart Rhythm Association (EHRA), European Society of Cardiology Working Group on Thrombosis, European Association of Percutaneous Cardiovascular Interventions (EAPCI), and European Association of Acute Cardiac Care (ACCA) endorsed by the Heart Rhythm Society (HRS), Asia-Pacific Heart Rhythm Society (APHRS), Latin America Heart Rhythm Society (LAHRS), and Cardiac Arrhythmia Society of Southern Africa (CASSA). Europace. 2019;21(2): 192–3.
3. Furugohri T, Isobe K, Honda Y, Kamisato-Matsumoto C, Sugiyama N, Nagahara T, et al. DU-176b, a potent and orally active factor Xa inhibitor: in vitro and in vivo pharmacological profiles. J Thromb Haemost. 2008;6(9): 1542–9.
4. Giugliano RP, Ruff CT, Braunwald E, Murphy SA, Wiviott SD, Halperin JL, et al. Edoxaban versus warfarin in patients with atrial fibrillation. N Engl J Med. 2013;369(22):2093–104.
5. Bounameaux H, Camm AJ. Edoxaban: an update on the new oral direct factor Xa inhibitor. Drugs. 2014;74(11):1209–31.
6. Sugidachi A, Asai F, Ogawa T, Inoue T, Koike H. The in vivo pharmacological profile of CS-747, a novel antiplatelet agent with platelet ADP receptor antagonist properties. Br J Pharmacol. 2000;129(7):1439–46.
7. Wiviott SD, Braunwald E, McCabe CH, Montalescot G, Ruzyllo W, Gottlieb S, et al. Prasugrel versus Clopidogrel in patients with acute coronary syndromes. N Engl J Med. 2007;357(20):2001–15.
8. Saito S, Isshiki T, Kimura T, Ogawa H, Yokoi H, Nanto S, et al. Efficacy and safety of adjusted-dose Prasugrel compared with Clopidogrel in Japanese patients with acute coronary syndrome. Circ J. 2014;78(7):1684–92.
9. Isshiki T, Kimura T, Ogawa H, Yokoi H, Nanto S, Takayama M, et al. Prasugrel, a third-generation $P2Y_{12}$ receptor antagonist, in patients with coronary artery disease undergoing elective percutaneous coronary intervention. Circ J. 2014;78(12):2926–34.
10. Otsuki H, Yamaguchi J, Kawamoto T, Yoshikawa M, Ebihara S, Tanaka K, et al. Safety and efficacy of low-dose prasugrel as part of triple therapy with aspirin and Oral anticoagulants in patients with atrial fibrillation undergoing percutaneous coronary intervention — from the TWMU-AF PCI registry —. Circ J. 2019;83(5):1000–5.
11. Vranckx P, Valgimigli M, Eckardt L, Tijssen J, Lewalter T, Gargiulo G, et al. Edoxaban-based versus vitamin K antagonist-based antithrombotic regimen after successful coronary stenting in patients with atrial fibrillation (ENTRUST-AF PCI): a randomised, open-label, phase 3b trial. Lancet. 2019; 394(10206):1335–43.
12. Mielke CH, Kaneshuro MM, Maher IA, Weiner JM, Rapaport SI. The standardized Normal ivy bleeding time and its prolongation by aspirin. Blood. 1969;34(2):204–15.
13. Mendell J, Lee F, Chen S, Worland V, Shi M, Samama MM. The effects of the antiplatelet agents, aspirin and naproxen, on pharmacokinetics and pharmacodynamics of the anticoagulant edoxaban, a direct factor Xa inhibitor. J Cardiovasc Pharmacol. 2013;62(2):212–21.
14. Kubitza D, Becka M, Mück W, Schwers S. Effect of co-administration of rivaroxaban and clopidogrel on bleeding time, pharmacodynamics and

pharmacokinetics: a phase I study. Pharmaceuticals. 2012;5(3):279–96.

15. Moll F, Baumgartner I, Jaff M, Nwachuku C, Tangelder M, Ansel G, et al. Edoxaban plus aspirin vs dual antiplatelet therapy in endovascular treatment of patients with peripheral artery disease: results of the ePAD trial. J Endovasc Ther. 2018;25(2):158–68.
16. Gibson CM, Mehran R, Bode C, Halperin J, Verheugt FW, Wildgoose P, et al. Prevention of bleeding in patients with atrial fibrillation undergoing PCI. N Engl J Med. 2016;375(25):2423–34.
17. De Caterina R, Lanza M, Manca G, Strata G, Maffei S, Salvatore L. Bleeding time and bleeding: an analysis of the relationship of the bleeding time test with parameters of surgical bleeding. Blood. 1994;84(10):3363–70.
18. Barber A, Green D, Galluzzo T, Ts'ao CH. The bleeding time as a preoperative screening test. Am J Med. 1985;78(5):761–4.

Permissions

All chapters in this book were first published by BioMed Central; hereby published with permission under the Creative Commons Attribution License or equivalent. Every chapter published in this book has been scrutinized by our experts. Their significance has been extensively debated. The topics covered herein carry significant findings which will fuel the growth of the discipline. They may even be implemented as practical applications or may be referred to as a beginning point for another development.

The contributors of this book come from diverse backgrounds, making this book a truly international effort. This book will bring forth new frontiers with its revolutionizing research information and detailed analysis of the nascent developments around the world.

We would like to thank all the contributing authors for lending their expertise to make the book truly unique. They have played a crucial role in the development of this book. Without their invaluable contributions this book wouldn't have been possible. They have made vital efforts to compile up to date information on the varied aspects of this subject to make this book a valuable addition to the collection of many professionals and students.

This book was conceptualized with the vision of imparting up-to-date information and advanced data in this field. To ensure the same, a matchless editorial board was set up. Every individual on the board went through rigorous rounds of assessment to prove their worth. After which they invested a large part of their time researching and compiling the most relevant data for our readers.

The editorial board has been involved in producing this book since its inception. They have spent rigorous hours researching and exploring the diverse topics which have resulted in the successful publishing of this book. They have passed on their knowledge of decades through this book. To expedite this challenging task, the publisher supported the team at every step. A small team of assistant editors was also appointed to further simplify the editing procedure and attain best results for the readers.

Apart from the editorial board, the designing team has also invested a significant amount of their time in understanding the subject and creating the most relevant covers. They scrutinized every image to scout for the most suitable representation of the subject and create an appropriate cover for the book.

The publishing team has been an ardent support to the editorial, designing and production team. Their endless efforts to recruit the best for this project, has resulted in the accomplishment of this book. They are a veteran in the field of academics and their pool of knowledge is as vast as their experience in printing. Their expertise and guidance has proved useful at every step. Their uncompromising quality standards have made this book an exceptional effort. Their encouragement from time to time has been an inspiration for everyone.

The publisher and the editorial board hope that this book will prove to be a valuable piece of knowledge for researchers, students, practitioners and scholars across the globe.

List of Contributors

Denizhan Ozdemir
Department of Internal Medicine, Yale New Haven Hospital, Yale School of Medicine, New Haven, CT, USA

Isabel Oliva Cortopassi
Section of Radiology and Biomedical Imaging, Yale New Haven Hospital, Yale School of Medicine, New Haven, CT, USA

Robert Lawrence McNamara
Section of Cardiovascular Medicine, Yale New Haven Hospital, Yale School of Medicine, Dana Clinic Building, 3rd Floor, 789 Howard Avenue, New Haven, CT 06519, USA

Kiyoaki Yamamoto, Atsushi Shinnakasu, Mihoko Kurano, Aiko Arimura, Hiroshi Arimura, Hiroshi Hashiguchi, Takahisa Deguchi and Yoshihiko Nishio
Department of Diabetes and Endocrine Medicine, Kagoshima University Graduate School of Medical and Dental Sciences, 8-35-1, Sakuragaoka, Kagoshima 890-8544, Japan

Tomoka Nagasato
Department of Systems Biology in Thromboregulation, Kagoshima University Graduate School of Medical and Dental Sciences, 8-35-1, Sakuragaoka, Kagoshima 890-8544, Japan
Fujimori Kogyo Co., Research Institute, 1-10-1, Yokohama, Kanagawa 236-0003, Japan

Ehteramolsadat Hosseini and Maryam Mohtashami
Blood Transfusion Research Centre, High Institute for Research and Education in Transfusion Medicine, Iranian Blood Transfusion Organization Building, Hemmat Exp. Way, Next to the Milad Tower, Tehran, Iran

Hideo Wada
Department of General Medicine, Mie Prefectural General Medical Center, Mie, Japan
Departments of Molecular and Laboratory Medicine, Mie University Graduate School of Medicine, Tsu, Mie 514-8507, Japan

Kazuo Kawasugi
Faculty of Medical Technology, Teikyo University, Tokyo, Japan

Mehran Ghasemzadeh
Blood Transfusion Research Centre, High Institute for Research and Education in Transfusion Medicine, Iranian Blood Transfusion Organization Building, Hemmat Exp. Way, Next to the Milad Tower, Tehran, Iran
Australian Center for Blood Diseases, Monash University, Melbourne, Victoria 3004, Australia

Noriaki Kawano
Department of Internal Medicine, Miyazaki Prefectural Miyazaki Hospital, 5-30 Kitatakamatsu, Miyazaki 880-8510, Japan

Ikuro Maruyama
Department of Systems Biology in Thromboregulation, Kagoshima University Graduate School of Medical and Dental Sciences, 8-35-1, Sakuragaoka, Kagoshima 890-8544, Japan

Toshimasa Uchiyama
Department of Laboratory Medicine, National Hospital Organization Takasaki General Medical Center, Gunma, Japan

Seiji Madoiwa
Department of Clinical Laboratory Medicine, Tokyo Saiseikai Central Hospital, Tokyo, Japan

Naoki Takezako
Department of Hematology, National Hospital Organization Disaster Medical Center, Tokyo, Japan

Yoshinobu Seki
Department of Hematology, Uonuma Institute of Community Medicine, Niigata University Medical and Dental Hospital, Niigata, Japan

Takayuki Ikezoe
Department of Hematology, Fukushima Medical University, Fukushima, Japan

Tsuyoshi Hattori
Department of Medical Affairs, Asahi Kasei Pharma Corporation, Tokyo, Japan

Kohji Okamoto
Department of Surgery, Center for Gastroenterology and Liver Disease, Kitakyushu City Yahata Hospital, Fukuoka, Japan

Iqtedar Ahmed Muazzam, Waqas Ilyas and George Bozas
Queen's Centre for Oncology and Haematology, Castle Hill Hospital, Hull University Teachng Hospitals (HUTH), Cottingham HU16 5JQ, UK

Anthony Maraveyas and Farzana Haque
Queen's Centre for Oncology and Haematology, Castle Hill Hospital, Hull University Teachng Hospitals (HUTH), Cottingham HU16 5JQ, UK
Joint Centre for Cancer Studies, Faculty of Health Sciences, The Hull York Medical School, University of Hull, Cottingham HU6 7RX, UK

Snehasis Pradhan, Kciku Gresa, Jan-Peter Röing genannt Nölke and Hans-Joachim Trappe
Department of Cardiology and Angiology, Marien Hospital Herne, Ruhr- University of Bochum, Hoelkeskampring 40, 44625 Herne, Germany

Jan Beyer-Westendorf
Thrombosis Research Unit, Department of Medicine I, Division Haematology, University Hospital "Carl Gustav Carus" Dresden, Dresden, Germany
Kings Thrombosis Service, Department of Haematology, Kings College London, London, UK

A. John Camm
Cardiology Clinical Academic Group, Molecular and Clinical Sciences Institute, St. George's University of London, London, UK

Keith A. A. Fox
Centre for Cardiovascular Science, University of Edinburgh, Edinburgh, UK

Jean-Yves Le Heuzey
Georges Pompidou Hospital, René Descartes University, Paris, France

Sylvia Haas
Formerly Klinikum rechts der Isar, Technical University of Munich, Munich, Germany

Alexander G. G. Turpie
McMaster University, Hamilton, Canada

Saverio Virdone
Thrombosis Research Institute, London, UK

Ajay K. Kakkar
Thrombosis Research Institute, London, UK
University College of London, London, UK

Ellen M. K. Warlo, Harald Arnesen and Ingebjørg Seljeflot
Center for Clinical Heart Research, Department of Cardiology, Oslo University Hospital, Pb 4956 Nydalen, 0424 Oslo, Norway
Faculty of Medicine, University of Oslo, Oslo, Norway
Center for Heart Failure Research, University of Oslo, Oslo, Norway

Laila Staerk and Anders Nissen Bonde
Department of Cardiology, Copenhagen University Hospital Herlev and Gentofte, Kildegaardsvej 28, 2900 Hellerup, Denmark
Department of Clinical Medicine, Faculty of Health and Medical Sciences, University of Copenhagen, Blegdamsvej 3B, 2200 Copenhagen, Denmark

Emma Kirstine Laugesen and Jonas Bjerring Olesen
Department of Cardiology, Copenhagen University Hospital Herlev and Gentofte, Kildegaardsvej 28, 2900 Hellerup, Denmark

Nicholas Carlson
The Danish Heart Foundation, Vognmagergade 7, 1120 Copenhagen, Denmark
Department of Nephrology, Copenhagen University Hospital Rigshospitalet, Blegdamsvej 9, 2100 Copenhagen, Denmark

Anne-Lise Kamper
Department of Nephrology, Copenhagen University Hospital Rigshospitalet, Blegdamsvej 9, 2100 Copenhagen, Denmark

Christian Torp-Pedersen
Department of Health Science and Technology, Aalborg University, Fredrik Bajers vej 5, 9100 Aalborg, Denmark

Gunnar Gislason
Department of Cardiology, Copenhagen University Hospital Herlev and Gentofte, Kildegaardsvej 28, 2900 Hellerup, Denmark
Department of Clinical Medicine, Faculty of Health and Medical Sciences, University of Copenhagen, Blegdamsvej 3B, 2200 Copenhagen, Denmark
The Danish Heart Foundation, Vognmagergade 7, 1120 Copenhagen, Denmark
The National Institute of Public Health, University of Southern Denmark, Studiestræde 6, 1455 Copenhagen, Denmark

Takahito Miyake, Hideshi Okada, Norihide Kanda, Fuminori Yamaji, Haruka Okamoto, Hiroaki Ushikoshi, Shozo Yoshida and Shinji Ogura
Advanced Critical Care Center, Gifu University Hospital, 1-1 Yanagido, Gifu 501-1194, Japan

Kei Noguchi and Hiroyuki Tomita
Department of Tumor Pathology, Gifu University School of Medicine, 1-1 Yanagido, Gifu 501-1194, Japan

J. Seelig
Department of Cardiology, Rijnstate, Wagnerlaan 55, 6815 AD Arnhem, the Netherlands
Department of Internal Medicine, Cardiovascular Research Institute Maastricht, Maastricht, the Netherlands

F. W. A. Verheugt
Department of Cardiology, Onze Lieve Vrouwe Gasthuis (OLVG), Amsterdam, the Netherlands

M. E. W. Hemels
Department of Cardiology, Rijnstate, Wagnerlaan 55, 6815 AD Arnhem, the Netherlands
Radboud University Medical Centre, Nijmegen, the Netherlands

L. Illingworth
Thrombosis Research Institute, London, UK

A. Lucassen
Department of Cardiology, St. Jans Gasthuis, Weert, the Netherlands

H. Adriaansen
Anticoagulation Clinic, Gelre Ziekenhuizen, Apeldoorn-Zutphen, the Netherlands

M. C. M. Bongaerts
Anticoagulation Clinic, Ziekenhuis Rivierenland, Tiel, the Netherlands

M. Pieterse
Stichting Cardiologie Amsterdam, Amsterdam, the Netherlands

J. P. R. Herrman
Department of Cardiology, Onze Lieve Vrouwe Gasthuis, Amsterdam, the Netherlands

P. Hoogslag
Department of Cardiology, Isala Diaconessenhuis, Meppel, the Netherlands

W. Hermans
Department of Cardiology, Elisabeth-TweeSteden Ziekenhuis, Tilburg, the Netherlands

B. E. Groenemeijer
Department of Cardiology, Gelre Ziekenhuizen, Apeldoorn, the Netherlands

L. V. A. Boersma
Department of Cardiology, St. Antonius Ziekenhuis, Nieuwegein, the Netherlands
Department of Cardiology, Amsterdam UMC, Amsterdam, the Netherlands

K. Pieper
Thrombosis Research Institute, London, UK
Duke Clinical Research Institute, Durham, USA

H. ten Cate
Department of Internal Medicine, Cardiovascular Research Institute Maastricht, Maastricht, the Netherlands
Anticoagulation Clinic Maastricht, Maastricht, the Netherlands

Ramin Rahmanzade
Faculty of Medicine, University of Basel, Basel, Switzerland
Biomedical Research and Training, University Hospital Basel, Basel, Switzerland

Francisco Cabrera Diaz and Claudia Zaugg
Hospital Pharmacy, Kantonsspital Aarau, Aarau, Switzerland

Philipp Schuetz
Faculty of Medicine, University of Basel, Basel, Switzerland
Department of Internal Medicine, Kantonsspital Aarau, 5000 Aarau, Switzerland

Ali Reza Salili
Department of Internal Medicine, Kantonsspital Aarau, 5000 Aarau, Switzerland
Department of Internal Medicine, Clinical Pharmacology, Kantonsspital Aarau, Tellstrasse 51, CH-5001 Aarau, Switzerland

Yayoi Yokoyama, Tomotsugu Yasuda, Hiroaki Furubeppu, Chinatsu Kamikokuryo and Yasuyuki Kakihana
Department of Emergency and Intensive Care Medicine, Kagoshima University Graduate School of Medical and Dental Sciences, 8-35-1 Sakuragaoka, Kagoshima 890-8544, Japan

Takashi Ito
Department of Emergency and Intensive Care Medicine, Kagoshima University Graduate School of Medical and Dental Sciences, 8-35-1 Sakuragaoka, Kagoshima 890-8544, Japan
Department of Systems Biology in Thromboregulation, Kagoshima University Graduate School of Medical and Dental Sciences, Kagoshima, Japan

Shingo Yamada
R&D Center, Shino-Test Corporation, Sagamihara, Japan

M. J. Moesker and M. C. de Bruijne
Department of Public and Occupational Health, Amsterdam Public Health Research Institute, Amsterdam UMC, Vrije Universiteit Amsterdam, De Boelelaan 1117, 1081BT Amsterdam, The Netherlands

C. Wagner
Department of Public and Occupational Health, Amsterdam Public Health Research Institute, Amsterdam UMC, Vrije Universiteit Amsterdam, De Boelelaan 1117, 1081BT Amsterdam, The Netherlands
Netherlands Institute for Health Services Research (NIVEL), Utrecht, The Netherlands

J. F. de Groot
Netherlands Institute for Health Services Research (NIVEL), Utrecht, The Netherlands

N. L. Damen
Department of Quality and Safety, Elizabeth Tweesteden Ziekenhuis, Tilburg, The Netherlands

N. R. Bijsterveld
Department of Cardiology, Flevoziekenhuis, Almere, The Netherlands

J. W. R. Twisk
Department of Epidemiology and Biostatistics, Amsterdam UMC, Vrije Universiteit Amsterdam, Amsterdam, The Netherlands

M. V. Huisman
Department of Thrombosis and Hemostasis, Leiden University Medical Centre, Leiden, The Netherlands

Jae C. Chang
Department of Medicine, University of California Irvine School of Medicine, Irvine, CA, USA

Noah Merin
Department of Medicine, Division of Hematology Oncology, Cedars Sinai Medical Center, Los Angeles, USA

Michael Tran, Chirag Sheth, Rohan Bhandari and Deborah Hornacek
Heart Vascular and Thoracic Institute, Department of Cardiovascular Medicine, Section of Vascular Medicine, Cleveland Clinic Foundation, Desk J-35, Cleveland Clinic Foundation, Cleveland, OH 44195, USA

Scott J. Cameron
Department of Cardiovascular and Metabolic Sciences. Cleveland Clinic Lerner College of Medicine, Cleveland, OH 44195, USA

Reto A. Schuepbach
Institute of Intensive Care Medicine, University Hospital Zurich, University of Zurich, Zurich, Switzerland

Dorothea M. Heuberger
Institute of Intensive Care Medicine, University Hospital Zurich, University of Zurich, Zurich, Switzerland
Surgical Research Division, University Hospital Zurich, University of Zurich, Zurich, Switzerland

Lama Abdelnour and Umut Selamet
Department of Medicine, Division of Nephrology, UCLA David Geffen School of Medicine, Room 7-155, Factor Bldg. 700 Tiverton Ave, Los Angeles, CA 90095, USA

Ramy M. Hanna
Department of Medicine, Division of Nephrology, UCLA David Geffen School of Medicine, Room 7-155, Factor Bldg. 700 Tiverton Ave, Los Angeles, CA 90095, USA
Department of Medicine, Division of Nephrology, University of California Irvine School of Medicine, Irvine CA, USA

Richard M. Burwick
Department of Obstetrics and Gynecology, Division of Maternal Fetal Medicine, Cedars Sinai Medical Center, Los Angeles, USA

Beshoy Yanny
Department of Medicine, Division of Digestive Diseases, UCLA David Geffen School of Medicine, Los Angeles, USA

Patrick Bui and Mary Fouad
Department of Medicine, UCLA David Geffen School of Medicine, Los Angeles, USA

Ira Kurtz
Department of Medicine, Division of Nephrology, UCLA David Geffen School of Medicine, Room 7-155, Factor Bldg. 700 Tiverton Ave, Los Angeles, CA 90095, USA
UCLA Brain Research Institute, Los Angeles, USA

Xiao Li, Yuanhua Zhuang, Lili Kang and Xiaoying Li
Qilu Children's Hospital of Shandong University, Jinan, Shandong, China

Xiaoyan Li
People's Hospital of Rongcheng, Weihai, Shandong, China

Xiuli Ju
Qilu Hospital of Shandong University, No107, Cultural west Road, Lixia District, Jinan, Shandong, China

Kei Suzuki and Hiroshi Imai
Emergency Critical Care Center, Mie University Graduate School of Medicine, Tsu, Mie, Japan

Takeshi Matsumoto and Kohshi Ohishi
Division of Blood Transfusion Medicine and Cell Therapy, Mie University Graduate School of Medicine, Tsu, Mie, Japan

Makoto Ikejiri
Central laboratory, Mie University Graduate School of Medicine, Tsu, Mie, Japan

Yoshiki Yamashita and Naoyuki Katayama
Department of Hematology and Oncology, Mie University Graduate School of Medicine, Tsu, Mie, Japan

Toshiaki Iba
Department of Emergency and Disaster Medicine, Juntendo University Graduate School of Medicine, Tokyo, Japan

Kohei Ohtsuka, Kimine Tanaka, Mayumi Yamanaka and Satoshi Fujii
Department of Medical Laboratory and Blood Center, Asahikawa Medical University Hospital, Midorigaoka-Higashi 2-1-1-1, Asahikawa 078-8510, Japan

Yoshie Kawahara
Department of Medical Laboratory and Blood Center, Asahikawa Medical University Hospital, Midorigaoka-Higashi 2-1-1-1, Asahikawa 078-8510, Japan
Present address: Japanese Red Cross Hokkaido Block Blood Center, Sapporo, Japan

Hiroyuki Kamiya
Department of Cardiac Surgery, Asahikawa Medical University, Asahikawa, Japan

Takayuki Kunisawa
Department of Anesthesiology, Asahikawa Medical University, Asahikawa, Japan

Mohammad Masoud Eslami, Mina Soufizomorrod, Saeid Abroun and Bahman Razi
Department of Hematology, Faculty of Medical Sciences, Tarbiat Modares University, North Kargar Av, Tehran 14117, Iran

Majid khalili
Department of Basic sciences, Maragheh University of medical sciences, Maragheh, Iran
Rahat Breach and Sleep Research Center, Tabriz University of Medical Sciences, Tabriz, Iran

Abu-Sayeef Mirza, Maher Elharake, Katherine Robinson, Thomas Brimer, Ashton Sequeira and Nathan Visweshwar
Department of Internal Medicine, H. Lee Moffitt Cancer Center and Research Institution, University of South Florida, 17 Davis Blvd., Suite 308, Tampa, FL 33606, USA

Seongseok Yun
Department of Internal Medicine, H. Lee Moffitt Cancer Center and Research Institution, University of South Florida, 17 Davis Blvd., Suite 308, Tampa, FL 33606, USA
Department of Malignant Hematology, H. Lee Moffitt Cancer Center and Research Institution, Tampa, FL, USA

Najla Al Ali, Rami Komrokji and Michael Jaglal
Department of Malignant Hematology, H. Lee Moffitt Cancer Center and Research Institution, Tampa, FL, USA

Hannah Shin, Joseph Luke O'Neil, Daniel Schwartz, Ethan Nowell, Grace Engle, Amra Kuc, Sabbir Mirza, Dhiraj Sikaria, Jesus Diaz Vera, Noah Hackney, Sammy Abusrur, Jose Jesurajan, Jameson Kuang, Shreyans Patel, Sabrina Khalil, Sonya Bhaskar, Alexander Beard, Toaa Abuelenen and Kevin Ratnasamy
Morsani College of Medicine, University of South Florida, Tampa, FL, USA

Ibraahim Badat
College of Arts and Sciences, University of South Florida, Tampa, FL, USA

Takeshi Yamashita
The Cardiovascular Institute, 3-2-19, Nishiazabu, Minato-ku, Tokyo 106-0031, Japan

Joji Hagii
Hirosaki Stroke and Rehabilitation Center, 1-2-1, Ougi-machi, Hirosaki, Aomori 036-8104, Japan

Ippei Ikushima
Sumida Hospital, 29-1, Honjo 1-Chome, Sumida-ku, Tokyo 130-0004, Japan

Takaaki Akasaka, Yoshiyuki Morishima, Takuyuki Matsumoto, Tomoko Motohashi and Tetsuya Kimura
Medical Science Department, Daiichi Sankyo Co., Ltd., 3-5-1 Nihonbashi Honcho, Chuo-ku, Tokyo 103-8426, Japan

Atsushi Takita
Biostatistics & Data Management Department, Daiichi Sankyo Co., Ltd., 1-2-58 Hiromachi, Shinagawa-ku, Tokyo 140-8710, Japan

Index

Printed in the USA
CPSIA information can be obtained
at www.ICGtesting.com
LVHW080048211023
761596LV00010BA/1963

9 781646 475674